Structure and Fabric
Part 2

Mitchell's Building Series

Structure and Fabric
Part 2

Seventh edition

Jack Stroud Foster
FRIBA

Raymond Harington
DIP ARCH

Roger Greeno
BA(Hons), FCIOB, FIPHE, FRSA

PEARSON
Prentice Hall

Harlow, England • London • New York • Boston • San Francisco • Toronto
Sydney • Tokyo • Singapore • Hong Kong • Seoul • Taipei • New Delhi
Cape Town • Madrid • Mexico City • Amsterdam • Munich • Paris • Milan

Pearson Education Limited
Edinburgh Gate
Harlow
Essex CM20 2JE
England

and Associated Companies throughout the world

Visit us on the world wide web at:
www.pearsoned.co.uk

First published 1973
Second Edition 1979
Third Edition 1983
Fourth Edition 1991
Fifth Edition 1994
Sixth Edition 2000
Seventh Edition 2007

ISBN 978-0-13-197096-0

British Library Cataloguing in Publication Data
A catalogue record for this book is available from the British Library

10 9 8 7 6 5 4 3
11 10 09

Typeset in 9/11pt Times by 35
Printed and bound in Malaysia (CTP-PJB)

Contents

Acknowledgements

The authors are indebted to many people and organisations who have given help and guidance in the preparation of this book and from whom they have received much information. The following, to whom thanks are especially due, were named in the precursor to this book *Mitchell's Advanced Building Construction – The Structure* by J S Foster where a note of their particular contribution was made: J F Crofts; Kenneth A Lock; A G Stone; Ivan Tomlin; Professor Z S Makowski; Leonard R Creasy; Kenneth W Dale; Peter Dunican; J W Tiller.

We are grateful to Concrete Limited; British Lift Slab Limited; Dollery and Palmer Limited and Putzmeister Limited; Halfen Limited; How-Kinnell Limited; Omnia Constructions Limited and many other firms and organisations who have freely given information and permission to base illustrations on material which they have readily provided.

With the permission of the Controller of HM Stationery Office we have drawn freely on *Principles of Modern Building*, Volumes 1 and 2 and on *Post-War Building Studies*, *Building Research Establishment Digests* and *Current Papers*, and have quoted from *The Building Regulations*. We have also drawn on *British Standard Specifications* and *Codes of Practice* with the permission of the British Standards Institution, from whom official copies may be obtained. We owe much in chapter 2 to the reading of Capper and Cassie's book *The Mechanics of Engineering Soils* and much of the material on the economic aspects of chapters 4 and 8 has been gathered from Leonard R Creasy's excellent paper on the subject to which reference is made in the text.

Thanks are due to the following for permission to quote from books or papers, to reproduce tables or to use drawings as a basis for illustrations in this volume: Architectural Press: *Building Elements*, R Llewellyn Davies and A Petty; *Guide to the Building Regulations*, A Elder for table 6.2; *Guide to Concrete Blockwork*, Michael Gage for figure 3.48 B; *Principles of Pneumatic Architecture*, Roger N Dent for figures 8.21 A, C, F and 8.49 A, B, E, F, G, H; *Structure in Building*, W Fisher Cassie and J H Napper for part of figures 2.5 and 2.13 A, B. Cement and Concrete Association: *Concrete Block Walls* for figure 3.48 C, D. Concrete Publications: *Reinforced Concrete Chimneys*, C Percy Taylor and Leslie Turner. Crosby Lockwood Staples: *Practical Problems in Soil Mechanics*, H R Reynolds and P Protopapadakis for figures 2.1 A and 2.5; *Design Problems of Heating and Ventilating*, A T Henley; *Oil Fuel Applications*, A T Henley. HMSO: *Air Structures – A Survey*, F Newby for figures 8.21 D, G, H, L and 8.49 C, D; DoE *Construction 3* and *7* for figures 3.20 and 3.16 respectively. Newnes-Butterworth: *Structural Steelwork for Students*, L V Leech for parts of figures 4.10, 4.16 and 8.25. Pitman: *Heating and Air Conditioning Equipment for Buildings*, F Burlace Turpin; *Soil Mechanics Related to Building*, J H G King and D A Creswell for figures 2.1 B, C, D, F and 2.8; *The Fabric of Modern Buildings*, E G Warland. Spon: *Mechanics of Engineering Soils*, P L Capper and W F Cassie for parts of chapter 2; *Walls and Wall Facings*, D N Nield. *Acier-Stahl-Steel 6/1974*, Centre Belgo-Luxembourgeois d'Information de l'Acier for figure 4.13 B. *Architects' Journal*, Architectural Press, 18 February 1971 for tables 6.4, 6.5, 15 September 1971 for figure 3.19. *Build International*, October 1969, Applied Science Publishers Ltd for figure 4.17. *Building Specification*, February 1970 for figure 3.34 right. *Building Research Congress 1951: Papers*, Division 1 for figure 2.5. British Gas Corporation: 'Flexibility with Flues' for figure 6.11, 'Fan diluted flues', text page 256; 'Gas Handbook for Architects and Builders' for table 6.3. Constrado: *Structural Steelwork Simplified* for figures 4.10 F, G, H, J, K, O, P and 4.16 G, H, J. *Proceedings of the Institution of Civil Engineers*, Institution of Civil Engineers, *RIBA Journal*, October 1973 for table 6.1 and figure 6.3. *The Structural Engineer*, Institution of Structural Engineers. The Australian Department of Labour and National Service – Industrial Services Division. Scaffolding Great Britain plc for figures 10.7, 10.8 and 10.12.

We also thank Messrs Anthony Collins, H J B Harding and R Glossop, F Kerr, Professor Z S Makowski, and H Werner Rosenthal.

We are grateful to George Dilks for his meticulous work in revising for the first edition of this work the original illustrations from *Advanced Building Construction* and for preparing new illustrations. We are also grateful to Jean Marshall for her work in revising many of the existing illustrations and for preparing new diagrams. We must also express our appreciation to the editorial team of the publishers and to Christopher Parkin, for their help and co-operation in seeing the work through to press.

JSF and RKH

I would like to thank Pearson Education, Raymond Harington and the estate of the late Jack Stroud Foster for allowing me the opportunity to continue the publication of this long established, authoritive industry standard. The original books by Charles F Mitchell were written in 1893 and continued by George A Mitchell. Many years later in 1963, Jack Stroud Foster updated and completely re-presented both *Elementary* and *Advanced Building Construction*. From these, the *Mitchell's Building Series* evolved into 'the' learning resource for students pursuing a variety of building, construction and architecture study programmes. The current revised and updated contents, along with much of the earlier material retained as relevant to existing buildings, preserves the concept of the book and will ensure its value as an industry reference for many years to come.

RG

Preface to 7th edition

In this edition account has been taken of relevant legislation which has come into force and of developments in constructional techniques; in particular, advantage has been taken of the opportunity to bring up to date much of the material. In this last task generous help has been received from many members of the construction industry and I would, especially, acknowledge the assistance received from Mr Alun Abraham, of Scaffolding Great Britain, in relation to developments in steel shoring.

I am indebted also to a number of academics from across the country who have submitted to the publishers helpful criticisms and suggestions which have had considerable influence upon the updating of this latest revision.

This volume is the second part of two volumes and the content develops the principles of building practice and procedures introduced in *Structure and Fabric Part 1*. It is also intended as complementary reading to the companion volumes in the *Mitchell's Building Series*.

Foreword

The two parts of *Structure and Fabric*, while being each complete in itself, are intended to form one work in which the second part extends and develops the material in the first.

The subject of the work has been treated basically under the elements of construction. Most of these are interrelated in a building and, as far as possible, this has been borne in mind in the text. Ample cross-references are given to facilitate a grasp of this interrelationship of parts. Contract planning and site organisation are both subjects relevant to constructional techniques and methods used on the site and to the initial design process for a building. These have been touched on in Part 1 and are developed in Part 2. The subject of fire protection by its nature is extremely broad but it is so closely linked with the design and construction of buildings that it has been covered on broad lines in Part 2 in order to give an understanding of those factors which influence the nature and form of fire protection as well as to give detailed requirements in terms of construction.

In view of the continual production of new and improved materials in various forms and the continuous development of new constructional techniques, using both new and traditional materials, the designer can no longer be dependent on a tradition based on the use of a limited range of structural materials, but must exercise judgment and choice in a wide, and ever-widening, realm of alternatives. This necessitates a knowledge not only of the materials themselves but also of the nature and structural behaviour of all the parts of a building of which those materials form a part. Efficiency of structure and economy of material and labour are basic elements of good design. They are of vital importance today and should have a dominating influence on the design and construction of all buildings.

In the light of this something is required to give an understanding of the behaviour of structures under load and of the functional requirements of the different parts; to give some indication of their comparative economics and efficient design, their limitations and the logical and economic application of each. In writing the two Parts it has been the aim to deal with these aspects. The books are not exemplars of constructional details. Those details which are described and illustrated are meant to indicate the basic methods which can be adopted and how different materials can be used to fulfil various structural requirements. The illustrations are generally not fully dimensioned; such dimensions as are given are meant to give a sense of 'scale' to the parts rather than to lay down definite sizes in particular circumstances. The function of the books is not primarily to give information on *how* things are done in detail, as this must be everchanging. Rather, the emphasis is on *why* things are done, having regard particularly to efficiency and economy in design. An understanding of the function and behaviour of the parts and of the logical and economic application of material should enable a designer to prepare satisfactory constructional details in the solution of structural problems.

The books are intended primarily as textbooks for architectural, building and surveying students, but it is hoped that students of civil and structural engineering will find them useful as a means of setting within the context of the building as a whole their own studies in the realm of building structures.

In books of this nature there is little scope for original work. The task consists of gathering together existing information and selecting that which appears to be important and relevant to the purpose of the book. The authors acknowledge the debt they owe to others on whose work they have freely drawn, much of which is scattered in the journals of many countries. An endeavour has been made to indicate the sources, either in the text or notes. Where this has not been done is due to the fact that over a period of many years of lecturing on the subject much material has been gathered, both textual and illustrative, the sources of which have not been traced. For any such omissions the authors' apologies are offered.

1 Contract planning and site organisation

This chapter is concerned with the process which is essential prior to the actual commencement of building operations on the site, dealing with the planning of the contract as a whole and the organisation and layout of the site. The preparation and use of programmes for the various stages of the building work, by means of which the work may be controlled, is discussed together with an outline of the factors which must be considered at this stage. The means by which a programme of work may be expressed for use on the site are also explained.

The planning and control of all resources necessary to the realisation of building production is of vital importance and at some point before commencing work on the site, thought must be given to the way in which the building operation will be organised. Most builders plan their work in some form or another but, in the past, only a few have done so in much detail and have committed their plan to paper. Buildings, and consequently their construction, have become increasingly complex and the proper management of a contract and the control of cost, on the part of the architect at design stage and the contractor during erection, are more than ever essential if building is to be carried out efficiently both in terms of time and money. Only by proper planning can aids to productivity, such as mechanical plant, incentives, and efficient use of labour, become fully effective. With the greater mechanisation of building operations and the increased use of expensive plant, the contractor must obtain maximum use of the plant and speed the construction of the job in order to keep costs to a minimum. The design/erection continuum must be seen as a production process from inception to completion and there must be a programme on which the job may be organised, against which performance may be assessed and within which control may be exercised.

As pointed out in Part 1 the building and civil engineering industry is peculiar in that the contractor who will be responsible for carrying out the work usually plays no part in the design of a project, and has no opportunity at this point to contribute from his experience on matters of construction, planning and the nomination of subcontractors and thus to assist in the work being carried out efficiently, quickly and economically. Although a negotiated contract or engagement of a design and build partnership is often suggested as a means of overcoming this lack of collaboration at the design stage, it is not always suitable or acceptable since the element of competitive tendering is absent. In these circumstances as well as an appreciation of the concept of buildability[1] it is essential that the architect should have sufficient knowledge of contract planning and of its implications to ensure a well organised job. Reference is made later to ways in which the architect can contribute to this end at the design stage.

Whether or not such contribution is made by the architect, the responsibility for actually carrying out the job in all its aspects is that of the contractor. In order to enable him to do this efficiently management methods common to other industries have become widely used in building.[2] The contemporary view of construction management embraces a great number of interrelated activities drawing on a vast range of resources: professional, manufacturing, different categories of contracting and supplying, off-site production of components involving an increased use of transport, and specialised assembly methods.

The subjects of this chapter, contract planning and site organisation, together with general control, are the construction aspects of production management which itself is a part only of overall management in building. **Planning** makes efficient and economical use of labour, machines and materials, **organising** is the means of delegating tasks and **control** enables planning and organisation to be effective. It is possible here to deal with them only in outline and for a more detailed consideration reference should be made to other works.[3]

1.1 Contract planning and control

This involves working out a plan of campaign or a programme for the contract as a whole and assembling the necessary data. The primary function of such a programme is to promote the satisfactory organisation and flow of the various building operations during the course of erection, by planning in advance the times and sequences of all operations and the requirements in labour, materials and equipment. In order to fulfil this function and also to provide important information required during the contract, a well planned programme will have certain clear objectives. It should:

1 show the quickest and cheapest method of carrying out the work consistent with the available resources of the builder
2 ensure continuous productive work for all the operatives employed and reduce unproductive time to a minimum, by the proper phasing of operations with balanced labour gangs in all trades
3 provide an assessment of the level of productivity in all trades to permit the establishment of equitable targets for remuneration
4 determine attendance dates and periods for all subcontractors' work
5 provide information on material quantities and essential delivery dates, the quantity and capacity of the plant required and the periods it will be on site
6 provide, at any time during the contract, a simple and rapid method of measuring progress, for the client and builder's information and the issue of interim certificates for the valuation of work in progress for accounting purposes.

If a builder's tender for any sizeable job is to be realistic, planning must start at the estimating stage and the following considerations must be taken into account: the most economical methods to be used for each operation and the sequence and timing of the operations, having regard to the resources at the contractor's disposal; whether hand or mechanical methods will be most economical and the most suitable type of plant to be used in relation to the nature and size of the job; the space available and the best positions for the various machines to be used; the best methods of handling materials and the most suitable places on the site for the storage of materials and for the placing of huts; suitable points of access to the site for lorries and machines. In deciding what methods to use

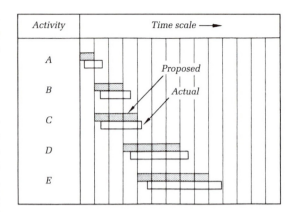

Figure 1.1   Gantt chart

for construction and the most suitable plant for different operations, the estimator would, when necessary, consult the contract planning and plant departments of the firm or plant suppliers.

Traditional methods of production control have tended to be based on criteria of usefulness to site managers and indeed it must be kept well in mind that the site is where the building is finally assembled and where managerial control must be effective.

A typical site-oriented control device is the *Gantt Chart* or *Bar Chart*, which allows a fairly simple and easily read plan of operations to be made available to all site personnel against which may be plotted actual performances (see figure 1.1). However, this excellent device only takes into account one of the resources, time, and unless further schedules of the resources needed for each operation are also available adjacent to the Bar Chart it does not inform on the critical relationships between the various activities depicted nor does it enable procedures involving a number of variables to be optimised since the complex interrelationships affecting the outcome of any plan (or alteration of plan) are not readily evident or quantifiable. This can, however, be achieved by means of a technique known as *Network Analysis*.

1.1.1 Network analysis

The essential difference between analysing a production problem by network as against linear or parallel linear methods lies in the identification of the dependency between operations. This approach leads to interrelated networks through which certain sequences can be seen to be 'critical' to the anticipated outcome in that they occupy the longest irreducible time required to execute the project (or part of the project) to which they are necessary. Figure 1.2 shows this in a simple set of five interrelated activities *A, B, C, D, E* of time values 1, 2, 3, 4 and 5 days.

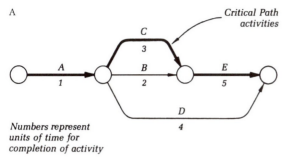

A

Numbers represent units of time for completion of activity

Dependencies:
E is dependent on activities
C, B and A being completed
C and B are
dependent on A being completed
D is dependent on A being completed

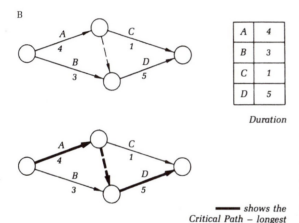

B

A	4
B	3
C	1
D	5

Duration

━━━ *shows the Critical Path – longest path through the network*

Figure 1.2 Network or Critical Path diagrams

In preparing a network the project is broken down into its operational parts (this can be at a strategic level or in extreme tactical detail) termed 'activities' which are represented by linear arrows (figure 1.2 A). The arrows are arranged to show the sequence of activities necessary to the occurrence of 'events' (shown by circles) which must precede further sets of activities. Activities (with the exception of 'dummy' activities) take up time, including waiting time, whereas events do not. An event cannot be said to have occurred until *all* the activities leading to it have been effectively completed.

Dummy arrows are used to show dependencies where activities are not directly sequential. For example in figure 1.2 B activity *C* is dependent on *A* being completed. *D* is dependent on *A* and *B* being completed – shown by use of a dummy arrow. Dummy arrows have no time value but

may still lie on the critical path if the activities they link are critical.

Figure 1.3 A, B, C shows how variations in time ascribed to the activities will result in different critical paths and how, in the case of C, all sequences can become critical. In theory the most perfect plan would result in an 'all critical' network but in life this would lead to a wholly inflexible situation lacking any time to manoeuvre or rethink situations. It is in this context that the 'float' or difference in time between the non-critical and the critical activity times becomes important to the production planner working in changing circumstances, in that he will be given options as to how he may deploy resources.

It is often convenient for the network to be set up initially using non-scaled linear arrows to represent the sequence interrelationship of activities and then to plot the network against a linear time scale as a prelude to examining the distribution of resources and to the preparation of a bar chart form of presentation. The generalised network statements shown in figure 1.3 A, B and C are each shown developed in this way in time scale form.

Although time is the planner's main parameter, other factors will affect the final assessment of times to be ascribed to the constituent activities of a network, e.g. cost, labour and material availability and the demands of other projects under the planner's control. It is convenient to think in terms of a 'normal' time when planning initially and figure 1.4 shows a statistical method of arriving at the time envisaged for an activity, given some historical data, using a Beta distribution in which the area under the curve represents the probability of the time expended on an activity lying between any given points along the baseline time scale. This particular distribution curve yields the formula:

$$t_e = \frac{a + 4m + b}{6}$$

where t_e = estimate of time to be used in network
a = optimistic or shortest estimate of duration time
b = pessimistic or longest estimate of duration time
m = most likely duration of activity judged from all available evidence.

However, there are occasions when the control of a project requires more sophisticated analyses to be carried out to allow the planner to respond to varying criteria for productivity: such as time/cost optimisation, levelling of resources and/or deployment of resources amongst a number of projects.

Time/cost optimisation This technique explores the possibilities of altering production time while optimising the costs of so doing. In building work it is generally the case that increased speed of production carries the penalty

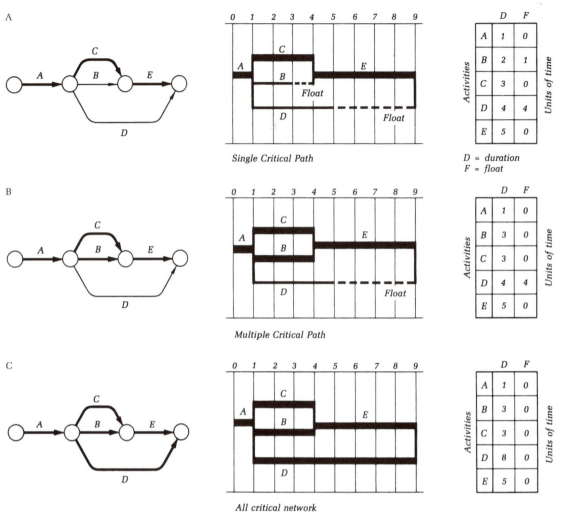

A

Single Critical Path

	D	F
A	1	0
B	2	1
C	3	0
D	4	4
E	5	0

D = duration
F = float

B

Multiple Critical Path

	D	F
A	1	0
B	3	0
C	3	0
D	4	4
E	5	0

C

All critical network

	D	F
A	1	0
B	3	0
C	3	0
D	8	0
E	5	0

NB Activities on Critical
Path have zero 'float'

Figure 1.3 Network or Critical Path diagrams

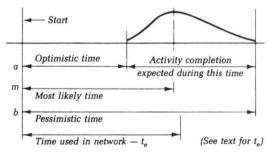

Figure 1.4 Activity times

of increased cost – usually due to having to use more operatives and/or machinery or to paying higher rates. Clearly
any reduction of the activity times on the critical path or
paths will reduce the overall production time but in so
doing will probably reduce the 'float' on other activities
to the point that they also become critical as shown in
figure 1.5. Given that activities A, B, C, D, E shown in case 1
can be carried out in 'normal' times shown in column X and
that activities A, C and E are capable of being carried
out by different means at 'crash' times for the increased
rates shown in column Y, it is then possible to define three

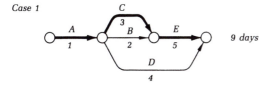

Case 1

9 days

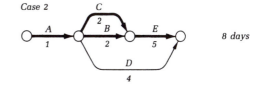

Case 2

8 days

Cost data

Activity	X		Y		
	Dur'n	Cost	Dur'n	Cost	
A	1	120	$\frac{1}{2}$	200	CC
B	2	80	2	80	
C	3	100	2	150	CC
D	4	60	4	60	
E	5	200	3	300	CC

CC = crash costs

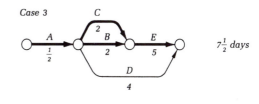

Case 3

$7\frac{1}{2}$ days

I.C. = rate of Indirect Costs – 70.00 per day

Activity	Duration (days)				
	Case 1 9	Case 2 8	Case 3 $7\frac{1}{2}$	Case 4 6	Case 5 $5\frac{1}{2}$
A	120	120	200	120	200
B	80	80	80	80	80
C	100	150	150	150	150
D	60	60	60	60	60
E	200	200	200	300	300
I.C.	630	560	525	420	385
Totals	1190	1170	1215	1130	1175

Case 4

6 days

Summary of cases 1–5 └─ Best time for least cost

Figure 1.5 Time/cost optimisation

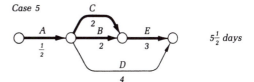

Case 5

$5\frac{1}{2}$ days

basic outcomes from the application of these figures: normal cost programme, all *crash* programme and best time/least cost programme, i.e. an optimisation of time and cost. These alternatives are shown in the networks of cases 1, 5 and 4. The effect of the pro rata indirect or overhead costs which are added to the direct cost variations should be noted.

Resource levelling and control   This technique enables a planner to assess the requirements of various resources to serve any given network of activities and to utilise 'float' in uncritical activities to optimise the use of resources or to reduce imbalances of resource demand. The technique ascribes the various resources to each activ-

ity and by comparison with established norms identifies excessive demands. It is possible to reposition activities requiring excessive use of resources and to balance the total requirements within the resources available or at least to reduce the time of excessive demand. The repositioning of certain activities will often render them critical when they are taken together with fixed waiting periods necessary to the planned use of resources.

This is illustrated in figure 1.6 where the network shown in A(i) yields the scaled network in bar chart form shown in A(ii). By allocating the resource units (RUs) for each activity (shown in the second column) a resource loading diagram can be prepared as in A(iii) which in this case shows an excessive demand of four RUs above the

A(i)

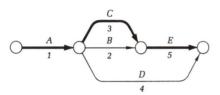

Network with Normal times (N)

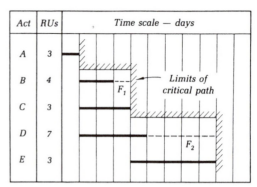

A(ii) *Bar Chart based on Network (N)*

Re-plan due to overload exposed by A(iii)

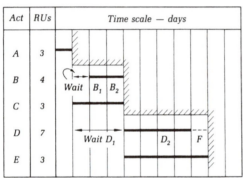

B(i) *Bar Chart still based on Network (N), but activities B and D moved within time spaces available (F₁ and F₂)*

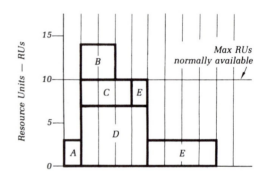

A(iii) *Resource Loading Chart for A(ii)*

Re-check Network as to new critical activities

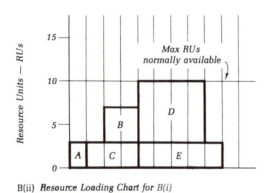

B(ii) *Resource Loading Chart for B(i)*

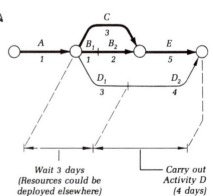

C

Figure 1.6 Resource levelling

resource units normally available during the second and third days, due to activities *B* and *D* coming together.

Figure 1.6 B(i) shows the repositioning of activities *B* and *D* in the excess times available for their execution and a resulting 'levelling' of the loading diagram to bring the requirements for resources within the limits of normal availability as in B(ii).

It will be noticed that this manoeuvre involves specific positioning of the waiting periods B_1 and D_1 and examination of the resulting network at figure 1.6 C shows that these constraints on the commencement of activities B_2 and D_2 leads to the former becoming critical and reduces the float of the latter to one day only.

The foregoing brief description of some of the uses of network analysis has been based on a simplified description of the networks involved. Readers who wish to study these techniques in depth should consult one of the many books dealing specifically with techniques of presentation which allow discrete descriptions of activities by numbers which facilitate input statements into computers, for which many standard programs exist for solving networks and analysing varying plans with a view to optimisation and control.[4]

It has been said with good reason that one of the most useful aspects of network analysis lies in exercising the logic used to set up the basic network of activities since the planner has a full knowledge of the practical consequences of any sequence of activities and the importance of their relationships. This aspect of a network approach to planning is illustrated in the method of presentation of the logic known as a *Precedence diagram*. Figure 1.7 B shows

a typical network restated in this form which eliminates the need for dummy activities normally used in conventional networks to indicate dependency (A).

1.1.2 The overall programme

On acceptance of the tender, contract planning commences and a working, or overall, programme is prepared by the contractor's planning staff together with the plant engineers and the site agent or foreman for the job. As already indicated this will be used as a guide for site activities, for detailed planning, for the buying and delivery of materials, for the co-ordination of sub-contractors' and main contractor's work and for assessing progress. Assumptions made at the estimating stage are borne in mind. It is essential at this point for the contractor to have full information from the architect in the form of a site survey, a full set of working drawings including, preferably, all details and full-size drawings together with those of all specialists, a specification, a copy of the bills of quantities and a complete list of all nominated sub-contractors.

Ample time should be allowed for planning before commencement of work on the site. For most projects, unless particularly small, at least four weeks should be allowed. The smaller jobs must be planned in detail at the outset, since there is not time for making adjustments during the course of a short contract. The larger projects may be planned on broader lines, since time will be available to carry out detailed planning as the work proceeds.

The preparation of the overall programme consists broadly of:

1 breaking the project down into a series of basic trade operations
2 establishing the quantities of work in each operation and the time content of each in terms of operatives and machines
3 arranging the operations in a sequence and balancing the size of gangs to give a maximum continuity of work for each trade and the minimum delay as one trade follows another
4 breaking down large projects into phases so that several operations may proceed simultaneously.

The programme is ultimately expressed in chart form which covers all the main operations throughout the contract, the phasing of the work on different parts where this is necessary and the duration of each operation, including the work of all sub-contractors and specialists. Together with this chart a written report or schedule is prepared, which includes a description of the methods to be used, schedules of plant giving the dates when each machine will be required, the labour requirements for each stage of the work, and information regarding site offices, storage

A

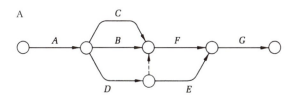

Dependencies:
C, B and D depend on A being completed
F depends on B, C and D being completed
E depends on D being completed
G depends on F and E being completed

B

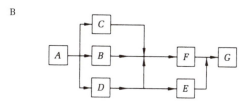

Figure 1.7 Precedence diagram

facilities, equipment and small tools. If, for any reason, the proposals in the contract plan differ from those assumed at the estimating stage, a written cost comparison is drawn up which shows the differences and the cost implications.

The overall programme shows the major operations and significant phases in the project, but detailed short-term planning at regular intervals on the site is necessary to ensure the satisfactory allocation of labour and materials to each individual operation as the work proceeds. This is usually carried out in two stages: (i) a reasonably detailed programme is prepared at monthly intervals, to cover the four weeks ahead, and (ii) a detailed programme is prepared each week, to ensure that labour, materials and plant will be available when required. To enable the site manager to give his full attention during the first few weeks to starting off the project, a detailed programme is prepared for him. This indicates in detail the materials and labour requirements of the first four weeks, together with the operational methods to be used.

The broad picture of the contract planning process given above will now be considered in greater detail.

Project breakdown   For the purposes of the overall programme the project must be broken down into groups of basic operations, each of which involves only one trade. For example, in housing, the cutting and fixing of the carcassing timber in first floor and roof or the first lift of the brickwork, or plastering throughout. For organised planning purposes the project is divided into stages.

As a general guide, these may be:

1 foundations and walling up to DPC
2 carcase to completion of roofing-in
3 finishings and all services
4 drains[5] and site works.

For larger projects and multi-storey work the breakdown stages can be:

1 sub-structure, or foundation work
2 frame, or basic structure
3 claddings, infillings, weather-proofing, etc.
4 finishings and services
5 drains[5] and site works.

Each stage is planned separately at first, to allow some flexibility in relating them on the site; delays due to bad weather or other causes can be provided for by varying the intervals between or overlapping the stages during the course of the work. Compensation for any variations from the programme arising within the stages can be made by increasing the gang sizes to speed up certain operations or, at times when productivity is greater than that assumed at the planning stage, labour can be put on to ancillary works and isolated jobs which, if omitted from the overall pro-

gramme, can be carried out at any time without interfering with the sequence of other operations.

Quantities of work and time content   In order to relate the various operations throughout the project, it is necessary to define the work content of each by means of a schedule of basic quantities, from which the operative time and machine hours required to complete each operation can be ascertained by the application of labour output rates and/or machine hours, or, as they are called, labour and plant standards. These standards in each case are established on the basis of information fed back from previous contracts or from work studies,[6] having regard to the type of labour which will be available and the likely demand on plant at the time of erection. The work content for each operation is then inserted on a schedule of basic operations which can be in the form of a series of *Data sheets*. These are lists of all operations in sequence under trade sections, each operation being numbered, against each of which is placed the quantity of work involved, the amount of labour, plant and materials in each, together with the estimated cost of each operation. Operations which can be carried out concurrently are noted.

These sheets together form a detailed analysis of the complete work and give information for all planning activities during the course of the contract. They provide a link between the overall programme and detailed work on the site and enable the site agent or manager to prepare the short-term plans accurately. In addition, they provide the basis on which materials can be ordered and the correct amount of labour can be applied to each operation. The sheets also provide definite operations against which operatives' time can be recorded for purposes of site bonus and costing procedures.

Sequence and timing of operations   In any section of work which contains two or more operations, one of the operations will govern the time required to complete the whole of the work. Similarly, in each stage into which a project may be divided, there will be one operation or a group of related operations governing the production time of the complete stage. This 'key operation' is the one which takes the longest time when the time cycles of all the operations are based on the use of the optimum size of gang for each. The longest of the key operations in each of the stages is termed the 'master operation' and fixes the rate of production for the whole project. The speed of the master operation is governed either by the time in which the work has to be completed, or the number of operatives undertaking this operation, in which case the amount of labour will fix the time required to complete the operation. In either case, it is essential to bring all other operations

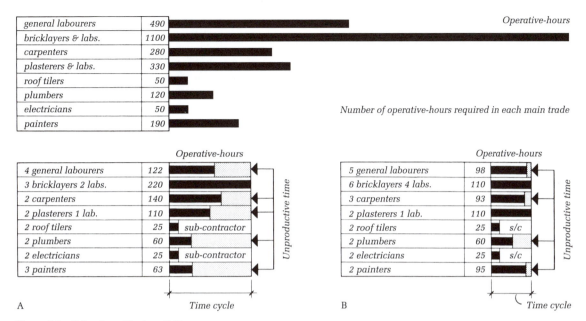

Figure 1.8 Balancing of trade activities

into phase with the master operation. This is necessary in order to ensure continuity of productive work for each specialist trade or sub-contractor and to minimise unproductive time in preparation and clearing up at the beginning and end of each operation.

The time cycles of individual trade operations in each stage are brought into phase by adjusting the labour content so that the working time for each operation is, as far as possible, the same as that of the key operation. This avoids one trade being idle while another related trade completes its work. Although some operations might be finished in a shorter period than the time cycle for the whole stage this would result in no difference in the overall building time, so that wherever possible trade operations should be balanced as shown in figure 1.8, which illustrates the effect of the balancing of working times upon continuity of work and unproductive time in the erection of a pair of semi-detached houses. A indicates the result of haphazard selection of operative numbers, resulting in unbalanced working times. In B it can be seen that planned and phased production using a balance of operatives results in balanced working times, a shortening of the complete time-cycle and a reduction in unproductive time.

However, it is not always possible to balance activities because in some circumstances there may be physical limits to the number of operatives that can be used on a particular operation, or in one trade there may be insufficient work to occupy even one person continuously throughout the complete time cycle. These operations, which must be carried out intermittently, are usually in the services installation and finishing trades, which are usually subcontracted so that arrangements can be made for the work to be done at intervals within the main cycle of operations (see figure 1.8 B).

Relationship between operations of plant and operatives It is essential, particularly in mechanical handling, that the number of persons working on any operation should be correctly related to the output of the mechanical plant serving them. This is necessary in order to avoid the plant being idle from time to time while operatives use the material already delivered to them. Concreting activities, for example, must be related in volume and operative numbers to the size of concrete mixer used or the volume of ready mixed delivery, so that each load of concrete can be received and placed by the time the next load is ready. The number of operatives that can work efficiently on any one site is, of course, limited by the size of the project, the nature of the structure and other considerations, so that this will set a limit to the size of mechanical plant capable of being used to advantage.

Phasing of work Where the project is extensive or consists of a number of blocks, it is usual to phase the work as a whole by dividing it into a number of sections, each of which is planned on the lines indicated previously and so related to the other sections that trade activities can proceed from one to another in a continuous progression.

Each operation should commence as soon as possible without necessarily waiting until the whole of the preceding one is complete, and each should employ the largest number of operatives practicable. In each stage and in all phases every operation should be planned for continuity until completion. Maximum production results when each member of a balanced trade activity is continuously engaged on the same work. It has been shown[7] that in such circumstances a definite increase in production takes place as the contract proceeds, up to a certain point, after which it tends to fall off slightly. In order to assist trade operatives, special instruction is sometimes given on working methods by means of large scale or full-size mock-ups of parts of the structure, particularly when new systems are involved.

The programme chart   The final step is to prepare a working schedule on the basis of the balanced production in each stage, from which programmes for the various stages are drawn up. The stage programmes are combined to give the final overall programme based on the methods and plant to be used and on the balanced production of work. A short interval may be left between the stages to provide for delays due to bad weather or other causes. The extent of these intervals will usually be governed by seasonal conditions and local circumstances. If the project is extensive and allows some freedom for the redisposition of activities in the event of delay at one point, no interval need be left between the stages.

The overall programme is usually expressed as a programme chart in the form of a Gantt or bar chart, on which the sub-division of work is shown on horizontal lines and of time on vertical lines (figure 1.1). Sometimes this is called a progress chart, since it is a useful means of recording the progress of the work. This overall or working programme is intended only as an outline of the site operations as a whole. It cannot be detailed because so many unknown factors which may affect the operations make it essential for detailed programming to be carried out at regular, short intervals during the course of the work. This gives flexibility and allows for rapid revision should progress fall seriously behind the overall programme.

Such a chart may be quite simple, or complex. As a simple progress chart it will consist of a list of the basic operations with a bar opposite each to indicate the length of time the operation is planned to take and at what point relative to the other operations. When the project as a whole is phased, the bars are hatched in sections or lettered to indicate the work in each phase. Usually, the chart also indicates the dates on which orders for materials must be placed, the dates for the delivery of the various pieces of plant and the total number of operatives required each week, with or without a breakdown into trades. A typical chart is shown in figure 1.9.

In addition to the data sheets and the overall programme, a *schedule of contract information* is prepared giving the recommended labour force for each stage of the contract under trades, details regarding the sequence of operations given on the data sheets, and details of equipment and methods of construction to be used. This schedule will also include full details concerning all subcontractors. A site layout plan and a site preparation programme will also be prepared at this stage, as well as the detailed programme for the first four-week period of the contract (figure 1.10).

1.1.3 Planning considerations

A number of factors which have a bearing on the decisions made during the contract planning stage are briefly considered here.

Site conditions and access   Site conditions will limit the type of plant that may be used. On wet sites it will be necessary to use tracked machines in the case of excavators and mobile cranes, and dumpers for transport. Sloping sites may make the use of rail mounted cranes unsuitable or uneconomical. On confined sites there may be insufficient room for a mixer or mixing plant and it may be necessary to use truck-delivered mixed concrete. Limitations of access may fix the maximum size of plant which can be brought on the site. A site closely surrounded by tall adjoining buildings may dictate the use of a derricking jib crane rather than a horizontal jib crane in order to be able to rise and clear the buildings as it turns from one position to another.

Nature of job   The type of structure and the general form, size and detailing of the building will all have an effect upon the way in which the contract is planned. Reference is made on page 15 to the significance of decisions made by the architect at the design stage. As far as the contractor is concerned, the nature of the structure in relation to the site must be considered so that a decision can be made on where best to place equipment and materials. It is desirable that all plant should be so placed on the site that the structure can be erected without moving the plant until most of it is completed. Plant should also be so placed that it can be removed easily after the project is completed.

In some circumstances the contractor may request the adjustment of the structure in some way, in order to permit the most efficient planning of the contract. For example, it may be desirable to enlarge a lift shaft slightly in order that a climbing crane may be accommodated within it, or for certain parts designed originally as in situ cast work to be carried out as precast work, or vice versa, in order

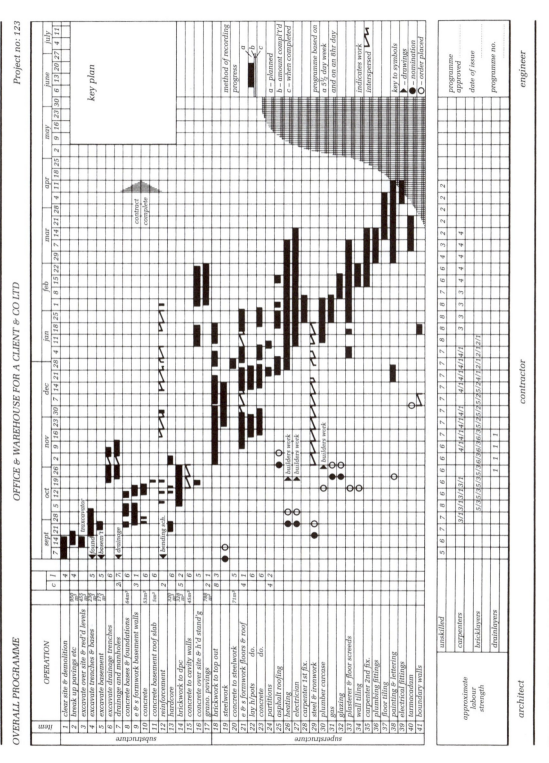

Figure 1.9 Contract programme

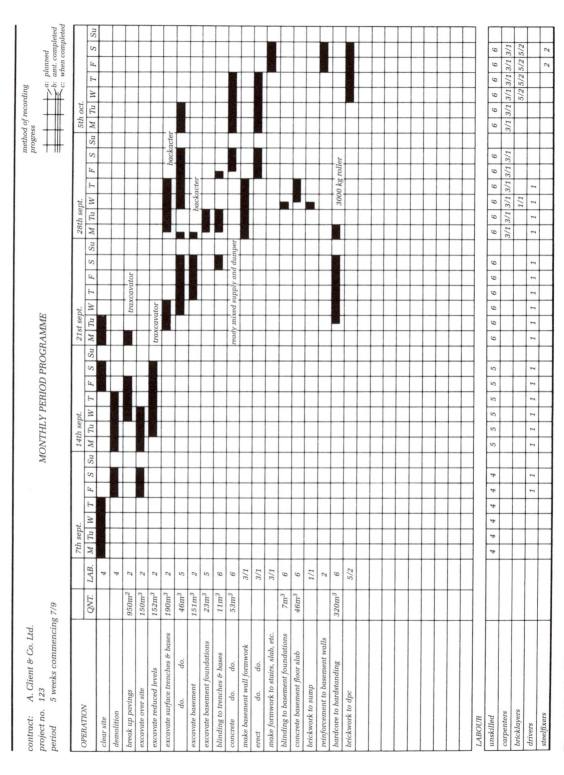

Figure 1.10 Monthly programme

maximise use of a crane. In addition to conditions round the site, the height and width of the building will influence the choice of the type of crane. Many mobile cranes have limitations of height and reach which make them unsuitable for buildings of three storeys and upwards (figure 1.11).

Plant The choice of the most suitable plant for any particular operation necessitates a consideration of the capabilities, limitations and outputs of different types of plant. The following paragraphs deal with a few aspects of the choice of plant at the planning stage.[8]

Excavation This can be carried out either by hand or mechanically by an excavator, which is a tracked or wheeled self-propelled chassis carrying a revolving platform, power operated jib and driver's cabin that can be used for a number of different excavating operations. It can also be rigged as a crane (figure 1.11). Excavated soil (spoil) can be transported in various types of truck vehicle. The length of haul to tip will vary with different projects, so that many combinations of excavator and transporting machines are possible. The contract planner, from experience and knowledge of plant, must in each case arrive at the most economic combination. The method adopted for excavating operations will be dependent upon:

1 the type of excavation to be carried out
2 the nature of the soil to be excavated
3 the volume of soil to be excavated
4 the length of haul to tip and the terrain over which the machinery has to dig and travel.

For small quantities, hand excavation is cheaper than mechanical excavation and the spoil can generally be spread locally. Where transport for the spoil is needed, the type used will depend on the distance to be hauled, the nature of the ground to be traversed and the cost of temporary roads, where necessary.

The total work to be carried out must be reviewed in order to establish whether or not it is possible to use one machine for a number of operations rather than a number of different machines. For example, an excavator rigged as a backacter will dig trenches and also, when rigged as a skimmer, could carry out the reduction of levels on the site if these were not too great in area, and thus avoid the use of another machine in addition to the excavator.

Work must also be phased in such a way that mechanical plant can be used. For example, if a run of drain trench sufficiently long to justify mechanical digging is situated near the building, work must be planned so that the trenches are dug before the building of the structure is commenced in order to provide room for the digger to work.

Handling The handling of structural units and materials in fabrication and erection can be carried out satisfactorily by crane or forklift truck (figure 1.12). If the use of a crane is to prove economical the work must be planned round the crane, the influence of which will, to a large extent, determine the production cycle. Careful consideration must be given to the quantity and nature of materials to be handled and whether or not there is sufficient to keep a crane fully occupied throughout the working day. The forklift truck has some advantages relative to the crane. It is designed specifically for handling materials and, like the crane, gives three-dimensional movement but with much greater horizontal movement although more restricted in the vertical direction. The rough terrain forklifts used on building sites have telescopic masts, some of which attain a lifting height of 6 m.

The delivery of incoming structural and fabricated elements should be phased with the building operations so that they can be off-loaded and placed immediately in their final positions by the crane wherever possible, thus avoiding double handling. In addition to the establishment of

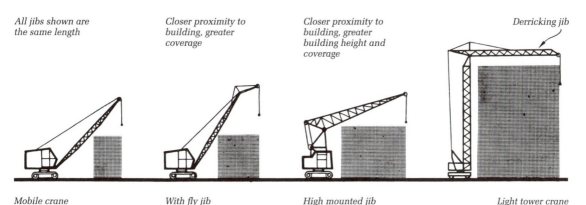

All jibs shown are the same length *Closer proximity to building, greater coverage* *Closer proximity to building, greater building height and coverage* *Derricking jib*

Mobile crane *With fly jib* *High mounted jib* *Light tower crane*

Figure 1.11 Mobile cranes – relative amount of working areas and coverage

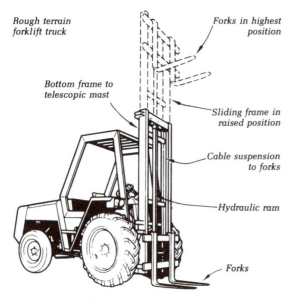

Rough terrain forklift truck

Forks in highest position

Bottom frame to telescopic mast

Sliding frame in raised position

Cable suspension to forks

Hydraulic ram

Forks

Figure 1.12 Forklift truck

balanced gangs the most important factor in planning for high productivity is the reduction of double handling. This involves the careful timing of materials deliveries and the delivery of all materials as near as possible to the point at which they will be used, together with the correct siting of hoisting plant, materials dumps and mixing plant in relation to the building and to each other. Materials should be grouped near cranes and hoists where these are employed so that they can be moved in order of requirement, and in such positions that the crane can hoist and place in one operation with the minimum change of position.

Bricks may be packaged by straps into multiples of fifty, or handled, together with blocks, on pallets, thus permitting the use of a forklift truck or providing a reasonable load for a crane to hoist. The sizes of precast elements or shuttering units should be related as far as possible to the lifting capacity of the crane in order to avoid excessive numbers of lifts and to speed up assembly. In this respect the architect is able to consider the size of precast units, such as claddings for example, on broad lines only, bearing in mind the likelihood or not of a crane being used. This is because the choice of crane depends on other considerations in addition to that of the size of any precast concrete units, and these, under normal contracting methods, become known to the contractor only after the design stage.

The most suitable type of crane will depend not only on the work it is required to perform but, as mentioned earlier, also on the nature of the building on which it is to be used. On large sites it is often necessary to introduce more than one crane and even on a single block, if it is long, a single crane may give insufficient coverage. During the planning stage, it may be necessary to investigate the advantages and disadvantages of different combinations of cranes by means of diagrams such as shown in figure 1.13.

It will be seen that when two cranes are used in order to obtain complete coverage of the building, the arcs of the jibs must intercept and adequate precautions must be taken to avoid collision of the jibs and hoist ropes as the cranes slew. There is less likelihood of such a collision when derricking jib cranes are used, and when horizontal jib cranes are employed the booms should be set at different levels.

It will also be seen from these diagrams that, as well as the amount of coverage given to the building, the

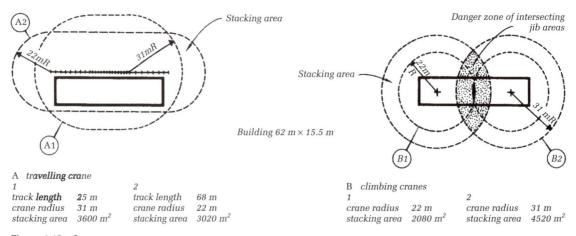

Stacking area

22mR

31mR

A2

A1

Stacking area

Building 62 m × 15.5 m

Danger zone of intersecting jib areas

Stacking area

22m R

31 mR

B1

B2

A *travelling crane*

1		2	
track **length**	25 m	track length	68 m
crane radius	31 m	crane radius	22 m
stacking area	3600 m²	stacking area	3020 m²

B *climbing cranes*

1		2	
crane radius	22 m	crane radius	31 m
stacking area	2080 m²	stacking area	4520 m²

Figure 1.13 Crane coverage

stacking area for materials covered by the arc of the jibs is a significant factor. In figure 1.13 A1, with a short track length and long jib, the stacking area is mainly on one side. With the shorter jib but longer track length in A2, stacking area is provided on one side and at the ends of the building. B2, with the longer jibs, provides a greater area than the smaller areas of B1 but results in a very large intersecting jib area.

Mixing and placing   Type and size of concrete mixer are dictated to a large extent by the quality and quantity of concrete required. To obtain highest efficiency, the concreting equipment must be carefully combined according to the kind of work to be done. When small to medium quantities, say up to 25 m^3 per day, are required, a mixer together with hand loading of the aggregate skip, some form of weight batching and hand barrow delivery can be economical, although scraper loading can be used, except for very small quantities. For site mixing, cement is provided in 25 kg bags. These bags will require a damp-free storage environment with protection from accidental damage. The alternative is a pump-fed gravity discharge system from bulk storage in an elevated cylindrical silo. Silo storage capacities range from about 12 to 50 tonnes. They may be an integral part of the site mixing plant and may have an independent weighing facility. The advantages of bulk purchasing loose cement are:

● the unit cost is considerably less than by the bag
● cement is drawn off in delivery sequence
● less site space is required for storage.

However, if there is insufficient labour available for site mixing or the site is congested with restrictions on space for mixing plant, then ready-mixed concrete is likely to be more expedient. Also, concrete supplied ready-mixed is produced under factory controlled conditions that are not exposed to the possibilities of site contamination of the constituents. Delivery lorries are usually of 4 to 6 m^3 volume with a rotating drum facility. In addition to specifying the mix composition and the compressive (characteristic) strength, the builder can determine whether to receive the concrete fully mixed during transit, partially mixed with additional water added at the site or dry mixed where all the water is added on site. The latter is a useful option where deliveries may be delayed, possibly due to traffic congestion or in areas that are difficult to access or are remote from the mixing depot.

Concrete pumps are an alternative to crane and skip for moving the mixed product from supply point to placement. To be cost effective, the volume of concrete required will have to be high and the delivery requirements continuous. Whether the production and supply is ready-mixed or site mixed, it will have to be organised on a well-managed system of rotation. Placing rates of up to 100 m^3 per hour are typical.[9]

When a project incorporates some cast in situ concrete work and some precast work, it is sometimes possible to arrange site precasting to occupy the idle time of a mixer and its associated handling equipment.

The decision on whether or not to set up a central mixing plant will depend on the amount of concrete to be produced, the relative costs of setting up a central plant and a number of individual mixers, and the relative costs of mixing by the two methods. In addition, the cost of transport from the central mixer to the various points of placing will also affect the decision.

Type of plant   In considering alternatives for the same operation, it must be borne in mind that the cost of mechanical operations is influenced by the nature of the particular activity, the sequence of work dictated by the design of the structure, the methods used for construction and by the amount of work to be done. The efficiency of any mechanised system cannot, therefore, be judged solely on the cost of equipment, but on the influence the system has upon all the related operations and thus upon the overall cost. It is often the case that a combination of plant which is dearer than an alternative combination permits the work to be carried out with greater continuity and in a shorter time, so that the overall cost is cheaper.

Design factors   The importance of a contribution from the architect at the design stage toward improvement in project organisation is stressed on page 1, and some indication is given here of the manner in which this might be made.

Relative to operations   If, in designing, account is taken of the operations which the craftsmen must perform in carrying out the work, and unnecessary labours are avoided, greater speed in construction will result.[10] Simplicity of construction and detailing leads to economies by enabling work to proceed quickly, thus reducing the contract period. Site operations can be simplified in many ways such as by reducing variations in the widths of foundations so that changes in trench-digger shovels are kept to a minimum; by maintaining floor slabs at a uniform level and allowing for variations in finishes by different thicknesses of screed rather than by variations in the floor slab; by designing openings and lengths of wall to brick dimensions in order to avoid the cutting of bricks; by avoiding breaks and returns in walls as far as possible, in order to simplify brick and block work, or shuttering in the case of concrete work, and to keep foundation runs straight to simplify excavation.

Relative to separation of trades   Interference and delay arise when more than one trade has to work on one item at the same time, or one trade has to wait for another before completing work it has partly finished. This can be avoided by the separation of trades at the design stage.[11] For example, by designing the brick walls to a single-storey building as panels running from floor slab to roof, with no openings in them, and the doors and windows as units between them running from floor to roof with no brickwork over, the bricklayer and carpenter can carry out their operations independently. If the roof is of timber construction, since no concrete lintels are required over openings, the concretor does not have to wait after having formed the floor slab in order to follow the carpenter and place the lintels. Finishing work is also simplified as the plasterer has only to work on plain surfaces.

Relative to continuity of work   Detailing that results in the division of work and the mixing of materials can prevent the most efficient organisation and the achievement of the most economical result. This can occur, for example, on small contracts of domestic work in which cross-walls in some blocks are of concrete and in others are of brick, so that the main contractor is not provided with sufficient concrete work to justify setting up an efficient mechanised concrete mixing plant. Similar circumstances can arise when there is a division of similar work between the main contractor and a nominated sub-contractor or when, for example, in precast concrete multi-storey construction the intermediate floors of maisonettes are designed in timber rather than in reinforced concrete. The former, although cheaper as a 'measured' item, disturbs the continuity of work and results in a greater total cost.[12]

The absence of continuity of work is an important factor which reduces the value of the mechanisation of most building operations because so much time is spent on preparatory work between operations rather than on productive work. Continuity is more likely if the operations are simple and repetitive. This should be dealt with at the design stage by detailing which results in (i) as few operations as possible for each aspect of the work, and (ii) the separation of the work of fabrication from that of assembly, since the problems of each are so different that any mechanisation must be independent.[13]

Relative to mechanical plant   As implied above, if full benefit is to be derived from the use of mechanical plant, all aspects of mechanisation and building methods must be considered at the design stage, so that when the contractor takes over the work it can be arranged in a manner suited to mechanical rather than to wholly manual operations. It is essential that the architect should be aware of the advantages of mechanised methods, of the nature and capabilities of various types of mechanical plant and of the factors which are likely to make their use of greatest value. These include continuity of operations and the use of plant at maximum capacity every time it is operated. For example, precast concrete cladding panels can be designed as large or small units. A crane will be more economically employed in hoisting them if the panels are designed as large units approaching in weight the maximum lifting capacity of the crane likely to be used, rather than if they are smaller in size and weight. The cost of hoisting one large or one small panel will be much the same so that the relative cost of hoisting a number of the small panels equivalent in area to one of the larger panels will be high.

1.2 Site organisation

1.2.1 Site planning

As described earlier, a programme covering operations during the first four weeks will have been drawn up at planning stage, in the preparation of which, where possible, the general foreman or site manager responsible for the contract will have assisted, so that he is in agreement with the proposals laid down. This programme will be generally in two parts:

1   Site preparation programme, which will cover the demolishing of any existing buildings, the setting out of the site and marking out of storage areas, the erection of huts and the construction of temporary access roads where necessary.
2   Period 1 programme, on the lines of that shown in figure 1.10, which will cover work during the first four weeks or so of the contract.

Together with these will be provided site layout plans to show (i) traffic routes, on which will be indicated any areas requiring particular attention, such as levelling-off or covering with temporary metal roadways or 'Summerfield' track, and any direction signs required; (ii) the location of offices, huts and stores; (iii) the position of bulk storage areas both during and after excavation together with the location of any equipment.

The general foreman or site manager will also be provided with copies of the overall programme, schedule of contract information and data sheets as well as all other necessary documents such as contract bills, specification, set of contract drawings, details of all material orders placed and to be placed at various dates during the contract, and finally details of the type and quantity of equipment to be used and the approximate periods when they will be required on the site.

Period planning   Work on the site will commence on the basis of the first monthly programme and during the third week of this period, and all subsequent stages, the next monthly programme will be prepared on the basis of the overall plan and data sheets. In the preparation of the monthly plan consideration must be given to the labour force desirable and practicable in the circumstances at the time, to plant requirements and availability, to the phasing and overlapping of operations to ensure completion of the work in the minimum time and to the planning of labour to maintain group identities. Steps must be taken to give adequate warning to all sub-contractors when they will be required on site.

Weekly planning Towards the end of each week progress will be reviewed and the next week's planned progress confirmed or modified if necessary. The following week's planned labour requirements will be reviewed and an estimate made of materials required for the next week but one and a note made of any action required to be taken regarding equipment. This weekly review will be prepared by the site manager in consultation with his trade foremen and any sub-contractors' foremen, and a written report will be submitted to the contractor's planning department.

In certain cases where close integration of fully mechanised operations is required over a short period, particularly in the case of reinforced concrete structures, a weekly programme would be drawn up in chart form by the planning department. Such a chart is illustrated in figure 1.14. In addition to weekly planning, the site manager will hold a brief meeting each day with his trade foremen and sub-contractor foremen to review the next day's work and to make the necessary preparations in regard to the placing of materials and equipment in readiness for the next day's operations.

The site manager should, at the beginning of the project whenever possible, indicate to the local employment office the anticipated 'build up' of labour force during the course of the contract.

Progress control Good site planning is a prior necessity to smooth and effective progress in construction work, but a regular review of the progress of all operations and its comparison with the programme or plan are essential.

Progress is maintained by the site manager, or on larger projects by a progress engineer, by the proper organisation of the delivery and placing of materials, by ensuring that all equipment and plant is in its correct position at the right time, and by adjusting the amount of labour when progress is likely to fall behind the programme because of unforeseen circumstances. Progress is checked during weekly planning by estimating or measuring the work completed, the percentage of each operation or group of operations completed being established and compared with the programme. Progress is marked on the charts as indicated in figures 1.1 and 1.9. When progress varies appreciably from the overall programme and where for this and any other reason it is considered desirable to alter the planned sequence of operations, the site manager would consult the planning department before making such changes.

Close co-operation between the site staff and the planning department is often maintained by means of regular and formal site production meetings between the site manager and the project planning engineer. When considering any changes, the effect on the supply of materials must be borne in mind, and when progress is faster than planned, the supply of materials in time for the work becomes the predominating factor. All sub-contractors must be notified immediately of any changes in the planned programme of work.

The site manager should maintain a record of current and planned labour strength in the form of a schedule or chart on which the following week's planned labour requirements will be entered during weekly planning. In addition, all incoming material will be recorded on a form, one for each main item, which should show among other information dates of order and receipt, quantity delivered and the balance of material outstanding.

The assessment of progress can be carried out expeditiously by computer planning software into which the overall programme has been entered and which is fed with the percentage completion of each operation periodically measured by staff as described above. The computer can calculate whether or not progress has fallen behind the overall programme and can plot on that programme the actual work done on the lines shown in figure 1.1. In addition a work programme for the next period can be produced together with an analysis of the work to be done in that period indicating where more resources may be required.[14]

As an aid to progress control on a project of any size, in addition to the site production meetings mentioned above, other regular site meetings should be held at which should be present the contract manager, site manager or general foreman, architect, clerk of works, quantity surveyor and any sub-contractors when necessary. At these meetings all aspects of the project requiring attention are discussed and decisions for future action made. To ensure that maximum benefit is obtained from such a site meeting an agenda should be prepared and circulated some days prior to the meeting and minutes should be prepared and circulated as soon as possible after the meeting.

1.2.2 Site layout

The need for an efficient layout of site has been referred to on page 13 of Part 1.

The layout of every site may be divided into an administrative area and a construction area. In the former will be

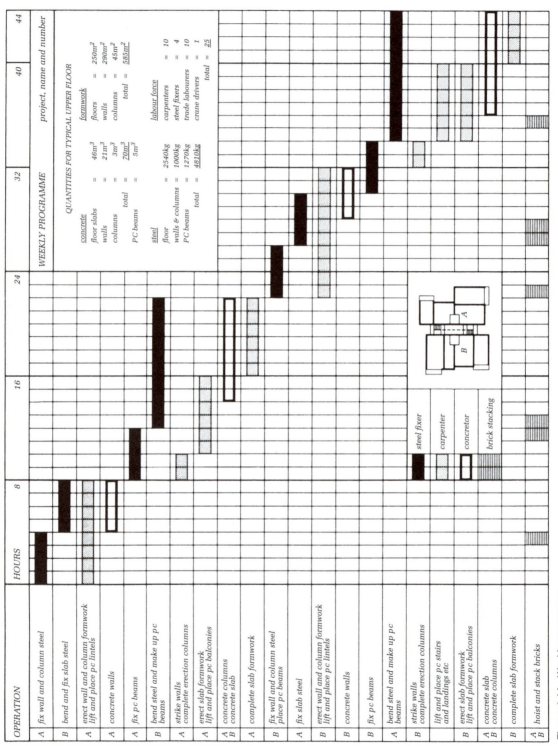

Figure 1.14 Weekly programme

located stores, offices, sub-contractors' accommodation, canteen and first aid facility. In the latter, which will be the actual site of the buildings being constructed, will be located consumable stores adjacent to the various buildings and all equipment required for construction purposes. The layout of both these areas forms an essential part of early planning in every contract, the neglect of which will lead to delay in the initial progress of the project, the tying up of more capital than necessary in materials and financial loss on the part of the contractor.

Proper access and departure routes for lorries should be provided and these should be clearly signposted. In determining the traffic routes attention must be paid to the position of all main services, such as water, gas, telephone, electricity, drains and excavations. Temporary roads must be positioned with sufficient distance between them and future buildings to allow for the movement or positioning of all mechanical plant.

Administrative area This should be located to give quick access to that area of the site which will require maximum labour control and the main storage area; sub-contractors' accommodation and canteen should be so located that accessibility for unloading materials is good and so that they are a minimum distance from the construction areas. The site office should be sited on the route into the administrative area and with as good a view as possible of the construction areas. The size of the site office will vary with the accommodation to be provided, which will depend in turn upon the size of the contract. It should provide accommodation for all or any of the following: site manager, quantity surveyor, timekeepers, bonus surveyor, checker, clerk of works, resident engineer, together with a small lockup store and conference room.

All contracts of any size in the present day require adequate telephone facilities for communication, electricity for power, compressed air for equipment, and lighting and heating facilities for offices and other accommodation. At a very early point in the planning stage the necessary arrangements for these services will be put in hand, particularly requests for telephone facilities and electricity.

The stores area should be situated near the site office and will consist of covered cabins for valuable or non-weatherproof stores, such as paint and ironmongery, and a locked pen for larger valuable stores which are weatherproof, such as metal window frames and pipes. Areas for sub-contractors' stores will be located near the sub-contractors' accommodation and sometimes they will be situated within the main stores area.

Where possible the moving of the administrative area to another part of the site during the course of a contract should be avoided by careful initial site planning, but where this is essential because of the nature of the site or the project, a further site layout plan would be prepared for each move involved.

Construction area This should contain the minimum practical quantities of materials and of necessary equipment and these should be so positioned that handling and movement are kept to a minimum. As the position of equipment, particularly mixers, hoists and cranes, will influence the position of materials such as sand, aggregates and bricks, the position of all plant should be planned before that of the materials. Materials arrive on the site in the order decided at planning stage, or in accordance with instructions issued from the site, and sufficient area must be provided to accommodate the size of batch ordered. In addition, overflow areas should be allocated. In planning the layout of the site, consideration must be given to the excavation stages as these may seriously restrict proposed storage areas.

Standardised materials, such as bricks, tiles and drain-pipes, should be stacked in unit dumps, the numbers in which remain constant although the length, breadth and height may be varied to suit site conditions. Aggregates, sited round the mixer, should be accommodated in light bays which will keep the materials separate, prevent waste and facilitate the checking of quantities in stock. If dumps are clearly identified by signboards, preferably put in position a day or two prior to the arrival of the materials, this saves time in giving instructions to lorry drivers on entering the site and avoids the possibility of dumping materials in the wrong position.

Notes

1 See Part 1, section 1.2.2.
2 See Part 1, *The industrialisation of building*, section 2.2 *et seq.*
3 *Introduction to Building Management*, 6th edition, R E Calvert *et al.*, Butterworth; Heinemann.
4 See *Project Management and Project Network Techniques*, 6th edition, K Lockyer and J Gordon, Prentice Hall.
5 On some sites it may be more expedient to excavate and lay drains during the excavation for foundations.
6 Work study is a tool of production management and is the name given to the study of work processes to find out if they are being done efficiently and, if not, to suggest means or alternative methods by which they may be carried out more efficiently. The process involves the examination of the way operations are performed, which is called *method* or *motion study*, and the time within which they are performed, which is called *time study*. Both of these studies are extensive but interdependent and are usually carried out concurrently by an executive trained in the technique of work study and called a 'Studyman' or 'Work Study Engineer'.

Although in normal building work a very large proportion of individual assembly operations are non-repetitive, in some of the trades there is considerable repetition in the work. In some cases 50 to 65 per cent of the work in the bricklayer and carpenter trades may be repetitive and some 40 per cent of work in concreting is repetitive. Work study, by establishing standard times and developing correct methods, gives considerable advantages in these spheres. In addition, methods for new work may be developed by this means and standards derived which are fair and can provide incentives to operatives to earn a bonus from repetitive work.

For fuller details see *Work Study* by R M Currie, Pitman.

7 See *Routine effect* in Part 1.

8 See also Part 1, page 14 and *Modern Construction and Ground Engineering Equipment and Methods*, 2nd edition, Frank Harris, Pearson Longman.

9 See also *Advanced Construction Technology*, 4th edition, Part 2.5, R Chudley and R Greeno, Pearson.

10 Under the CDM – Construction (Design and Management) Regulations 1994 – the architect, as project coordinator, must also take account of the safety of the craftsman in carrying out the operations.

11 See also Part 1, section 2.2.2.

12 See also Part 1, page 14 on the use of prefabricated components in otherwise traditional construction.

13 See also Part 1, page 14 on continuity of work.

14 The Building Research Establishment has developed a computer program for assessing progress on these lines. It is known as BREMONITOR. There are also numerous commercially produced software packages available to suit varying site situations.

2 Foundations

This chapter introduces site exploration as the means by which knowledge of the soils underlying a site is obtained for use in foundation design. Methods used for carrying out this work are described. Classification of soils according to their various characteristics and their likely behaviour under load is outlined followed by a consideration of some aspects of foundation design and the choice of foundation type, having regard to these factors. The various types of foundation are then described and the chapter concludes with a description of methods used in underpinning existing foundations.

A foundation has been defined in Part 1 as that part of a building which is in direct contact with the ground and its function is that of transmitting to the soil all the loads from the building in such a way that settlement is limited and failure of the underlying soil is avoided. Wind loads, in this context, are assumed to cause pressure on the soil but they can, in fact, result in uplift forces at the foundations due to wind suction on the roof, particularly in extensive light single-storey buildings with flat or low-pitched roofs (see Part 1, chapter 7 for effects of wind), or due to lateral wind pressure on slender, tall structures tending to cause overturning (see Part 1, chapter 3). In such circumstances the foundations may be required to function in holding the structure down against wind uplift. Reference to this is made in chapters 4 and 8 of this volume.

Superstructure, foundations and soil act together. As indicated in Part 1 the design of the foundations cannot therefore be satisfactorily considered apart from that of the superstructure they carry. Nor should the superstructure be designed without reference to the nature of the soil on which the foundations rest. In the case of probably 75 per cent of new buildings, similar in size and character and founded on the same type of soil as the surrounding exist-

ing buildings, no problems arise in this respect. The structure and foundations of a new building can be similar to those already erected. But when a new building is much higher than those already existing or varies radically in some other way, or is built on a previously, undeveloped 'green-field' site, its design should be with regard to the type of foundation likely to suit the ground. Only in this way can a satisfactory and economic solution to the design as a whole be obtained.

It is therefore necessary for the designer to have not only some knowledge of the nature and strength of the materials to be used for the foundations and superstructure, but also some knowledge of the nature, strength and likely behaviour under load of the soils on which the building will rest. An overall picture of the condition below the surface of the site is an essential factor in the selection of the type of superstructure, since it may affect fundamentally the whole planning of the building. In addition, it may affect the placing of a building or a group of buildings on a site. The science of soil mechanics, by means of site explorations and tests, provides this essential information.

2.1 Soil mechanics

In the past the foundation for a building was designed on the basis of experience and although the majority were successful there were many failures due to unknown factors which experience on previous buildings had not brought to light.

Until the end of the first quarter of the twentieth century the soil had not been the subject of analytical study by engineers, although in the nineteenth century many observations were made on soils and the data recorded during the extensive construction work carried out in the development of railways and canals and in the erection of industrial buildings.

Towards the end of the nineteenth century the width of foundations was adapted to the nature of the soil by applying permissible bearing values to the main categories of soils, i.e. gravel, sand, silt and clay. It was assumed that no settlement occurred unless the unit load on a base was greater than the bearing value of the soil on which it rested. In the first quarter of the twentieth century it was realised that every load produces settlement regardless of the permissible bearing value of the subsoil, the settlement depending on many factors other than the pressure at the foundation including the nature of the section of soil to a considerable depth and the dimensions of the loaded area. This led to the concept of allowable settlement rather than permissible bearing value, and stimulated research to determine the relationship of loads to the stresses and strains set up in the soil and the behaviour of various soil types under load and to establish satisfactory methods of obtaining, classifying and testing soil samples in order to provide adequate data for foundation design. These investigations established the science of what is now known as soil mechanics.

Soil mechanics provides the engineer with information on the properties of soils which geology alone is unable to provide but which is essential in the design of any structure bearing on or against the soil and which permits a more rational approach to such problems than is possible by empirical methods based solely on experience. The practical means of providing the engineer with this necessary factual information is by thorough investigation, or exploration, of the soils on a building site and the laboratory examination of soil samples.

2.1.1 Site exploration

The aim of a site exploration,[1] or sub-soil survey, is to provide a picture of the nature and disposition of the soil strata below the level of the ground and to obtain samples of the soils in the different strata for subsequent laboratory tests and examination.

Such an exploration should be carried out at a very early stage of a building project, commencing at the same time as the preliminary design of the structure. Because of the close connection between the soil conditions and the design of the superstructure it is inadvisable to complete the exploration before the design has been considered. This avoids unnecessary expenditure in obtaining unwanted information and the possibility of obtaining insufficient information for design purposes.

The extent of an exploration will depend on the size and type of structure, the nature of the site and the availability of local geological information. When the proposed structure is light and simple it may be cheaper to check site conditions with a few trial pits or boreholes and to use a high factor of safety rather than to carry out an extended survey (see Part 1, page 52).

Should initial borings indicate considerable horizontal variations in the sub-soil a greater number would be required and in the case of clay soils an extended exploration would provide information to permit a suitable distribution of load on the foundations to be arranged so that the effects of differential settlement were minimised. Full information from an extended exploration will also permit the use of a low factor of safety with consequent economies in the foundations, which can be considerable in the case of large, heavily loaded buildings.

The exploration should be taken deep enough to include all strata likely to be affected significantly by the loading from the building and this depth will depend on the type of structure, on its weight and, particularly, on the size and shape of the loaded area. Usually investigations must be made to a depth of at least one and a half times the total width of a pad foundation or three times that of a strip.[2]

The cost of an exploration will, of course, vary with the type of structure and the nature of the soil but is low when compared with the total building costs: it will range from 0.1 to 1 per cent of the cost of the structure, being lower for large projects because the fixed charges for setting up for an exploration will be constant.

As a preliminary to the sub-soil survey a desk study of the site and surrounding area should be made as described, and for the reasons given, in section 4.3 of Part 1.

Trial pits and boreholes   On the site itself the sub-soil strata may be examined by means of trial pits or boreholes. As mentioned already the number and disposition of these will depend on the type of structure to be erected, its function, occupancy and the site conditions. The number must be sufficient to provide an adequate picture of the sub-soil conditions over the site and sufficient in depth to include all strata likely to be affected to an appreciable extent by the imposed building load. A spacing of 10 to 30 m is usually adequate depending on circumstances.

Where practicable trial pits (see Part 1) are preferable to boreholes because the exposed rock or soil may be examined in situ, irregularities noted and any inclination of strata measured. Beyond a depth of about 3 m the cost of sinking a pit increases rapidly relative to that of boring, so that for depths of over 6 m boreholes are invariably used.

Hand boring   To depths up to 6 to 9 m in reasonably soft soils unlined boreholes may be put down by hand by means of a simple post-hole auger and extension rods. The equipment consists of a clay or gravel auger, according to the nature of the soil being penetrated, screwed to a length of steel pipe to which further lengths of pipe are added as the auger sinks (figure 2.1 A). The auger, which for this purpose is usually 100 to 150 mm in diameter, is turned by means of a tee-piece handle of steel pipe.

Other types of borers are available which facilitate boring in stiffer soils. One is shown in figure 2.1 B. The hinged upper end of the drill allows loosened soil or stones to rise upwards as boring proceeds and when the drill is raised for clearing the flap closes and retains the material on the drill. The side cutters can be adjusted to cut a hole wider than the drill to allow liners to follow the drill as it descends in loose gravel and sand where the walls of the borehole may fall in. The cutting action extends the cutters but on withdrawal a slight turn to the left draws them within the radius of the drill and allows it to be drawn upwards. In clay and other cohesive soils they admit air or water thus preventing a vacuum below the drill as it is withdrawn and facilitating its withdrawal.

A cylinder attachment can be fitted to avoid frequent clearing when boring deeper than 3 m; the hinged flap on the drill and a hinged door on the cylinder for clearing result in a form of gravel shell (figure 2.1 C). Special points can be fitted to facilitate boring in damp sand and soft rocks (D). Hard rocks must first be broken by chisel bits (E), then raised and dropped to break the rock, after which the drill or auger is used to extract the chips.

In free running dry sand or wet sludge the cylindrical borer shown in F can be used, the sand or sludge falling into the top of the cylinder in which it is brought to the surface.

For boring to depths greater than about 9 m it is usually necessary to use boring tackle and winch (G). Boring procedure generally is by hand as already described but the raising of the auger or gravel shell with its load of soil is facilitated by the use of the winch. In hard rocks or compact gravels the winch rope can be used to raise and drop the chisel bits. Lining tubes fitted with cutting shoe and driving cap can be sunk in loose sand and gravel by driving with a 'monkey' and leader[3] suspended from the winch, the core being removed by auger or a shell similar to those used for bored piles (see figure 2.31). The tubes are ultimately removed by the hand-winch or by screw-jacks. Liners are essential with soils such as loose sands and gravels but even with stiffer soils a liner for the first metre or so prevents loose top soil being knocked into the hole and acts as a guide to the auger when being inserted after clearing.

A hand rig is suitable for holes up to 200 mm in diameter and up to 24 m in depth, beyond which a power winch is usually necessary.

Mechanical boring Mechanical mobile borers speed up the boring process and reduce costs when a large number of holes is required. This type of borer is used for drilling the holes for short-bored piles and is illustrated in figure 2.30 A. Continuous-flight augers (see page 62) are suitable for use in cohesive soils and a soil sampler (page 26) may be lowered down the hollow stem into the soil below the auger. Holes up to 250 mm diameter may be formed up to 50 m in depth.

For piercing very hard soil or rock various types of rotary drills may be used with hollow core-bits for the recovery of sample cores of rock.

When bedrock or other easily identified strata such as sand are to be located a quick method is by wash-boring in which the cutting bit is carried at the end of jointed tubes and as it is power rotated water is pumped down the tube and out of the bit. As it returns up the borehole, which is lined with tubes, the water carries with it particles of the soil cut by the bit from which a particular stratum may be identified (figure 2.1 H). In softer soils a bit is not required, the jet of water itself breaking up the soil to permit the liner to be sunk. The method is not suitable for obtaining a detailed section of the subsoils below a site, although driving can be stopped at intervals and samples taken below the wet, disturbed soil at the bottom of the hole by means of an extra long-reach sampling tube.

Geophysical methods of exploration For economic reasons boreholes, particularly if deep, frequently cannot be placed close enough to give an accurate picture of the subsoil conditions. Geophysical methods may be adopted in such cases as a means of providing data between borings or of establishing the most useful positions of such borings and at the same time reducing the number necessary.

These methods can be used to obtain rapidly the depth and position of changes in strata over large areas but are only successful when the soil formations have marked differences in characteristics.

The most suitable geophysical methods for providing the information on thicknesses and depths of strata required for work on foundation studies are seismic and electrical. The seismic method of recording the reflection and time of travel of vibrations set up in the soil is useful for locating bedrock and for evaluating the stiffness of various types of soil, but is not suitable for use in built-up areas or on small sites. The electrical method is used for locating the position and depth of different subsoils. The most generally used of the electrical methods is the earth resistivity method, which involves introducing a known electric current into the ground between two electrodes driven into the soil and then measuring the potential difference between two inner electrodes. The spacing of these electrodes controls penetration below the ground surface and calculations based on the observed currents and potentials enable the resistivity of the soil at different

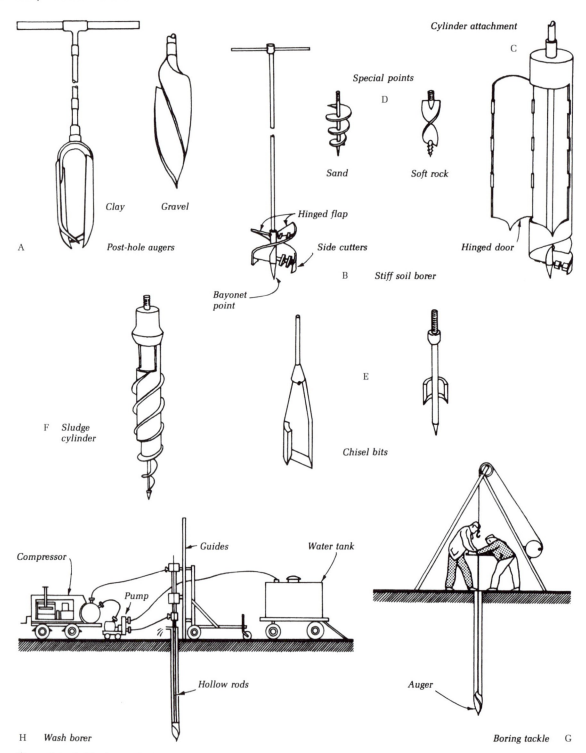

Cylinder attachment

C

Special points

D

Sand

Soft rock

Clay

Gravel

Hinged flap

Side cutters

Hinged door

A

Post-hole augers

B Stiff soil borer

Bayonet point

E

F Sludge cylinder

Chisel bits

Compressor

Guides

Water tank

Pump

Hollow rods

Auger

H Wash borer

Boring tackle G

Figure 2.1 Soil boring methods

depths to be established. From a knowledge of the differences in resistance of various soils it is possible to determine the depth and thickness of the subsoil strata. Some boreholes are an essential part of this method in order to prove the accuracy of the estimates and to provide data for adjusting them if necessary. They also provide the soil samples required for testing purposes.

The correct interpretation of the data obtained calls for much skill and experience and this work is invariably carried out by specialists.

In situ field tests[4]   These are tests carried out in the field as distinct from those carried out in the laboratory and form part of the procedure for site exploration.

Vane test   This is used in soils such as soft clay and silty clay from which it is difficult to obtain good undisturbed samples for testing purposes. The apparatus consists of a small four-bladed vane attached to a high tensile steel rod as shown in figure 2.2 A. This is pushed about 750 mm into the soil at the bottom of a borehole and the force required to twist the vane is measured. This force is required to overcome the shear strength of the soil acting over the surface of the cylinder of soil turned by the vanes so that from it the shear strength of the soil can be calculated. It can be used up to a depth of about 30 m.

Penetration test   This is used in non-cohesive sands and gravels from which undisturbed samples can be obtained only with very great difficulty. The penetrometer consists of a rod on the bottom end of which is screwed a bullet-shaped point 50 mm in diameter and on the top end a driving cap over which slides the tube of a 65 kg drop-hammer which drops through 760 mm and may be operated by hand or by a hand-winch working over a rig. Extension rods are added as driving proceeds and at intervals the number of blows of the hammer required to drive the point a given distance is recorded in order to determine the relative density of successive strata. This 'Standard Penetration Test' consists of recording the number of blows required to produce a penetration of 300 mm, the density of packing of the soil being classified as:

Loose	– less than 10 blows
Medium dense	– 10 to 30 blows
Dense, or compact	– 30 to 50 blows.
Very dense	– more than 50 blows.

A variation used in more cohesive soils has a split barrel sampler screwed into an annular driving shoe and an adaptor on to the extension rod. When removed from the ground, the barrel is detached and separated to provide an undisturbed soil sample. For cohesive soils the following classification applies:

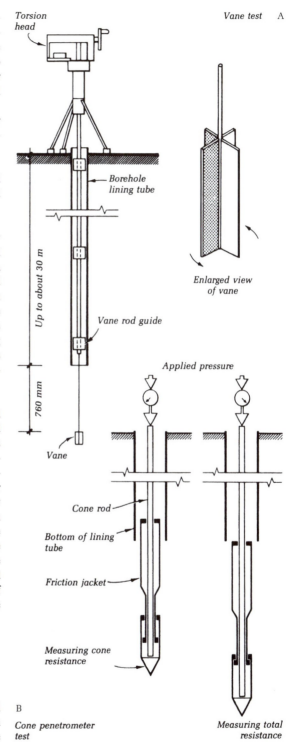

Figure 2.2   In situ soil tests

Very soft – 0 to 2 blows
Soft   – 2 to 4 blows
Medium   – 4 to 8 blows
Stiff   – 8 to 15 blows
Very stiff – 15 to 30 blows
Hard   – more than 30 blows

Dutch or cone penetrometer test This is used to determine the bearing capacity of piles or as a rapid means of
preliminary exploration. The apparatus is shown diagrammatically in figure 2.2 B. The force on the 35.6 mm diameter cone is measured as it is pushed down by the cone
rod. When the cone retaining sleeve engages with the friction jacket measurement is made of the force required to
push both down; by deducting the force on the cone from
the latter the friction may be calculated.

All of these tests are used only to supplement borings
from which fuller evidence of the nature of the subsoil is
obtained.

Loading test This consists of applying loads to a steel
plate at approximately the proposed foundation level and
measuring the amount and rate of settlement which occurs
under progressively increasing loads. At the beginning
settlement is rapid after the application of the load but it
soon ceases; as the loading increases settlement will continue for a longer period between each increment of load
until a point will be reached when settlement continues
indefinitely. The ultimate bearing capacity of the soil is
judged by the maximum load which can be applied before
this occurs.

This test is of short duration and measures primarily the
settlement due to direct compression and lateral displacement of the soil. It is, therefore, suitable for uniform non-
cohesive soils and soft rocks in which settlements are
almost entirely due to these causes, but the ultimate bearing capacities obtained should be treated with reserve
since the area of the test plate is smaller than the foundations and the depth of the soil affected is small; they
should never be applied to deeper strata likely to be
stressed by large or closely spaced foundations. The test is
not suitable for clay soils in which consolidation settlement, taking place over a very long period of time, may be
the critical factor.

Soil samples and laboratory analysis[4] Samples of the
different soils encountered are required for testing and
analysis in the laboratory where this is appropriate to the
nature of the soils. For tests such an mechanical analysis
to determine particle sizes samples of all types of soil can
be taken as extracted by the boring tool or from the bottom of a trial pit and placed in tins, bottles or polythene
bags. For compressive and shear strength tests samples

undisturbed by the boring or digging process are required
because the strength properties can be altered by disturbance. For example, the shear strength of some clays may
drop to one half to one quarter of their original strength
when disturbed or 're-moulded'.

Samples with structure and moisture content unaltered
can be taken from trial pits by a hand sampling tube when
the soil is sufficiently firm and cohesive. This consists of
a short length of metal tube sharpened to a cutting edge
at one end and fitted to a removable handle. The tube is
pushed into the soil, given half a turn to shear off the sample and extracted. The handle is removed and the tube
sealed at both ends by rubber bungs or plastic caps.

Various sampling tools or soil samplers for obtaining
soil specimens from boreholes have been developed,
based on a thin metal tube with a cutting edge and
designed to minimise disturbance as much as possible;
these replace the auger as each new stratum is reached and
extract an undisturbed core of the subsoil. When taken
from the boring rods the cylinder is capped or sealed at
both ends. The type commonly used for cohesive soils is
shown in figure 2.3. Standard compression machines use
38 mm diameter cores but for other tests a core of not less
than 75 mm diameter and about 450 mm long is required.[5]

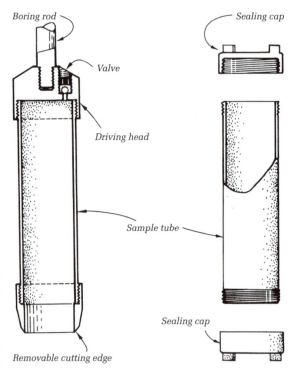

Boring rod

Valve

Driving head

Sample tube

Removable cutting edge

Sealing cap

Sealing cap

Figure 2.3 Soil sampling tool

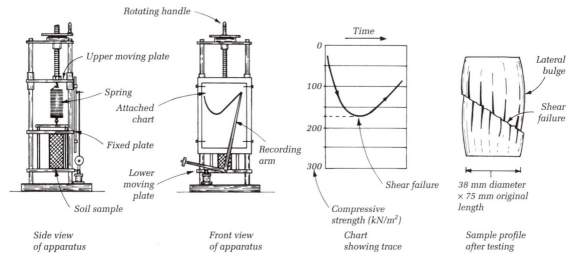

Figure 2.4   Unconfined compression test

Unconfined compression test   This very simple portable soil testing apparatus can be applied to samples as they are extracted directly from the ground, or later in the preferable surroundings of a laboratory. If samples are transferred to a laboratory, they should be retained and sealed in the sampling tube to keep them moist before testing.

The test is limited to self-supporting, cohesive non-fissured soils (clays). Although very basic in its principle of compression, given several test samples, the equipment provides quite adequate guidance on the shear strength potential of a soil. Samples are a standard 38 mm diameter and, when carefully extruded from the sampling tube, cut to lengths of 75 mm. Figure 2.4 shows a sample under test, with shear failure defined as the lowest point on the trace recording.

Accessing samples   It is extremely difficult to obtain undisturbed samples of most non-cohesive soils and in situ field tests must be made. Reasonably undisturbed cores can, however, be taken from moist sand above ground water level and for sand below ground water level undisturbed samples may be obtained by a compressed air sampler by means of which the core can be retracted into an air chamber and brought up without contact with the water in the borehole.

Sample cores of rock are obtained by the use of hollow core bits on a rotary drill. The size of core necessary to obtain a continuous sample will increase with the softness of the rock.

Undisturbed samples should be taken at each change of strata and at every 1.5 to 2 m depth of borehole and samples for classification purposes should be taken at every metre depth from the cores as they are laid out in sequence for inspection.

When ground water is encountered a sample should be taken and its level recorded.

Reports   A site exploration report should contain information about the following:

1 Topographical features to include site levels, slopes and hollows
2 Site dimensions
3 Orientation, i.e. N, S, E or W
4 Trees – preservation orders
5 Existing buildings – preservation/listing orders/demolition
6 Proximity of neighbouring properties, shared/party walls and boundaries
7 Natural land drainage, potential for flooding
8 Evidence of mining operations
9 Location of services/utilities, e.g. gas, drains, water, etc.
10 Exposure levels, i.e. potential wind loading.

The procedures and methods for obtaining the above are well documented elsewhere.[6] The purpose of this text is to consider soil investigation, and in this context the site exploration report should include the following:

1 A complete record of the depth and thickness of the various strata in each trial pit or borehole and a note of the level of any ground water
2 A copy of the site plan showing the position where each soil sample is taken

3 A full description of each stratum based on the classification of soils by grain size. Apart from boulders, these categories are gravel, sand, silt, clay and peat, the latter not considered to be a satisfactory bearing soil.[7]

4 On 'brown-field' sites an analysis of the subsoil to determine whether it contains any contaminants from previous industrial use

5 On sites previously used as landfill, measures to be taken to determine the gas hazard potential from methane, carbon dioxide and volatile organic compounds generated from decaying matter

6 Consultation with the local planning and building control authorities to ascertain the possibility of naturally occurring radioactive gases (radon) from uranium and radium deposits in the ground.

Factors 4, 5 and 6 are considered in more detail under section 2.1.3. Soil categories in 3, together with the bearing capacities of the soils and simple field tests by which these types may be identified on site, are given in Part 1, table 4.3.

A note should be made regarding the degree of compaction of gravels and sands and of the presence of clay or silt. Some clays break into irregular polyhedral pieces when a lump is dropped. This is most important because these 'fissured' clays can carry only about half the load they could in an unfissured state. The weight, colour and smell of all the soils may also be recorded.

Site exploration reports are based on the records of the findings in each trial pit or borehole, an example of which is given in figure 2.5.

Reference should be made to Part 1, chapter 4, where the subject of soils and their characteristics is introduced. Particular attention should be given to the different properties of the two extreme groups of soils, the fine-grained, cohesive soils and the coarse-grained, cohesionless soils and to the effects these have upon the behaviour of the soils under load and under different climatic conditions.

The information obtained from the exploration of a site must be extended by a laboratory examination of the soil samples collected from the various strata. This involves a number of tests to classify the soil in greater detail and to establish certain properties which govern the design of foundations.

2.1.2 Soil classification

In order to be able to judge the likely behaviour under load of any particular soil, it is necessary to be able to identify it so that it may be compared with similar soils the behaviour of which is already known and recorded.

It has been found that the physical properties of a soil, which are those most relevant to foundation design, are closely linked with the size and nature of the soil particles and with its moisture content, and soils are classified on the basis of these characteristics.

Classification by grain size The types of soil already mentioned, excluding peat, are defined in terms of particle or grain size as follows:

Gravel: particles larger than 2 mm (up to 60 mm, above which the term 'boulders' is applied)
Sand: particles between 2 mm and 0.06 mm
Silt: particles between 0.06 mm and 0.002 mm
Clay: particles smaller than 0.002 mm

The further sub-division made within these limits in gravels, sands and silts is shown on the graph in figure 2.6.

The size distribution of soil particles is established by means of sieving and sedimentation. The particle sizes in millimetres for gravels and sands closely correspond to certain BS sieve sizes[4] and these sieves are used in the classification of these soil types (see figure 2.6). In the case of silts and clays with particles below 0.06 mm in size, sieving cannot be used and the method known as sedimentation is usually adopted, in which the soil particles are mixed with water, generally containing a dispersing agent to keep the particles apart, and samples of the particles in suspension at a given depth and at regular intervals of time are taken. As the larger particles sink more quickly than the smaller the distribution of particle size in the silt range and the amount of clay particles may be calculated.

Many soils are, of course, a combination of particles from the main groups of soils and this necessitates a further classification relating them according to the percentage of each which they contain. This is shown in the classification or grain size triangle in figure 2.7, in which there are eleven sub-divisions and from which it can be seen, for example, that a silty clay loam is defined as a soil containing up to 30 per cent sand, 50–80 per cent silt and 20–30 per cent clay.

The distribution of particles according to size in a sample of soil may be plotted graphically to give a 'grading curve' from the shape of which can be judged the type of soil being considered. In order to spread the smaller sizes on the graph adequately the logarithm of the particle size instead of the actual size is plotted along the horizontal axis. Some typical grading curves are shown in figure 2.6. Curve A indicates a clay sample (52 per cent being in the clay range, 36 per cent in the silt range and 8 per cent in the fine sand zone – see also figure 2.7). D shows a well-graded coarse sand. The latter can be compared with curve C which is a fine sand less well-graded as shown by the steep part of the curve which indicates a preponderance of the small sizes between the limits of the steep section of the curve. Curve B indicates a silty loam, the flat part of

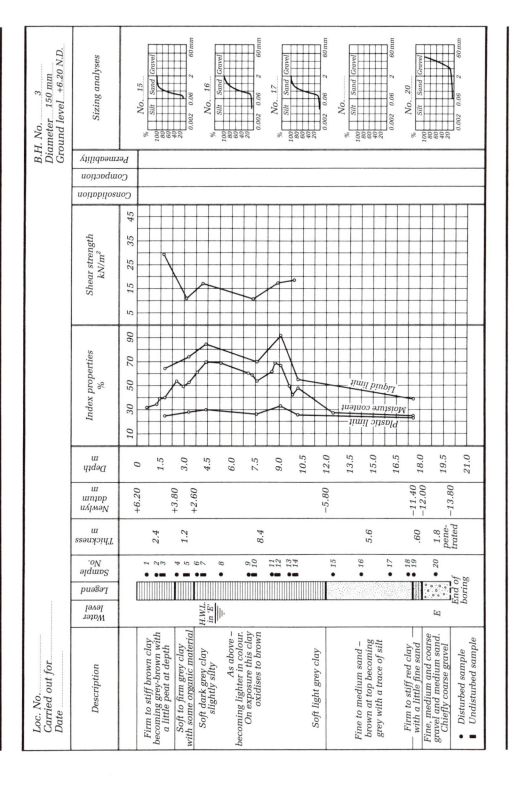

Figure 2.5 Borehole record

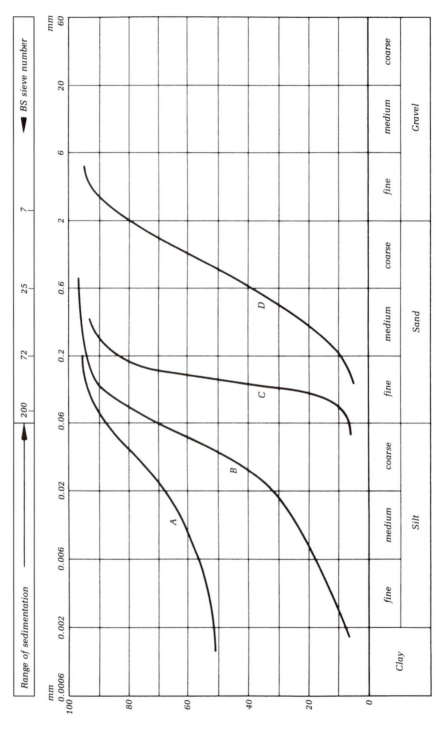

Figure 2.6 Particle-size distribution curves

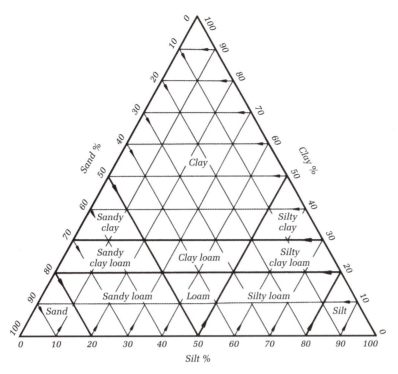

Figure 2.7 Soil classification triangle

the curve at the top indicating a low percentage of the larger particles lying between the limits of the flat section of the curve. It will be seen from these curves that the higher on the graph and the farther to the left a curve lies, the finer is the material of which the soil consists; conversely, the lower and the farther to the right a curve lies the coarser is the material represented.

Classification by limits of consistency Classification by grain size distribution alone provides insufficient data for classifying fine grain soils because different soils in this category may have the same grading curve but exhibit different properties because of variations in the shape and nature of their particles. These soils change in consistency and, in the case of clays particularly, in volume with changes of moisture content and a further classification is adopted based on the limits of consistency. These are, at one extreme the moisture content at which a soil in liquid state on drying out becomes plastic and can be moulded in the fingers, which is termed the *Liquid Limit* and, at the other extreme, the moisture content at which the now plastic soil, on further loss of moisture, begins to solidify and crumble between the fingers, which is termed the *Plastic Limit*. At a lower moisture content the soil completely solidifies and ceases to lose volume with further loss of moisture.

The liquid limit is measured in a standard Casagrande Liquid Limit apparatus[8] and the plastic limit is measured by rolling the soil on glass with the palm of the hand until it is a thread 3 mm in diameter. The moisture content at which it just fails to do this and breaks into pieces is the plastic limit. The difference between the two limits, that is liquid limit less plastic limit, is called the *Plasticity Index* and indicates the degree of plasticity of a soil, the more plastic the soil the greater the plasticity index. Typical values of some soils are given in table 2.1.

In addition to classifying a soil in terms of its particle-size distribution and plasticity it is necessary to establish certain other essential characteristics which govern design. These are compressibility, permeability, strength and density. The latter may be measured on the site.

Compressibility Soil is made up of solid particles with voids between filled with either air or water, and the ratio

Table 2.1 Plasticity index

Soil	Liquid Limit %	Plastic Limit %	Plasticity Index
Silt	36	20	16
Silty clay	48	24	24
London clay	75	25	50

of the volume of voids to the volume of solids is called the voids ratio. When soil is under pressure, any compression which may take place does so by a decrease in the volume of the voids since the volume of the solid particles cannot vary.

The property of a soil to deform under load by the closer wedging together of the soil particles due to the expulsion of air and water from the voids is known as compressibility. When compression is due to the expulsion of air, as under roller or tamping, it is called compaction and when due to the expulsion of water it is called consolidation.

Consolidation tests are carried out in an oedometer or consolidation machine and a curve is constructed showing the relation between the voids ratio (e) and the applied pressure (p) at successive stages of consolidation, known as a p–e curve, from which values can be obtained for an equation for the calculation of settlement (see figure 2.14 and page 39).

Permeability   The passage of water through the voids of a soil is known as permeability and the rate of settlement under pressure depends on the ease with which any water present can flow out of the soil, that is on its degree of permeability. In sands and gravels the voids are so large that any water in the soil rapidly flows out and permits the soil particles to consolidate quickly but in the case of clay soils consolidation is a much slower process. This is because the voids in clay are minute in size compared with those in sand, for example, and are usually full of water which must force its way slowly through the fine spaces as pressure is exerted on the soil. The degree of permeability, expressed as a coefficient in metres per second, is obtained by measuring the percolation of water through a soil sample in a permeameter.

Strength of soils   The resistance to failure offered by a soil depends on its shear strength, failure occurring by surface slip. Shear resistance is ascertained in a *shear box* by means of which is measured the force required to shear a sample in half horizontally under varying degrees of vertical loading. The force applied at failure divided by the cross-sectional area of the sample gives the ultimate shearing stress.

Shear strength is considered to be made up of (i) internal friction, the resistance due to friction between the particles, and (ii) cohesion, the resistance due to the tendency of the particles to hold together. Granular or coarse-grained soils derive their shear strength almost entirely from internal friction so that as vertical pressure squeezes the particles closer together the shear resistance increases. These are called cohesionless or non-cohesive soils. Fine-grained soils depend on cohesion only; however much they are loaded the particles of a soil such as clay develop no friction so that the shear resistance remains constant and equal to the cohesion of the soil. These are called cohesive soils. Intermediate types of soil exhibit both forms of resistance. The graph in figure 2.8 illustrates this clearly. Sample *A*, a sandy soil, exhibits no cohesion but shows a large angle of internal friction, that is to say there is a considerable increase in shear resistance with increase in vertical load. Sample *B*, a sandy clay, because of the

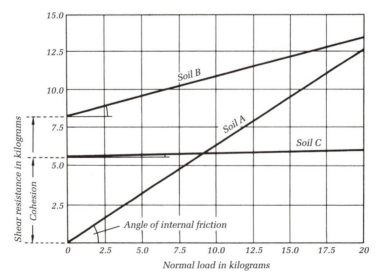

Figure 2.8   Shear box tests

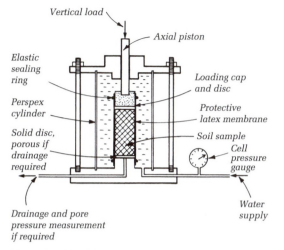

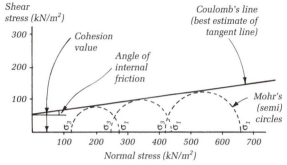

Figure 2.9 Triaxial compression test

clay content has considerable cohesion and some frictional strength which is indicated by the relatively small angle of internal friction. The shear strength of sample C, a silty clay, depends almost entirely on cohesion although a slight amount of frictional strength is present. In the case of a clay the graph line would be horizontal.

This difference between the sands and clays is most important because it is in clay soils and those with small angles of internal friction that shear failure is most likely to occur. An increase in pressure on the soil due to the weight of a building will result in no increase in the strength of a clay soil, whereas in the case of a sand its shear strength will rise as the weight of building increases.

Shear strength may be measured indirectly by compression tests, of which there are two. One, the unconfined compression test described previously on page 27 and shown in figure 2.4, is suitable only for clays. Although principally a compression test, the shear strength of clays is known to be half the compressive strength. The other test is the triaxial compression test, which can be used for either sand or clay.

The equipment illustrated in figure 2.9 is applied to three 38 mm diameter × 75 mm long soil samples extracted from the same sampling tube. Each is subjected to a higher lateral water pressure before vertical loading. The calculated results are plotted as the best tangential line across the three semi-circles (Mohr's circles) shown.

The formula for stress failure (σ_1) is:

$$\sigma_1 = \sigma_3 + (L \div A)$$

where σ_3 = hydraulic/water pressure
L = load at failure
A = cross-sectional area of sample at failure.

2.1.3 Contaminated land

Subsoil analysis in not just limited to determining the physical composition, loadbearing capacity and shear strength of soils. Analysis of soil samples for contaminants is also of importance. The Environment Agency estimate that there could be some 300,000 hectares (approximately 740,000 acres) of land in England, Scotland and Wales affected by some form of contamination. Land contaminants are not necessarily man-made from former industrial or agricultural uses; they may occur in the natural formation and composition of the subsoil. Contaminants occur as solids, liquids or gases and include substances that are corrosive, explosive, flammable, radioactive or toxic. For example, the effect can be measured directly by the deterioration of concrete foundations in soil contaminated with sulphates, or it may be less obvious, such as the insidious threat to personal health from radioactive gas.

As well as assessing the potential for damage to the substructure of a building, a site investigation must provide an assessment of the risk that ground contaminants could have on the occupants or users of any proposed buildings.

Hazard identification Information must be obtained on the past and present function to which the site and its buildings have been used. Guidance may be found from historical maps, local knowledge, local authority and library archives and Land Registry records. A detailed analysis of soil samples from a geotechnical survey will be required to establish the presence or otherwise of acids, salts, asbestos, heavy metals, cyanides and coal tar, to name but a few of the potentially harmful substances that could have permeated the ground from industrial premises and commercial processing plant.

Further analysis of the soil from 'brown-field' locations will determine the organic content and its potential for decomposition and production of explosive methane gas. The presence of methane is very likely in former infill sites. These have been a common means for refuse disposal since the early part of the Industrial Revolution. Tips are often uncharted and, with town and city expansion, inevitably these sites become absorbed in the urban spread. Former tips can be perceived as quite harmless 'green-fields'. However, closer investigation during a walkover survey often reveals the absence of natural growth and patchy discoloured vegetation.

Ground water levels and local underground flow conditions should also be determined, as contaminants can migrate from one area to another through the liquid intermediary. Methane will normally rise and dissipate naturally into the atmosphere with no apparent effect. If a building is constructed on gas-contaminated land, the methane can accumulate under and within the structure. Given sufficient concentrations, methane can reach explosive levels if accidentally triggered by a spark, e.g. boiler pilot light.[9]

Radon gas occurs naturally by radioactive decay of uranium and radium deposits in the ground. Subsoil comprising granite deposits may contain these radioactive elements. Where present, the products of radioactive decay can migrate through the ground and then disperse naturally and harmlessly into the atmosphere. However, where buildings occur, the products can penetrate through and into the structure by attaching to microscopic particles and droplets of moisture. Radon is invisible and odourless, and a build-up within a building is known to be hazardous to health. It has been estimated that radon exposure may be responsible for about 5 per cent of all cases of lung cancers in the UK.[10] Radon concentrations can be measured; the average concentration is about 20 Bq/m^3 (Becquerels per cubic metre) within a building. In specific parts of the UK, notably some areas of the West Country, Northamptonshire, Derbyshire and Scotland, the figure can be at least ten times higher. Practical measures for reducing levels of radon (and methane) are considered in Part 1, section 8.3.6.

Remedial measures for processing ground contaminants
Removal A low-technology, simple but effective measure, limited to relatively small volumes of contaminated soil at depths up to about 5 m. The number of landfill sites licensed to accept this type of waste is limited and there may be resistance by some ground operatives to handle it.

Containment or encapsulation Cover systems or containment can be achieved by isolating contaminated soil by filling a peripheral trench with an impervious agent such as bentonite clay slurry and capping the site with an impervious horizontal layer between the trenches. Trench excavation will need to be taken down to uncontaminated sound strata such as rock. These 'tanking' techniques can be incorporated within the sub-structural design of a building (see section 3.10).

Vapour extraction This is used to remove fuel and solvent deposits from the ground, in addition to gases from organic and radioactive sources. Bore holes are located at convenient locations around a site. Attached to these are extension pipes of several metres height, terminated with a cowl or propeller fan to assist with dispersal of gases.

Electrolysis This is appropriate where deposits of metal are found in wet ground. A low-voltage direct current is applied between an anode and cathode within a sump, from which the attracted metal accumulation can be removed.

Biological This is otherwise known as phytoremediation. Specific plant growths are used to absorb harmful chemicals from the ground. Another biological technique known as bioremediation applies naturally occurring microbes to consume petrochemicals and convert them to water and carbon dioxide. Nutrients and oxygen must be plentiful and the temperature maintained to at least 10 °C. Below ground this is easily achieved, particularly with the energy generated by the bacterial activity.

Chemical Liquid hydrogen peroxide or potassium permanganate is pumped through boreholes into the contaminated soil to convert fuel/oil/petrol deposits to water and carbon dioxide. An alternative chemical process can be applied to excavated soil. The soil is mixed with solvents to break down oil and petrochemicals and then replaced.

Thermal Steam is pressure injected through the subsoil to evaporate any chemicals. Variations include use of electric currents and radio waves to heat the soil water content.

Incineration This is not normally practical on site, therefore it involves extraction, processing elsewhere and replacement.

2.2 Foundation design

Two main considerations enter into the design of foundations: (i) the soil must be safe against failure by shear, which may cause plastic flow under the foundation, and (ii) the structure must be safe against excessive settlement due to the consolidation of soil under the foundation, and particularly against differential or unequal settlement under various parts of the building. In order to investigate the stability of any foundation it is necessary, therefore, to know first something of the distribution and intensity of pressure between the foundation and the soil and the intensity of pressure and shearing stresses at various points within the soil. Secondly, it is necessary to know something of the mechanism of failure of the soil when over-loaded.

2.2.1 Pressure distribution

Distribution of contact pressure   The assumption usually made that a uniformly loaded foundation will transmit its load so that the soil is uniformly stressed is generally incorrect. The manner in which the load is distributed to the soil depends on the nature of the soil and on the rigidity or stiffness of the foundation. This is shown in table 2.2. It will be seen that there is a considerable variation of pressure under rigid foundations, which is illustrated in figure 2.10, although in practice the pressure distribution tends to become more uniform as indicated in A and B which also show the fundamentally different pattern of distribution in cohesive and cohesionless soils. The pressure will also vary with the relative density of the soil. Foundations, of course, vary between the perfectly rigid and the perfectly flexible type, in addition to which the soil may combine both cohesive and frictional properties in varying degree.

This knowledge permits preliminary decisions to be made before the detailed design of a foundation is undertaken and also acts as a guide to the choice of the type of foundation by which the load will be transferred to the soil. The distribution of contact pressure has little effect on the stresses in the soil at depths greater than the width of the foundation. Its greatest significance lies in establishing the stresses which are set up in the foundation itself. In practical foundation design, based on a uniform pressure distribution, the factor of safety normally used generally covers the under-estimate of bending moments on cohe-

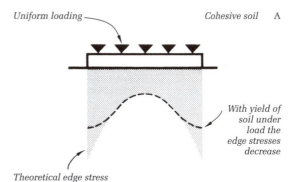

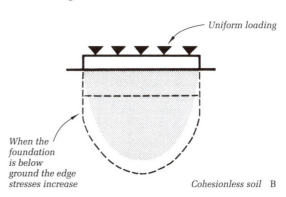

Figure 2.10   Contact pressure distribution

sive soils. On cohesionless soils the estimate will usually be greater than the actual bending moments in the foundation.

Distribution of vertical pressure   A knowledge of the distribution of the normal vertical stress in the soil is required for the solution of settlement problems.

The intensity of vertical pressure at any point and at any depth in the soil may be established by means of *Boussinesq's formula*. This gives the intensity of pressure at a point O as

$$q = \frac{3PD^3}{2\pi R^5}$$

where q equals the intensity of pressure, P equals the applied point load, D equals the depth of the point O under consideration, R equals the distance between the point of application of the load and the point in the soil, O, as shown in figure 2.11 A. $R = \sqrt{(d^2 + D^2)}$ for which d may be obtained as $\sqrt{(a^2 + b^2)}$. The stress intensities at any point in the soil due to a number of concentrated loads may be added. Thus a foundation transmitting a uniformly distributed load may be divided into a number of smaller areas and the load on each regarded as a concentrated load (B). From this the general pattern of pressure distribution under the whole of the foundation may be established.

Table 2.2   Distribution of contact pressure

Foundation	Soil type	
	Cohesive	Cohesionless
Approaching fully flexible	Tendency to uniform distribution of pressure	Tendency to uniform distribution of pressure
Approaching fully rigid	Tendency to high stresses at edges – becoming more uniform as ultimate load is approached	Tendency to high stresses in centre – at all loads

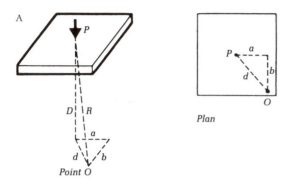

Plan

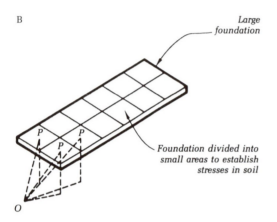

Figure 2.11 Pressure intensity in soil

Division into 600 mm squares gives sufficient accuracy for depths below 3 m.

In practice this process involves an excessive amount of calculations, which may be reduced by the use of appropriate computer software.

For general preliminary purposes, however, and for establishing soil pressures in relation to foundations imposing small loads and creating small soil pressures, the average pressure at any level may be ascertained reasonably accurately by the method given in Part 1, section 4.3 based on the spread of load from the edges of the foundation.

Increased pressure at any level due to the overlapping pressures from a group of closely spaced independent foundations may also, for preliminary purposes, be determined in the same way. The spread is taken from the edges of the outermost foundations in the group and the applied load is the sum of the combined foundation loads.

A diagram may be drawn by joining points of equal pressure in the soil to produce what is known as a bulb of pressure diagram as shown in figure 2.12 A. It will be seen that the larger the loaded area, the deeper is its effect upon the soil (Bi, ii). If a number of loaded foundations are placed near each other the effect of each is additive and

if they are closely spaced one large pressure bulb is produced similar to that produced where the whole area is uniformly loaded (Biii).

If the increase in pressure at a given depth due to a building is not greater than 10 per cent of the original over-burden pressure (see page 38) at that depth, the effect on the soil at that particular depth will usually be negligible. It has been shown in many cases that beyond the bulb of pressure bounded by the line joining points stressed to one-fifth of the applied pressure, consolidation of the soil is negligible and has little effect upon the settlement of a foundation. For practical purposes this is considered to be the significant or effective bulb of pressure. For a uniformly loaded square or circular foundation this bulb extends to a depth of approximately one and a half times the width of the foundation (see A). For a strip foundation it extends to as much as three times the foundation width. As already noted, the pressure bulb, together with the geology of the site, indicates the depth to which investigations of the soil should be made. This is particularly important in the case of wide foundations where deep, underlying weak strata, which would not be affected by the pressures from narrow or isolated foundations, would be stressed by the wider foundations as shown in Bii.

Distribution of shear stress The distribution of shear stress and the position of maximum shear stress is important when there is a possibility of shear taking place in the soil. The maximum shear stress in the soil may be established analytically or graphically by Mohr's circle. The maximum shear stress set up by a single strip foundation carrying a uniform load of p N/m^2 is p/π N/m^2. This occurs at points lying on a semicircle having a diameter equal to the width of the foundation. Thus at the centre of the foundation the maximum shear occurs at a depth equal to half its width. The bulb of shear stresses is shown in figure 2.12 C.

2.2.2 Ultimate bearing capacity of soil

Should the load on a foundation be excessively high, plastic failure of the soil would occur and the foundation would sink into the ground. When a cohesive soil is surface-loaded this type of failure occurs by a wedge of soil directly under the foundation being forced downwards. This pushes the soil on each side outwards causing it to shear along a curved slip plane, so that heaving of the surface on each side takes place, as shown in figure 2.13 A. The depth of a foundation therefore must be such that the weight of soil above the base on each side together with the shear resistance of the soil is sufficient to prevent this. In practice, it is unusual for both sides of a foundation to be equally strong and failure is likely to occur on one side

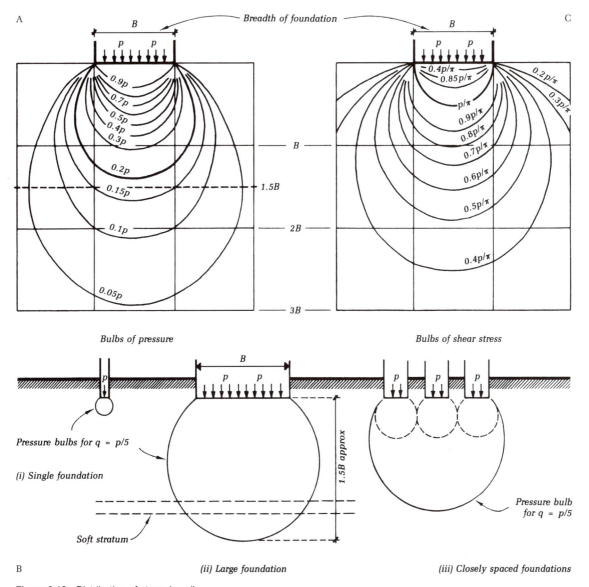

Bulbs of pressure

Bulbs of shear stress

Pressure bulbs for q = p/5

(i) Single foundation

Soft stratum

1.5B approx

Pressure bulb for q = p/5

B

(ii) Large foundation

(iii) Closely spaced foundations

Figure 2.12 Distribution of stress in soil

only, as shown at B. The larger section of the slip plane is assumed to be a circular arc.

The intensity of loading at which failure occurs is known as the ultimate bearing capacity of the soil and is related to the shear strength of the soil. The bearing capacity may be found by an analysis of plastic failure or by a graphical method. The circular arc method, illustrated in figure 2.13 C, is useful when the strength of the soil varies with depth. The critical slip circle is found by trial and error, the centre being located approximately by the use of tables or graphs based on accumulated data. Moments of

the load W, the weight w of the soil within the arc and the total shear resistance, S, along the slip surface are taken about the centre of the arc (S = length of slip plane × shear strength of soil per unit area). For stability the sum of the moments of soil weight and shear resistance must be at least equal to We, the moment of the foundation load. The factor of safety against failure is the sum of the resisting moments divided by the moment of the foundation load:

$$\text{Factor of safety} = \frac{(S \times r) + (w \times y)}{W \times e}$$

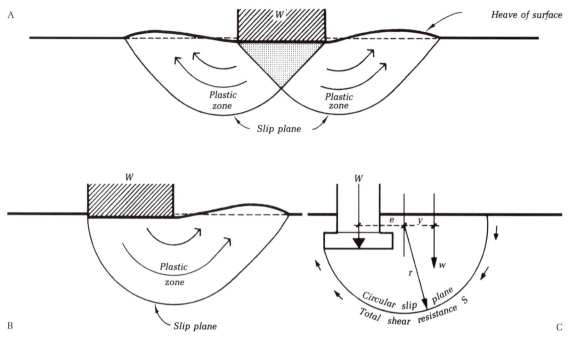

Figure 2.13 Mechanism of soil failure

It is generally assumed that when a clay is loaded uniformly on the surface by a strip load, the ultimate bearing capacity is about $5\frac{1}{2}$ to 6 times the shear strength of the clay. In practice, foundations are not always in the form of long strips, nor do they always apply their load to the surface of the ground. Allowances must be made for this. In addition, in order to avoid plastic failure and to keep settlement within small limits, a factor of safety must be adopted. This factor is usually taken as three and thus the safe uniform bearing pressure is about twice the shear strength of the soil when surface loaded.

Normally the foundation is below the surface of the ground and a volume of soil, called the *over-burden*, must be removed to form the pit. The soil below is thus relieved of the pressure due to its weight (see figure 2.15). This pressure may be allowed to offset wholly or partly the pressure due to the building load, so that for subsurface loading, the safe bearing pressure becomes twice the shear strength of the soil per unit area plus the weight per unit area of the excavated overburden.

Higher bearing pressures may be allowed on independent rectangular foundations. Although there is some difference of opinion, it is generally considered that the safe bearing pressure for a strip foundation on cohesive soils may be increased by about 30 per cent for a square foundation.

There is less risk of plastic failure with cohesionless soils than with clay, and with these the limiting conditions depend on settlement rather than on shear failure. In most cases serious settlement takes place before the ultimate bearing capacity of the soil is reached. The ultimate bearing capacity for granular soils may be calculated from *Ritter's formula*, involving the depth of the foundation below the surface, the breadth of the foundation and the unit weight and angle of internal friction of the soil. In practice considerable regard is paid to the results of Standard Penetration Tests, which give an indication of the relative density of the soil strata (see page 25).

2.2.3 Settlement of foundations

The vertical downward movement of the base of a structure is called settlement and its effect upon the structure depends on its magnitude, its uniformity, the length of time over which it takes place, and on the nature of the structure itself. As explained in Part 1, uniform settlement over the whole area of the building, provided it is not excessive, does little damage. Unequal settlement at different points under the building, producing what is known as relative or differential settlement, may, however, set up stresses in the structure through distortion. These may be relieved to some extent in the case of a brick structure in weak mortar, for example, by the setting up of a large number of fine cracks at the joints, but in more rigid structures overstressing of some structural members might occur. The maximum amount of settlement which should be permitted, that is, the allowable settlement in any particular circumstances, is discussed on page 41.

Settlement may be caused by:

1 the imposed weight of the structure on the soil
2 changes in moisture content of the soil
3 subsidence due to mining or similar operations
4 general earth movement.

Settlement under the load of the structure may be due firstly to elastic compression, which is the lateral bulging of the soil and which takes place without a change in volume. It is usually small and occurs as construction proceeds. Secondly, it may be due to plastic flow, which has already been discussed and which must be prevented by adopting sufficient depth and bearing area for the foundations. Thirdly, settlement may be due to consolidation.

Cohesive soils Settlement in cohesive soils is dependent on the bearing pressure, the compressibility of the soil and on the depth, width and shape of the foundation. For the same bearing pressure a wide foundation will settle more than a narrow one.

Settlement due to the consolidation of cohesive soils may continue for years after the completion of the building. The compressibility of clay is appreciable, but as the reduction in volume takes place by the expulsion of some pore water, and as the permeability of clay is low, consolidation takes place very slowly. The amount of settlement in a layer of clay of thickness H equals

$$\frac{e' - e''}{1 + e'} \times H$$

where e' is the voids ratio (see page 32) from the initial consolidation due to the overburden p', and e'' is the voids ratio due to the total pressure p'', set up by the foundation load plus the overburden. The values for both e' and e'' are taken from the p–e curve (such as that shown in figure 2.14), constructed from a consolidation test carried out on a sample of the soil. It is possible to establish the time required for a given degree of consolidation to take place as the time is directly proportional to the square of the thickness of the soil layer, inversely proportional to the coefficient of permeability and directly proportional to the slope of the p–e curve. Therefore, if the thickness of a clay stratum is known, together with the nature of the drainage conditions at its boundaries, it is possible to prepare a curve such as that shown in figure 2.15, showing the probable progress of settlement over a period of time.

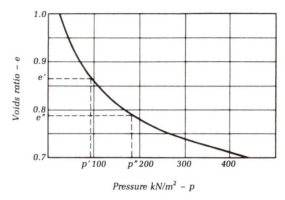

Figure 2.14 Laboratory compression curve (p–e curve)

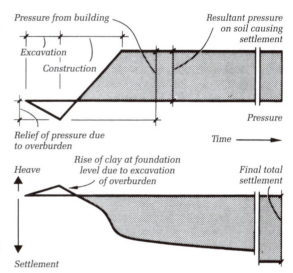

Figure 2.15 Settlement of clay soils

Cohesionless soils In sands and gravels the amount of consolidation under load is relatively small and the rate of settlement keeps pace with the construction of the building. On completion of the building no further settlement takes place unless caused by unforeseen circumstances such as the erosion of sand by water outside the limits of the building, the removal of surcharge by adjacent excavations or consolidation of the soil by vibration. This is because these types of soil are to a large extent incompressible, the particles resting on one another, so that under load they do not move very far and settlement is not large. What settlement does occur takes place quickly because the permeability is high. The actual settlement depends, of course, on the bearing pressure and the relative density of packing of the soil particles. It also increases with the width and depth in the soil of the foundation.

Bearing pressures on soils   In the design of foundations a distinction must be made between the bearing pressure which a soil is safely capable of withstanding and the pressure which it is advisable not to exceed in practice.

The *bearing capacity* or *presumed bearing value* is the maximum pressure that a soil will bear without risk of shear failure, irrespective of any settlement which may result. It is obtained, as described above, by dividing the ultimate bearing capacity by an appropriate factor of safety, and is intended for preliminary design purposes.[11]

The *allowable bearing pressure* is the maximum pressure which should be applied to the soil taking into account shear failure, settlement and the ability of the particular building to withstand settlement. The allowable bearing pressure thus depends on both the soil and the type of building concerned, and is generally less than the bearing capacity.

As explained earlier, if the soil is overloaded shear failure of the soil will result. To avoid this type of failure the soil pressure immediately beneath the foundations must not at any point exceed the bearing capacity of the soil. If the foundations rest on soil such as rock or deep beds of compact gravel (so that settlement is not a significant factor) it is sufficient to ensure that the bearing capacity is nowhere exceeded. If the soil is such that possible settlement must be considered, the soil pressure immediately beneath the foundations must not exceed the allowable bearing pressure and should, as far as possible, be equal at all points.

Uneven loading of the soil should be avoided, under both individual and combined foundations, by arranging the centre of gravity of applied loads to coincide with the centroid of area of the foundation as explained in Part 1. Where some eccentricity is unavoidable the design must ensure that the allowable bearing pressure is not exceeded at any point so that relative settlement is not excessive.

Equality of applied soil pressure cannot be achieved beneath a number of independent bases if the loads on some are almost entirely dead loads but the loads on others have a high proportion of live load, since the dead loads are applied all the time but the live loads, which must be provided for, may never occur. The use of a suitably low allowable bearing pressure may, in such cases, reduce the differential settlement to an acceptable amount. It is, however, a better arrangement, particularly on soils of low bearing capacity, to plan the building so that the dead to live load ratio is approximately uniform in the various columns.

Settlement due to mining subsidence is confined to certain localities and this, together with general earth movement, is discussed on pages 43 and 44.

The significance and effects of moisture changes in the soil are explained in Part 1, section 4.2, to which reference should be made.

2.2.4 Choice of foundation type

This subject is introduced in Part 1, chapter 4, particularly in relation to small-scale buildings, and table 4.4 in that volume indicates the suitability of foundation types to the various types of soil. The underlying reasons for the recommendations made are discussed here.

The foundations and sub-structure together with the superstructure of a building are interrelated and interdependent. In order to obtain maximum strength and economy in the total structure they must be considered and designed as a whole. The larger and taller the building the more important this becomes.

The type of foundation adopted very often depends largely upon the form of construction used for the structure above. But with soils such as clay or silts, and in subsidence areas, the building itself must often be designed to react to differential settlement with no ill effect.

It is a comparatively easy matter to provide adequate foundations to small buildings, but as the size, particularly the height, of the building increases, so does the need for economy in cost of foundations increase. An increase in height increases the load on the foundations in two ways: first, because the total weight of the building is greater, and secondly, because the greater horizontal forces due to wind pressure shift the resultant load on the foundations to one side, causing a local increase in soil pressure as explained in Part 1. Generally, unless conditions are exceptional, the dead weight of the building will keep the resultant within the middle third of the base. Although methods such as power boring for cylinder piles make it economically possible to build much higher on soils the nature of which previously limited the height of building, problems of settlement are intensified. Overall settlement may be greater with the liability of greater differential settlement and the possibility of tilting of the whole structure.

In terms of overall building costs the problem of dealing with settlement must be considered from the point of view of the structure as well as of the foundation on which it rests. In extreme cases it may be essential to use a high factor of safety against ultimate failure but for economic reasons, as far as the foundations are concerned, a factor of not more than three is often desired and this may lead to considerable settlement. Although forms of structure can be designed which are flexible enough to resist such settlement safely, those forms of construction which are economic in material often involve rigidity of structure, resulting in sensitivity to differential movement. The cost of structure must therefore be balanced against the cost of foundations and in some circumstances it may prove cheaper to limit settlement by using a high factor of safety so that a rigid and economic superstructure may be used.

Many factors are involved in the choice of a foundation type. The actual arrangement required in any particular case will depend upon the nature and strength of the sub-soil, the type of foundation indicated by considerations of economy in the structure as a whole, the nature of the structure, the distribution of the loads from the superstructure and the total weight of the building and its parts.

Foundations relative to nature of soil A knowledge of the general manner of pressure distribution in different soils according to the type of foundation acts as a guide in making a preliminary choice of foundation. By a careful distribution of the loads from the superstructure it is possible to reduce and sometimes eliminate uneven pressure.

When it appears that settlement would be large, the bearing pressure may be reduced by adopting a greater spread of foundations, or by using a raft. It should be remembered, however, that although this procedure may be beneficial when the foundation rests directly on the compressible soil, it is not effective in preventing settlement when consolidation can take place in a deep-seated compressible stratum. This is because the wider the foundation the greater will be the depth at which the soil will be stressed. This can be seen in the comparative pressure bulb diagrams shown in figure 2.12 B.

When the building rests on a comparatively shallow compressible stratum, it may be cheaper to carry the foundations down to firmer soil at a lower level by means of piers or piles.

Where the compressible layers are deep-seated it is possible to limit consolidation by excavating basements so that the pressure imposed by the weight of the building is wholly or partly offset by the reduction in overburden pressure resulting from the removal of the excavated soil and water. The provision of deep basements may also be used as a means of reducing differential settlement under buildings in which some parts are heavier than others, such as a tall tower block with lower surrounding parts of the building forming a 'podium'. A deep basement under the tower block will have the effect of reducing the pressure in the soil at that point due to the relief of overburden pressure, thus reducing the difference in the pressures under the whole building (figure 2.16 A). The application of this method is, however, limited by the expense of the excavation which, if to be effective, may need to be very deep, and by the effect of soil movement during excavation (see page 54).

For greatest economy both in superstructure and foundations, all loads should, as far as possible, be carried directly from the point of application to the point of support. This is particularly important when the loads are great. A raft foundation may be an economic way of transferring to a weak soil the load from a large number of lightly loaded columns but when heavy load concentrations are to be transferred to the soil, deep piles passing through the weak soil to bed rock may be the cheapest solution.

Foundations relative to nature of structure The consideration of the behaviour of soils under load shows that rigidity or flexibility of structure can influence the design of slab and raft foundations. Similarly, the behaviour of the structure under load depends not only upon the dead and imposed loads, but also upon the nature of the foundations and the manner in which the structure is supported upon them, since some of the forces acting upon the structure are those developed at its junction with the foundations (see chapter 8). Accordingly, as indicated earlier, consideration must be given on the one hand to the choice of a foundation appropriate to the type of structure likely to be most economic, and on the other hand to the possibility of using a structure suited to the soil conditions and to foundations which appear most appropriate. For example, in some circumstances on sites with weak soils the most economic building might result from the use of a few expensive types of foundation, rather than from the use of a larger number of more lightly loaded and cheaper foundations. In such circumstances the structural frame would be designed with this in mind, the columns being widely spaced in order to reduce the number of foundations required.

Design of structure to resist settlement Since a certain amount of settlement is usually inevitable it must be reckoned with and provision made for it in the design of the structure. The effects of differential settlement depend to a large degree upon the rigidity of the structure. For example, in structures in which beams rest simply on brick walls, unequal settlement unless excessive will not affect the internal stresses in the structural elements because these are not distorted (figure 2.16 B), although finishes may be affected. But in rigid continuous structures even a small amount of differential settlement of the supports will cause secondary shearing forces and bending moments in the beams and columns due to the distortion of the frame arising from the rigid joints (see figure 2.16 C). Buildings with rigid frames may be supported upon a stiff raft foundation (D), or the foundations of individual columns can be connected by a series of continuous beams so that the structure settles as a whole, the frame being designed to act as a very deep girder to resist any secondary stresses set up by settlement. Although secondary stresses may be set up in rigid structures by settlement of the foundations, such structures do permit forms of construction which require less resistance from the soil than non-rigid forms. This applies especially to single-storey structures where

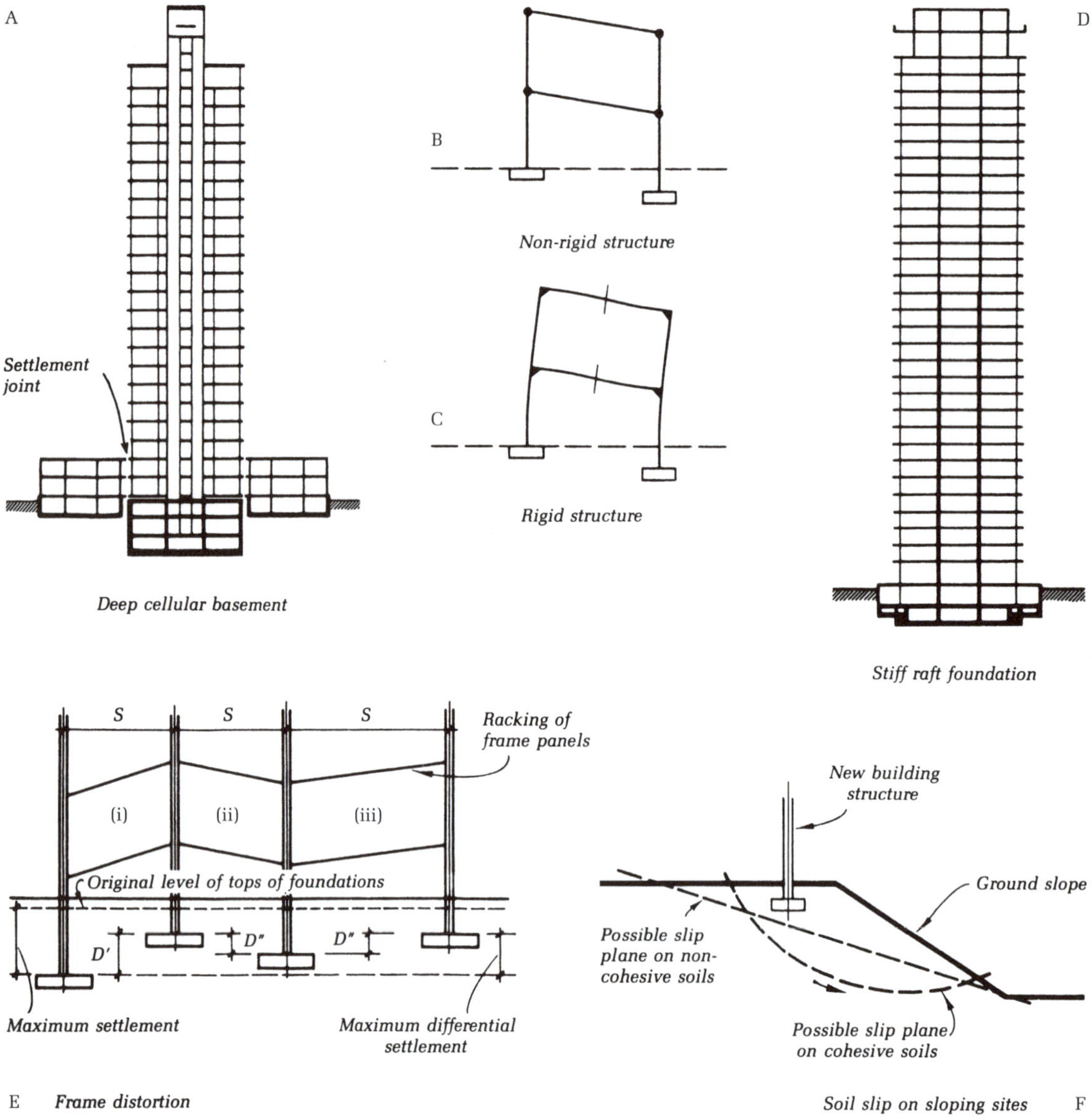

A

Settlement joint

Deep cellular basement

B

Non-rigid structure

C

Rigid structure

D

Stiff raft foundation

S   S   S   Racking of frame panels

(i)   (ii)   (iii)

Original level of tops of foundations

D' D'' D''

Maximum settlement   *Maximum differential settlement*

E   *Frame distortion*

New building structure

Ground slope

Possible slip plane on non-cohesive soils

Possible slip plane on cohesive soils

Soil slip on sloping sites   F

Figure 2.16 Foundations, soil and superstructure

rigid frames with hinged bases may be used to prevent the transfer to the foundations of bending stresses set up by wind pressure on the superstructure above. This is discussed more fully in chapter 8.

Although the structure of a building supported on a non-rigid foundation may be of comparatively flexible construction, so that its stability is not impaired by movement, differential settlement is usually limited in order to avoid damage to claddings and finishes and excessive unevenness in the floors. The limit of this movement will depend not only upon the type of structure and the nature

of the finishes, but also upon the extent to which damage and subsequent repair to finishes is acceptable.[12] The acceptable limit may vary with different buildings. Where large movements such as occur on subsidence sites must be accommodated, special forms of construction to deal with this problem must be devised and this is discussed in section 2.3.4.

It has been suggested that the differential settlement of uniformly loaded continuous foundations and of equally loaded spread foundations of approximately the same size is unlikely to exceed half of the maximum settlement, and that normal structures such as office buildings and flats can satisfactorily withstand differential settlements of 19 mm between adjacent columns spaced about 6 to 7.5 m apart. It is suggested that an allowable pressure should be selected such that the maximum settlement of any individual foundation is 25 mm.[13] These figures relate only to foundations on cohesionless soils.

Since most damage arising from differential movement is due to distortion of the rectangular panels between beams and columns, and since the angle of 'racking' depends on the distance between the columns as well as on the actual amount of settlement, it is possible to formulate rules on the same principle as the deflection limitations used in beam and slab design, in which the maximum differential settlement between adjacent columns is limited to a certain fraction of the span between them. Thus the difference in the settlements of two adjacent columns, that is, the differential settlement between them, divided by their distance apart, is a measure of the severity of the settlement. There is considerable risk of damage in a normal framed building when this fraction exceeds about 1/300, so that at design stage it is advisable to so limit the estimated differential settlement so that this fraction does not exceed 1/500. This is illustrated in figure 2.16 E. It will be seen that the distortion in frame panel (i) is greater than that in the centre panel (ii). This is obvious from inspection and also from the fact that D/S in panel (i) is greater than D/S in panel (ii). The differential settlement in panels (ii) and (iii) is the same, but as the distance between the columns of panel (iii) is greater than that between those in panel (ii), the severity of 'racking' in panel (iii) is less and therefore the likelihood of damage is less than in panel (ii). This again is clear from a comparison of the fraction D/S in each case. Having established in this way the safe estimated differential settlement, the building frame should be designed so that the strains due to the settlement are not excessive.

Although this method of relating estimated settlement to the structure is useful, the relative uncertainty and approximate nature of settlement calculations compared with structural calculations must be borne in mind. This limits their application – more refined calculations will not make the soil more uniform. Further, the flexibility of structure, generally speaking, is not continuous throughout a building due to the stiffening effect of structural walls and lift shafts. This tends to localise distortion in different parts and the resultant damage may be greater than that anticipated on the basis of the ratios quoted above.

Foundations relative to ground movement Reference to ground movement has been made in Part 1, chapter 4, where the effect of seasonal moisture variations and of tree roots in causing settlement on clay soil is discussed and

where reference is made to the lifting effect on foundations by frost heave. These volume changes cause local movements of the soil, but mass movement of ground takes place in unstable areas: these occur as a result of subsidence in mining areas and in areas of brine pumping, landslips on unstable slopes and creep on clay slopes.

Subsidence When subsidence takes place, in addition to the total vertical movement there is a horizontal movement which exerts forces on any structure resting on or built into the subsiding ground. This can best be understood by considering the subsidence caused by the extraction of a coal seam. As the working face moves forward and support is removed from the ground behind, settlement will take place after a certain time depending on the nature of the overburden and the method of extracting the coal. This will cause a depression of the ground at the surface which will move in advance of the working face causing a tilt and curvature in the ground which is termed the 'subsidence wave'. This is illustrated in figure 2.17.

As the wave passes beneath a building it will subject it to changing forces according to the position of the building on the wave:

(i) At the crest of the wave a cantilever effect will be produced in the foundations and tensile forces will be set up in the ground. This tension will be transmitted to the foundations (2)

(ii) On the flank of the wave the building will be subjected to eccentric loads because it is tilted out of the vertical (3)

(iii) In the trough of the wave support will be removed from the centre of the foundation and a beam effect will be produced setting up tension in the foundation. Compressive forces will be set up in the ground (4)

(iv) When the subsidence wave has finally passed, the building will resume its vertical position but at a lower level (5).

Although the total subsidence may be considerable the length of the wave is usually so very much greater than the maximum length of the building that day-to-day settlement is slight.[14] Precautions against undue damage of the structure by these movements must, however, take account of this behaviour of ground and structure during the passage of the subsidence wave. Methods used to minimise the effects of subsidence are discussed later in section 2.3.4.

Unstable slopes and creep Unstable slopes subject to landslip may often be recognised by the characteristic uneven surface of the ground, and where possible such areas should be avoided as building sites. The tendency for the upper strata of clay soils to move downhill on

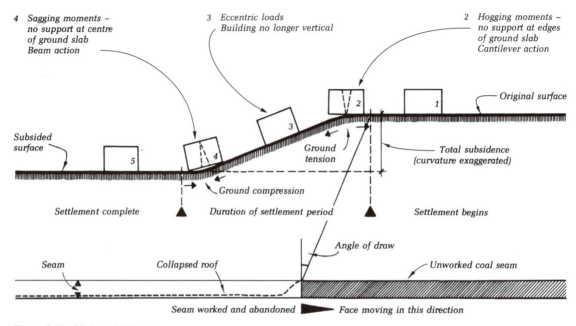

4 Sagging moments – no support at centre of ground slab Beam action

3 Eccentric loads Building no longer vertical

2 Hogging moments – no support at edges of ground slab Cantilever action

Original surface

Subsided surface

Ground tension

Total subsidence (curvature exaggerated)

Ground compression

Settlement complete

Duration of settlement period

Settlement begins

Angle of draw

Seam

Collapsed roof

Unworked coal seam

Seam worked and abandoned ◢ *Face moving in this direction*

Figure 2.17 Mining subsidence

sloping sites is always present and will be governed by the angle of the slope, the characteristics of the soil and other factors. Landslips on slopes previously just stable may be started by work carried out on adjoining land. Sometimes only the surface layers creep downhill and where this has occurred evidence of this will be seen in tilted fences and boundary walls and curved tree trunks. The creeping layer may vary in depth from a few millimetres to some metres, and may move on slopes as shallow as one in ten. Where buildings must be erected on such sites large scale and expensive works may be required to stabilise the ground.

Building on any sloping site, whether signs of slip or creep are evident or not, should be preceded by very careful site exploration and careful design of the foundations. Non-cohesive soils will tend to slip at different angles according to the angle of internal friction of the particular soil and this must be carefully established. The calculated shear on any inclined plane passing under the foundations, and with the upper and lower edges cutting the ground surface, should not be greater than half the total resistance available due to friction on the worst surface (see figure 2.16 F). In cohesive clay soils slip may take place on a curved surface, as described earlier, and the calculated shear stress on any such surface passing under the proposed foundations and with its upper and lower edges cutting the ground surface should not be greater than half the shear strength likely to be available under the worst anticipated future conditions.

When building on sloping sites it is most important that water should be drained away from the uphill side of foundations.

Generally, any slope tends to be unstable and placing a load at the top increases this tendency. If site exploration and analysis indicate an inadequate factor of safety against slip this must be increased by one of the following:

- reducing the slope, including forming retaining walls and placing a load at the base of the slope
- reducing the load
- placing the load below the line of failure of the slope
- maintaining or increasing the ground strength by drainage or water diversion, having regard to ground water movements which will obtain when the building is erected
- maintaining ground strength by placing filter layers to prevent erosion.

Differential settlement which would cause little damage on level sites tends to be more serious on sloping sites, as the parts which settle tend to move downhill, causing large cracks. Longitudinal tensile reinforcement in simple footings is very useful in such cases.

2.3 Foundation types

So far in this chapter consideration has been given to the resistance of the soil to the loads imposed by building, to its behaviour under load, and to the factors which affect the choice of foundation. It now remains to continue from Part 1 the examination in detail of the actual types of foundations used in various circumstances to transfer the building loads to the soil.

The broad classification into *shallow* and *deep* foundations has been explained in Part 1 and some indication has been given of the situations in which spread, pile and pier foundations are likely to be used. This is extended here under the headings of the different foundation types now to be discussed.

The foundation types defined in Part 1 may be broken down further into the following sub-divisions, each of which will be considered in detail.

Strip and slab	Wide strip foundation
foundations	Isolated column foundation
	Continuous column foundation
	Combined column foundation
	Cantilever foundation
	Balanced base foundation
Raft foundations	Solid slab raft
	Beam and slab raft
	Cellular raft
Pile foundations	Friction piles
	End bearing piles
Pier foundations	Masonry
	Mass concrete
	Cylinders and monoliths

Spread foundations and short-bored pile and pier foundations suitable for small-scale buildings have been described in Part 1, chapter 4.

2.3.1 Spread foundations

As explained in Part 1, when loads are heavy or the soil is weak, necessitating wide spread foundations, the required thickness of mass concrete may make them excessively deep. To minimise the thickness in such cases it is normal to introduce reinforcement to take up the bending stresses and to keep the thickness of the concrete to that required to give sufficient resistance to shear. In extreme cases only is shear reinforcement also provided. If the thickness of the foundation can be enough to obviate the need for shear reinforcement, the risk of cracks and the consequent corrosion of the steel by ground moisture is avoided. The detailed design of such foundations is covered in textbooks on reinforced concrete. They will be considered here in terms of their types and applications.

In spite of the greater depth of foundation considerable benefit may be derived from the use of mass concrete bases in waterlogged ground. These may often prove cheaper than the continuous pumping needed to permit the fixing of the reinforcement and placing of concrete around it in the dry.

Wide strip foundation   This foundation may be used to carry loadbearing walls, including brick or concrete walls to multi-storey buildings. Since the loads will be heavy the necessary width of foundation will be greater than that required for a small-scale building (figure 2.18 A). The bending stresses set up by the double-cantilever action of the foundation across the base of the wall are, therefore, likely to be high and to resist these reinforcement is placed at the bottom of the slab, which will be the tensile zone, usually in the form of rods at right-angles to the length of the wall.

Isolated column foundations   This is the most commonly used foundation type in framed buildings (figure 2.18 B).

Reinforced concrete pad   This is a square or rectangular slab of concrete carrying a single column. Reinforcement is placed at the bottom in both directions to resist the bending stresses set up by the double-cantilever action of the slab about the column base (figure 2.19 A, B). Shear reinforcement is normally not provided. The critical plane for bending in the case of a reinforced concrete column is at the face of the column, but in the case of a steel stanchion it is at the centre of the base-plate. When bending moments due to wind pressure on the structure above will be transferred to the foundation slab through a rigid connection, allowance must be made for this in the design of the slab. The critical plane for shear is assumed to be at a distance from the face of the column equal to the effective depth of the slab (A).

The thickness of the slab may be reduced towards the edges to economise in concrete either by stepping or tapering the top face. Slopes greater than 25 degrees require top shuttering to prevent the concrete building up at the edges when poured. Stepped bases should not be provided unless site supervision is sufficient to ensure that the concrete in each base will be placed in one operation.

In order to protect steel and concrete from the soil during construction a blinding layer of concrete, 50 to 100 mm thick, is usually laid over the bottom of the excavation.

BS 8110-1: *Structural use of Concrete. Code of Practice for Design and Construction* gives a table of the required minimum concrete cover to the reinforcement for different grades of concrete and different conditions of exposure to ground water.

Steel grillage   This form of foundation makes use of steel beams to take up the shear and bending stresses in the foundation (figure 2.20).

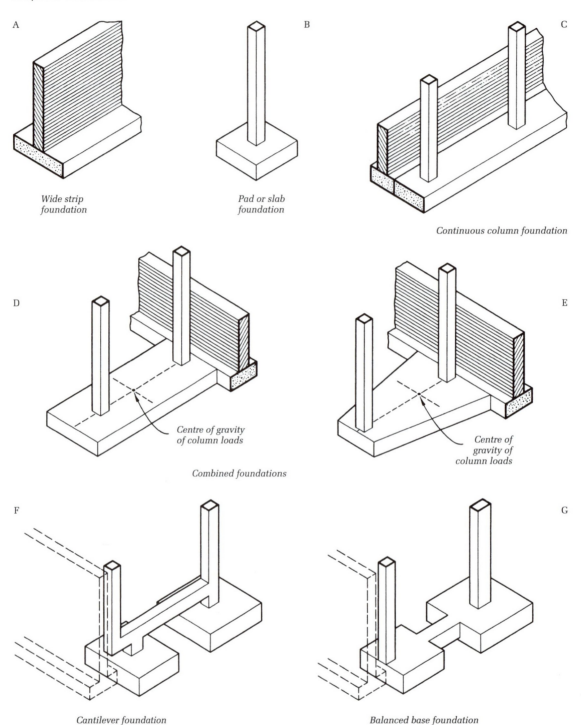

A

Wide strip foundation

B

Pad or slab foundation

C

Continuous column foundation

D

Centre of gravity of column loads

Combined foundations

E

Centre of gravity of column loads

F

Cantilever foundation

G

Balanced base foundation

Figure 2.18 Foundation types

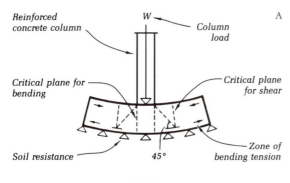

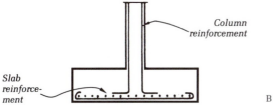

Figure 2.19   Pad or slab foundation

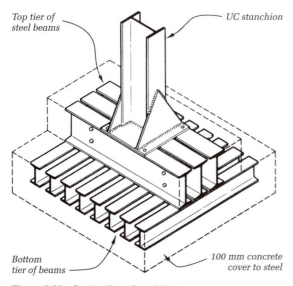

Figure 2.20   Steel grillage foundation

The grillage foundation is expensive in steel and is not often used except when very heavy loads from steel stanchions must be carried on wide slabs and where the depth of the foundation is restricted in order to keep it above the subsoil water table.

Continuous column foundation   This type of foundation supports a line of columns and is one form of combined foundation (figure 2.18 C). Although it is, in fact, a strip foundation the term 'continuous column foundation' is used here to distinguish it from the normal strip foundation

carrying a wall which is subject only to transverse bending, whereas the continuous column foundation is subject to both transverse bending and longitudinal bending due to the beam action between the columns as shown in figure 2.21 B. Circumstances in which it might be used are: (i) where the spacing in one direction and loading of the columns are such that the edges of adjacent independent column foundations would be very close to each other or would overlap, and (ii) where there exists some restriction on the spread of the foundations at right-angles to a line of columns, such as a site boundary or an existing building adjacent to a line of outer columns, which would exclude the possibility of an independent foundation of adequate size to each (figure 2.21 A). This is common in urban areas where a new building is to be erected between the existing party walls of adjoining buildings, and it is inadvisable or unnecessary to cut under the party walls to form foundations.

Assuming that sufficient width of foundation, relative to the spacing of the columns, is available to provide the necessary area of foundation, the strip is designed as a continuous beam on top of which the columns exert downward point loads and on the underside of which the soil exerts a distributed upward pressure (B). The stresses set up are the reverse of those in a normal continuous floor beam. The main tensile reinforcement is therefore required near the upper face between columns and near the bottom under the columns to resist the negative bending stresses as shown in C. Transverse reinforcement must always be provided to locate the main bars during concreting, whether or not it is required to resist transverse bending.

This type of foundation is sometimes formed as an inverted Tee-section in order to provide adequate longitudinal stiffness.

Combined column foundation   It has been suggested that a continuous column foundation may be used when there exists some restriction on the spread of independent foundations at right angles to a line of columns. This assumes that the columns can be placed far enough away from the restriction to provide sufficient area of foundation since, in order to obtain a symmetrical disposition of column relative to foundation (to ensure even stressing of the soil), the maximum width available is twice the distance from the centre line of the columns to the restriction (see figure 2.21 A). When this distance is too small to provide adequate width of foundation the columns close to the restriction may be linked to an inner line of columns on what is called a combined foundation, by means of which sufficient area and an even distribution of pressure may be obtained (figure 2.18 D, E).

One or more pairs of columns may be combined in this manner, but the problem of design is the same in each

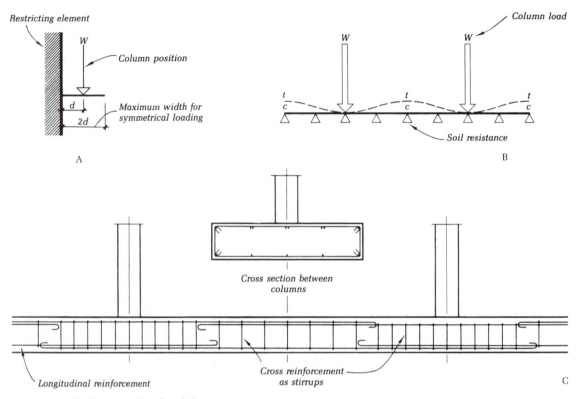

Figure 2.21   Continuous column foundation

case. This is the provision of a slab with an area sufficient to prevent overstressing of the soil and excessive settlement under the combined column loads, and of such a shape and proportion that its centroid lies as nearly as possible on the same vertical line as the centre of gravity of the column loads.

In the case of a pair of columns which are equally loaded or of which that next to the restriction is more lightly loaded, a simple rectangular shaped slab is used (figure 2.22 A). The position of the centre of gravity of the column loads is first established. This will lie midway between equally loaded columns or, if these are unequally loaded, nearer the more heavily loaded column. In the latter case the position is found by taking moments about a point. This will determine the length of the foundation slab since this length must be twice the distance from the centre of gravity of the loads to the line of restriction (assuming that the slab is to extend to this) in order to make the centroid of the slab coincide with the centre of gravity of the loads. The width of the slab is then determined according to the area required for the foundation.

Trapezoidal base   When, in the case considered above, the necessary projection of the slab beyond the more heavily loaded column is restricted, for example by a lift pit close

to the column (figure 2.22 B), or when the loading is reversed and the more heavily loaded column is nearest the restriction (C), a trapezoidal base must be used. This is necessary in order to permit the centres of gravity of loads and slab to lie on the same vertical line, since the position of the centroid of a trapezium along its axis may be made to vary with changes in the proportions of the ends. The length of the slab in the first case will be limited by the restriction at each end, and in the second case is arbitrarily fixed by extending the base just beyond the more lightly loaded column. The widths of the two parallel ends are then determined by the following equations, in which the known factors of the area required, the length of the slab and the calculated position of the centre of gravity of the column loads, are used to give a trapezium, the centroid of which will coincide with the centre of gravity of the loads (D):

1   $A = \dfrac{(b' + b'')}{2} \times L$

or, $b' + b'' = \dfrac{2A}{L}$

where $b' + b''$ equals the sum of the lengths of the parallel sides,
A equals the required area,
L equals the determined length.

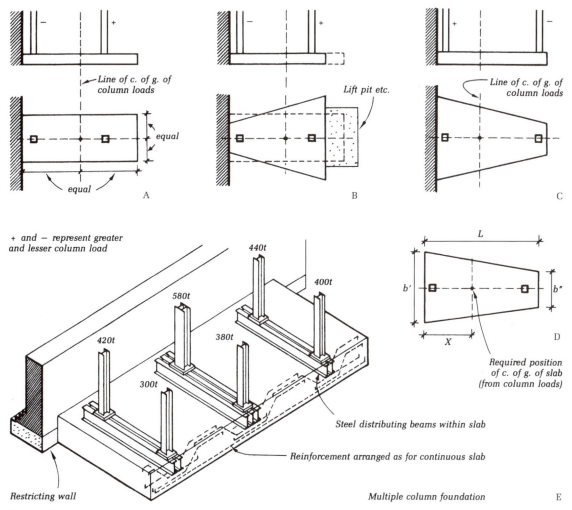

Figure 2.22 Combined column foundations

$$2 \quad X = \frac{L}{3} \times \frac{(b' + 2b'')}{(b' + b'')}$$

$$\text{or, } b' + 2b'' = \frac{3X}{L} \times (b' + b'')$$

where X equals the distance of the centre of gravity of the column loads from the longest parallel side b'. Combining $b' + b''$ from equation (1) with $b' + 2b''$ from equation (2) provides:

$$3 \quad (b' + 2b'') - (b' + b'') = b''$$

from which can be found b'.

The combined foundation slab, whether rectangular or trapezoidal in shape, will be reinforced along its length on the line of the columns, in a similar manner to the continuous column foundation, to resist the tensile stresses set up by the beam action between the columns and any cantilever action in the ends. Transverse reinforcement will also be provided.

As indicated earlier, this type of foundation may be in the form of a multiple column slab, carrying two or three pairs of columns, in which, as before, it is desirable that the centres of gravity of column loads and slab coincide (figure 2.22 E). This, however, is not always practicable and a certain amount of eccentricity may have to be accepted within the safe limits of soil strength and settlement. Combined foundations may be constructed as a single or double tier steel grillage in conjunction with steel stanchions (E). The loading and spacing of a single pair of stanchions may be such that sufficient width of slab will be obtained when the necessary number of steel beams in a

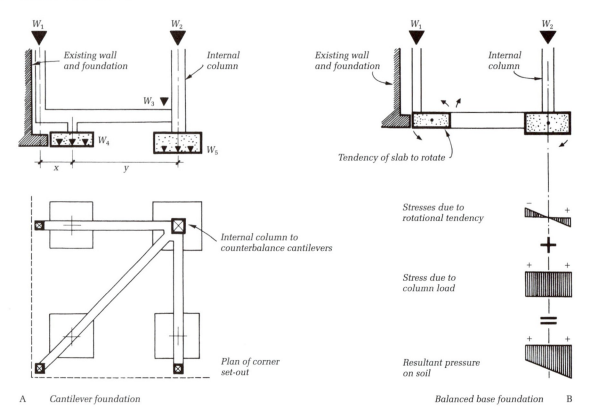

A   *Cantilever foundation*

Balanced base foundation   B

Figure 2.23 Balanced foundations

single tier linking the stanchions has been encased with concrete to give the required 100 mm cover. If this is not the case the alternative to a bottom tier of beams at right-angles to the line of stanchions is to reinforce the concrete with transverse rods as in the case of the multiple column slab shown in E. The use of combined foundations is not limited to the circumstances visualised above and a common or combined foundation may be provided for a number of adjacent columns where the size of independent foundations would be such that they would overlap.

Balanced foundations These consist of the cantilever foundation and the balanced base foundation which may be used in the following circumstances instead of a combined column foundation:

● As an alternative to a trapezoidal combined foundation slab.
● When some obstruction at a column position prevents an adequate foundation being placed directly under the column. This occurs, for example, when the column is placed close to the wall of an adjoining building, the foundations of which project under the column, or when a sewer passes directly under a column. In these circumstances a cantilever foundation would be used.

Cantilever foundation The foundation consists of a ground beam one end of which, cantilevering beyond a base set a short distance in from the obstructed column and acting as a fulcrum to the beam, picks up the foot of the column while the other end is tailed down by an internal column as shown in figure 2.18 F.

The loads on the beam and the two base slabs are quite simply ascertained from the loads on the two columns, and from these the beam size and slab areas may be calculated (see figure 2.23 A):

$W_1 \times x = W_3 \times y$ therefore

$W_3 = \dfrac{W_1 \times x}{y}$ (this counterbalancing force must be provided by the column load W_2)

$W_4 = W_1 + W_3$ (the load on the fulcrum slab)

$W_5 = W_2 - W_3$ (the load on the internal slab: the 'uplift' at the end of the cantilever equal to W_3 reduces W_2 by this amount)

The counterbalancing force W_3, supplied by the internal column, must be provided wholly by the dead load on this column. This must be at least 50 per cent greater than the uplift due to the combined dead and live loads on the outer column in order to provide an adequate margin of safety.

Care must be taken to maintain this margin during construction. When the dead load on the internal column is insufficient for counterbalancing purposes, added weight must be provided by increasing the size of the internal base as necessary or alternatively, anchorage must be provided in the form of tension piles or ground anchors under the slab. These, or a mass concrete counterweight, must be used when no suitably placed column is available for counterbalancing the cantilever beam. When a corner column is involved, a diagonal cantilever beam may be used as shown in figure 2.23 A.

It should be noted that the short fulcrum column shown in the illustrations is not essential. The cantilever beam may rest directly on the top of the fulcrum slab.

As indicated earlier, maximum economy is achieved if loads are transferred axially. Cantilever foundations should not, therefore, be resorted to unless it is impossible to realign a column to permit axial support of its load.

As in a combined foundation, the cantilever foundation may be constructed in reinforced concrete or in steel. In the latter case the steel cantilever beam in conjunction with steel stanchions may bear on steel grillages or on normal reinforced concrete slabs.

Balanced base foundation In this form of balanced foundation the beam is dropped to the level of the base slabs (figure 2.18 G). It can be used when a base can be placed under, although eccentric to, the outer column and may be viewed as a beam balancing, or resisting the tendency to rotate on the part of the eccentrically loaded foundation slab (see figure 2.23 B).

In design, the foundation slab to the outer column is made large enough to take the load W_1 from the column. The tendency of this slab to rotate due to the eccentric loading is resisted by a balancing beam linking it with the inner foundation slab which, in turn, will tend to rotate under the action of the balancing beam. The spread of this slab must, therefore, be sufficient not only to distribute safely to the soil the load from its own column, but also to prevent overstressing of the soil at its far edge due to the rotational tendency (B).[15] If, for this reason, the spread became excessive, it would be necessary to use a cantilever foundation. The balancing beam must be stiff enough to fulfil its function satisfactorily. In both the cantilever and balancing beams the tensile zones would be at the top where the main reinforcement would be placed.

Raft foundations A raft foundation is fundamentally a large combined slab foundation designed to cover the whole or a large part of the available site. Reasons for the use of a raft foundation have already been mentioned in the section on the choice of foundation type and they may now be considered in greater detail.

A raft may be used when the soil is weak and columns are so closely spaced in both directions, or carry such high loads, that isolated column foundations would overlap or would almost completely cover the site. In these circumstances it may be economic to use a raft when it appears likely that more than three-quarters of the site will be covered by isolated column foundations. When a raft is indicated the following considerations should be borne in mind before a final decision is made:

- When the depth of the weak strata down to firm soil is not much greater than about 4.5 m, it may be cheaper to use foundation piers, as these can be economically constructed to this depth.
- When the weak soil extends to a depth greater than this bearing piles might, in some circumstances, be a less expensive foundation than a raft.
- When the weak soil extends to such a great depth that bearing piles would be uneconomic, a raft can be used provided that the building is reasonably compact in plan and fairly evenly loaded. When the loading is very uneven, friction piles could be used as an alternative and possibly simpler solution to the problem.
- Treatment of the ground to improve its bearing capacity, with techniques such as vibro-compaction of sand, may be economical (see page 71).

The choice of method depends on the extent of the individual loads, the size of project and the ground conditions. Each method must be compared economically with the others on the basis of a cost/benefit analysis.

When a raft is used this reduces the ground pressure immediately beneath the foundation, but the pressures at greater depths are little less than they would have been under closely spaced isolated column foundations (see page 35). Care must be taken to ensure that the pressures at such depths are acceptable.

Use of rafts generally A raft used to distribute the building loads over a large area of weak soil may be flexible or rigid in form, as required by the characteristics of the soil, the distribution of loading and the form of structure used for the building it supports. As already explained, the rigidity of a raft has an important effect upon the distribution of pressure and upon settlement. It is possible to use a stiff raft to minimise differential settlement. For example, in the case of a building bearing on a cohesionless soil and having high load concentrations round the edges, the use of a rigid raft would result in a greater uniformity of pressure under the raft than if a flexible form were used. However, a high degree of rigidity results in a most expensive structure and a certain amount of differential settlement usually has to be accepted, the safe limits depending upon the construction of the building which is supported.

A raft may also be used as a means of bridging over weak areas on a site which otherwise is generally fairly firm. A raft should not, however, be carried directly over a local hard area in an otherwise fairly weak site. In such circumstances the hard area should be excavated to a suitable depth and backfilled with a material compacted so as to bring its bearing capacity to an amount not exceeding that of the remainder of the site.

The use of basements in foundation problems has been mentioned earlier where it is shown that a basement may be used as a means of reducing the net intensity of pressure on the soil by a relief of overburden pressure, in order to reduce the overall settlement under the whole area of a building.

In saturated and soft subsoils, such as very wet mud or silt, it is possible to devise a foundation which will enable the building to float on the subsoil. On a liquid soil having no angle of repose, the building on its foundations would sink into the ground until a volume of soil and water equivalent in weight to that of itself is displaced with the result that the net load applied to the supporting soil is zero. In other words, stability is achieved when the building is floating. Such foundations are coloquially termed 'floating foundations' but since the principle is applied to soils other than those described a preferable term is 'fully compensated foundations'. For this purpose deep and stiff foundations are required. In the case of high and heavy buildings it will be necessary to sink the structure to considerable depths in order to displace sufficient soil, thus leading to considerable pressures on the walls and base. In order to resist these the basement or basements are constructed on a cellular basis as described on page 54.

A basement, as explained on page 41, may also be used under the heavier parts of a building in which large variations of loading occur, as in a building with a tall tower block surrounded closely by lower blocks, in order to produce a uniform distribution of pressure. In such circumstances, if the basement is deep, the whole of the basement must be constructed as a stiff rigid structure (figure 2.16 A).

A raft may also be used to provide a stiff foundation under a building the structure of which is sensitive to differential movement (see page 88).

Stiff rafts may also be used in some circumstances to overcome the difficulties of subsidence sites or, alternatively, these difficulties may be overcome by the use of an extremely flexible raft carrying a building of sufficient flexibility to deal with the movements arising during actual subsidence. This is discussed more fully on pages 65–6. The flexibility of a raft foundation varies, as in the case of normal beams, with the depth and the degree of stiffness between the parts and, where great stiffness is required, a deep beam or cellular structure is essential.

Design of rafts generally   In order to distribute the soil pressure evenly, the centre of gravity of the building loads as a whole should lie on the same vertical line as that of the raft. This is particularly important since the raft form is used on weak yielding soils where a slight unevenness of pressure will cause considerable differential settlement. Although, as seen in combined slab foundations, equal loading of the columns is not essential in order to make the centre of gravity of the loads coincide with the centroid of the raft, this is facilitated if the plan of the building is symmetrical and without projecting elements. The ideal is a simple, regular shaped raft carrying symmetrically arranged, equally loaded columns and on which any heavier parts of the structure are grouped symmetrically about the axis. If practical limitations on the extent of the raft at the sides make it impossible to bring about the coincidence of the two centres of gravity, it may be necessary to make the raft irregular in shape in order to reduce the eccentricity to a minimum. In circumstances where some part of the raft must be unevenly loaded to an excessive degree, or where a considerable projection or arm in the plan form is unavoidable, the unevenly loaded or the projecting section should be on a separate raft with the building structure above also separated. This will avoid adverse effects on the structure arising from differential settlement due to the uneven loading on the subsoil.

The layout of drain and service pipes should be considered at design stage so that the effects of holes and ducts passing through the raft may be taken into account. When founding on very soft clay or silt, in which long-term settlement may occur, provision should be made for relative movement between the building and any services entering the building, if necessary by means of flexible connections as suggested in Part 1, section 4.4.1.[16]

The weight of a building structure is not uniformly distributed over the raft, but is concentrated at the wall or column points and, since a completely rigid raft is not practicable in most cases, there is the tendency for greater pressure to occur under these points, giving a non-uniform distribution of pressure under the raft. Nevertheless, to simplify design procedure, other than in the case of an extremely flexible raft, the assumption is frequently made that the pressure distribution is uniform or varies uniformly under the raft, and that the upward reaction of the soil on the raft is uniform. To find the reaction of the soil per m^2, the total load from the structure is divided by the area of the raft and from this may be calculated the thickness of the slab and the reinforcement required for the raft, the design process being similar to that for a floor, but inverted.

Problems due to subsoil water arise during the construction of basements on account of the upward pressure of the water. The total dead weight of the basement and building above must, when completed, exceed the maximum upward pressure of the water unless tension piles or ground anchors (pages 55 and 99 respectively) are used to prevent flotation. If neither of these is being used, the basement when under construction will be without the superstructure load and must be prevented from floating. If the dead weight of the basement alone is insufficient for this, one of a number of methods can be used:

1 The subsoil water level may be lowered by pumping or other means. This is usually essential in the initial stages of construction at least.
2 The basement may be temporarily filled with water to a height such that its weight, together with the dead load of the basement, resists the upward pressure of the subsoil water.
3 Holes may be formed in the floor through which the subsoil water may enter the basement and prevent the build up of external pressure. When construction has advanced enough to provide sufficient dead load, the holes are sealed.

Pumping to keep the water level down is most commonly used, but when construction is sufficiently far advanced the use of the other methods, particularly (2) which is simpler than (3), enables pumping to be stopped. If a basement was to be used only to reduce pressure on a weak soil, it might prove more economic to use piles in order to avoid the construction of a basement in such conditions.

Rafts may be divided into three types according to their design and construction: (i) solid slab, (ii) beam and slab, and (iii) cellular (figure 2.24).

All are basically the same, in consisting of a large, generally unbroken area of slab covering the whole or a large part of the site. The thickness of slab and size of any

beams will be governed by the spacing and loading of the columns and the degree of rigidity required in the raft.

Solid slab raft   This type of raft consists of a solid slab of concrete reinforced in both directions. Light solid slab rafts are used for the small loadbearing wall type of building, such as outbuildings, or for light framed structures where the bearing pressures are relatively low and these are described in Part 1, section 4.4.2. For larger and more heavily loaded buildings, this type of raft is often economic only up to a thickness of about 300 mm. Unless the column and wall loads are very heavy, the slab is reinforced top and bottom with two-way reinforcement. When loads are heavy, more reinforcement is required on the lines of the columns to form column bands similar to a plate floor. The reinforcement to the panels between would then be two-way reinforcement mainly at the top. If the slab is situated at ground level, it is generally desirable to thicken the edge of the slab or to form a downstand beam (figure 2.24) of sufficient depth to prevent weathering away of the soil under the perimeter of the raft (see Part 1). In some cases deep edge beams are combined with the raft to contain the soil under the slab in order to transfer the load to a lower level. This is illustrated in the three-storey building shown in figure 2.24. The ground floor slab, 300 mm above ground level, is stiffened by cross and edge beams to act as a raft, the pressure from which is transferred to the ground below through soil filled within and contained by the deep edge beams before the floor slab is cast.

A thin solid slab raft divided into small sections, held together by steel mesh at mid-thickness, may be used for light framed buildings on sites liable to subsidence. For design reasons, discussed later, the raft rests on the surface of the ground, acting as the ground floor slab and no downstand edge beams are provided, soil erosion being avoided by a few feet of paving round the building.

Beam and slab raft   When loading or requirements of rigidity necessitate a slab thickness greater than 300 mm it may be economic to use beam and slab construction, which may be visualised as an inverted floor, the slab bearing directly on the ground and the beams projecting above it (figure 2.24). The reinforcement is arranged accordingly with regard to areas of tensile stress. On sites where the weak strata are overlaid by a thin layer of comparatively stiff soil at building level, economy in construction could be effected by placing the beams below rather than above the slab. The beams could then be cast in excavated trenches of the appropriate size without the use of shuttering, and the necessity of a sub-floor would be avoided. The slab at the bottom, however, has the advantage of providing stiffness to the base of the beams, which

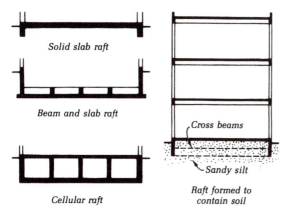

Solid slab raft

Beam and slab raft

Cellular raft

Cross beams

Sandy silt

Raft formed to contain soil

Figure 2.24   Raft foundations

will be in compression in the zones of negative bending moment between columns. Where possible, the raft should be cantilevered beyond the outside lines of columns so that the bending moments in the inner slabs are reduced.

When there is some unevenness of loading on the columns of a frame, but where loading and strength of soil are such that a full raft is not required, beams may be used to link isolated column foundations to give some degree of rigidity in order to limit differential settlement.

Cellular raft   When stresses in the raft are high, and particularly when great rigidity is required, the beams must be deep and when the overall depth is likely to exceed 900 mm a cellular form of construction is adopted. This consists of top and bottom slabs with edge and intermediate beams in both directions forming a hollow cellular raft (figure 2.24). When such a raft is extensive in area and great rigidity is required to reduce differential settlement, the depth may need to be as much as a full basement storey or more using reinforced concrete cross walls constructed monolithically with the floors. The cellular basement is also necessary when very deep basements are used to reduce overall settlement by a relief of overburden pressure (figure 2.16 A and D). As already explained cellular basement construction is one means of dealing with these problems in the case of very tall buildings, as the needs both to limit overall settlement and to reduce differential settlement often arise.

A normal cellular raft is completely cellular, but, in the case of a basement, reinforced concrete walls constructed monolithically with the floors and dividing the basement area into small compartments, make it unsuitable for many purposes. An alternative method may be used in which heavy reinforced concrete columns and floor beams are designed to form Vierendeel girders[17] in both directions. To obtain sufficient rigidity, the junctions of top and bottom of the columns with the beams will normally need large haunches. This restricts the floor area of the basement and in some circumstances it may be preferable to use pile foundations instead of a cellular basement as an alternative, although possibly more expensive method.

In soft soils a cellular raft is constructed and sunk as a monolith as described on page 64. The soil is supported by the sides as excavation proceeds and on completion a base slab is laid at the bottom. In the case of large rafts it may be necessary to divide the raft area into a number of separate monoliths which, after sinking, are joined together by in situ cellular construction to form a single unit.

In deep excavations there is an upward movement of the excavated surface called 'heave' which is caused by the pressure of the overburden at the sides of the excavation forcing up the base of the excavation (see figure 2.15). If a very deep basement is needed in soft clay the heave

and, subsequently, the ultimate settlement of this, may be the determining factor in deciding the depth to which the foundation should be taken. If sufficient relief of overburden pressure cannot be obtained the excess pressure may be taken by piles and if the piles are driven prior to the excavation, the amount of heave can be reduced.

2.3.2 Pile foundations

These may be defined as a form of foundation in which the loads are taken to a low level by means of columns in the soil on which the building rests.

It is a method of support adopted (a) on sites where no firm bearing strata exist at a reasonable depth and the applied loading is uneven, making the use of a raft inadvisable, (b) when a firm bearing stratum does exist but at a depth such as to make strip, slab or pier foundations uneconomical, that is at depths over 3 to 4.5 m, but not so deep as to make use of a raft essential. Piles to depths of 18 m are common and in exceptional circumstances they are used to depths of 30 m or more, but piles over this length are considered long, (c) in shrinkable clay soils as a means of founding below the zone of seasonal moisture movement and of dealing with the problems of moisture change in the soil caused by vegetation, especially trees and shrubs.

The short bored piles used at depths below the zone of seasonal moisture movement in clay soils are described in Part 1, section 4.4.3. The effects of trees on this type of soil are also described in section 4.2. The methods of protecting buildings from the consequences of these effects are indicated in table 4.4 in Part 1 and are described later in that chapter.

Piles are also used when pumping of subsoil water would be too costly or support to excavations too difficult to permit the construction of spread foundations.

Irrespective of the type, piles may be divided into two categories according to the manner in which they lower the level of the applied pressure: (i) friction piles, and (ii) end bearing piles (figure 2.25 A, B). Most piles do, in fact, carry the load by a combination of friction and end bearing.

Friction piles   These transfer their load to the surrounding soil by means of the friction between their surfaces and the soil, and to a slight extent by end bearing, and can be used in deep beds of clay and silt as an alternative to a raft. They may also be used in conjunction with a raft, as already described, when the latter cannot be taken deep enough to obtain sufficient relief of overburden pressure to keep settlement within acceptable limits. Such foundations should be designed with great care as they may result in unacceptable differential settlement.

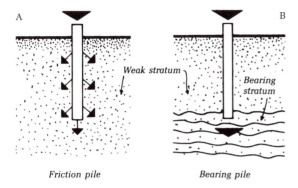

Friction pile   *Bearing pile*

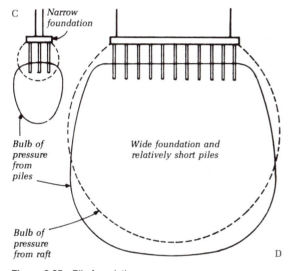

C |Narrow foundation

Bulb of pressure from piles

Wide foundation and relatively short piles

Bulb of pressure from raft   D

Figure 2.25 Pile foundations

End bearing piles These carry their load through weak strata and transfer it to a firm stratum on which their ends rest. They may be viewed as simple columns receiving lateral restraint from the soil through which they pass.

Pile groups The bearing capacity of a pile depends upon its length, its cross-sectional area and the shear strength of the soil into which it bears and it is frequently necessary to use a group of piles, rather than a single pile, in order to obtain adequate bearing capacity. Such a group is termed a 'pile cluster' (figure 2.26 A). For reasons of stability a group of at least three piles is often used under any heavy load rather than a single pile. This, in addition, also provides a margin of safety should one be defective in some way. It is, however, not always necessary to use three piles if displacement and rotation are limited by beams connecting the pile caps (see below).

End bearing piles in a cluster should be placed not less than twice the least width of the piles centre to centre, and friction piles not less than the perimeter of the piles centre to centre.

In the case of non-cohesive soils the soil tends to consolidate during driving but in cohesive soils it tends to be remoulded by displacement upwards and outwards, with a

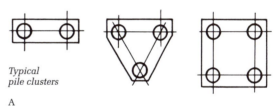

Typical pile clusters

A

The friction pile has the effect of carrying the bulb of pressure to a low level so that the high stresses are set up in the soil at a level where it is strong enough to resist them rather than near the surface where it is weaker. It may be assumed that friction piles form a raft imposing the load upon the soil about two-thirds down their depth. In order to obtain an effective lowering of the bulb of pressure it is essential that the ratio of pile length to building width should be high. The wider a foundation is the deeper will its bulb of pressure penetrate and unless the piles in a wide building are long relative to its width they will make little difference to the actual depth of the bulb of pressure (compare C and D in figure 2.25).

Friction piles may also be used to act in tension in holding a building down against the wind uplift referred to at the beginning of this chapter and in holding down against flotation (see page 52).[18]

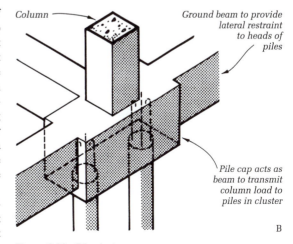

Column

Ground beam to provide lateral restraint to heads of piles

Pile cap acts as beam to transmit column load to piles in cluster

B

Figure 2.26 Pile clusters

loss of strength. If this remoulding is not excessive the soil partly regains its strength after rest and by adequately spacing the piles the remoulding process is minimised.

The tops of the piles in a cluster are connected by a block of concrete termed the 'pile cap', reinforced to transmit safely the column load to the heads of the piles in the cluster (see figure 2.26 B). Adjacent caps are usually linked to each other by ground beams of reinforced concrete, this always being done when there are fewer than three piles under a load. The beams provide lateral rigidity to the tops of the piles and must be able to resist any bending due to eccentric loading on the piles. This is important when the piles are in soft ground giving little lateral support. The beams will normally also carry the lower floor slab and walls of the structure above.

The safe bearing capacity of a pile or cluster of piles may be established in a number of ways:

1 By a loading test on a trial pile or cluster, by means of kentledge bearing directly on the cap, by jacking against kentledge or, more commonly, by jacking against a reaction beam tied down at its bearings to tension piles or ground anchors. The test load is usually up to $1\frac{1}{2}$ times the working load
2 By measuring the 'set' of the pile, which is the distance it sinks into the ground under a given number of blows of a hammer of given weight dropped through a given distance. With this data the bearing capacity is calculated by means of dynamic formula. Dynamic formulae are reliable in non-cohesive soils and hard cohesive soils but in cohesive soils other than 'hard' a check by calculation should be made as in (3)
3 In cohesive soils, by calculation based on the measured shear strength of the ground at various levels
4 By means of a small diameter penetration test, making use of the cone penetrometer developed to measure the soil resistance at its foot and the frictional resistance along its outside surface (page 25). These measurements can be taken at varying depths over a building site and the lengths and bearing capacities of the piles predetermined before work begins
5 By previous experience with similar piles nearby.

As will be seen from the earlier part of this chapter, tests will not give a complete picture of the final settlement of the completed structure as a whole, since it will produce a deeper bulb of pressure than the single pile or cluster of piles (figures 2.12 and 2.25), and this must be considered separately having regard to the factors on which this depends.

Types of pile and applications   Piles are considered as driven or formed in situ according to whether they are preformed and driven into the ground by blows of a hammer or are formed on site by placing concrete and reinforcement in a borehole.[19]

Some types of pile are, in fact, basically formed in situ but a shell or former into which the concrete is poured is driven into the soil and from the point of view of soil mechanics are considered as driven piles. To distinguish them from driven precast concrete, timber or steel piles they may be termed 'driven tube' piles.

Driven piles   These may be of timber, steel or precast concrete. The latter, either normally reinforced or prestressed, is most commonly used in building work. Timber piles are rarely used in building work except for minor purposes, such as supports to manholes in running sand for example. The use of timber piles is restricted by the limitations in cross sections and lengths available. Splices are impractical for this application. In addition the supply of large numbers of timber piles is difficult at short notice. Timber tends to rot if not always submerged and difficulties are encountered in the use of timber piles if there is hard driving in the early stages. Steel piles in the form of open ended box or H sections, or pipes, are used when the length is very great, especially when driven into rock to carry heavy loads. These conditions would require large section heavy concrete piles, difficult to handle and liable to damage, and in such circumstances they would be used instead of steel only in situations where the latter would be subject to heavy corrosion.

Precast concrete piles   These may be square, hexagonal or round in section and may be solid or hollow. 18 m long precast piles are normal and lengths up to 15 to 18 m can be obtained from stock. Large section piles over 30 m long have been used; they are usually cast on or near the site. The toe of a precast pile is usually fitted with a cast iron or steel protective shoe and the top is protected during driving by a steel cap or 'helmet' in order to prevent spalling. Shoes are not essential in soft ground but special shoes are provided for entering rock.

Precast piles may be obtained cast in units of varying but standard lengths up to 10 m, which can be joined by special steel caps cast in each end to permit any length of pile to be formed. They may also be obtained cast in segments 1 m long, assembled by spigot and socket joint, for light loadings. These have a central hole for a reinforcing or post-tensioning bar.

The reinforcement in precast concrete piles is designed to resist the stresses set up in transport and slinging and in driving, as well as those set up by the working load. Prestressing may be used, in which case the whole pile may be prestressed as cast, or long piles can be formed of factory made precast sections transported to the site where they are made up into full pile lengths and post-tensioned prior to driving.

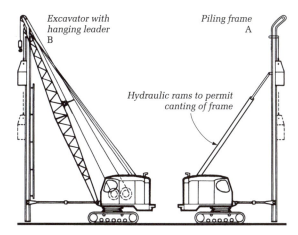

*Excavator with
hanging leader
B*

*Piling frame
A*

*Hydraulic rams to permit
canting of frame*

Figure 2.27 Piling rigs

Pile driving The pile is supported between guides or
'leaders' in a vertical steel piling rig or frame (figure 2.27)
sufficiently high to take the pile and to allow room above
the cap for a hammer, or for the drop of a weight, by
means of which the pile is driven into the ground. Pile
driving frames (A) are normally mobile and arranged to
rake over to drive piles at an angle when required. For
piles less than 12 m in length an excavator may be
equipped with hanging 'leaders' to act as a pile driver, a
hydraulic ram between the bases of the leader and jib permitting canting of the leader for raking piles (B). Driving
by a weight or drop hammer is performed by allowing a
weight or 'monkey', attached by cable to a steam or diesel
operated winch, to fall on to the head of the pile in successive blows. The winch raises the 'monkey' each time to
the top of the drop and is then released to allow the 'monkey' to fall. The distance of drop is usually about 1.50 m
although drops up to 3 m are sometimes used.

Pile driving hammers rest directly on the pile and travel
down with it. Driving is more rapid than by a drop hammer because of the large number of blows delivered per
minute. They are operated by steam or compressed air as
single- or double-acting hammers. Petrol operated hammers are also available, acting on the principle of earth
rammers. With a single-acting hammer the motive power
simply lifts the hammer after which it falls by its own
weight. With a double-acting hammer the steam or air
assists in forcing it down on to the pile after having raised
it. The weight of the striking parts of a double-acting hammer is much less than that in a single-acting hammer but
the former gives a more rapid succession of blows, as does
the more modern hydraulic impact hammer which works
in a similar way with the same advantages.

Driven piles are most suitable for open sites where the
length required will be constant and where there is ample
headroom for the driving frames. Considerable vibration
and noise is set up during driving and on some sites this
would preclude the use of this type of pile, although the
use of electric or hydraulic vibratory pile drivers considerably reduces noise.

In certain types of soil, such as fine, clean sands and
fine gravels, driving is facilitated by the use of a jet of
water discharged under pressure at the toe of the pile to
displace the soil underneath. Various arrangements of jets
are used and the water is taken to the toe through a pipe or
pipes running through the pile from the top. Towards the
end of driving the water is cut off so that the pile is driven
to rest for the last 1.20 to 1.40 m through undisturbed soil.

Driven tube piles These are more suitable than solid
driven precast concrete piles on sites where there is likely
to be a considerable variation in the lengths of piles. They
may be formed in various ways using steel or concrete
tubes. In some forms driving is from the top, in others
from the toe by means of a mandrel or by a drop hammer
falling through the length of the tube. In some forms the
tube or shell is left permanently in position, in others it is
withdrawn as concreting proceeds. In common with driven
piles vibration occurs under the blows of the hammer, the
degree of which depends upon the particular system used.

Top driving Where driving is from the top a steel tube of
required diameter is fitted with a cast-iron conical shoe,
placed in a driving frame and driven down by a drop or a
steam hammer in the same way as a precast concrete pile
(figure 2.28 A). Reinforcement, in the form of a cage made
up of longitudinal rods linked by helical or horizontal
binding with spacers at intervals to ensure adequate concrete cover, is lowered into the tube. Concrete is then
poured in through a hopper or skip as the tube is gradually
withdrawn leaving the driving shoe behind.

Tamping of the concrete may be done by a drop hammer falling on to each charge of concrete as the tube is
slowly withdrawn, consolidating it and forcing it into
close contact with the surrounding soil. Alternatively, the
concrete is consolidated by vibrating the tube as it is
withdrawn. In other systems, using a tube with a thickened
bottom rim, the concrete is tamped by the tube which, as it
is withdrawn, is subjected to rapid up and down blows
from a steam hammer, causing it to consolidate the concrete by its thickened edge (figure 2.28 B).

The normal tube length is from 12 to 15 m, but this may
be extended when necessary by screwing on a further
length when the first has been driven or, alternatively, the
tube may be used as a leader for a precast concrete extension pile driven through the tube and taking the conical
shoe with it. This has the advantage of reducing frictional
resistance since the driving force has to overcome only the

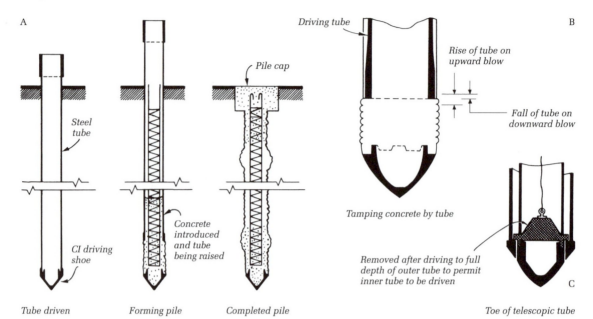

A

Driving tube

B

Pile cap

Rise of tube on
upward blow

Fall of tube on
downward blow

Steel
tube

Concrete
introduced
and tube
being raised

CI driving
shoe

Tamping concrete by tube

Removed after driving to full
depth of outer tube to permit
inner tube to be driven

Tube driven   Forming pile   Completed pile

Toe of telescopic tube

C

Figure 2.28 Top driven tube piles

resistance of the extension pile as it is being driven, the upper tube remaining fixed in position. The same result may be obtained by the use of telescopic tubes in which the upper length after it has been driven to its full extent acts as a leader to an inner tube which is driven to the full required depth (figure 2.28 C). Both these methods can be used where piles must pass through the water of a river or lake, the upper tube being left in position as a casing to the concrete filling.

Toe driving Driving from the toe substantially reduces the amount of vibration set up during the driving operations. One type of pile driven in this way uses a steel tube of the appropriate diameter held in a driving frame with the open bottom end resting on the ground. The first 600 or 900 mm of the tube are filled with dry gravel or dry mix concrete which is compacted by blows from a drop hammer falling within the tube (figure 2.29 A). When the friction between the gravel and the inside of the tube is sufficiently great subsequent blows pull the tube down into the soil. By means of marks on the hammer cable a check is kept on the depth of the plug and more gravel is introduced when necessary so that the tube is continually sealed to prevent the entry of soil or subsoil water. Reinforcement and a semi-dry concrete mix may be placed as described earlier and tamping carried out by the drop hammer as the tube is gradually withdrawn, or a wet mix concrete may be used and consolidation be carried out by vibrating the tube as already described.

In this type and in all types in which the tube is withdrawn it is possible to form an enlarged bulb base of concrete which increases the bearing area of the pile as shown in figure 2.29. Light steel tube piles driven in this way may be used when vibration must be limited but the nature of the subsoil, such as waterlogged soil or running sand, might make boring impossible. The tubes are left permanently in position and are filled with reinforcement and concrete. Any required lengthening of the tubes is carried out by site welding.

Precast concrete tube piles Another type of pile driven from the toe uses a tube or shell of concrete round a steel mandrel, the tube being left in the soil to act as a permanent casing to a core of reinforced concrete cast inside (see figure 2.29 B). The concrete tube is built up from precast reinforced concrete sections threaded over the mandrel, at the bottom of which is a solid concrete shoe. The hammer blows are transmitted to the shoe through the mandrel and by means of an adaptor at the top of the mandrel. Sufficient force is also transmitted to the tube to overcome skin friction and to enable it to follow the shoe. After being driven to a set the mandrel is withdrawn, any surplus sections of tube are removed from the top and the reinforcement and concrete are placed inside.

Bored piles On sites where piling is to be carried out close to existing premises, or for underpinning buildings with the minimum of disturbance, driven or driven tube

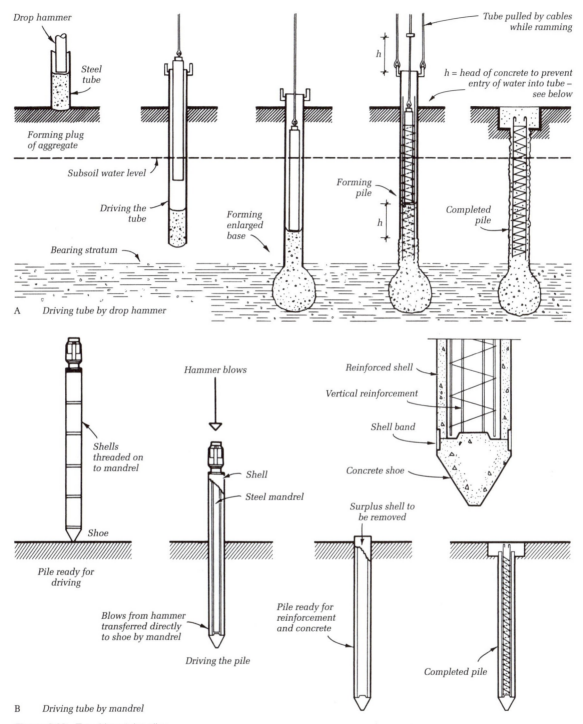

A   Driving tube by drop hammer

B   Driving tube by mandrel

Figure 2.29   Toe driven tube piles

piles are unsuitable because of the amount of vibration set up. In such cases some form of bored pile is used in which the vibration is much less. The use of bored piles is usually cheaper than any form of driven pile when a few only are required.

The pile hole may be bored either by rotary drilling or by percussive boring.

Rotary drilling This may be most advantageously used on large contracts in cohesive soil where there is good access for the drilling machines. These incorporate a drilling auger which is fixed to the end of a long, square vertical drilling rod, or kelly bar, which passes through a turntable by means of which it is rotated. This type of machine can drill up to a depth of 6 m with diameters up to 900 mm and it may be mounted on a lorry (see figure 2.30 A) or be part of a specially constructed crawler-mounted machine or, very commonly, be mounted on an excavator crane specially rigged for drilling (B). Temporary support for the borehole may be provided by steel lining tubes where necessary, but this method of boring is most economically used in deep cohesive soils where an upper lining only might be required for the non-cohesive overlying soil. Where it is used in other types of soil and where the soil strata vary a bentonite suspension, as described under

diaphragm retaining walls on page 98, may be used instead of a lining, the boring being carried out through the fluid.

Percussive boring This method requires only a light shear-leg type of rig instead of a large driving frame and is useful, therefore, on sites where levels vary and where space or headroom is restricted.

In forming the pile by percussive boring a steel tube is sunk by removing the soil from inside it by means of a coring tool (figure 2.31 A, B), the tube then sinking under its own weight or being driven down by relatively light pressure, generally by means of the coring tool itself acting through a steel channel passed horizontally through the clearing holes in the sides (C).

The tube is made up from screw-coupled sections varying from 1 to 1.35 m in length, so that work can be carried out with the headroom restricted to as little as 1.80 m, fresh sections being coupled on until the final pile length is reached. The coring tool is raised and dropped inside the tube by a winch driven by a petrol or diesel engine to which it is attached by a steel cable running over a pulley at the top of the shearlegs. A clutch enables the operator to drop the tool as required.

The type of coring tool used varies according to the nature of the subsoil. In clay a steel cylinder with a bottom cutting edge is used. It has rectangular holes in the sides to enable stiff clay to be more easily removed by a spade or crowbar when the cutter is brought up full from the tube (A). In sands and other granular soils a cylinder similar to the clay cutter is used, but without the extracting holes in the side, with a hinged flap at the bottom opening upwards to allow the soil to enter the cylinder as it drops and then closing to retain it when the cylinder is raised. This is called a sand and ballast shell (B).

Forming the pile When the required depth is reached, a cage of reinforcement is lowered into the tube and concrete is introduced in batches through a hopper as the tube is gradually withdrawn either by block tackle operated by the winch or by hydraulic jacks operating against lugs on the top of the tube (F). Tamping can be carried out by a drop hammer or by compressed air.

In the latter method, which eliminates vibration during the tamping process, after each batch of concrete is introduced a pressure cap is screwed on the top of the tube and compressed air is admitted which forces the concrete down and out against the surrounding ground (D). With this method a bulb foot, similar to that formed by a drop hammer, may be produced in concrete by forcing cement grout into the granular subsoil surrounding the foot of the pile.

When ground water is present the concrete may be placed under the water by means of a tremie pipe or the

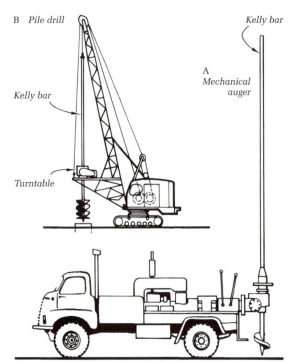

Figure 2.30 Rotary drills

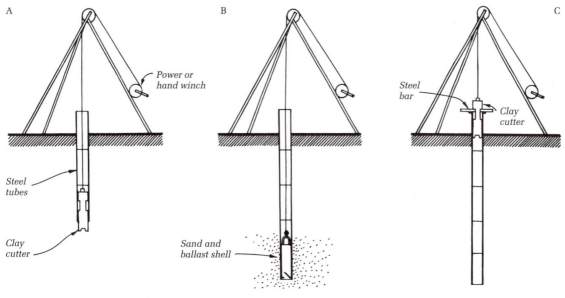

A

Power or hand winch

Steel tubes

Clay cutter

Sinking tubes through clay subsoil

B

Sand and ballast shell

Sinking tubes through granular subsoil

C

Steel bar

Clay cutter

Tube driving (when necessary)

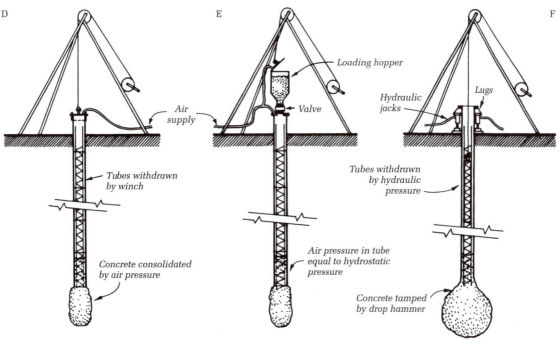

D

Air supply

Tubes withdrawn by winch

Concrete consolidated by air pressure

Forming pile with compressed air

E

Loading hopper

Valve

Air pressure in tube equal to hydrostatic pressure

Placing concrete through an air lock

F

Hydraulic jacks

Lugs

Tubes withdrawn by hydraulic pressure

Concrete tamped by drop hammer

Forming pile with a drop hammer

Figure 2.31   Bored piles

water can be kept out of the tube during concreting by compressed air, in which case the concrete is placed through an air-lock attached to the top of the tube in place of the normal pressure cap as shown in (E). This consists of a cylindrical steel hopper with a feed hole at the top and a large valve at the bottom. Admission of air first blows out the water through a pipe lowered down the tube. This pipe is then closed, and by maintaining the air at sufficient pressure to balance the head of subsoil water, the tube is kept free of water. Concrete is then placed in batches through the air-lock, each batch being forced down and out by a temporary increase in the air pressure as the tube is being raised.

In using bored piles or driven tube piles where the tube is withdrawn, the possibility of 'necking' must be borne in mind. This is liable to occur when the external ground water pressure is more than the pressure of the concrete where it flows out of the bottom of the tube. It may also occur when the pile is in very soft clay which will try to press back into the pile when the tube is withdrawn if the concrete pressure is insufficient. Compressed air on the concrete helps to overcome this if it is maintained at the critical time. To minimise this difficulty it is possible to introduce into the tube, instead of poured concrete, a core pile made up of short precast concrete units assembled initially on a central steel tube, with reinforcing bars passing through holes round the units. As the casing tube is withdrawn grout is introduced into the central hole under compressed air, filling all voids in the pile and passing through holes to fill the space between the core pile and the sides of the bore hole.

Cylinder piles   These are large diameter bored piles which may be from 600 mm to 3 m or more in diameter and possess the advantages of the traditional cylinder foundation which has been used for many years by engineers to carry very large loads. Developed essentially for use in cohesive soils, an auger on a heavy duty machine is used for drilling the hole. These piles are generally unreinforced except where they pass through a soft stratum near the ground level and may be sunk to a depth of 60 m in suitable conditions using a telescopic kelly bar. A lining is not normally used except in cases where a granular stratum overlays a clay stratum, when the hole may be lined to a depth of from 10 to 15 m. At greater depths bentonite slurry may be used where necessary to support the sides of the borehole. Drilling is rapid and shafts can be sunk into firm clay at a speed of up to 3 m per hour. The base of the piles may be expanded up to two or three times the diameter of the shaft by means of an under-reaming tool consisting of a cylinder from which cutting wings are made to extend when at the base of the shaft.

Cylinder piles have a number of advantages over normal types of piles in terms of cost and construction time, particularly when heavy column loads have to be carried. A group of normal piles can be replaced by a single large diameter cylinder pile which reduces drilling costs and construction time. In many cases a group of friction piles can be replaced by a single under-reamed cylinder pile which carries its load mainly in end bearing. For example, loads of 3000 tonnes can be carried on a single cylinder in suitable soil, such as the lower levels of the London clay. In most cases, because a column can be carried by a single cylinder, pile caps can be greatly reduced or eliminated, thus shortening the construction time and reducing the cost. In addition, the soil strata in the boring can be inspected as the work proceeds and concreting can be carried out more efficiently than in the case of normal in situ piles, and can be properly inspected. These piles, because of their capacity to carry very heavy loads, are valuable as foundations to very tall buildings founded on soft soils where the alternative to cylinder piles would be a large number of long friction piles.

In practice little is known about the relation between base and shaft resistances, but tests indicate that if the base resistance is to be fully developed large settlements are to be expected. This must be carefully considered when under-reamed piles are used. The under-reamed pile can resist large tensile forces and is therefore useful where tension piles are required.

Auger-fed piles   These are bored piles up to 600 mm diameter formed with a *continuous flight auger*, which is an auger up to 20 m in length, the helix of which extends the full length of the stem, the latter being hollow. The auger is supported from an excavator crane equipped with a set of leaders which guide the auger during boring. This is similar to the rig used for driving piles (figure 2.27 B) except that the leaders extend well above the crane jib because of the length of the auger.

The auger is bored into the soil to the full depth required for the pile. Concrete specified to a higher than normal water/cement ratio with rounded aggregates, or a sand/cement grout, is then pumped down the hollow stem to form the pile as the auger is withdrawn. In the process the auger takes with it the spoil from the hole. Since the spoil is removed only as the concrete is fed into the borehole it continues to support the sides of the hole until the latter has been completely filled with concrete. No lining tube or bentonite slurry is, therefore, required for this purpose in unstable soils and the stopping and starting often associated with bored piling is eliminated.

Reinforcement may be pushed in manually if it extends only a few metres down the pile. Longer reinforcement is

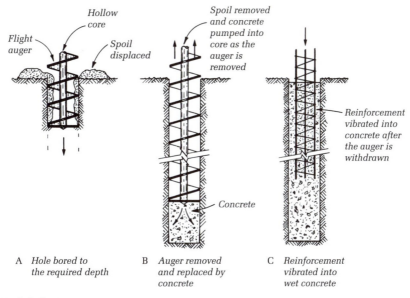

A   *Hole bored to the required depth*

B   *Auger removed and replaced by concrete*

C   *Reinforcement vibrated into wet concrete*

Figure 2.32   Auger fed piles

introduced with the aid of vibration. Figure 2.32 shows the principles.

Jacked piles   Where complete freedom from vibration and noise is essential a jacked-in pile can be used in which precast sections of reinforced concrete are successively jacked down one after the other. The jack works either directly on solid precast sections joined together by lengths of tubular steel grouted in a central hole as they sink into the ground (figure 2.36 D), or on the top of a built-up steel mandrel which forces down a concrete shoe and tubular precast sections, the completed tube being filled with reinforcement and concrete.

When these piles are used for underpinning work as shown in figure 2.36 D the jack operates against the underside of the foundation, otherwise it works against a travelling kentledge. The expense of jacked-in piles limits their use.

Rectangular concrete piles   These are large rectangular reinforced concrete piles formed in the same manner as diaphragm retaining walls (page 98), making use of a bentonite suspension to support the sides of the excavation. They can be formed up to 2.2 m × 1.8 m and are useful when very heavy loads must be taken down to a deep bearing stratum lying below water-logged or unstable soils.

2.3.3 Pier foundations

Brick and concrete piers   When a good bearing stratum exists up to 4.5 m below ground level, brick, masonry or mass concrete foundation piers in excavated pits may be used. This type of pier foundation is described in Part 1, section 4.4.4. At this depth, unless the nature of the ground necessitates considerable timbering or continuous pumping, piles are not always economical.

Cylinder and monolith foundations   These are a form of pier foundation used when very heavy loads must be carried through water-logged or unstable soil down to bed rock or to a firm stratum.

A *cylinder foundation* is constructed by means of a caisson, a cylinder of steel or concrete which provides support to the sides of the excavation and by means of which soil and water can be excluded. The caisson is from 1.80 m upwards in diameter and is built up as high as is practicable before excavation begins (see figure 2.33 A1). The soil is excavated from within by grabbing, hand clearing being used usually only at the cutting edge (A2), and the caisson sinks to the required depth under its own weight or with the aid of kentledge. The bottom edge of a concrete caisson is formed into a cutting edge usually protected by a steel shoe. After sinking, the caisson is left

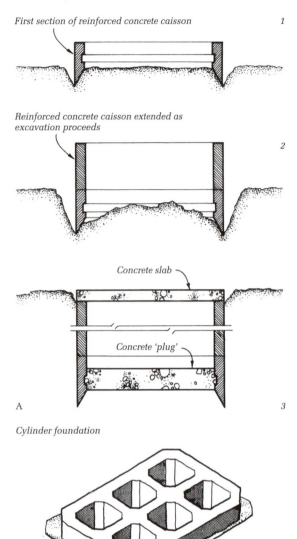

First section of reinforced concrete caisson   1

Reinforced concrete caisson extended as excavation proceeds   2

Concrete slab

Concrete 'plug'

A   3

Cylinder foundation

Reinforced concrete monolith

B

Figure 2.33   Cylinder and monolith foundations

order to permit the air within the caisson to be maintained at sufficient pressure to prevent the entry of water. Open caissons can be taken to depths up to 45 m or more; theoretically there is no limit to the depth. In stiff clay soils considerable weight is required to overcome the skin friction as the caisson sinks and in these conditions cylinder piles, already described, form a suitable alternative. Bentonite slurry, however, pumped between the caisson and the soil can now be used as a lubricant. This is especially necessary when very large diameter caissons, 30 to 60 m in diameter, forming the total base of a building structure, are being sunk. Pneumatic caissons may be used for any type of soil, but the depth to which they may be sunk is limited to about 34 to 37 m below the water surface level. This corresponds to an air pressure of about 345 kN/m^2 which is the greatest pressure in which people can work. In practice pneumatic caissons are generally not sunk much beyond 25 m.

As an alternative to large diameter cylinders in suitable circumstances the diaphragm retaining wall technique described on page 98 could be used.

A **monolith** is a rectangular open caisson of steel or concrete with a number of wells for excavation, used when loading necessitates a very large area of foundation (see figure 2.33 B). Heavy mass or reinforced concrete monoliths have the advantage of greater weight. These can be very large depending upon the nature of the project. The initial lift of a monolith might be as high as 3 m or more constructed on a temporary concrete slab which would be smashed and removed prior to sinking. Further lifts about 1.80 m high would be constructed until the required depth had been reached. As with cylinders monoliths may be sealed at the top and bottom or be filled in solid according to the requirements of a particular job.

The advantages of cylinder and monolith foundations over normal pile foundations are that the soil may be inspected at the base and during the course of excavation and the concrete can reliably be placed in position. The area of a caisson should be so related to that of the superstructure above that it is great enough to allow for some deviation from its correct position during sinking.

permanently in position. It may either be filled with concrete or sealed at the bottom with a concrete 'plug' and at the top with a concrete slab (A3). The interior may be left empty or filled with sand or water.

Cylinders which are open at top and bottom, known as open caissons, are used where ground conditions permit. When it is necessary to keep the excavation free of water and silt, a pneumatic caisson is used. This has a working chamber with a closed top fitted with an airlock through which operatives, equipment and materials must pass in

2.3.4 Foundations on subsidence sites

The behaviour of soil and buildings under the action of subsidence has been described on page 43. A number of methods are used to minimise the effects of mining subsidence upon a building; of these the following are commonly used.

● The foundation of the building may be designed as a strong reinforced concrete raft, generally in cellular form to give adequate stiffness, or as a stiff system of beams with three-point support on the soil through spherical bearings on piers or pads. The foundation structure, together with the building supported by it, will tilt as the ground moves, but the building will not distort and will, therefore, be free from damage. These methods are expensive, particularly the construction of stiff cellular rafts, and are generally used only for large structures, since the cost of the foundations is likely to be much in excess of the value of any damage caused in a small building.

- The building itself may be designed as a rigid frame structure or be strengthened by reinforcement to act as a single structural element, capable of resisting without cracking all the stresses set up as it tilts and as the support from the ground reduces in area and varies in position with the movement of the ground. With both these methods it is desirable to break up long buildings into short blocks, connected only by some form of flexible joint.

- In contrast to the above methods, the building and its foundation can both be designed as flexible elements, the construction of the structure and its cladding being detailed to accommodate the differential movements of the soil without damage either to structure or finishes.

- Hydraulic jacks may be used to raise or lower the building at different points to keep pace with the subsidence movement and thus avoid tilting. With this method the building structure would rest on reinforced concrete or prestressed concrete foundation beams bearing on concrete bases, provision being made for the jacks to be inserted at the points of support where adjustment is required. The jacks can be individually and automatically adjusted as the settlement takes place by connecting them to a central control system.[20]

Precautions in long buildings When the method adopted requires a long building to be broken down into short independent units, these should not exceed about 18 m in either direction in order to keep differential settlement within reasonable limits. If the independent units must be connected this is accomplished by movement or slip joints in walls, roof and floor slab. These gaps must be sealed in such a manner that freedom of movement is permitted. At the walls thin corrugated metal can be used, or a soft filler joint of cellular plastic or expanded rubber finished with a silicone pointing. These softer fillers can be used as a form of sliding element at roof and floor. As an alternative to corrugated metal, rendering on expanded metal has been used. When it becomes known that movement will take place the rendering is cut down to permit 'concertinaing' to take place, and when the movement is complete and the gaps have again opened up they are made good permanently.

Protection within the limits of these small units should provide resistance against the forces set up in the ground and against those set up by the cantilever and beam actions of the foundation (see page 43). This is done by providing steel reinforcement disposed according to the stresses set up in the foundation slab by the different external ground forces. The amount of the ground forces transmitted to the slab may be limited by reducing as much as possible the friction between the slab and the ground. This is accomplished by laying the ground or foundation slab on waterproof paper or polythene sheet on a bed of sand, shale or other granular material 150 to 300 mm thick.

Resistance to the forces due to the flexure of the foundation at the crest or trough of the wave can be provided only at considerable expense by means of a stiff and rigid foundation, often in heavy and deep cellular form. In small and light structures the cost of such foundations, sometimes as much as 10 per cent of the cost of the building, is likely to be greater than the cost of repairing any damage caused by unrestricted flexural movements, and foundations of this type would be economical only in large, multi-storey structures in which the high cost of the foundations would be distributed over a total floor area much greater than that of the foundation itself.

Use of flexible structure Damage to the structure arising from the flexural movements of the foundation when this is not rigid are minimised by using, as far as possible, flexible forms of construction such as timber framing and brickwork bedded in weak mortar to prevent serious cracking. Light reinforcement introduced in brickwork at top and bottom and around openings also assists in this respect, especially when the differential movements are likely to be severe. In addition to these precautions flexural movement can be reduced by siting the building with the shorter axis in the direction of the likely maximum curvature when this can be ascertained.

These methods accept as inevitable a considerable amount of damage to the structure, requiring repair after subsidence is complete, or they make use of heavy, rigid and costly foundations designed to resist the flexural stresses. Even so in many cases experience has shown these to be inadequate to prevent extensive damage to the structure. This is probably due to the fact that the deeper the structure is in the ground the more likely it is to be affected adversely by the horizontal ground forces.

An alternative method suitable for buildings of light construction, in which the foundation and superstructure are integrated into a wholly flexible structure, uses a foundation designed as a thin, flexible slab which accommodates itself to the curves and tilt of the subsidence wave. The building structure is formed as a light, completely pin-jointed frame capable of adjusting itself to the differential vertical movements transferred to it through the foundation slab (see figure 2.34). Claddings designed to permit free movement, such as tile hanging, weatherboarding or hung concrete slabs, are used and windows are designed in timber with sufficient clearances to accommodate the maximum anticipated distortion.

The foundation slab, laid on sand or shale, consists of a 125 mm thickness of concrete reinforced at the centre of the thickness and divided into panels not greater in area than 18.5 m². The panel edges are painted with bitumen so that the slab will bend fairly freely with the ground. The primary function of the reinforcement is to hold these slabs together. The sand or shale bed, together with the lightness of structure, reduces friction and minimises the ground forces transmitted to the slab.

Where the loading on columns is such as to require spreading over a relatively large area local top and bottom reinforcement is placed in the slab but no thickening of the slab occurs. Alternatively, reinforced precast concrete

foundation pads are used to which the slab reinforcement is tied when the slab is cast up to them.

Since no flexural stresses are set up in the slab as a whole the need to limit its length in order to keep differential settlements within small limits does not arise. The length of the building is not, therefore, dictated by this consideration but rather by that of the maximum length which can economically be held together by tensional reinforcement. This appears to be about 55 m for a single-storey building reducing to 43 m for one of three or four storeys, up to which height it is considered that this form of foundation is applicable.

Rigidity is provided in the pin-jointed frame structure by specially designed diagonal steel braces incorporating coiled springs at the ends. These do not move under normal wind and dead loads but react to the greater stresses due to subsidence. One of the springs always acts in compression, the other remaining inoperative. When the subsidence wave has passed the springs bring the frame back to its normal position and continue to control it under normal loads.[21]

This type of flexible foundation and building structure is, of course, suitable for soft or made up ground, on which differential settlements are likely to develop after the building is completed, as well as for normal sites. There is no reason why its use for lightly loaded buildings, for which it was designed, should be limited to subsidence sites.

A form of roller bearing has been used for heavy single-storey structures, in which two sets of rollers at right angles to each other in a tray are situated at the foot of portal frames to permit free horizontal movement of the soil and foundations in any direction without transfer to the structure.

2.4 Underpinning

The term underpinning is applied to the process of excavating under an existing foundation and building up a new supporting structure from a lower level to the underside of the existing foundation, the object being to transfer the load from the foundation to a new bearing at a lower level. This may be necessary for any of the following reasons:

- When excessive settlement of a foundation has occurred.
- To permit the level of adjacent ground to be lowered, for example where a new basement at a lower level is to be formed.
- To increase the loadbearing capacity of a foundation.

Before underpinning operations commence, the structure of the building should first be examined for weaknesses such as poor brickwork or masonry and for effects

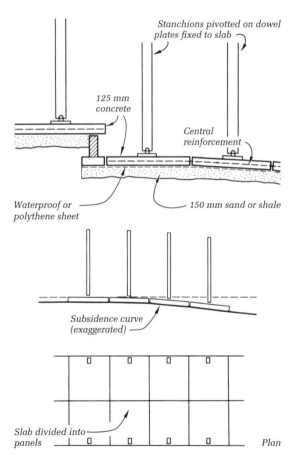

Figure 2.34   Flexible structure and foundation

of settlement which may be accentuated during the course of the work. In very old or badly damaged buildings, the structure may require strengthening by grouting up cracks and loose rubble masonry, for example, or by tying-in walls by tie rods or prestressing cables. Temporary support should be provided by adequate shoring and by strutting up of openings (see chapter 10). In some circumstances a wall or column must be relieved of all loads bearing on it by strutting floors and beams down to a solid bearing clear of the underpinning. During the course of underpinning high structures it may be advisable to keep a check on any possible movement by taking readings of plumb bobs suspended from high points on the structure and by making checks on levels from time to time. Strips of glass or 'tell-tales' should be resin bonded across all cracks as a check on any further movement where it has previously occurred.

2.4.1 Continuous underpinning

Where possible walls should be underpinned with the minimum of disturbance to the structure and to the occupants of the building, and in most cases the work can be carried out without the support of needles passing through the wall to the interior. The work should be carried out in sections or 'legs' in such a sequence that the unsupported lengths of existing foundation over excavated sections are equally distributed along the length of the wall being underpinned (figure 2.35 A). In buildings other than small lightly loaded structures, the sum of these unsupported lengths should not at any one time exceed one-quarter of the total length of the wall, or in the case of a very weak wall, or one carrying heavy loads, one-fifth to one-sixth of the total length. Unless unavoidable, a section should not be excavated immediately adjacent to one which has just

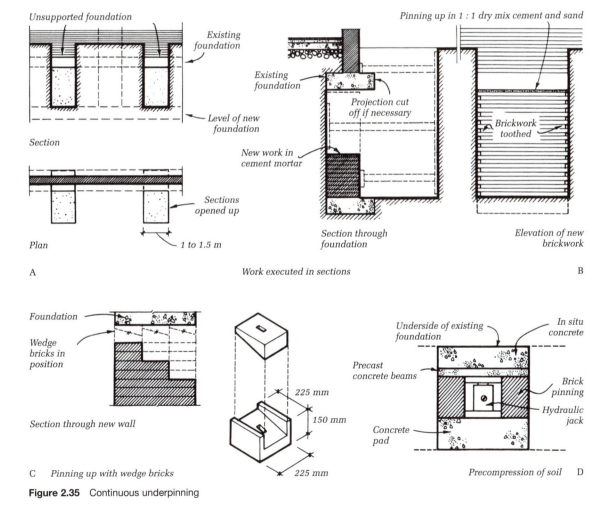

Figure 2.35 Continuous underpinning

been completed. On a low, lightly loaded building such as a two-storey house with the walls in good condition, it is common practice to work to a sequence in which not more than one-third of the length is unsupported at any one time. The lengths of wall which can be left temporarily without support over the excavated sections will depend upon the thickness and general state of the wall and its foundation, the load which the structure will be carrying during the course of the work and an assessment of the risk that the situation imposes on any operative working in the excavation. Advantage is taken of the 'arching action' of normally bonded brick and stone walls over openings and the length of sections can usually be from 1 to 1.50 m.

As each section is excavated, support will generally be required, the depth of each stage of excavation depending upon the firmness of the soil. Poling boards or trench sheeting or, in loose soils permitting only small depths to be excavated at a time, horizontal sheeting, will be used to support the soil faces under the wall. When the nature of the soil or other circumstances precludes the withdrawal of poling boards or sheeting as underpinning proceeds, these are usually formed of precast concrete about 1 m by 300 mm by 50 mm thick. Holes through the thickness permit grout to be pumped through to fill solidly any voids between the soil face and the back of the boards. This is essential in order to prevent the voids closing up afterwards and causing settlement at the back of the wall.

When the full depth of the excavation is reached, the new concrete foundation will be placed if brick underpinning is being used or, if concrete is being used either for a normal wall or a retaining wall, the first lift of the concrete. (For details of retaining wall construction see *Support for excavations*, chapter 10.) The ends of the brickwork in each section should be left toothed ready for bonding with the next section (figure 2.35 B); all concrete work should have grooves formed to provide a key for the next section of concrete. Any horizontal reinforcement must project, being turned up against the sides until the adjacent section is excavated. It is then turned down to be spliced with that in the next section. All brickwork should be to at least semi-engineering quality, laid in a cement and sand mortar of 1:3.

After cleaning off the underside of the existing foundation, brickwork is built up to within about 25 to 38 mm of the foundation. When time has been allowed for the mortar to set and shrink, the brickwork is pinned up with half-dry 1:1 cement and sand rammed in hard with a 25 mm board to make solid contact with the foundation above. In thick walls it is necessary to step down the brickwork from back to front in order to pin up satisfactorily, each step being pinned up in turn, commencing at the back. It is possible to pin up with specially made two-piece wedged-shaped engineering bricks, bedded in stiff mortar

at the top of each step which, as they are driven home, reduce the mortar thickness to a minimum. The two pieces are keyed together by mortar squeezed into a vertical slot in each as the wedge is tightened (see figure 2.35 C).

Concrete underpinning This must be placed over front shutters and can generally be taken only to within about 75 mm of the existing foundation. This, after the concrete has had time to set and shrink, is filled with half-dry 1:1 cement and sand or a dry mix fine aggregate concrete, rammed home with a mechanical tamper. Alternatively, pressure grouting can be used to pin up. Daywork joints in concrete walls which are to be water resistant must be freed of laitance. This can be washed away from freshly placed concrete by high-pressure water spray within $1\frac{1}{2}$–2 hours of placing, leaving a good, clean key. When the adjacent sections are opened up the exposed concrete faces may need to be hacked to form a good key and wire brushed to remove any remaining laitance and loose material. The keyed surface should be thoroughly wetted and brushed over with a thin coat of 1:2 cement and sand mortar or a proprietary bonding agent immediately before fresh concrete is placed against it.

Prestressing can be applied to continuous in situ concrete underpinning carried out in the normal way to form a continuous beam under the existing foundations. This is done by sinking some sections to the full required depth to act as supporting piers to a beam formed between them by the intermediate sections which are not sunk so deep. These are formed into a continuous beam by post-tensioning with cables, passed through holes formed in the sections as they are cast and anchored at the ends. In addition to providing continuous support under the foundations, which may be essential when the superstructure is in a very poor condition, this method keeps to a minimum the amount of deep excavation which must be carried out.

Underpinning in each section should commence as soon as possible after excavation, and be carried out as quickly as possible. No earth face should be exposed overnight. If delay is likely the base of the excavation should be protected by a blinding layer of concrete or the last few inches of excavation should be held over.

Temporary support When the walling is weak or when sections must be opened up in lengths greater than those previously suggested or where the safe bearing pressure of the soil is likely to be exceeded during the underpinning operations, it may be necessary to provide temporary support in the form of needles passing through the wall or under the foundation. Needles of timber or steel should be of ample size to avoid deflection and should be kept in close contact with the structure by means of folding wedges or jacks. Supports must be well clear of the area of underpinning work and taken on to a solid bearing. When the needles pass through the wall above foundation level, subsidiary needles or springing pieces hung from the main needles may be required to carry the section of wall below the level of the main needles, particularly if the walling is weak (see figure 2.37 A).

In order to minimise the number of needles penetrating within the building it is possible to use widely spaced main needles which pass through the wall and carry a beam parallel to the outside face of the wall. This acts as a fulcrum to intermediate counterweighted cantilever needles which penetrate only the thickness of the wall (figure 2.37 B). When support by needles is impracticable, temporary steel struts bearing on concrete pads may be inserted beneath the existing foundation. The underpinning work may then be built up or cast between them before they are removed or, alternatively, they may be left in position and eventually cast in with new concrete work.

Precompression of soil The method of underpinning and final pinning up described, if carefully executed, will result in negligible movement due to settlement of the new work, but in large and heavy buildings a system of precompressing the soil on which the new work will bear is adopted. By this means consolidation of the soil is effected before the load from the underpinning above is applied and subsequent settlement is avoided. In this method a pad of reinforced concrete is cast on the bottom of the excavation and when it has matured a hydraulic jack of the normal type or a Freyssinet flat jack is fixed to it at the centre of the underpinning section. Precast concrete beams are then placed on top of the jack and concrete is placed on them up to the underside of the existing foundation (figure 2.35 D). Before this has set it is jacked up so that all voids in the underside of the foundation are completely filled with the concrete. When the concrete on the beams has matured a predetermined load is applied by the jacks so that the soil is compressed and its bearing capacity is 'pretested' in advance. When all the work has been carried out in this way, the spaces between the jacks in adjacent sections are pinned up with engineering bricks in cement mortar which, when set, permits the jacks to be removed and the remaining spaces to be filled up tightly.

Hydraulic jacking may also be used to ensure a positive contact between the face of the excavated soil at the back of each section and any supporting poling boards or sheeting. Each stage of excavation is supported at the back by 100 to 150 mm of in situ reinforced concrete acting as a permanent poling board. Temporary struts back to the opposite excavated face incorporate hydraulic jacks by means of which a predetermined pressure is applied to the concrete poling board about 24 hours after casting, while it is still 'green'. At this stage the concrete is still flexible

and this ensures with greater certainty close contact with the soil face and the elimination of any voids behind the concrete.

2.4.2 Piled underpinning

Continuous underpinning as described may in some circumstances be impracticable or uneconomic. The presence of subsoil water with certain types of soil might make the work extremely difficult, or when the load has to be transferred to a great depth the cost of continuous underpinning might be excessive. In such circumstances piles may be used with the load from the wall transferred to them by beams or needles. In order to avoid vibration which would be undesirable in many cases of underpinning, bored or jacked-in piles are used.

Arrangement of beams and piles This depends on many things, including the state of the structure and the necessity or otherwise of avoiding disturbance to the use or contents of the building. The most straightforward method is probably to sink pairs of piles at intervals along the wall, one on each side, connected by a needle or beam passing through the wall just above or immediately below the foundation, whichever best suits the circumstances (see figure 2.36 A). Reinforced concrete or steel beams can be used for the needles. The latter occupy less space and are generally simpler to fix. An alternative method which may sometimes be more suitable is to stagger the piles on each side of the wall (B). This has the effect of halving the number of piles required, although each will take a greater load, and of lengthening the needles. If the positions most convenient for sinking the piles do not coincide with the most suitable positions for the needles, two beams, one on each side of the wall and parallel to it, may be carried by the piles and will support the needles which will pass through the wall at the most suitable points.

These methods, because of the internal piles, necessitate work within the building being underpinned. Where this is not possible, groups of two or more piles along the outside of the building may be used to support the wall by means of cantilever capping beams projecting under the wall (C).

As an alternative to cantilevered supports, jacked-in piles can be used immediately under the existing foundations in order to avoid internal work. These piles have been described on page 63 and the method is illustrated in figure 2.36 D. It is, of course, essential that the weight of the structure be sufficient to provide adequate dead weight as a reaction to the jacks as the piles are being forced down.

Quite a different method of piling[22] makes use of small diameter (120 to 280 mm) drilled friction piles passing through the actual foundation (figure 2.36 E), thus

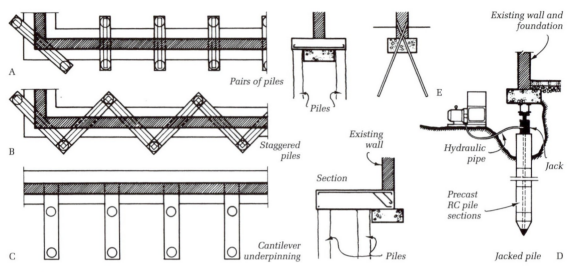

Figure 2.36 Piled underpinning

eliminating the necessity for needles to transfer the load to the piles. This system of mini piling is also known as root piling as it provides the structure with roots into sound strata. The piles are bored at an angle by means of a rotary drilling rig, either directly through the foundation or from a point up the wall as shown in figure 2.36. As the hole is drilled a steel lining tube is pushed into the borehole. Reinforcement is introduced, a cement and sand grout is pumped in under pressure to form the pile and the lining tube is extracted. This type of angle piling can be end bearing on solid subsoil, in which case the lining will often be left in place. Formation of the piles is quiet and free of vibration and can be carried out with limited headroom.

Continuous beam support When the superstructure and its foundation are in a weak condition, continuous beam support to the wall may be essential. In small, lightly loaded buildings such a beam, projecting beyond the wall face, can sometimes be supported adequately on a line of single piles only outside the building. In other cases the beam must be carried by needles bearing on piles on each side of the wall or on cantilever capping beams as described above. The supports may, however, be spaced at greater intervals and the beam designed to span between them. When support to the wall may be given above the level of the foundation the beam can be formed in a number of ways which simplify construction, first by the use of 'stools'. These are concrete blocks with holes in them through which reinforcing bars can be passed. When very deep beams are required the stools are formed of a number of vertical steel bars welded to top and bottom plates. Holes are cut in the wall at approximately 1 m centres at

the required level and the stools are inserted (see figure 10.11 C, page 353). These act as props to the wall above when the intervening sections of the wall between them are cut away. When this has been done for the full length of the underpinning, reinforcement may be threaded through the stools and tied in position. Shuttering is then erected and the concrete placed to form a continuous beam incorporating the stools as part of it. The gap at the top of the beam is pinned up as described earlier.

Precast prestressed beams Another similar method makes use of prestressed concrete beams and involves no in situ casting of concrete. Precast blocks or segments, with holes cast through them for a prestressing cable, are inserted in turn next to each other in holes cut in the wall, each being pinned-up on top and wedged underneath until the whole of the wall is supported by a line of these blocks (see figure 10.11 D). The joints between them are filled with dry mix mortar rammed tight and openings are left at each end for anchorage and stressing of the cable. A post-tensioning cable is then threaded through the full length, tensioned and anchored at each end to form a single, homogeneous beam. Alternatively, stressed steel underpinning beams may be used made up of 600 mm to 1.50 m lengths of steel beams with steel diaphragm plates welded to each end and drilled for high tensile torque bolts. Each length is inserted in turn and pinned up in the same way as the segments of a post-tensioned concrete beam, so that the wall above is continuously supported (see figure 10.11 E). The beam is inserted in the wall with the joints between the lengths of beam left slightly open on the tension side, and after packing solid at the top, the bolts are tightened by a torque wrench. In closing the joints, the lower flange of the beam is stressed so that the beam takes up its load without deflecting.

In low-rise masonry buildings, such as dwellings not exceeding three storeys, continuous beams of reinforced brickwork (see section 3.2) can be formed in situ by cutting slots 50–60 mm deep in the joints into which are inserted reinforcing bars of stainless steel which are then grouted into position with high strength mortar.[23]

The methods just described are also useful when new openings must be formed in existing walls, permitting the work to be carried out without the dead shoring and strutting which is otherwise necessary as described in the section on shoring in chapter 10.

As an alternative to cutting into the wall, concrete beams have been cast as a continuous ring round and against the outside face of a building just above foundation level. These are then post-tensioned and have the effect of tying the walls of the building into one rigid unit, enabling it to act as a whole under the action of differential settlement, thus avoiding cracking of the walls.[24]

2.4.3 Underpinning to column foundations

In underpinning framed structures, the main problem is to provide satisfactory support to the columns while they are being underpinned. Before excavation is begun these must be relieved of load by dead shores under all beams bearing on them.

Reinforced concrete columns and brick piers can be supported by means of a horizontal yoke formed of two pairs of rolled steel beams set in 25 to 50 mm deep chases in the sides of the columns (figure 2.37 C). The bottom pair are large enough and long enough to act as needles to transfer the load to temporary support and the upper pair are at right-angles to the needles and bear on them. The pairs of beams are tied together by transverse tie rods or angles. An alternative method, which avoids chasing into the sides of the column and which can deal with greater loads, is to grip the column or pier in a heavy steel cramp designed to grip more tightly as it takes up the weight of the column (D). The base of the cramp bears on needles on opposite sides of the column which transfer the load to supports well away from the column base. Reinforced concrete columns may be supported by a reinforced concrete collar (E). Steel stanchions can be supported on steel needles by steel angle or channel cleats welded to the stanchion flanges and bearing on the needles (G). Lightly loaded stanchions can sometimes be carried by a large diameter steel pin, passed through a hole drilled in the web and bearing on the needles.

In all cases the supports to the needles must be far enough away from the column or pier to avoid collapse by

dispersion of the pressure on to the sides of the excavation, and also to allow sufficient working space. The needles must have good solid bearings, strong enough to take the loads transferred to them from the columns and this may sometimes necessitate the use of piles at the bearings. If underpinning is due to the excavation of a new basement under the columns, the temporary pile supports may have to penetrate below the level of the new basement to a depth sufficient to permit them to act as free standing columns when the main excavation is carried out and while the column is being underpinned.

Circumstances may make it impossible to provide support to the needles on two sides of a column or stanchion and a counterweighted cantilever support must then be used. This would be formed of a pair of long, deep steel needles bearing on a fulcrum support some distance from the column in the form of a base slab or piled foundation (see figure 2.37 F). One end of the needles would pick up the yoke or cleats attached to the column and the other would be tailed down by kentledge or tension piles. A hydraulic jack placed between the needles and the fulcrum support enables the latter to be 'precompressed' and any settlement taken up to prevent movement of the column.

2.4.4 Ground treatment

Soil stabilisation Suitable soil such as gravel and sand may be strengthened, or its permeability decreased, by pressure grouting, and water in sand and silts can be frozen by refrigeration. The task of underpinning (and of normal foundation construction) can often be simplified by the application of these processes to the soil and in some circumstances they can be used as an alternative to underpinning. In the latter case, where permanent effects are required, grouting would be used, since freezing is only practicable as a temporary measure. Grouts may be solutions of various chemicals which gel in the soil or suspensions of cement, fly-ash or bituminous emulsion.[25]

Soil compaction Piles may be used in non-cohesive soils such as silt, loose sand or gravel, as a means of compacting it by the vibration of repeated driving and extraction in order to increase its bearing capacity. This necessitates the use of a large number of piles over the area of a site and alternative methods of compacting have been developed. One is a mechanical technique known as vibroflotation or vibrocompaction, in which compaction is effected by a large, heavy poker vibrator about 5 m long supported by a mobile crane. Water jets incorporated at the lower end produce a saturated mass of soil into which the vibrator sinks under its own weight until it has reached the depth required for compaction. A column of compacted soil up to 3 m in diameter is produced and by

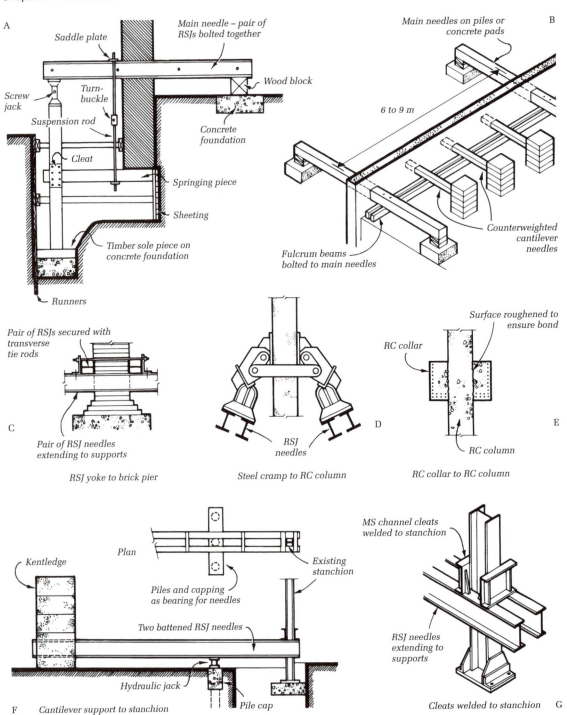

A

Saddle plate

Main needle – pair of RSJs bolted together

Main needles on piles or concrete pads

B

Screw jack

Turn-buckle

Wood block

Suspension rod

Concrete foundation

6 to 9 m

Cleat

Springing piece

Sheeting

Timber sole piece on concrete foundation

Counterweighted cantilever needles

Runners

Fulcrum beams bolted to main needles

Pair of RSJs secured with transverse tie rods

Surface roughened to ensure bond

RC collar

C

RSJ needles

D

RC column

E

Pair of RSJ needles extending to supports

RSJ yoke to brick pier

Steel cramp to RC column

RC collar to RC column

MS channel cleats welded to stanchion

Kentledge

Plan

Existing stanchion

Piles and capping as bearing for needles

Two battened RSJ needles

RSJ needles extending to supports

Hydraulic jack

Pile cap

F   Cantilever support to stanchion

Cleats welded to stanchion   G

Figure 2.37   Underpinning

arranging the compaction points so that these columns overlap full compaction over any area is achieved. In weak cohesive soils or loose fills the same technique is adopted, but coarse granular fill is introduced as the vibrator is withdrawn and is compacted by the vibrations to form dense stone columns in the soil at intervals of one to two metres. Another technique for loose, wet sands compacts by means of an electric spark across electrodes at the end of a pipe sunk by water jetting, which produces spheres of compaction about 3 m in diameter. These can be made to overlap vertically and horizontally across a given area.

Dynamic consolidation, in which a weight of up to 20 tonnes with a base of about 4 m² is dropped through 20 m or more on a pre-selected grid pattern, may be used to improve the loadbearing capacity of poor granular soils and of loose fills. Soft cohesive soils may be compacted, in suitable circumstances, by temporarily pre-loading by a surcharge of soil or fill some metres high.[26]

Notes

1 The term exploration is used here rather than site investigation as the latter implies a broader study of a site than just the investigation of the subsoil. For the scope of site investigations, and for fuller details of site explorations, see BS 5930: *Code of Practice for Site Investigations.*

2 See pages 35–7 where this and other matters related to distribution of pressure in the soil are discussed.

3 Lining tubes are supported between a guide frame or leaders. The term 'monkey' refers to a cast iron weight raised and allowed to fall about 1.5 m on to a hardwood or dense plastic driving cap. Figure 2.27 shows a crane-suspended hanging leader.

4 See BS 1377: *Methods of Test for Civil Engineering Purposes. General Requirements and Sample Preparation.*

5 For variations of this tool, see BS 5390: *Code of Practice for Site Investigations.*

6 The BRE have produced a series of Digests entitled *Site Investigation for Low-rise Building*:
Ref. 318: *Desk Studies*
    322: *Procurement*
    348: *The Walk-over Survey*
    381: *Trial Pits*
    383: *Soil Description*
    411: *Direct Investigations.*

7 See Part 1, section 4.2 and section 2.1.2 of this book.

8 The apparatus and methods of carrying out these laboratory tests are not described in detail. Most books on soil mechanics include full details; see *The Mechanics of Engineering Soils* by P Leonard Capper and W Fisher Cassie, Spon, for example.

9 Case studies: Loscoe (Derbyshire), 1986 and Kenilworth (Warwickshire), 1989.

10 National Radiological Protection Board, *Health Risk from Radon*, 2000.

11 Values for bearing capacities of various types of soil are given in Part 1, table 4.3.

12 See BRE Digest 386, *Monitoring Building and Ground Movement by Precise Levelling*, and Digest 475, *Tilt of Low-rise Buildings with Particular Reference to Progressive Foundation Movement.*

13 See *Soil Mechanics in Engineering Practice*, 3rd edition, by K Terzaghi, R B Peck and G Mesri, Wiley-IEEE, 1996.

14 See BRE Digest 251, *Assessment of Damage in Low-rise Buildings, with Particular Reference to Progressive Foundation Movement.*

15 See Part 1, *Eccentric loading on foundations*, pages 40 and 56.

16 See also *MBS: Introduction to Building*, 4th edition, section 17.4.6.

17 See page 147.

18 Alternatively, ground anchors may be used for this purpose. See page 99.

19 These are also referred to as *displacement* and *non-displacement* (or *replacement*) *piles*, respectively.

20 See end note 12, re. BRE Digest 475.

21 Brockhouse Steel Structures Limited has marketed this system of construction under the name of *Clasp.*

22 A patented system known as *Pali Radice (Root Pile)*, Fondedile Foundations Limited.

23 A patented method known as *Matrix System*, Bersche-Rölt Limited.

24 Known as *Hoopsafe*, see *Advanced Construction Technology*, 4th edition, by R Chudley and R Greeno, Pearson.

25 See BS 8004: *Code of Practice for Foundations* for a description of these and other geotechnical processes and also *Modern Construction and Ground Engineering Equipment and Methods*, 2nd edition, F Harris, Pearson.

26 See BRE publications: *Specific Dynamic Compaction*, K Watts, and *Design Guide: Soft Soil Stabilisation. EuroSoilStab: Development of Design and Construction Methods to Stabilise Soft Organic Soils*, various authors.

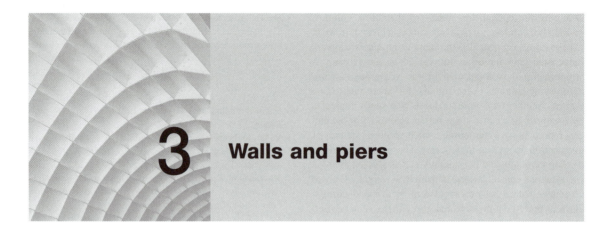

3 Walls and piers

Following a consideration of the behaviour of walls under load and the need for and means of providing lateral support, the various types of masonry wall construction are described together with plain monolithic concrete walls. Retaining walls are discussed as a separate type followed by a description of methods of waterproofing walls. External facings and claddings to walls are examined at some length and the methods used to control movements in walls in order to prevent damage are explained.

Types of walls and the functional requirements of walls are discussed in Part 1, chapter 5, where the construction of masonry and frame walls is described in detail. The behaviour of the wall under various conditions of loading is discussed in chapter 3 of that volume. Some further aspects of the strength and stability of walls are examined here.

Eccentric loading Floor and roof loads on walls are frequently applied on one side, rather than on the axis of the wall or bearing. In the case of concrete floor slabs with a span of less than thirty times the wall thickness it may be assumed that the load is applied through the centre of the bearing. However, the deflection of a floor tends to concentrate the load on the bearing edge of the wall and this tendency will be greater with flexible floors, such as normal timber or longer span concrete floors. BS 5628: *Code of Practice for the Use of Masonry* recommends that at the discretion of the designer it may be assumed that the load from a single floor (or roof) acts at one-third of the depth of the bearing area from the loaded face of the wall or leaf in the case of a cavity wall. Even if the bearing of such a floor covers the full width of the wall it is desirable to assume this eccentricity.[1] Where timber joists are

supported by joist hangers the load may be assumed to act at a distance of 25 mm from the face of the wall.

As explained in Part 1, chapter 3, such eccentric loading tends to cause bending in the wall, having the effect of increasing the compressive stress in the wall on the loaded side and of decreasing it on the opposite side (figure 3.1 A where *p* represents the stress due to the load applied axially). The simple formula by means of which the intensity of these stresses is established is given on page 38 of Part 1. This variation in stress must be borne in mind for the increased stress on the loaded side could become greater than the safe compressive strength of the wall and, if the eccentricity is too great, tensile stresses will be set up in the side opposite that on which the load is applied (figure 3.1 B). This has a marked effect upon the strength

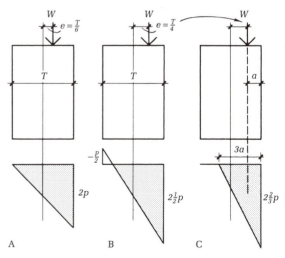

Figure 3.1 Effect of eccentric loading

of the wall, particularly in masonry walls in which the tensile strength is small and is, in practice, usually ignored.

The tensile strength of the wall is governed not by the tensile strength of the mortar used but by the adhesional strength or bond between the units and the mortar. Failure invariably occurs as a result of a breakdown of this bond rather than from a failure of the mortar or units in tension. Bond is extremely variable and average ultimate values range from 0.2 to 0.9 N/mm^2 depending on the mortar mix, type of masonry unit, workmanship and weather conditions at the time of building. Normally no direct tension should be permitted in unreinforced masonry.

It is assumed that cracking occurs when tensile stresses are set up. Thus when the eccentricity is greater than one-sixth of the thickness and tension occurs, part of the wall will cease to function structurally. It is also assumed that on cracking a redistribution of the compressive stress takes place such that the resultant passes through the centroid of the stress triangle (figure 3.1 C). The working portion of the wall is thus reduced to three times the distance of the point of application of the load from the bearing face of the wall and the maximum stress on this reduced working area is established from $(2 \times W)/(3 \times a)$ and for the eccentricity shown would be $2\frac{2}{3}p$. In the calculation of walls this is allowed for by the application of reduction factors to the compressive strength of the walling material.

Overturning Lateral loads, such as the pressure of wind or of stored materials on normal walls and of earth on retaining walls, tend to bend and overturn the wall as shown in Part 1, chapter 3. This causes the line of resultant pressure to become eccentric and sets up a stress distribution similar to that caused by eccentric loading, with a reduction in working area when cracking occurs. In free-standing walls this may be overcome by making the wall thick enough to bring the resultant well within the base or by buttressing. In buildings account is taken of horizontal and vertical lateral supports, which may be present in the form of floors and cross walls or a structural frame, to provide added resistance against overturning to that of the wall itself (see Part 1, figure 3.20).

Buckling As explained in Part 1, chapter 3, short walls or piers ultimately fail by crushing, but as the height increases they tend to fail under decreasing loads by buckling. For purposes of calculation the height of a wall is generally based on the distance between floors, assuming these give adequate lateral support to the wall. The actual proportion of the floor to floor dimension to be considered for this purpose and the methods of providing adequate lateral restraint by means of floors are discussed later.

The manner in which these possible causes of failure are accounted for in the regulations governing the minimum thickness of walls is discussed under the section on *Determination of wall thickness.*

Loading This is considered in terms of the dead load, the live load and the wind load.

The *dead load* is the weight of all parts of the structure bearing upon the wall together with its own self-weight. Weights of building materials used as a basis for estimating such dead loads are given in BS 648: *Schedule of Weights of Building Materials.*

The *live load* is the weight of superimposed floor and roof loads including snow. Minimum loadings are given in BS 6399: *Loading for Buildings.* For floors see BS 6399-1, and for roofs BS 6399-3; reference is made to these on pages 193 and 252 respectively.

In most types of building it is unlikely that all floors would be fully loaded at the same time. In the design of walls, piers and columns BS 6399 allows for this by permitting varying percentage reductions of the total imposed floor loads according to the number of floors carried (see table 3.1). Where a floor is designed for loadings of 5 kN/m^2 or more these reductions may be made provided that the loading assumed is not less than it would have been if all floors had been designed for 5 kN/m^2 with no reductions. No reduction is permitted for plant or machinery specifically allowed for, or warehouses, storage buildings and garages.

When a wall or pier supports a beam, the reduction permitted in respect of a single span of beam supporting not less than 40 m^2 of floor (see *Floor loading*, page 193) may be compared with that given by table 3.1 and the greater of the two taken into account in the design of the wall or pier.

The *wind load* may be assessed by methods given in BS 6399-2 based on a 'basic' wind speed appropriate to the district in which the building is to be situated, taken from a map of isopleths which is adjusted to take account of topographical and other factors such as height above ground and size and shape of building, and is then converted by formulae to the actual pressure or suction exerted on a particular surface of the building (e.g. wall or roof).

Table 3.1 Reduction of floor loads

Number of floors including roof carried by member under consideration	Per cent reduction in total imposed load on all floors carried by member under consideration
1	0
2	10
3	20
4	30
5 to 10	40
over 10	50

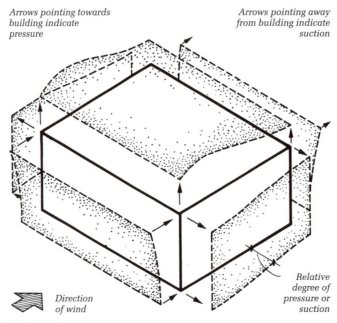

Arrows pointing towards building indicate pressure

Arrows pointing away from building indicate suction

Direction of wind

Relative degree of pressure or suction

Figure 3.2   Pressure and suction due to wind

The distribution of wind pressure on walls is not uniform. It is greater near the eaves, and near the ends of a wall high suctions are possible as shown in figure 3.2 where the distance of the curved 'planes' from the faces of the building indicates the degree of pressure or suction on the surfaces of the building.

3.1 Solid masonry walls

For many years loadbearing brickwork has provided the cheapest structure for houses and blocks of flats from three to five storeys in height, and, broadly speaking, loadbearing brick or block wall construction will still be economical for small-scale buildings of all types where planning requirements are not limited by its use.[2] Until masonry walls were calculated on a scientific basis great heights required great thickness of wall, but with the development of adequate design methods it became possible to build fourteen to fifteen storeys high with only 280 mm external cavity walls and 175 mm solid internal loadbearing walls, and five storeys using only 102.5 mm internal walls. This necessitates careful consideration of such design aspects as (i) the choice of a suitable plan form, (ii) maintaining a suitable proportion of height to width of building to keep wind stresses to a minimum, and (iii) running concrete floor slabs through to the outer face of external walls to reduce the eccentricity of floor load (see page 74).

A suitable plan form is one in which the floor area is divided into rooms of small to medium size and in which the floor plans repeat on each storey. This results in the walls running through in the same plane from foundations to roof and the floors, being of moderate span, transferring moderate and distributed loading to the walls. Maisonettes, flats and hostels are building types in which these requirements may be fulfilled.

For high-rise buildings the disposition of the walls must be such that they provide structural rigidity in the direction of both axes of the building so that some form of cellular plan (figure 3.3 A) or one with cross walls in both directions (B) is essential.

As a design concept, the use of masonry relative to its material strength, wall thickness, height and length is considered in sections 3.1.1 to 3.1.3. In practice, the need for a regular plan form with floor size limitations will restrict spatial layout, flexibility of design and future changes of use. Other criteria such as provision for adequate thermal insulation will also influence the thickness of structure, and when considering the economics of construction practice, loadbearing masonry is generally limited to modest heights. However, where a traditional façade or cladding is desirable in high-rise construction, non loadbearing masonry can be applied to a structural frame of steelwork or reinforced concrete (see section 3.11.2).

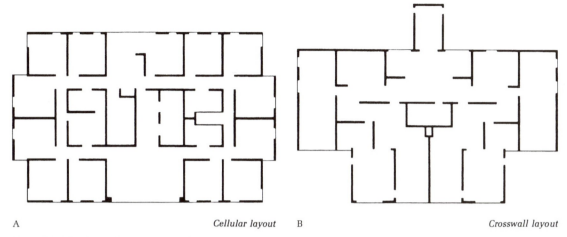

A   *Cellular layout*   B   *Crosswall layout*

Figure 3.3   Plan forms for masonry construction

3.1.1 Strength of masonry walls

The strength of a masonry wall depends primarily upon the strength of the units and of the mortar. In addition the quality of workmanship is important. Variations of 25 to 35 per cent in the strength of brickwork have been found due to bad workmanship, mainly on account of badly mixed mortar and imperfect bedding of the bricks. This is an important consideration in high-strength masonry walls where all units should be properly bedded and all vertical joints flushed up solid.

Mortar and its effect   The effect of the strength of the units and of the mortar is illustrated by the graphs in figure 3.4 which are based on tests on brickwork. It will be seen that the strength of the wall increases with the strength of the individual bricks, but not in direct proportion to the unit strength. This is due to the significant effect of the mortar, which is shown in the adjoining graph, where the effect of varying mortar strengths is related to the strength of the wall.[3] It can be seen that little advantage is gained, from the point of view of ultimate wall strength, in increasing the strength of the mortar beyond a certain point for any particular grade of brick. In terms of actual mortar mixes this is illustrated in table 3.2 which shows that, although a cement mortar will give greatest strength, a cement-lime mortar with 50 or 60 per cent of the cement replaced by lime gives only a slightly lower strength of brickwork.[4] A straight non-hydraulic lime mortar will generally give only a brickwork strength of less than half that attained with a cement or cement-lime mortar. It should be noted that the optimum strength of brickwork is obtained with a 1:3 basic proportion for the mortar.

The use of these relatively weak mortars has a significant effect upon the weather resisting properties of solid masonry walls as explained in Part 1, chapter 5. In addition, they improve to some extent the resistance of the wall to cracking, which is usually caused by differential movement rather than by excessive loading, by their ability to 'give' slightly to take up movements of the wall and thus minimise cracking and distribute it through the joints.

Bricks   These are graded according to their 'average crushing strength', which is the average strength of a random sample of ten bricks tested in accordance with BS 3921: *Specification for Clay Bricks.*[5] The prices of bricks normally available fall into groups which closely correspond to these grades of strength. The crushing strengths for clay bricks generally range from about 7 to more than 100 N/mm².

In order to keep the thickness of walls within reasonable limits and, preferably, of the same thickness for the full height of the building, particularly in the case of cross wall construction, variations in the types of bricks and in the mortar mixes are made according to the stresses at different heights. Excessive variation is uneconomic and leads to difficulties in supervision on the site. Sufficient flexibility in strength can, however, be obtained in most buildings by the use of three to four grades of bricks with one or two mixes of mortar. Ordinary fletton bricks, which are relatively inexpensive, when used in 215 mm thick imperforate cross walls are adequate for flat blocks up to four or five storeys in height when the floor spans are short. Concrete blocks may be used instead of bricks. When these are specified advantage may be taken of the greater basic strength of blockwork (see page 81).

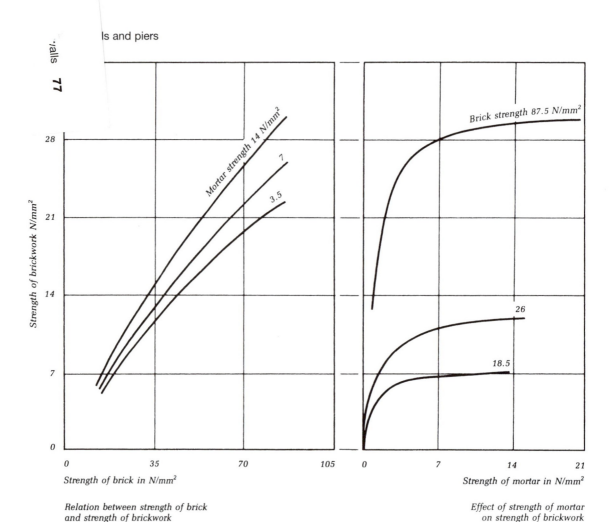

Strength of brickwork N/mm²

Strength of brick in N/mm²

*Relation between strength of brick
and strength of brickwork*

Strength of mortar in N/mm²

*Effect of strength of mortar
on strength of brickwork*

Figure 3.4   Strength of brickwork

Table 3.2   Effect of mortar proportions on strength of brickwork

Proportion of cement and lime to sand (by volume)	Strength of brickwork expressed as percentage of the strength of brickwork built in 1:3 cement mortar, for the following ratios of lime:cement (by volume)						
	All cement	*50:50*	*60:40*	*70:30*	*80:20*	*90:10*	*All lime*
1:1	–	72	70	66	58	47	–
1:1½	–	87	84	77	68	56	–
1:2	96	94	90	84	74	60	–
1:3	100	96	92	87	79	65	48
1:4	–	92	87	81	71	59	–

3.1.2 Determination of wall thickness

Reference is made in Part 1, chapter 5 to the empirical method of determining wall thicknesses by reference to the height and length of wall and to other conditions laid down for this purpose in the Building Regulations,[6] and it is explained why this method is generally used for small traditional types of building. The method limits, amongst other things, the height and width of the buildings to which it may be applied, the span and area of the floors relating to the wall, the imposed loads on various parts of the structure and the width of any openings in the wall.

None of these limitations, however, normally presents a problem in the case of the types of building to which the method is applicable.

In the case of larger and taller buildings the wall thicknesses are determined by means of calculations. This is also often necessary with some smaller types of loadbearing wall structures especially when a large proportion of openings is required which would reduce the effective wall area to a minimum. The reasons for this are that in using the empirical method it is not possible to take into account stronger materials than those laid down as a minimum, thus to gain the advantage of reduced thicknesses or areas of wall, nor the actual loads likely to bear upon the walls in various parts. In addition the limitation on the width of openings necessitated by an assumed maximum loading and a specified minimum strength of material tends to be restricting in design. The determination of wall thickness by calculation, having regard to actual loads and varying strengths of material, can lead to more economic construction in such cases and to a greater freedom in the disposition of openings. The thickness of a wall or column determined in this manner is based upon the calculated load to be carried, the actual strength of the materials used and the slenderness of the wall or column under consideration.

The Building Regulations accept the recommendations of BS 5628-1 (*unreinforced masonry*) and BS 5628-3 (*materials and components, design and workmanship*), as a basis for determining by calculation the necessary thickness of walls and columns for all types of building. This standard defines certain terms in relation to the process of calculation:

- slenderness ratio
- effective height
- effective length
- effective thickness
- working or design strength.

Slenderness ratio   Reference has already been made to the failure of walls and columns[7] under decreasing loads with increasing height. The ratio of height to thickness as a measure of this tendency is termed the slenderness ratio. This is defined as the ratio of effective height to effective thickness, but in the case of walls it may be based alternatively on the effective length if this is less than the effective height. This takes account of the stability which is provided by vertical as well as by horizontal lateral supports.

The lower the slenderness ratio the less the likelihood that a loaded structure will tend to buckle. Maximum slenderness ratios vary with application and the British Standard should be consulted for data relating to specific situations. For general purposes, a maximum of 27 is acceptable for unreinforced masonry, except in the case of

walls less than 90 mm thick in buildings of more than two storeys where it should not exceed 20. In many buildings where loading is light and the necessary wall thickness is small the slenderness ratio becomes the controlling factor, limiting as it does the height for any given thickness of wall.

Effective height   This is based on the distance between adequate lateral supports[8] provided by floors and roof and depends upon the degree of support they are assumed to provide. The greater the degree of support the smaller is the proportion of the distance between supports taken as the effective height. This is illustrated in figure 3.5 A to C in respect of walls and D, E in respect of columns.

BS 5628-1 defines two degrees of resistance to movement provided by lateral supports according to the nature of the junction between the wall and the supporting element. Firstly, 'enhanced resistance', which may, in the case of horizontal lateral support, be assumed where (i) floor and roof structures span on to the wall or column from both sides at the same level (figure 3.5 B), (ii) an in situ or precast concrete floor or roof bears at least 90 mm on the wall or column, irrespective of the direction of its span (A), or (iii) in houses up to three storeys, a timber floor spans on to the wall from one side with a bearing of at least 90 mm. In the case of vertical lateral support, 'enhanced resistance' may be assumed where the intersecting or return walls referred to under *Effective length* are properly bonded to the supported wall. Secondly, 'simple resistance' which, in the case of horizontal lateral support, is to be assumed where the support is connected to the wall or column by anchors and ties (figure 3.5 C), and, in the case of vertical lateral support, where the intersecting or return walls are connected to the wall by metal anchors evenly distributed throughout their height at not more than 300 mm centres.

Columns must be considered about both axes. If lateral support is provided in one direction only, as indicated by the beam in D, the effective height relative to that direction will be as shown, but in the other direction, where there is no top support, it must be twice its height above the lower support. In the absence of any top support (E) the latter value must be taken relative to both axes.

Where the wall between two openings is by definition a column as in G and F and the support to the wall provides 'enhanced resistance' to lateral movement its effective height is $\frac{3}{4}H$ plus $\frac{1}{4}Z$; where 'simple resistance' only is provided it is H, the distance between the lateral supports.

Where the thickness of a pier (J) is not greater than $1\frac{1}{2}$ times the thickness of the wall of which it forms part (K) it is to be treated as a wall in establishing the effective height; where it is greater it is to be treated as a column in the plane at right angles to the wall.

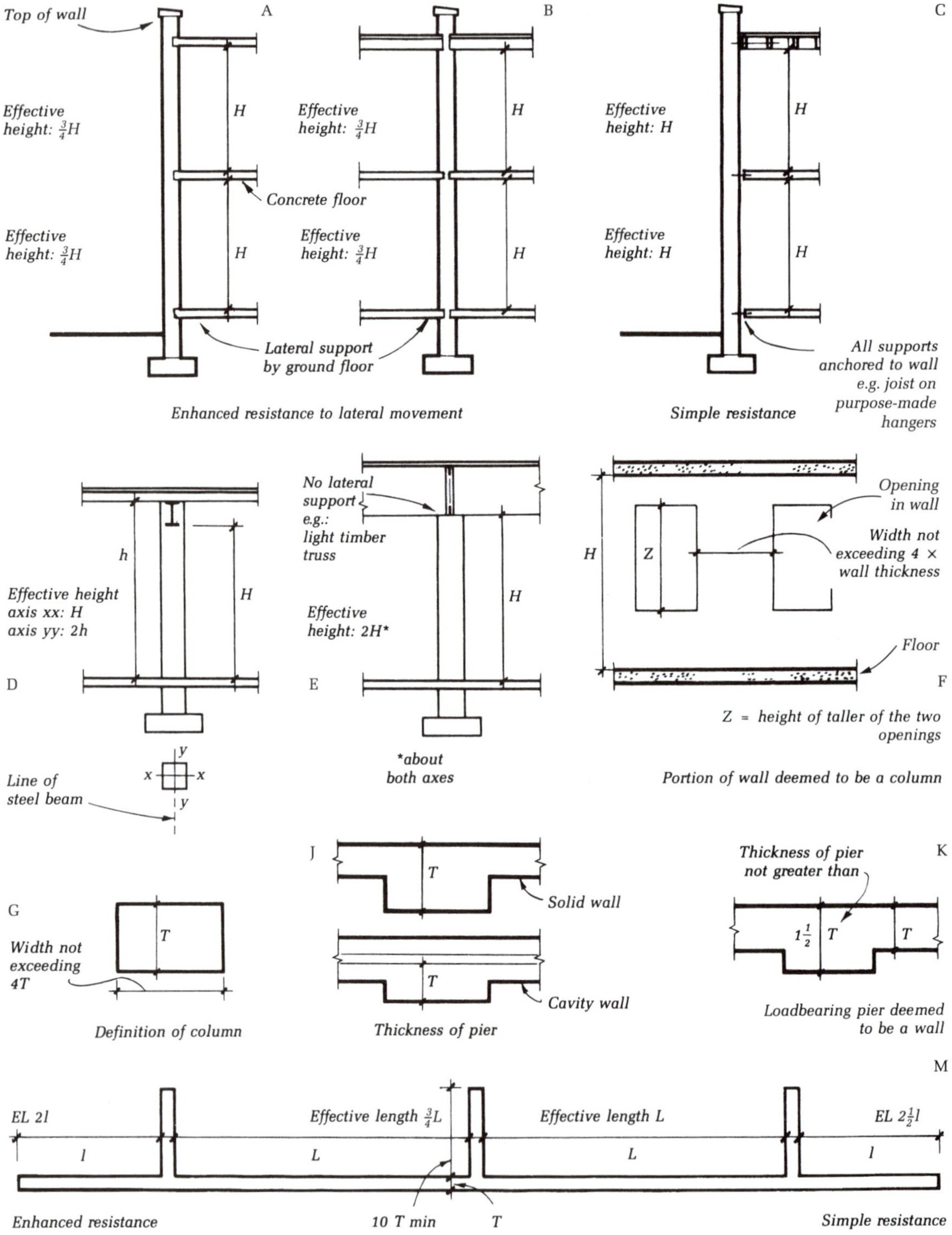

Top of wall

A

Effective height: $\frac{3}{4}H$

H

Effective height: $\frac{3}{4}H$

H

Concrete floor

Effective height: $\frac{3}{4}H$

H

Effective height: $\frac{3}{4}H$

H

Lateral support by ground floor

Enhanced resistance to lateral movement

B

Effective height: $\frac{3}{4}H$

H

Effective height: $\frac{3}{4}H$

H

C

Effective height: H

H

Effective height: H

H

All supports anchored to wall e.g. joist on purpose-made hangers

Simple resistance

h

Effective height axis xx: H axis yy: 2h

H

D

Line of steel beam

y

x —|—x

y

Definition of column

No lateral support e.g.: light timber truss

H

Effective height: 2H*

H

E

*about both axes

G

Width not exceeding 4T

T

Opening in wall

H

Z

Width not exceeding 4 × wall thickness

Floor

F

Z = height of taller of the two openings

Portion of wall deemed to be a column

J

T

Solid wall

T

Cavity wall

Thickness of pier

K

Thickness of pier not greater than

$1\frac{1}{2}$ T

T

Loadbearing pier deemed to be a wall

M

EL 2l

l

Enhanced resistance

Effective length $\frac{3}{4}L$

L

10 T min   T

Effective length L

L

EL $2\frac{1}{2}l$

l

Simple resistance

Figure 3.5 Calculated walls

Effective length This is related to the distance between vertical lateral supports in the form of intersecting or return walls not less in thickness than the supported wall or the loadbearing leaf of a cavity wall, which extend from the junction not less than ten times the thickness of the supported wall.

The effective lengths appropriate to both 'enhanced' and 'simple' standards of resistance to lateral movement are shown in figure 3.5 M.

Effective thickness This is the actual thickness of a solid wall excluding plaster, rendering, or any other applied finish or covering. Allowance is made for any stiffening piers which may be bonded to the wall by multiplying the actual thickness by a factor which varies with the size and spacing of the piers, resulting in an effective thickness greater than the actual thickness. Table 3.3 gives these factors. In the use of this table, buttressing or intersecting walls may be considered as piers of width equal to the thickness of the intersecting wall and of a thickness equal to three times the thickness of the stiffened wall. Some examples of effective thickness are shown in figure 3.6.

If a column has no lateral support or has support in both directions the effective thickness will be based on the least dimension. When the column has support in one direction only slenderness ratio values must be established using effective thicknesses based on both dimensions, and the larger be adopted.

Working or design strength The stresses permitted in a wall or column are regulated according to the strength of

Table 3.3 Determination of effective thickness of wall stiffened by piers – stiffness coefficients

Pier thickness divided by wall thickness	Pier spacing divided by pier width				
	6	8	10	15	20
1.0	1.0	1.0	1.0	1.0	1.0
1.5	1.2	1.15	1.1	1.05	1.0
2.0	1.4	1.3	1.2	1.1	1.0
2.5	1.7	1.5	1.3	1.15	1.0
3.0	2.0	1.7	1.4	1.2	1.0

Note: See also Building Regulations, Approved Document A, section 2C.

the bricks or blocks, the type of mortar to be used and the slenderness ratio of the wall or column. Basic or characteristic compressive strengths of masonry, arising from combined uniformly distributed dead and superimposed loads and related to the strength of the units and the type of mortar used, are given in tables in BS 5628-1 for brickwork and blockwork. The tables for blockwork take into account the increase in strength resulting from the shape of the units referred to in Part 1, page 94. The design strength is established by multiplying the basic strength by a reduction factor, which takes account of the slenderness ratio of the member and any eccentricity of loading, and dividing by an appropriate factor of safety taking into account the likely quality of the mortar used and the quality of control in manufacture of the walling units and in site supervision. For members with a slenderness ratio up to eight and with

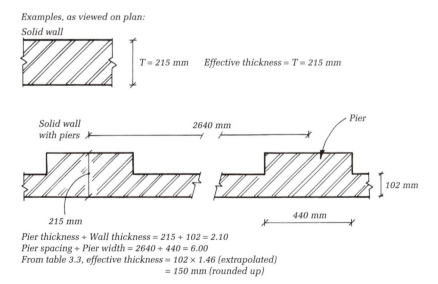

Examples, as viewed on plan:

Solid wall

$T = 215$ mm   Effective thickness $= T = 215$ mm

Solid wall with piers

2640 mm

Pier

102 mm

215 mm

440 mm

Pier thickness ÷ Wall thickness = 215 ÷ 102 = 2.10
Pier spacing ÷ Pier width = 2640 ÷ 440 = 6.00
From table 3.3, effective thickness = 102 × 1.46 (extrapolated)
= 150 mm (rounded up)

Figure 3.6 Measurement of effective thickness of solid walls

an eccentricity of loading not exceeding one twentieth no reduction factor is applied.

In the phraseology of BS 5628 the 'design vertical load resistance' of a wall per unit length or of a column is found by multiplying the design strength so established by the thickness of the wall or by the area of the column respectively. This Standard makes allowance, with certain provisos, for increased local stress under the bearings of concentrated loads.

Walls built of different materials   Walls built of materials of differing strengths bonded together are less important now as loadbearing structures since the general practice is to use a thin non-structural facing material attached to a structural backing, but provision is made in BS 5628-1 for dealing with such a combination in two ways, which is still relevant to the modern cavity wall (see section 3.3.1). The weaker material may be considered to be used throughout the full thickness and the design strength established on that basis. Alternatively, the area of that portion of the wall built of the stronger material only may be considered as carrying the load, in which case the design strength is established using a slenderness ratio calculated on the thickness of that material alone.

The thickness of random rubble walling should be based on design strengths of 75 per cent of the corresponding strengths for coursed walling of similar materials.

The traditional practice of building in a timber wall plate to support floor joists should be avoided. Although a convenient means for providing level support for joists, the integrity of construction is affected by the significant reduction in effective wall thickness. Building in the ends of joists will also reduce the effective wall thickness at these specific points, resulting in higher stresses at the floor levels. Therefore, with this practice, care must be taken to ensure that there is sufficient wall strength and that the brick- or blockwork is of a high standard at joist locations. Where joists bear on both sides of the wall, as in cross wall construction (figure 3.10 A), and the continuity of masonry bonding is disturbed, the infilling between the joists may be of in situ concrete.

The alternative use of metal joist hangers built into the wall, as shown in figure 3.7 D, may be considered. This must be with due regard to the *resistance*, as indicated in figure 3.5 C.

3.1.3 Lateral support

Adequate horizontal lateral support or restraint to a wall as defined in note 8 on page 134 is provided by floors and roofs and these must be capable of acting as a stiff frame or diaphragm to transfer the lateral forces to suitable buttressing walls, piers or chimneys. This may necessitate the provision of extra strutting in timber floors and flat roofs and lateral and diagonal bracing in pitched roofs to enable them to fulfil this function. This is especially so with light timber trussed rafter roofs which may require wind bracing in addition to the stability bracing normally incorporated in the roof structure (see Part 1, page 156).

Requirements for lateral restraint   The Building Regulations, Approved Document A, Structure, require all external, separating and compartment walls of residential buildings up to three storeys and small single storey buildings, whatever their length, to be laterally supported by the roof and, in addition, those greater than 3 m long by every floor. Any other internal loadbearing wall of any length requires roof or floor support at the top of each storey.

Connections   The connections between the wall and the supporting floors and roof must be such that the lateral forces may be effectively transmitted both in compression and tension.

In the case of a concrete floor or roof this is achieved if the slab bears at least 90 mm on the wall irrespective of the direction of span. Restraint is less certain with a timber floor the joists of which bear on the wall, but experience has shown that the partitions in small houses have a stiffening effect, in the light of which the Building Regulations are satisfied if the floor joists in houses of not more than two storeys have a minimum bearing of 90 mm on the supported walls or 75 mm if on a timber wall plate at each end, provided the joists are at not more than 1.2 m centres.

Adequate lateral support to loadbearing partitions and separating walls is provided by continuous contact between each face of the wall and a floor or roof structure on each side at or about the same level or by contact at not more than 2 m intervals, the points of contact on each side being as near as possible in line. This requires the tight fit of any joist hangers and of joist ends against the wall, or, if the joists run parallel with the wall, packing between the wall face and the first joist with noggings between the first three joists as shown in figure 3.7 A.

Positive anchorage   In all situations other than those described above, positive connections must be made by means of metal straps or anchors. Where the bearings of a concrete slab are less than 90 mm or where the slab only butts against the wall metal anchors must be used, cast in the slab at 300 mm centres. Precast concrete floors abutting the wall should be anchored to it by metal straps at least 30 mm by 5 mm cross section, with the ends bent down 100 mm, extending at least 800 mm from the wall to take in one or two lines of beams or blocks. These may be placed at the top within the concrete topping to the floor and be similar to that shown in figure 3.7 A, but bent down at both ends.

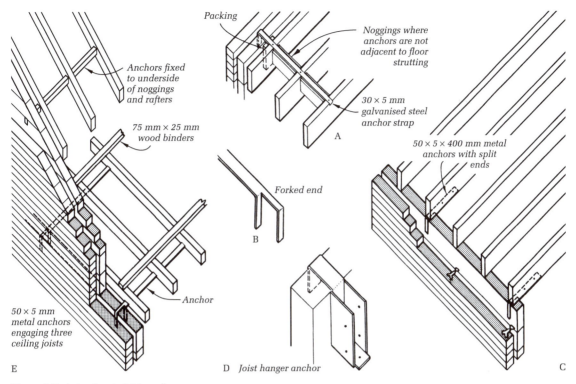

Figure 3.7 Lateral restraint to walls

Adequate restraint may be obtained with timber floors and roofs by means of metal anchors. These should be at least 30 mm by 5 mm cross section[9] and 1200 mm long, securely fastened to the joist ends at 1.25 m intervals except in houses up to three storeys where the spacing may be up to 2 m. These would have split ends for building in to a bed joint of a solid wall. For cavity walls 50 mm anchors, split and bent up or down at least 100 mm to form a forked end, should be used (figure 3.7 B, C). If joist hangers are used to support the joists 30 mm wide anchors with the end bent down only over the cavity face of the inner leaf may be used provided the hangers are bedded tight against the wall face and there is a close fit between the joist end and hangers. These methods ensure the satisfactory transmission of lateral forces both in tension and in compression. Alternatively, joist hangers designed to act also as anchors may be used, with the same provisos (D).[10]

To obtain the stiffness necessary to provide lateral support where the joists run parallel to the wall similar strap anchors engaging at least three joists are used, with any joist strutting being adjacent to the anchors. The strutting between these joists may with advantage be solid; where the anchors do not occur at floor strutting noggings must be placed between the joists (A) and if anchors with the

end bent down only over the cavity face of the inner leaf of a cavity wall are used it is essential to pack out the space between the first joist and the wall face at the anchor positions to make a tight fit between the two. This, again, is to ensure the satisfactory transmission of lateral forces both in tension and in compression.

Stiffening of floor joists is also necessary should a sound insulating floating floor be used. In this case the floor boards will not stiffen the floor structure and the necessary stiffness should be provided by strutting right across the floor on the line of the anchors. 150 mm by 25 mm boards should also be notched into and spiked to the ends of the joists where they bear on loadbearing walls.

Gable walls Lateral support to gable walls should be provided by anchors at the verges fixed to the rafters at 2 m centres, commencing as near the ridge as will give a satisfactory connection (figure 3.7 E). If the height from the floor immediately below the gable to half-way up the gable is greater than sixteen times the thickness of the supported wall, anchors at 2 m centres along the base of the gable, connecting with the topmost ceiling, must also be provided (E).

Interrupted support   Should the lateral support along a wall be interrupted by an opening in the floor this must not exceed 3 m, and the spacing of the anchors on each side should be closer than 2 m to provide in total the same number as if there were no opening.

Piers and walls providing vertical lateral support to a loadbearing wall must be of sufficient height and thickness to provide efficient support and should be effectively bonded at the intersection. Block bonding should not normally be used but always a fully bonded junction. Where large blocks are used or where block bonding is essential, satisfactory bond at the intersections may be obtained by the use of 38 mm by 5 mm metal ties, about 600 mm long with the ends bent up 50 mm. These should be spaced not more than 1.2 m apart.

3.2 Reinforced masonry walls

Reinforced masonry, usually in the form of reinforced brickwork, has been used to a considerable extent in the United States of America and in other earthquake prone countries such as India and Japan. It has been used to a limited extent in Great Britain since the nineteenth century. The reinforcement enables the masonry to withstand tensile and shear stresses in addition to the compressive stresses which it is capable of bearing alone.

Horizontal reinforcement may be bedded in the bed joints of the masonry as shown in the brick lintels described and illustrated in Part 1, chapter 5, but vertical reinforcement necessitates the use of slotted, recessed or perforated bricks or the use of special bonds, such as rat-trap bond (see Part 1, figure 5.7), within the cavities of which the reinforcement may be grouted, as it may be within the cavities of hollow concrete blockwork as shown in Part 1, figure 5.29.

Reinforced masonry is designed in accordance with the recommendations of BS 5628-2 (*reinforced and prestressed masonry*) in which is laid down the types of steel to be used and the sizes and cover for reinforcement and mixes for mortar and concrete. Basic or characteristic compressive strengths to be used in design are given for brick and block masonry in tables which take account of the strength of units and type of mortar used.

Brick walls, with the appropriate reinforcement within them, can be made to act compositely with reinforced concrete floors above and below them to form deep box girder cantilever beams, the floor slabs acting as flanges and the walls as webs. This is a reasonable technique when the component parts are already being used in a system as in, say, masonry cross wall construction, but it should be emphasised that care should be taken not to use brickwork for work which clearly calls for reinforced concrete. BS 5628-2 indicates that it may be more appropriate where the proportion of concrete infill is large to design the element as reinforced concrete in accordance with BS 8110-1: *Structural use of Concrete. Code of Practice for Design and Construction*, or where the area of main reinforcement is small as unreinforced masonry in accordance with BS 5628-1.

Prestressing can be used for tall brick and stone masonry columns to overcome the tensile stresses caused by eccentric loading and lateral wind pressure. This permits the overall size to be kept to a minimum. It can be done by building a hollow column, the central space accommodating post-tensioning cables or bars (see chapter 4, page 182) which are anchored to the foundation and to a steel plate or reinforced concrete pad at the top. This is simpler than building up a combination of solid masonry and normal reinforcing bars. Design recommendations for prestressed masonry are also given in BS 5628-2.

3.3 Cavity loadbearing walls

The advantages of the cavity wall in terms of weather resistance and thermal insulation have been described in Part 1, chapter 5, and the function of the wall ties in linking the two relatively thin leaves to provide mutual stiffening is explained. This is particularly important where one leaf only is loaded, the other then acting as a stiffener. Details of the necessary spacing of the ties in the wall are given and reference is made to the need to avoid the use of lime or the weaker cement-lime mortar mixes in order to ensure adequate bond of ties to wall.

Differential movement between the two leaves is likely to occur in large areas of external cavity walling, which over a period of time may weaken the bond of the ties in the joints. This, together with slight deformations of light wire ties, will reduce their effectiveness as a stiff link between the leaves. In the case of buildings not exceeding four storeys or 12 m in height, whichever is less, it is considered that the outer leaf may be uninterrupted for its full height. In taller buildings these movements can be sufficiently limited by supporting the outer leaf at vertical intervals of not more than three storeys or 9 m whichever is less and by dividing the wall horizontally into separate panels not exceeding 12 m wide. In such circumstances vertical twist metal ties are preferable to the lighter wire ties. The vertical subdivision may conveniently be formed by running floor slabs through to the outer face at appropriate levels, since in most buildings over three storeys in height some of the floors, even in maisonette blocks, will be of reinforced concrete. The detailing of the vertical joints between the wall panels will depend on the type of structure (see figure 4.37).

3.3.1 Wall thickness

BS 5628-1 provides for calculation on the same basis as for solid walls, provided that where both leaves share the load the effective thickness of the wall is taken as two-thirds of the sum of the thickness of the two leaves. This takes into account the fact that a cavity wall has less lateral stiffness than a solid wall equal in thickness to the sum of the thicknesses of the two leaves. Alternatively the effective thickness may be taken as the thickness of either leaf, if this is greater (see figure 3.8).

Allowance may be made for the stiffening effect of any piers bonded to the wall by the application of the stiffness coefficients given in table 3.3 on page 81, to the thickness of the stiffened leaf, using this increased thickness in establishing the effective thickness of the wall as given above for a wall without piers.

Where one leaf only supports the load, advantage may be taken of the stiffening effect of the other leaf by using in the calculation of the slenderness ratio an effective thickness as given above. The bearing capacity of the wall is, of course, based only on the area of the loadbearing leaf.

When the load is carried by both leaves, one of which is of weaker material, as is the case when a lightweight inner leaf is used with a brick outer leaf, the design strength to be used for both leaves should be based on the weaker of the two materials and determined by using an effective thickness as described above. This is in accordance with the provision made in BS 5628-1 for solid walls built of two materials.

Examples as viewed on plan:

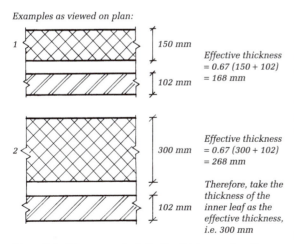

1

150 mm

102 mm

Effective thickness = 0.67 (150 + 102) = 168 mm

2

300 mm

102 mm

Effective thickness = 0.67 (300 + 102) = 268 mm

Therefore, take the thickness of the inner leaf as the effective thickness, i.e. 300 mm

Figure 3.8 Measurement of effective thickness of cavity walls

Means of establishing wall thickness The Building Regulations provide for establishing the thickness of load-bearing cavity walls either by the empirical method or by calculation in accordance with the provisions of BS 5628.

Under the empirical method the Regulations relate the required thickness of a cavity wall to its height and length by requiring the sum of the thickness of the two leaves and 10 mm to be not less than the thickness which would be required for a solid wall of the same height and length. Each leaf must be at least 90 mm thick and, subject to the provisions below, the cavity must be at least 50 mm and not more than 300 mm wide. An exception is a cavity wall with a minimum leaf thickness of 75 mm where the cavity width is restricted to a maximum of 75 mm. Lengths of wall ties vary to provide for an embedment depth of at least 50 mm. The practical reasons for these limitations on cavity width are given in Part 1, section 5.3.2, together with the number and types of wall ties required for different combinations of leaf thickness and cavity width.

3.4 Diaphragm and fin walls

These are external walls of loadbearing brick or block construction developed to deal with the structural problems inherent in the design of walls to tall singlestorey buildings, such as sports halls and factories. In these buildings the vertical loading is comparatively small and is not a great problem, but because the walls are anything from 6 to 10 m high between lateral supports slenderness becomes a significant factor, especially in view of the fact that the absence of internal walls to provide lateral restraint throws upon the external walls the whole burden of resisting lateral wind pressure and transmitting the resulting forces to the foundations.

Diaphragm wall This is, in effect, a wide cavity wall in which two 102 mm brick or 100 mm block leaves are tied together by 102 mm brick or 100 mm block diaphragms or ribs at intervals of 1 to 1.25 m along the wall (figure 3.9 A). The two leaves thus act as the flanges and the ribs as the webs of a stiff box section. The ribs or diaphragms are preferably bonded to the leaves but may, alternatively, be tied to them by stainless steel or non-ferrous anchors. The whole wall may be founded on normal strip foundations, generally reinforced and 175 to 250 mm thick. In weak soils and/or heavy structural loading, the reinforced strip will be designed as a beam transferring load to piled foundations. The bases of the cavities are filled with weak concrete to ground level, open drainage perpends being provided in the outer leaf as in a normal cavity wall. The overall thickness of the wall will vary from about 300 mm to 700 mm depending mainly on its height.

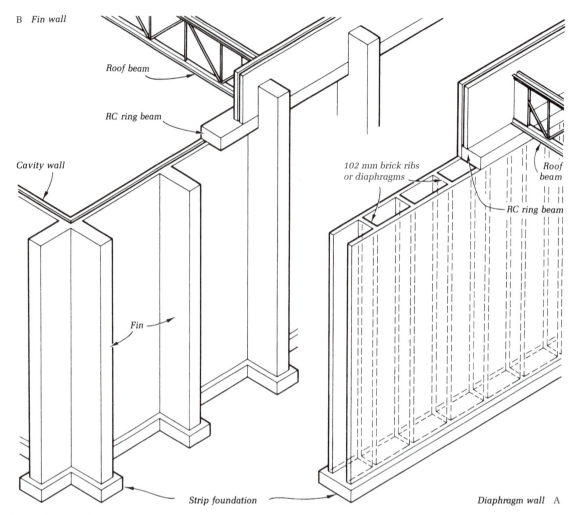

B *Fin wall*

Roof beam

RC ring beam

Cavity wall

Fin

Strip foundation

102 mm brick ribs or diaphragms

Roof beam

RC ring beam

Diaphragm wall A

Figure 3.9 Diaphragm and fin walls

The roof structure, which usually bears on a concrete ring or capping beam at the head of the wall, is designed to act as a stiff horizontal wind girder capable of transmitting the wind forces on any wall to the two flanking walls which, acting as shear walls, resist the lateral pressure of the wind and transmit the forces to the foundations. The roof, in addition, ties in and props the heads of the walls which, therefore, act as propped cantilevers resulting in greatly reduced bending moments at the base of the wall.[11] This, together with the stiffness deriving from the box section of the wall, results in a considerably thinner wall than if it acted as a freestanding vertical cantilever.

Wide openings in the wall may require a thickening of the ribs on each side to take the extra vertical loading from the lintel or, alternatively, a pair of 102 mm ribs with a small space between may be used for this purpose at the jambs to permit a vertical damp-proof course to be run in the one next to the opening on the line of the door or window frame. As in all cavity walls a cavity damp-proof course must be provided in the spaces between the ribs above the lintel of every opening. Tests and time have shown that, in walls not less than about 400 mm thick, damp is not likely to pass through the ribs to the inner leaf before drying out, provided that bricks and mortar of appropriate quality are used to suit the conditions. When very severe conditions of driving rain are expected, or in thinner walls, engineering bricks may be used for the ribs or a vertical damp-proof course should be applied between all ribs and the outer leaf, with metal anchors tying ribs to leaf. These vertical DPCs are best formed by bituminous paint.

Apart from cavity filling the methods of thermal insulation described for cavity walls in Part 1, section 5.3.3

may be applied to diaphragm walls. The ribs in walls of this type may provide a route for thermal bridging unless uniformly insulated internally.[12]

Fin wall   This is a conventional cavity wall with deep piers or fins on the external face in which the outer leaf acts as the flange and the fins as the ribs of stiff structural T-sections (figure 3.9 B). The cavity in this wall is bridged only by the cavity ties which should be vertical twist type at the normal spacings.

The spacing of the fins is from 3 to 5 m, normally arranged to coincide with the spacing of any roof beams. The depth and thickness of the fin must be such as to avoid buckling at the outer edge under compressive stress (see Part 1, section 3.3.3) as it acts as a vertical cantilever. Fin depths vary from 1 to 2 m and thicknesses from $1\frac{1}{2}$ to 2 bricks. For wide spacings of the fins the thickness of the outer leaf may be increased when necessary. The wall may be founded on normal strip foundations with projections under the fins or as described for diaphragm walls.

The considerations regarding the function and design of the roof structure are the same as those in relation to the diaphragm wall and, for the reasons given, the fin wall is also assumed to act as a propped cantilever.

Wide openings in the wall, as in the case of the diaphragm wall, may require a thickening of the fins on each side. Damp-proofing and thermal insulation are dealt with as in normal cavity wall construction. When desired for architectural reasons the depth of the fins may be reduced up their height by sloping back the outer edge to produce a trapezoidal shape on side elevation or the outer edge may be stepped up its height or be simply splayed off at the top.[13]

3.5 Plain monolithic concrete walls

A plain monolithic concrete wall means a wall of cast in situ concrete containing no reinforcement other than that which may be provided to reduce shrinkage cracking, together with a certain amount of round openings. The monolithic reinforced concrete wall is considered in chapter 4 and in this chapter reference will be made only to the plain concrete wall of either normal, no-fines or lightweight concrete.

As with reinforced concrete walls they are most economic when used both to support and to enclose or divide, provided they are at reasonably close spacing, that is to say, up to about 5.5 m apart. They are, therefore, used mainly for housing of all types, both as external and internal loadbearing walls, when low building costs can be attained.

Dense concrete is generally used for high buildings although no-fines concrete has been used for heights up to ten storeys in this country. In Europe blocks as high as 20 storeys have been constructed with no-fines loadbearing walls more economically than with a frame.

Plain monolithic concrete walls suffer certain defects which, in some respects, make them less suitable as external walls than other types. With normal dense aggregates the thermal insulation is low[14] and the appearance of the wall surface may be unsatisfactory, requiring some form of finishing or facing. In addition the unreinforced concrete wall, and particularly the no-fines wall, is unable to accommodate itself to unequal settlement as does a reinforced wall by virtue of the reinforcement or as a brick or block wall in weak mortar does to a certain extent by the setting up of the fine cracks in the joints. Thus, as a result, large cracks tend to form in the wall. Nevertheless, where foundations are designed to reduce unequal settlement to a minimum such walls can be used successfully.

Aggregates used for dense plain concrete are natural aggregates conforming to the requirements of BS EN 12620: *Aggregates for Concrete*, air-cooled blast furnace slag and crushed clay brick. Aggregates for light-weight concretes are foamed slag, pulverised fuel ash (fly ash), clinker, pumice and any artificial aggregate suitable for the purpose. No-fines concrete may be composed of heavy or lightweight aggregate.

The thickness of plain concrete walls is determined in accordance with the provisions of BS 8110-1 which defines effective height in relation to lateral supports and limits the slenderness ratio for walls without lateral support to 30 and, for laterally supported walls, rising to 45 according to the type of support received. Methods of taking into account eccentricity of loading and concentrated loads are given together with the method of establishing the design strength of the wall.

Dense concrete walls   These are constructed from concrete made with a well-graded aggregate giving a concrete of high density.

In most buildings the thickness of any type of plain concrete wall must, by reason of other functional requirements, be thicker than the minimum dictated by loadbearing requirements. An example of this is the dense concrete separating wall, which must be 190 mm thick in order to provide an adequate degree of sound insulation between houses and flats (see Building Regulations, Approved Document E, section 2).

Lightweight aggregate concrete walls   These will give better thermal insulation than dense concrete when used for external walls but care must be taken in the choice of aggregate for external use because of the danger of excessive shrinkage and moisture movement occurring with certain types. Untreated clinker has a corrosive action on steel and should not be used if shrinkage reinforcement is to be incorporated.

All types of lightweight aggregate concrete are more permeable than dense concrete, and where the wall is exposed to the weather a greater content of cement in the mix and increased cover to the steel is required, with possibly the further protection of rendering or cladding. Concrete with a wide range of density and compressive strength can be obtained by the selection of appropriate aggregate and mix (see *MBS: Materials*, tables 8.1 and 8.21).

No-fines concrete walls These are constructed with a concrete composed of cement and coarse aggregate alone, the omission of the fine aggregate giving rise to a large number of evenly distributed spaces throughout the concrete. These are of particular value in terms of rain exclusion as explained in Part 1, chapter 5. No-fines concrete is suitable for external and internal loadbearing walls or for panel wall infilling to structural frames.

The weight of no-fines concrete is about two-thirds that of dense concrete made with a similar aggregate. Aggregates graded from 19 mm down to 9.5 mm are used with mixes of 1 to 8 or 10 for gravel aggregate and 1 to 6 for light-weight aggregates. The aggregate should be round or cubical in shape and no more water should be used than that required to ensure that each particle of aggregate is thoroughly coated with cement grout without the voids being filled. The hydrostatic pressure on formwork is only about one-third of that of normal concrete. This is an advantage since horizontal construction joints should be minimized and formwork one or two storeys high can be employed without it being excessively heavy. Any normal type of shuttering can be used or open braced timber frames faced with small-mesh expanded metal are suitable. These permit inspection of the work during pouring. The economies which can be effected in the formwork are important in view of the high proportion of the total cost of concrete work accounted for by the formwork alone.

Weaknesses of no-fines concrete No-fines concrete walls should not be subjected to bending stresses nor to excessive eccentric or concentrated loads. Slender piers and wide openings are, therefore, unsuited to no-fines construction. Isolated piers should not be less than 450 mm in width or one-third the height of adjacent openings.

The bond strength of no-fines concrete is low but for openings up to about 1.5 m wide the walling itself may be reinforced to act as a lintel, provided there is a depth of wall not less than 300 mm above the opening. As a precaution against corrosion the steel should be galvanized or coated with cement wash and bedded in cement mortar. For wider openings an in situ or precast reinforced lintel of dense concrete is generally necessary. Even when the wall

above openings is not required to act as a lintel to carry floor or roof loads, horizontal reinforcement equivalent to at least two 12 mm steel bars should be placed above and below all openings. In buildings with timber floors the steel above the openings in external walls is usually made continuous.

For small scale domestic buildings a wall thickness of 200 mm is usually structurally sufficient. Wall thicknesses of 300 mm are used for multi-storey loadbearing wall construction. For external walls some form of thermal insulation is required such as described for solid brick walls in Part 1, section 5.3.3. Lateral support to the walls must be provided by positively anchoring the floors to the walls by some means such as vertical steel dowels set in the wall and passing through the ends of the anchor straps.

Because of its weakness in tension, walls of no-fines concrete are sensitive to differential settlement. Particular attention must, therefore, be paid to the design of the foundations. For small buildings the lower part of the walls and the strip foundation should be of dense concrete, reinforced if necessary. For high buildings adequate stiffness is usually obtained by the use of rigid reinforced dense concrete cellular foundations.[15,16]

Shrinkage reinforcement This may be required in in situ cast concrete walls, other than those of clinker aggregate or no-fines concrete, particularly in external walls, in order to distribute the cracking due to setting shrinkage and thermal movement, and thus minimise the width of the cracks. Where this reinforcement is considered to be necessary in external walls greater than 2 m in length, BS 8110-1 recommends that it should be in each direction not less than 0.25 to 0.30 per cent of the cross-sectional area of the concrete according to the grade of concrete used. It also makes recommendations in respect of internal walls, the positioning and distribution of the reinforcement and the provision of extra reinforcement round openings where shrinkage effects are greatest.

As the drying shrinkage of no-fines concrete is low, reinforcement for this purpose is not usually necessary except, perhaps, with some lightweight aggregates, because the stresses set up by the slight shrinkage are relieved by the formation of fine cracks round the individual particles of aggregate. Shrinkage reinforcement may also be omitted from dense concrete walls where the mix is lean and of low shrinkage and where end restraints on the walls are small and work can be carried out continuously.[17]

3.6 Cross wall construction

This is a particular form of loadbearing wall construction in which all loads are carried by internal walls running at right angles to the length of the building (see figure 3.10 A).

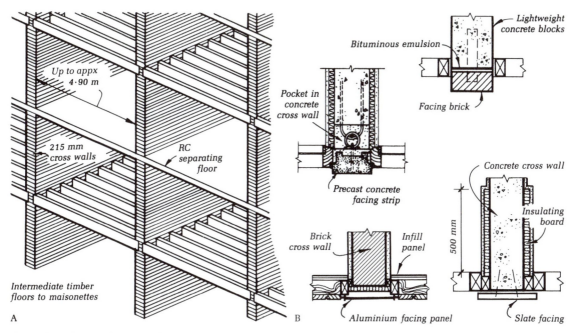

Figure 3.10   Cross wall construction

The majority of buildings require dividing up by internal partitions or separating walls. In certain types in which these occur at sufficiently close and regular intervals and where a high degree of fire resistance and sound insulation is required, necessitating relatively thick and heavy walls, an economic structure results when the walls are made to carry the loads from floors and roof. Typical examples are terraces of houses, maisonette, flat and hotel blocks, and schools.

The advantages of this form of construction are:

- Simplicity and uniformity of construction – the walls consist of simple unbroken runs of brick- or blockwork or in situ concrete.
- Projecting beams and columns are eliminated.
- It lends itself to repetition and standardisation of both structural and non-structural elements and thus to the prefabrication of the latter.
- The external walls, being free from load, may be designed with greater freedom in the choice of materials and finish.
- As a result of these factors, construction costs are relatively low.

Concrete cross wall construction in multi-storey buildings over four to five storeys in height is normally reinforced and is called box-frame construction. It is discussed later in chapter 4. At this point smaller buildings only will be considered, built of bricks, blocks or plain concrete.

Although not a disadvantage in the type of building to which cross wall construction is most suited, planning is restricted by the fact that the walls must run up in the same plane on all floors, from foundations to roof. It is desirable for economic reasons that the walls should also run in an unbroken line on plan from front to back of the building although this is not structurally essential. For maximum economy the walls should also be free of openings as far as possible. A further restriction on planning is the limitation which must be placed on the spacing of the walls in order to obtain an economic combination of walls and other elements of the structure. Of these, the floors and external walling are the most important.

Spacing of walls   Cross walls should be spaced at regular or regularly repeating intervals along the building in order that a limited number of floor spans can be standardised in terms of thickness, reinforcement and formwork. Thus also external wall elements, other than those of brick or blocks, may be standardised in a limited number of units.

The optimum spacing is about 4 to 4.25 m but spacing may range from 3.4 to 5.5 m without varying the cost of the structure itself by more than a few per cent. Within this range any increase or decrease in the cost of the floor structure is balanced by the respective decrease or increase in the cost of the walls. Although within the limits of 3.4 to 5.5 m the cost of the structure remains reasonably constant, the cost of external cladding increases directly with the increase of wall spacing. Analysis shows that while there is usually little to be gained from the point of view of cost and structure in using intermediate cross walls within plan units of up to 5.5 m in width, they should be introduced for greater widths.

Design generally   The design of the walls, if of masonry, is carried out on the basis of BS 5628-1, the loading and the stresses in the walls being determined in the manner described earlier. If the walls are of plain in situ concrete, design is carried out on the basis of BS 8110-1 as indicated on page 87.

A cross wall structure will normally be stable against the pressure of wind parallel to the walls but will, in most cases, require some longitudinal bracing against wind pressure at right angles to the walls. This is because there is little rigidity at wall and floor junctions and the brick, blockwork and plain concrete walls have little tensile strength. The floor loads and the self-weight of the walls set up 'precompression' in the walls which enables the latter and the floor-to-wall joints to resist tensile stresses up to the limit of this 'precompression'.[18] In many cases this may be insufficient to prevent the development of tensile stresses and the following methods may be adopted, either singly or in conjunction with each other, to provide longitudinal bracing to ensure overall stability:

- The use of a staircase, lift tower or services duct of which the walls at right angles to the cross walls act as buttresses and are assumed to take all lateral forces.
- The provision of longitudinal walls in the plan to take the lateral forces.
- The return of the ends of the cross walls as an 'L' or 'T' for the same purpose.
- The introduction of certain piers and walls in reinforced concrete, or the forming of the joints between walls and floors in reinforced concrete.
- The introduction of longitudinal reinforced concrete beams.

External cavity panel walls of brick and lightweight concrete block alone can provide sufficient rigidity to the structure of buildings up to five storeys high and of reasonable length, provided the area of window is not excessive and the workmanship is sound, especially at the bonding of panels to cross walls.

Comparison of suitable materials   Bricks, blocks or plain in situ concrete are all suitable materials for this form of construction. The advantage of blocks in terms of stress allowances has already been mentioned. They have the added advantage of providing a faster means of wall construction than with bricks. Plain concrete walls have in some cases proved as cost effective as comparable areas of block walls, particularly where the opportunity has been taken to use a crane for rationalising the building operations.

As an alternative to unreinforced masonry and concrete, reinforced concrete framed construction will normally only be cost viable for medium-rise and above, repetitive construction typical of blocks of flats. The reinforced concrete wall[19] even at its most economic spacing of about 5.5 m is unlikely to be cost effective for low-rise cross wall constructed buildings.

In cross wall construction the in situ solid reinforced concrete floor gives greatest rigidity. Hollow block floors and precast floors may be used if the junction with the wall can be cast in situ, their effectiveness depending upon the degree of rigidity which can be obtained at this point. This rigidity can be increased, whatever the type of floor, by introducing vertical reinforcement in the wall at the floor level. This, however, is more practicable for concrete walls than for those of bricks or blocks where the rods would complicate the laying of the units. When timber floors are used, for example in three-storey houses or as the intermediate floors in maisonettes (see figure 3.10 A), lateral support to the walls must be obtained from these by close contact with each wall face (see page 83) if the joist ends are not built-in. Where the joists run parallel to the walls and bear on precast concrete cross beams, the latter, if spaced not too far apart, may provide adequate restraint. No longitudinal rigidity to the structure as a whole is provided when timber floors are used.

Weather protection   When the walls terminate in returned ends for reason of stability or are completely covered by the outer leaf of a cavity panel wall the ends have adequate protection. When, however, the wall ends are exposed on elevation between thin infilling panels as shown in the details in figure 3.10 B, some positive weatherproofing is essential in order to prevent moisture penetration across the short distance between the exposed and internal faces of the wall. The methods adopted involve the use of a damp-proof barrier behind a brick facing or the use of other forms of facing with a minimum number of joints, as shown.

Thermal insulation   In addition to the possibility of water penetration at the ends of cross walls, that of thermal transfer by cold bridging must also be considered. If unchecked, heat energy transfer through dense concrete cross walls, in particular, may cause condensation on the inner wall faces near to the external walls. The condensation risk is considerably reduced by installing thermal insulation on the return faces, and by constructing the cross walls of low-density materials such as lightweight concrete blocks. Depending on the relationship of the infilling panels and the wall ends, insulation may be applied to the inner faces or to the external face of the cross wall as shown in figure 3.10 B. However, it should be noted that the stringent insulation requirements for the UK are likely to limit the use of this construction technique.

3.7 Panel walls

Masonry and monolithic walls may be used as non-loadbearing filling to the panels formed by the structural columns and beams of a framed building or to areas of calculated masonry walling. These are referred to as 'panel walls'.

In a framed structure most of the wall elements are assumed to sustain no loads from floors and roof. Compressive strength is therefore of less importance than transverse strength to withstand pressure from wind or stacked materials. Although heavy construction can provide good fire resistance, durability and sound insulation, the compressive strength of materials such as brickwork and concrete is not, therefore, generally fully utilised when they are used for panel walls, and considerable weight is added to the structural frame. To overcome this disadvantage, in addition to the use of lightweight concrete for masonry and monolithic panel walls, other means of providing the infilling or enclosing element have been devised to reduce the deadweight while still fulfilling the functional requirements of a non-loadbearing external wall. These are dealt with later under 'External facings and claddings' (page 103).

Combined action of panel and frame   Investigations by the Building Research Establishment have shown that panel walls and the supporting frame act together giving considerable rigidity and strength to the frame. In some tests the supporting beam was wholly in tension, the wall acting with it as the compression element. If this composite action is taken into account in order to economise on the actual beam elements or to provide lateral rigidity to the frame against wind pressure on the building as a whole, then the compressive strength of a panel wall does become a significant factor.[20]

The strength and stability of a panel wall depends upon its stiffness, which in an unreinforced wall will vary with its thickness, upon the nature of the edge support and, in the case of brick or block walls, on satisfactory bond between mortar and unit. Good edge fixing on all four sides is essential in order to develop maximum strength against lateral forces. Although greatest pressure can be sustained when the edges are actually built into the frame to give a rigid fixing, satisfactory edge support can be achieved by the use of metal ties built in wall and frame. In masonry walls the lateral strength is related to the bond existing between the mortar and the walling units.[21] This

does not imply, however, the need to employ a mortar rich in cement. It has been shown clearly that an increase in the tensile strength of the mortar by the use of a high cement content does not result in a proportional increase in the strength of the wall.[22]

The use of bed joint reinforcement laid at intervals up the wall improves lateral strength. BS 5628-2 suggests 14 mm^2 minimum cross-sectional area at vertical intervals not exceeding 450 mm.

Design considerations generally   The Building Regulations make no specific reference to panel walls. These would be designed in accordance with the appropriate sections of BS 5628-1 or 3 having regard to the dead weight of the wall, the lateral pressure to be resisted and the high pressures and suctions caused by wind near the eaves and corners of a building (see page 75). For normal conditions and panel size, a 215 mm solid or 255 mm cavity wall is usually sufficient.

BS 5628-1 defines types of edge support conditions and gives limiting dimensions of panel walls in terms of maximum area and of thickness related to height according to the degree of edge support they receive and gives the data required for designing these walls by calculation. BS 5628-3 gives empirical guidance for proportioning panel walls in buildings up to four storeys high in locations providing some protection from wind, such as in towns or in well wooded areas. This is given in tables which relate the maximum permitted area of wall to the height of the wall, the degree of edge support, the construction of the wall (solid or cavity) and the wind zone in which the building is situated. The height or length of solid panel walls must not exceed forty times the thickness, which must be not less than 190 mm, and of cavity walls thirty times the total thickness of the masonry. In the latter the minimum thickness of leaf must be 100 mm and the maximum cavity width 100 mm (figure 3.11 A, B). Limits are set for the size and position of openings in the wall.

BS 5628-3 limits the extent to which a panel wall may project beyond its supporting beam (see figure 3.11). A solid panel must not overhang by more than one-third of its thickness and a cavity panel by not more than one-third of the thickness of the overhanging leaf.

When it is desired to carry a brick panel over the structural frame to give a continuous brick facing to the building, a galvanised steel angle bracket may be fixed to the structural beam and accommodated in the thickness of a brick bed-joint. This permits a 102.5 mm brick cover to the beam and at the same time keeps the projection over the support down to one-third of the leaf thickness. A variation is shown in figure 3.12.

Monolithic concrete panel walls may be used and if constructed of lightweight concrete they will provide

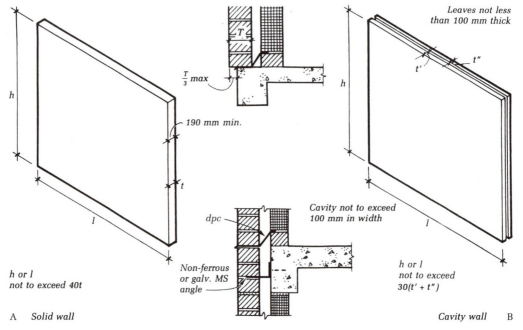

$\frac{T}{3}$ max

190 mm min.

h

l

t

h or l
not to exceed 40t

A   Solid wall

dpc

Non-ferrous
or galv. MS
angle

Leaves not less
than 100 mm thick

t' t''

h

Cavity not to exceed
100 mm in width

l

h or l
not to exceed
$30(t' + t'')$

Cavity wall   B

Figure 3.11   Panel walls

some degree of thermal insulation in external walls. Satisfactory edge fixing is given by the normal bond with structural members either of concrete or encased steel, and by tying in the shrinkage reinforcement to the structural members.

Provision for differential movement   When brick panel walls are used in a reinforced concrete frame it is essential to make provision for vertical differential movement between panel and frame caused by the irreversible expansion of the brickwork (see note to table 3.5, page 132 and *MBS: Materials*, chapter 6, *Bricks and blocks*) and the shrinkage and creep of the concrete frame. As a result of these opposing movements stresses may be set up in the panels which may cause distortion and failure of the brickwork, such as displacement of the bricks next to the support and, occasionally, buckling of the panels (figure 3.12 A).

In practice, adequate support for the outer leaf must be ensured by careful construction of the frame and provision for movement be made by means of a compression joint at the top of each panel. In most cases the normal bed joint thickness of 10 to 12 mm will be sufficient for this. Such a joint removes the top restraint to the panel. This can be provided, however, by some form of anchor which gives lateral restraint while permitting vertical movement (figure 3.12 B).[23]

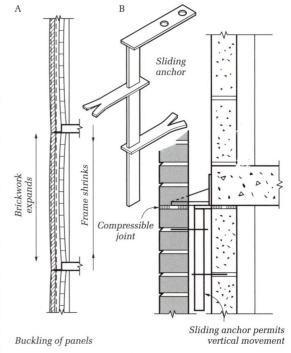

A   B

Sliding
anchor

Brickwork
expands

Frame shrinks

Compressible
joint

Buckling of panels

Sliding anchor permits
vertical movement

Figure 3.12   Movement joints to panel walls

Support against wind   When a cavity panel wall will be subjected to high wind pressures lateral support can be given by narrow angle or zed section steel wind posts set within the cavity and fixed top and bottom to the structural floors, the top fixing being sliding. Anchors built into the leaves of the wall at vertical intervals are secured to the posts through vertically elongated holes which give lateral restraint and support while permitting vertical movement as does the sliding anchor referred to above.

3.8 Compartment, separating and other walls

A *compartment wall* is a wall the function of which is to sub-divide a building into compartments of limited floor area or, in the case of storage buildings, cubic capacity, in order to restrict the spread of fire. The required fire-resistance must be appropriate to the class of building of which it forms part; in hospitals, if this is one hour or more, the wall must be constructed of materials of limited combustibility.[24]

A *separating wall* is one which separates adjoining semi-detached and terrace houses or separates a flat or maisonette from any other part of the building. Its primary function is to prevent the spread of fire between adjoining buildings or between different occupancies in the same building and to provide an adequate degree of sound insulation. A separating wall in any of these contexts should be constructed as a compartment wall.[25]

A *party wall* means a wall separating adjoining buildings belonging jointly to the owners on each side. It may or may not be loadbearing and its functions are the same as those of a separating wall.[26]

3.9 Retaining walls

The function of a retaining wall is to resist the lateral thrust of a mass of earth on one side and sometimes the pressure of subsoil water. In many cases the wall may also be required to support vertical loads from a building above.

3.9.1 Stability

A retaining wall must be designed so that (i) it does not overturn and does not slide, (ii) the materials of which it is constructed are not over-stressed, and (iii) the soil on which it rests is not over-stressed and circular slip is avoided on clay soil.

The pressure on the back of the wall is called *active pressure* and tends to overturn the wall and push it forward. The force exerted by the earth on the front of the wall in resisting movement of the wall under the active pressure is called the *passive earth resistance* (figure 3.13 A). The actual thrust of retained earth on a wall and its

direction and point of application can only be determined approximately. There are several theories for assessing the active pressure and the passive resistance which are explained in detail in most textbooks on the theory of structures. A popular application uses Rankine's formula, shown in this instance in figure 3.14. Active water pressure is avoided by the provision of a vertical rubble drainage layer, or drainage counterforts, discharging through weep holes in the wall or, if the wall encloses a basement, through lateral drainpipes at the back of the wall as shown in figure 3.13 B, C, which conduct the moisture to a main drain. In the case of clay soils, this will prevent saturation leading to increased pressure due to the reduction in the shear strength of the soil arising from the increase in moisture content. An impermeable covering to the retained soil is also useful for this purpose.

The tendency of the wall to slide is resisted by friction on the underside of the base and by the passive earth resistance at the front of the wall as explained in Part 1, page 39. When the frictional resistance is insufficient, the passive earth resistance must be used to increase the total resistance to sliding to the required amount. This may necessitate the provision of vertical ribs on the base of the wall to increase the passive resistance.

Overturning will occur (i) if the line of the resultant pressure falls outside the base of the wall, or (ii) if the eccentricity of the resultant is such that the maximum pressure at the toe is great enough to cause settlement leading to rotation of the wall (Part 1, page 40). As the resultant will normally pass through the base at some eccentricity and, in fact, in reinforced concrete retaining walls will usually fall well beyond the middle third of the 'stem' or wall thickness, a triangular distribution of pressure on the soil must be assumed as in the case of normal eccentrically loaded walls.

The pressure created by retained soil can also be represented graphically by a triangular distribution as shown in figure 3.14. The resultant thrust or active pressure (P) occurs on a plane located at one-third of the height (H) of the wall, the centre of gravity of the pressure diagram. This gives an overturning or bending moment of $P \times H/3$, expressed in units of kilo Newton metres (kNm).

Soil slip   In addition to possible movement due to sliding and overturning, when on clay soil the wall may also move because of the tendency of a mass of clay to slip and carry the wall with it. This movement, which can occur under any foundation on clay, is described in chapter 2 and occurs on a circular arc. Because wet clay soils have a low angle of repose, any bank of clay has a tendency to slip in this manner. The safe angle of slope is a function of the height and it is possible for the height of a retained mass of clay soil to be such that the arc of circular slip is

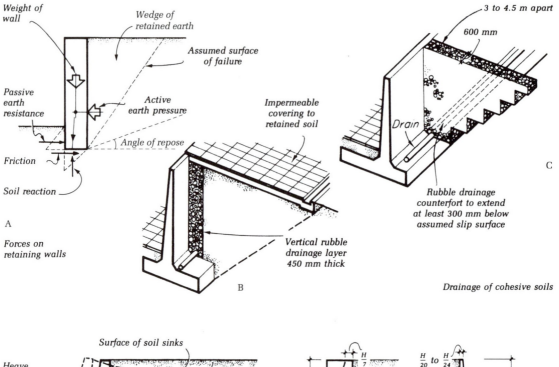

A

Forces on
retaining walls

Drainage of cohesive soils

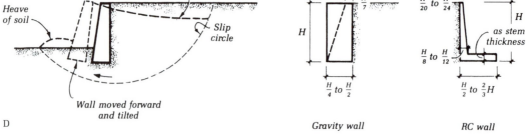

D

Circular slip failure under retaining wall

Approximate proportions of retaining walls

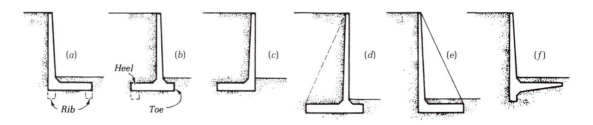

F Typical reinforced concrete retaining walls

Figure 3.13 Retaining walls

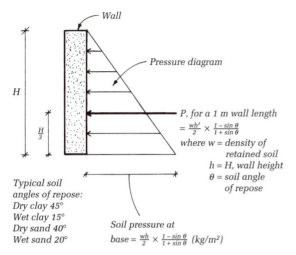

Typical soil
angles of repose:
Dry clay 45°
Wet clay 15°
Dry sand 40°
Wet sand 20°

Figure 3.14 Retaining walls – soil and active pressure

situated well below the base of the wall. The strength and stability of a retaining wall have no bearing upon such soil movement beneath it and no variations in the detailed design of the wall would affect its overall stability in this respect (see figure 3.13 D). When this type of failure appears likely, sheet piling may be used, taken to a depth below the slip circle sufficient to prevent movement taking place.

In basement retaining walls, sliding or rotation can be overcome when necessary by making the active pressures on each side of the basement counteract each other through the floors. The weight of the structure over often assists the weight of the wall in resisting overturning.

3.9.2 Types of retaining wall

There are two main types of retaining wall: (i) the gravity wall, constructed of brickwork, masonry, mass concrete or enclosed soil and (ii) cantilever or L-shaped walls constructed of reinforced concrete.

Gravity retaining walls In building work gravity retaining walls are commonly used for heights up to 1.8 m and depend on mass for their strength and stability. They are designed so that the width is such that the resultant of lateral earth pressure and the weight of the wall falls in such a position that the maximum compressive stress at the toe does not exceed the maximum safe bearing capacity of the soil, nor the permissible bearing stress of the material of which the wall is constructed. An endeavour is usually made to keep the resultant within the middle third of the base so that no tensile stresses are set up at the back of the wall at its bearing on the soil or in any of the lower joints.

This would result in high compressive stresses at the front of the wall. A width of base between one-quarter and one-half of the height is usually satisfactory (figure 3.13 E). For high walls the rectangular section is uneconomic in design, since the material at the front of the wall operates against the resultant passing within the middle third and also adds to the load imposed on the soil. The front face, therefore, may usefully be sloped or stepped back as explained in Part 1, page 40.[27]

Crib walls These are a form of gravity retaining wall constructed of precast reinforced concrete or preservative treated timber units built up to form a series of open boxes into which suitable free-draining soil is placed to act as an integral part of the retaining wall (figure 3.15).

The wall consists of a concrete foundation strip on which the first layer of units is accurately levelled and into which they key. The units consist of stretchers laid parallel with the face of the wall and headers, with ends formed to interlock with the stretchers, laid at right angles to them. The face of the wall should be battered to a slope not greater than 1 in 6 or less than 1 in 8, although it may be vertical where the height is less than the length of the headers. The filling should be of soil able to develop high friction such as coarse sand, gravel or broken rock and should be extended about 900 mm behind the crib to form a drainage layer.

The crib wall is suitable for terraces and embankments and in single width as shown in figure 3.15 for heights up to 5 m. For greater heights up to 9 or 10 m, and when poor quality common infill is used, double or triple width cribs are required in order to obtain adequate mass. These use longer headers and three or four rows of stretchers.

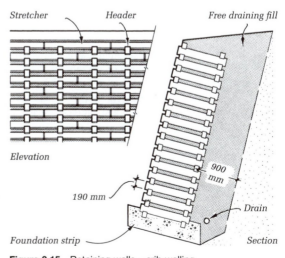

Figure 3.15 Retaining walls – crib walling

Some forms are designed with deep stretchers to give a solid face to the wall, but with open joints between to provide free drainage to the infill.

Where precast concrete is used to form this type of wall, less concrete is used and it is quicker to construct than a mass concrete gravity wall in circumstances where it is suitable.

Cantilever retaining walls In the cantilever retaining wall the practice of sloping the front face of a gravity retaining wall is carried further so that full benefit may be derived from the advantages which arise in doing so. This is particularly necessary in high retaining walls where the size of a gravity wall would be excessive. As the resultant of the lateral pressure and weight of wall falls well outside the thickness of the wall, or vertical stem, high tensile stresses are induced and it is necessary to use reinforced concrete so that the stem can act as a vertical cantilever. These walls are more economical in the use of materials, occupy less space and weigh less than gravity walls.

Different forms of reinforced concrete walls are illustrated in figure 3.13 F. That shown in (a) is most commonly used in building structures where it is not possible to excavate behind the stem of the wall. When some excavation can be carried out behind the wall, advantage should be taken of this to form a base projecting partly in front and partly behind the stem as shown in (b) so that the weight of the soil on the heel, or back portion of the base, can assist in counterbalancing the overturning tendency. The most economical arrangement is usually that in which the length of the heel is approximately twice that of the toe. The form shown in (c) has the whole of the base under the retained soil. Because of the increased stability given by the weight of the retained soil, the base may be shorter than in type (a). The inherent economy of types (b) and (c) may, however, be offset by the cost of excavation when the soil must be removed in order to construct the base slab, and type (c) may not be a suitable form when the stem carries superimposed loads from a structure above, since these would be concentrated on the toe.

Counterfort walls When the height of the wall is over 7.5 m, and the thickness of the wall might be excessive, it is usually more economical to use a counterfort retaining wall as shown in (d) and (e), in which vertical ribs called 'counterforts' act as vertical cantilevers. Type (e) with the counterforts at the front of the stem is also called a 'buttressed retaining wall'. The pressure of the soil on the wall is transferred to the counterforts by the wall slab which spans horizontally between them. Similarly, the base slab is designed to span between the counterforts. In very tall counterfort walls it is expedient to use horizontal secondary beams spanning between the counterforts with the slab spanning vertically between them. If the spacing of these beams is varied down the height of the wall, the bending moments in each span of the slab can be kept the same so that the same thickness of slab can be maintained throughout the full height of the wall.

In the case of deep basements formed as cellular rafts, the cross walls or cross frames can act as the buttress counterforts, and in normal single-storey basements columns from the superstructure above can sometimes be used in a similar way, if suitably spaced and running down on the plane of the wall. Alternatively, in some circumstances the wall may span between the basement and ground floor.

Resistance to sliding When frictional resistance to sliding is insufficient, it may be necessary to form a projecting rib on the underside of the base in order to increase the depth of earth providing passive resistance. This may be in any convenient position from the extremity of the toe to the heel as shown dotted on type (a). The best position is at the heel so that the great bearing pressure under the toe of the base can prevent the spewing of the soil in front of the rib. A rib is usually essential when the vertical load is small compared with the lateral thrust of the earth as in a freestanding wall such as that in figure 3.16 B. This sometimes occurs also in a building with a high basement retaining wall such as the multi-storey wall illustrated in A. The construction of the base at an angle assists in increasing the resistance to sliding and at the same time will result in a more even distribution of pressure on the soil. The sloping up of the base in this way is most useful when incorporated with a projecting rib under the stem of the wall as in (f) in figure 3.13 F.

The approximate proportions of the type of wall commonly used in basements are shown in figure 3.13 E.

Disposition of reinforcement When the base of a cantilever retaining wall projects entirely in front of the stem as a toe, the main reinforcement is arranged vertically on the tension side of the stem, that is, nearest the retained earth, and is carried round on the under side of the base slab (figure 3.16 A, B, D). Shear stresses are not often so great as to require the provision of shear reinforcement. If any part of the base projects under the retained soil, the downward pressure of the soil will set up a reverse bending moment and the tension side of the slab under the soil will be at the top where the horizontal reinforcement must be placed (B).

When a retaining wall carries a substantial superstructure load (C) the stem may need to be thicker and, due to the concentrated load at the heel, the base can sometimes be shorter. Since the stem acts as a column, reinforcement is required at both faces.

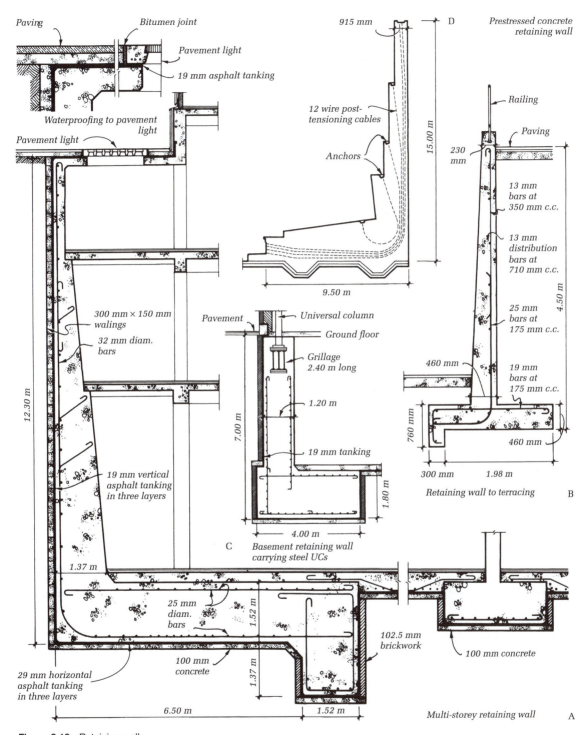

Figure 3.16 Retaining walls

In the counterfort wall the counterforts are designed as vertical cantilevers fixed at the base. When they are on the same side as the soil retained, the compressive edges are stiffened by the stem slab but there is no such stiffening when they act as buttresses on the front of the stem, so the edges may need thickening to resist the tendency to buckle. The stem and base slabs are designed as slabs continuous over the counterfort or buttress supports, the base slab distributing the pressure on to the soil. The disposition of the steel reinforcement in the base will depend on whether counterforts or buttresses are used. In the first case there will be tension at the bottom of the slab between the counterforts due to the downward pressure of the soil it carries. In the second case there will be tension at the top due to the upward pressure of the subsoil below.

Prestressing can be applied to high retaining walls and permits the thickness to be kept to a minimum. Set backs, at suitable intervals on the base and stem, provide positions for the anchorage of the post-tensioning cables as indicated at D.

The stem and base of the cantilever retaining wall may be tapered as the bending moments reduce, but it is often cheaper not to do so in order to simplify shuttering and the placing of the concrete.

Diaphragm retaining walls   This type of wall was used originally for purposes of water cut-off, especially to intercept seepage below dams. In construction work it is now taken to mean a reinforced concrete wall cast in a deep trench for use as a retaining wall (figure 3.17 A).[28]

The trench is excavated to the required depth in alternate sections up to about 6 m long, depending on the soil, either by hydraulic grab or by rotary or percussive methods (B). As excavation proceeds the soil is replaced by a bentonite mud slurry which is produced from a clay having thixotropic properties which cause the slurry to gel when at rest, thus giving support to the sides of the excavation and keeping out subsoil water. During excavation the mixture of excavated material and slurry is pumped out and sieved, leaving most of the slurry to be returned to the trench.

On completion of each panel prefabricated cages of reinforcement are lowered through the slurry and concrete is placed through a tremie pipe. As the concrete rises the bentonite slurry is displaced and can be re-used until its gelling properties reduce to the point when it requires re-conditioning or replacing. The intermediate panels are then excavated and the wall completed following the same procedure.

Wall thicknesses vary from 450 mm to 1 m by grab excavation and to 1.5 m by rotary or percussive methods. Depths from 30 to 45 m are possible.

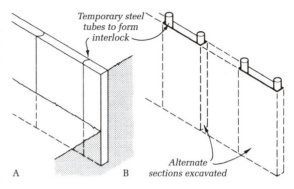

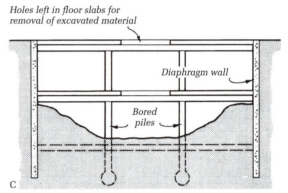

Figure 3.17   Diaphragm retaining walls

Bored pile walls   Diaphragm retaining walls may also be constructed with contiguous bored piling in which bored piles are formed in line touching each other by first completing alternate piles and then forming the intermediate piles between them. If the piles are more closely spaced and the intermediate piles are bored while the concrete in the adjacent piles, already cast, is still green, the auger will cut into the concrete and the piles will overlap giving continuity and a better seal between them. Pile diameters of 375 mm and upwards are used. Figure 3.18 illustrates the principles of application to basement construction.

With these methods the surface exposed on excavation will require some facing up and any weaknesses in the joints may need to be sealed by injecting or spraying with cement grouts, or by lining with a metal mesh and rendering with cement and sand. Alternatively, a lining to the wall with a cavity and a hollow floor may be provided as described on page 102.

Stabilisation of diaphragm walls   Since these walls have no toe as in the cantilever retaining wall, stability is provided either by struts in the form of the floor structures

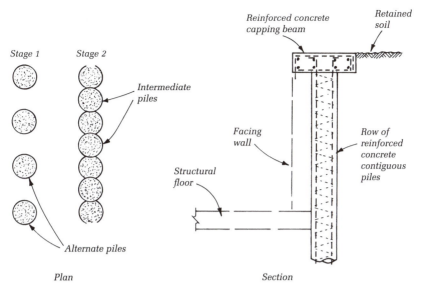

Stage 1 *Stage 2*

Intermediate piles

Alternate piles

Reinforced concrete capping beam

Retained soil

Facing wall

Structural floor

Row of reinforced concrete contiguous piles

Plan *Section*

Figure 3.18 Contiguous piling diaphragm retaining wall

(figure 3.17 C) or by ties in the form of ground anchors (figure 3.19 B).

In the former method initial stability may be provided at the top by temporary girders while the excavation is carried out between the walls and the basement structure is constructed upwards from the lowest level. The diaphragm walls must be sufficiently deep beyond the lowest basement floor to achieve adequate passive resistance from the soil. Alternatively, construction may follow the excavation downwards, the floor structures being constructed at each stage of excavation as shown in figure 3.17 C. If internal columns are part of the structure these must be formed as piles rising the full depth of the basement before excavation commences.

In the latter method no temporary top strutting nor permanent strutting by floors is required. Ground anchors consist of a prestressing cable or bar with a fixed anchorage which holds the cable in the ground at one end, and a tensioning anchor at the other (figure 3.19 A). The cable is inserted in a hole formed by rotary or rotary percussive drilling or by vibro driving depending upon the nature of the soil, the hole being lined where necessary. Anchors are formed in different ways according to the nature of the soil and the loads to be sustained as shown in figure 3.19. In their application to diaphragm retaining walls anchors are inserted as excavation proceeds thus providing stability as the wall is exposed (B).

Ground anchors may also be used in construction work to overcome uplift due to flotation of deep basements and

relatively light structures in high groundwater (C), either as a permanent measure or as a temporary measure during construction, or to overcome uplift of foundations due to wind forces (see page 21).

3.10 Waterproofing

Methods adopted to prevent the lateral penetration of rain and the penetration of ground moisture through walls are described in Part 1, chapter 5. Methods of providing protection against entry of subsoil water under pressure and of dealing with the problem of rising damp in existing walls are discussed here.

3.10.1 Waterproofing of basements

When walls and floor form a basement below ground level, the penetration of moisture through the sides of the wall as well as through the floor is commonly prevented by the application of an unbroken membrane or coating of suitable impermeable material over the whole of the walls and floor. When the basement is below the level of subsoil water, the latter will be forced through the walls and floor under pressure and special precautions must be taken, either by completely enclosing the basement in a waterproof 'tank' of impermeable material or by the use of high-grade dense concrete, with or without an integral waterproofer, for walls and floor. Prestressed concrete can also be used for this purpose.

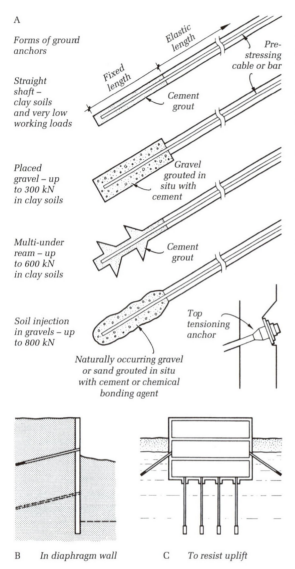

A

Forms of ground anchors

Straight shaft – clay soils and very low working loads

Elastic length

Fixed length

Pre-stressing cable or bar

Cement grout

Placed gravel – up to 300 kN in clay soils

Gravel grouted in situ with cement

Multi-under ream – up to 600 kN in clay soils

Cement grout

Soil injection in gravels – up to 800 kN

Top tensioning anchor

Naturally occurring gravel or sand grouted in situ with cement or chemical bonding agent

B   *In diaphragm wall*   C   *To resist uplift*

Figure 3.19 Ground anchors

Waterproof tanking This may be carried out with asphaltic bitumen brush applied on a fabric base in three layers with all joints lapped and sealed or with tough plastic sheeting similarly sealed at all joints. A more recent development utilises the property of sodium bentonite, a clay of volcanic origin, to swell on contact with water to form an impervious gel. A layer of this is sandwiched between two linked propylene textile fabrics to form a composite waterproofing material.[29] Another modern technique uses a two-layer liquid application of urethane pre-polymers over a primed, dry, clean and reasonably smooth concrete or masonry surface. Primer and subsequent coat-

ings are thixotropic (promptly gel) and can be applied with a brush or roller. On exposure to atmospheric moisture, the material cures to form a robust rubber-like coating.[30] All these materials are flexible and rapid in installation. The traditional alternative to these is asphalt, which has been widely adopted for this purpose. The methods described here are specifically for asphalt, but in principle can be interpreted to suit other liquid applications.

These membranes may be placed either on the outside or the inside face of the structure. When placed on the inside it may be necessary to provide 'loading' walls and floors of sufficient strength to prevent the waterproofing membrane being forced off the structure by the pressure of water.

Asphalt tanking Asphalt in tanking work is laid in three coats to a total thickness of not less than 30 mm on horizontal surfaces and not less than 20 mm on vertical faces. All internal angles are reinforced by means of a fillet 50 mm on the face, formed in two coats. All joints in the coats must be broken or spaced by at least 150 mm in horizontal work and 75 mm in vertical work.

External tanking The advantages of placing the asphalt on the outside face as shown in figure 3.16 A and figure 3.20 A are that the structure itself provides the necessary resistance against the pressure of water on the asphalt and that the asphalt keeps the water out of the structure. Faults due to poor workmanship or to settlement are, however, difficult to locate and remedy, because the point at which the water enters the internal face may be a considerable distance away from the fault in the asphalt. In spite of this the advantages of the external membrane are such that it is usual to adopt this method in all new buildings.

The asphalt to the floor is laid on a 75 to 100 mm blinding layer of concrete and is usually protected immediately by a 50 mm fine concrete screed. The concrete floor slab which is then laid must either be thick and heavy enough to resist the upward pressure of any subsoil water or be suitably reinforced to fulfil the same function. The vertical asphalt to the walls should be protected externally by a protective skin. This is usually 102.5 mm of brickwork, but the project requirements may necessitate the use of thin in situ reinforced concrete or precast concrete walings (figure 3.16 A), for reasons given on page 68.

The asphalt may be applied either direct to the retaining wall or to the protective skin, according to the circumstances. When working space is available behind the wall, the wall is first built and the asphalt then applied direct to the face of the wall which, if of brickwork, should have the joints raked out to a depth of 13 mm to form a key. If of concrete, it should be treated in some way to give a rough surface for the same purpose (figure 3.20 A). Where

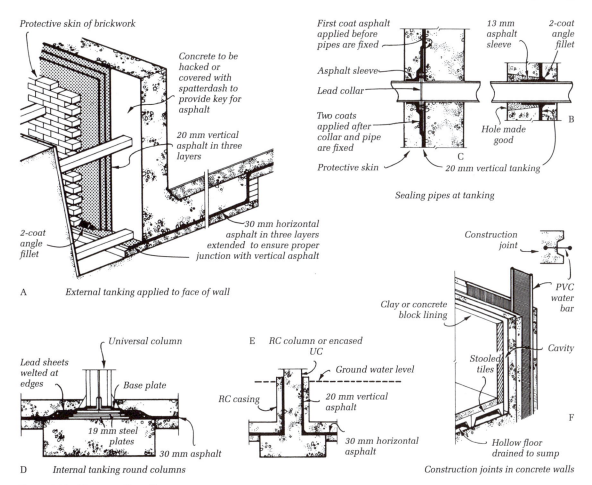

Protective skin of brickwork

Concrete to be hacked or covered with spatterdash to provide key for asphalt

20 mm vertical asphalt in three layers

2-coat angle fillet

30 mm horizontal asphalt in three layers extended to ensure proper junction with vertical asphalt

A   External tanking applied to face of wall

First coat asphalt applied before pipes are fixed

Asphalt sleeve

Lead collar

Two coats applied after collar and pipe are fixed

Protective skin

13 mm asphalt sleeve

2-coat angle fillet

Hole made good

B

C

20 mm vertical tanking

Sealing pipes at tanking

Universal column

Lead sheets welted at edges

Base plate

19 mm steel plates

30 mm asphalt

D   Internal tanking round columns

E   RC column or encased UC

Ground water level

RC casing

20 mm vertical asphalt

30 mm horizontal asphalt

Construction joint

PVC water bar

Clay or concrete block lining

Stooled tiles

Cavity

Hollow floor drained to sump

F

Construction joints in concrete walls

Figure 3.20   Waterproofing of basements

possible the excavation on the outside of the wall should be sloped back to limit the use of timbering, but if strutting is essential this must be arranged so that the position of the struts can be changed to permit the asphalt to be applied at the strut points. The protective skin would most suitably be of 102.5 mm brickwork built up as the asphalting is carried out, so that the struts may be repositioned to bear on the brick skin and thus avoid possible damage to the asphalt (A). When no space at the back of the wall is available, for example in underpinning work or on a confined site where the basement extends to the boundary of the site, the asphalt is applied to the protective skin. In these circumstances temporary support to the soil face behind the protective skin must be maintained until the new retaining wall has been constructed, and the asphalt work must be carried out in short lifts as each section of the wall rises as described on page 356.

In the case of a reinforced concrete wall which is cast direct against the asphalt, care must be taken that the asphalt is not damaged or pierced by reinforcing bars or tamping rods. In some circumstances when the basement wall is of concrete, even when working space is available behind the wall, it may be more economical to build up the brick skin as a thin, self-supporting wall, stiffened by piers at intervals, to which the asphalt can be applied. This then functions as permanent formwork for the concrete wall to be cast between the asphalted skin and inside formwork.

The pits for column foundations and retaining wall bases are covered with a blinding layer of concrete and the sides lined with 102.5 mm brickwork built up off the edge of the blinding layer. The whole is then tanked with asphalt, the upper edges being joined to the asphalt layer in the floor (figure 3.16 A).

Internal tanking Application of the asphalt to the inside face of the structure is largely confined to existing buildings where it would normally be required as a damp-proofing remedial measure rather than as a tanking against water under pressure. When used as tanking, apart from the disadvantages of not protecting the structure from water and of requiring loading walls and floors to protect and hold the asphalt in position against the pressure of water, there are problems involved in waterproofing round columns.

If the internal tanking is carried over the top of the column foundation slab, the column will pierce the asphalt skin. It is thus necessary to form a sleeve of asphalt round the column rising from the horizontal layer to some distance above the highest surface level of the subsoil water as indicated in figure 3.20 E. A reinforced concrete casing will then be required round the sleeve to prevent it being forced off the column face. As in the case of the walls and floor, the reinforced concrete column or steel universal column is permanently, or at least periodically, saturated. An alternative method which may be used for steel columns in a new building is to sandwich two or three lead sheets between steel base plates, large enough to project at least 230 mm beyond the plates all round (D). The edges of the lead sheets are welted all the way round and the exposed faces painted with bitumen, after which the asphalt is worked between them up to the edges of the base plates and to the base of the stanchion. This can make a reasonably watertight junction provided the water pressure is not excessive. However, neither of these methods is really satisfactory and in new buildings it is better to drop the asphalt skin completely under the foundation slab.

Although asphalt is not likely to be over-stressed under a foundation slab, the pressure on the asphalt due to heavy loads over confined areas should be considered and be limited to 65 to 86 kN/m^2.[31] Manufacturers of other waterproofing materials should be consulted on the suitability of their products for use in specific situations.

Passage of services Service pipes and drainpipes must often pass through asphalt tanking and some provision must be made to prevent water entering at these points. The usual method when the water pressure is not high is to form an asphalt sleeve round the pipe about 300 mm long and extending an equal distance on both sides of the line of the asphalt tanking (B). This sleeve is applied before passing the pipe through the wall after it has been thoroughly cleaned, scored and painted with a coat of bitumen. The asphalt tanking is then worked up to the sleeve and the joint is reinforced with a fillet. When the water pressure is high, and is likely to force water between the sleeve and the surface of the pipe, a metal collar is incorporated at the junction of the pipe and the tanking to form a seal (C). The collar may be formed by a 3 to 6 mm plate welded on the pipe or by a lead sheet sandwiched between a flanged pipe joint as shown. The collar should project a minimum distance of 150 mm all round the pipe and both faces should be painted with bitumen before the asphalt is worked round it. Waterproofing round pavement lights is shown in figure 3.16 A.

Waterproof structure High-grade and thoroughly consolidated concrete can be highly impervious to moisture, but in practice it is not easy to obtain an impervious structure owing to the difficulty of working round strutting and forming proper construction joints, which are usually weak points. A number of precautions are generally taken to overcome this weakness. PVC or synthetic rubber water-bars may be incorporated at the joints (figure 3.20 F). When long lengths are being cast it is preferable to leave a gap of 450 to 610 mm between adjacent sections and to fill these later after the edge faces have been prepared in the manner described on page 68. This minimises the extent of shrinkage. Vibration should be used in order to obtain maximum density of concrete and an integral waterproofer may be incorporated with the mix, although it is inadvisable to rely on this in the absence of high quality concrete.

When dependence is on the impermeability of the concrete rather than on waterproof tanking, the possibility of damage and inconvenience through leakage may be overcome by constructing a hollow floor consisting of a concrete topping over special half-round or stooled flat tiles, and building up a lining of clay or concrete blocks 50 to 75 mm thick and 50 mm in front of the face of the basement walls (F). Should any leaks occur the water will drain into the hollow floor from which it will run into a sump constructed for the purpose and fitted with a float controlled electric pump which will come into operation when the sump fills. In some circumstances, this method, even allowing for the cost of periodic pumping, can be more cost effective than full waterproof tanking.[32]

The prestressing of concrete, because it maintains the whole of the concrete in compression, prevents cracks occurring under load and also has the effect of closing up shrinkage cracks at working joints. Since high quality concrete must be used for prestressed concrete work and prestressing overcomes the weakness at construction joints, it is possible to obtain a waterproof construction with greater certainty than with ordinary reinforced concrete. However, prestressed concrete is not likely to be used solely on this account but where for structural reasons it appears appropriate, advantage can be taken of its impermeable qualities.

3.10.2 Rising damp in existing walls

The commonest cause of rising damp in existing walls is the absence of a damp-proof course.[33] In some cases this can be reduced by forming a narrow external trench or dry area against the base of the wall. This will permit the damp to evaporate outwards before it rises to a level where it can penetrate to the inside of the building.

Capillary tubes When the wall is thick, evaporation from the inside of the wall may be assisted by inserting porous high capillary tubes along the base of the wall. These are tubes about 50 mm in diameter made of high-capillary earthenware inserted in holes formed in the wall and sloping upwards slightly from the outside face. They should penetrate about two-thirds the thickness of the wall and must be bedded in a weak mortar in order to provide sufficient capillarity. This is used very dry to minimise shrinkage and breaking away from the surrounding wall. The spacing of the tubes depends upon the degree of dampness in the wall but will usually be in the region of 450 to 900 mm apart. The functioning of the system depends on the fact that evaporation of the moisture through the tube causes a fall in temperature and an increase in weight of the air in the tube. This damp air slips out from the bottom of the tube and is replaced by fresh air: thus a continuous circulation of air is set up and evaporation is continuous.[34] This method can also be used without cutting a trench as an alternative to inserting a normal damp-proof course above ground line, although it is now generally used with water-repellent barriers as a means of drying out the dampness already in the wall.

Insertion of damp-proof course The traditional method of inserting a damp-proof course in an existing wall is to cut out at intervals short sections of the wall at the appropriate level on the lines of normal underpinning work and rebuild them to incorporate a damp-proof course of engineering bricks or other suitable material. This is a lengthy and expensive operation, and methods have been developed which involve sawing either by hand or power-driven saw a narrow slot in a mortar bed joint into which is driven a metal damp-proof membrane. By this method the work can be carried out much faster and at less than half the cost of the traditional method. It is, however, only suitable for walls in which the courses are reasonably straight and in which the walling is sound. Unsound brickwork and loose rubble in the core of thick stone walls will fall and block the slot.

Water-repellent barrier Silicone water-repellent may also be used to form a water barrier at the base of a wall.

Holes of 12 mm diameter penetrating to two-thirds thickness of the wall are drilled about 150 mm apart and a silicone-rubber latex solution is pressure injected through them until the full thickness of the wall is saturated and a damp-proof barrier formed. This method is particularly suitable for walls the nature of which would make difficult the insertion of a damp-proof course.[35]

Electro-osmosis An electrical method of drying out damp walls and maintaining them in a dry state is based on the phenomenon of electro-osmosis. The principle of this is that if an electric current is passed between two electrodes buried in a capillary material any free water in the material will flow towards the cathode or negative electrode. In practice, a number of positive electrodes, consisting of spirals or loops of copper strip, are embedded in holes drilled to a depth of about two-thirds of the wall thickness from 200 to 500 mm apart. These are connected by a copper wire embedded in a chase in the wall. Negative electrodes connected by insulated cable are buried in the soil at or below foundation level. The leads connecting the wall and soil electrodes are connected to accessible junction boxes so that an electric current can be applied if required. This is seldom necessary because by short circuiting the electrodes sufficient current to operate the system is obtained due to the potential difference naturally existing between a damp wall and the soil on which it rests and from which derives its dampness. One particular advantage of the system is that it is not necessary to fix the wall electrodes below the floor because the wall below the electrodes will be dry down to the soil electrodes at the foundation level.

Waterproof rendering, corrugated bituminised fibre based lathing or plastic mesh attached to the wall by separating studs, both of the latter forming a base for plaster with air spaces behind, may be applied to the internal face of damp walls to protect finishes. These methods will not, however, dry out the wall.

3.11 External facings and claddings

The traditional masonry technique of bonding stone with a brick backing produces a structure which combines strength and satisfactory appearance and in which the facing material acts structurally with the backing material. The advent of the framed structure with solid infilling panels led to the use of thin applied slabs of stone or pre-cast concrete as an external facing rather than ashlar work, and the increasing use of framed structures naturally resulted in attempts to reduce the dead load of the non-structural enclosing construction. This factor, together with the development of lifting devices for use on building

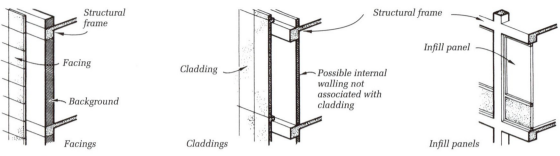

Figure 3.21   External finishes

sites, led to the technique of constructing walls with large, light elements or 'claddings' attached to the structural frame. This separation of the finishing process from that of the basic structure helps to reduce the number of operatives on the site at any given time and reduces the risk of finishes being damaged by following trades.

Fixing techniques will vary according to the nature of the background structure, the type of facing or cladding system used and whether it is applied as the work proceeds or after the completion of the structure. In discussing the various methods by which building structures may be finished externally it is convenient to subdivide them under the following headings:

Facings   Methods of finishing such as rendering and wall tiling which require a continuous background structure to give the necessary support and fixing for the materials forming the external face of the building.

Claddings   The term 'cladding' is here taken to mean a method of enclosing a building structure by the attachment of elements capable of spanning between given points of support on the face of a building, thus eliminating the necessity for a continuous background structure. A cladding element will often be large enough to take a large part of the wind force acting on the building and must be strong enough to transfer this load to the basic structure. Claddings may be heavy elements such as precast concrete slabs, or lightweight elements such as glass reinforced plastic panels.

Infilling panels   This is a method of providing enclosure by large, fairly light elements which are generally based on some light framing of timber or metal. These elements differ essentially from claddings in that they are fixed between the members of the structural frame of the building, rather than applied to the face of the frame to form a skin. In addition to supporting their own weight, infilling panels must be strong enough to support wind loads and

transfer them to the main structure through properly designed fixings.

These three categories of external finishes are shown diagrammatically in figure 3.21.

3.11.1 Facings

Wall tiles   Tiles for external use may be glazed or unglazed and in either form must be capable of resisting the action of frost. They are bedded on a rendered backing with a proprietary adhesive recommended by the tile manufacturer, a suitable mastic or strong mortar. Concrete surfaces, particularly soffits, should be well hacked to provide a key. When covering a large surface tiles are best contained in panels, or broken up into separate panels to reduce the risk of cracking due to movement of the background. The edges of the panels should have open weathered or mastic/silicone filled joints to allow for movement (see also *MBS: Finishes*, chapters 4 and 5).

Stone   Small thin slabs of slate and quartzite are bedded solid and supported by flat metal cramps every third or fourth course. Thin slabs of marble or slate may also be fixed by screwing and pelleting.

Terracotta and faience   These materials are described briefly in *MBS: Materials*, chapter 5. For external work they are produced in the form of slabs 300 mm × 200 mm, 450 mm × 300 mm and 600 mm × 450 mm by 25–32 mm thick (figure 3.22 A, B). They are scored or dovetailed on the back to give a good key and may be fixed with non-ferrous metal cramps in a manner similar to thin stone slabs. Hollow blocks are also produced which are filled with fine concrete and bonded into brickwork. The blocks are limited to 0.084 m^3 volume and are usually used for such elements as cills and copings (A and C). The protective glaze on faience only covers the face and a small strip of the return surfaces. Arrises are rounded. Joints should be 6 mm wide.

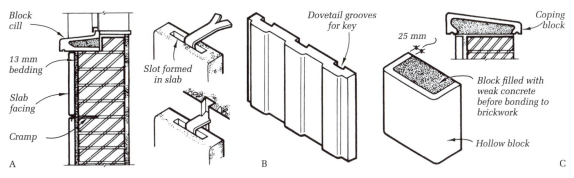

Figure 3.22 Terracotta and faience facings

Mosaic and flint Ceramic, glass or marble mosaic is made up of tesserae usually $19 \times 19 \times 5$ mm in size, supplied gummed face down on paper which is soaked off when slabbing up is completed or, alternatively, it is supplied with nylon net or strip glued to their backs which is embedded in the bedding mortar.

Floated coats should be provided as for tiles to which a Portland cement bedding mortar, 1 to 3 or 4 mix, is applied not more than 10 mm thick. Up to $\frac{1}{2}$ volume of hydrated lime may be added to the richer cement mix. The backs of paper faced mosaic should be pre-grouted to fill the joints between the tesserae and differences in level in the backs of marble mosaic should be made out with 1 cement: 2 fine silver sand. The mosaic is then beaten into the bedding without delay. Nylon-backed and similar mosaics are grouted after fixing is completed.

Movement joints should be provided as for tiles.

Flints may be natural or 'knapped' (see Part 1, *Rubble walls*, page 98). They require a sand and cement bed some 125 mm thick to enable the stones to be set in well over half their length. The bedding between the stones is brushed down while still 'green'.

Glass Opaque glass forms colourful and easily cleaned surfaces. The different types available and the methods of fixing sheets as external facings are discussed in *MBS: Materials*, chapter 12. Typical fixing details are shown in figure 3.23 for facings up to 2.4 m high. Above this fixing clips to project over the face of the glass or cover strips must be used.

Fully supported metal sheeting Steel, stainless steel, copper, bronze, aluminium and lead in sheet form may be applied as a wall facing using techniques similar to those used for roofing, the sheets being joined together with welts and standing seams and secured by cleats nailed to horizontal timber insets in concrete or battens in screed as shown in figure 3.24. Lead sheet is too heavy to rely on cleat fixing and the sheets are held by a number of brass

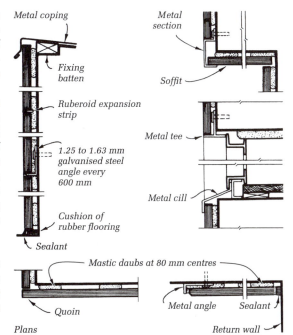

Figure 3.23 Glass facing

screws driven into lead plugs in the backing and finished-off with lead welded dots.

As an alternative to these traditional techniques very thin sheeting bonded to 18 mm high density chipboard backing is available. The panels are nailed or screwed to a framework of 50 mm timber vertical battens and cross noggings fixed to the structural wall. Vertical joints at 600 mm centres may be recessed with metal top-hat section cover strips or be formed as traditional roll joints or standing seams. Cross joints are welted. The sheeting is not bonded close to the edges of the panels to allow for fixing through the chipboard and for forming the joints.

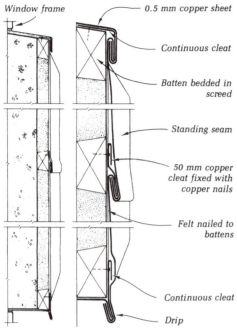

Window frame

0.5 mm copper sheet

Continuous cleat

Batten bedded in screed

Standing seam

50 mm copper cleat fixed with copper nails

Felt nailed to battens

Continuous cleat

Drip

Fully supported metal sheeting

Figure 3.24   Metal facings

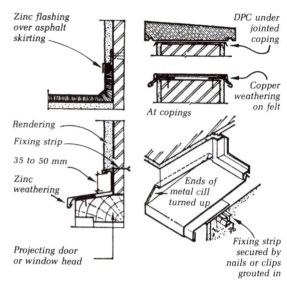

Zinc flashing over asphalt skirting

DPC under jointed coping

Copper weathering on felt

At copings

Rendering

Fixing strip

35 to 50 mm

Zinc weathering

Ends of metal cill turned up

Projecting door or window head

Fixing strip secured by nails or clips grouted in

Figure 3.25   Protection of renderings

Renderings   Renderings are used to enhance the water resistance of external walls (see Part 1, table 5.1 and section 5.2.2) and to modify their colour and texture. Large areas of cement–lime rendering should be avoided by the provision of movement joints at intervals, but this is not necessary with acrylic and silicone resin renderings.

The penetration of water behind the rendering must be prevented in order to preserve the water resistance of the wall and to avoid the possibility of sulphate attack on the rendering should the bricks in a masonry wall contain soluble salts. This is ensured by providing adequate edge protection by means of flashings, copings and cills which throw water well clear of the wall face and by placing damp-proof courses below mortar jointed copings and cills as shown in figure 3.25.

The external face of solid masonry parapet walls and freestanding walls should not be rendered since water can penetrate through the other face resulting in the possibility referred to above. When rendering is required on a parapet wall cavity construction should be used (see Part 1, chapter 5). Renderings should not be carried across damp-proof courses since the rendering below the DPC may be pushed off the wall by salts from the soil, permitting water to rise by capillary movement between rendering and wall, by-passing the DPC and causing dampness in the walling above. Chimney stacks, being usually under severe expo-

sure, are best left un-rendered. (See also *MBS: Finishes,* chapter 5.)

3.11.2 Claddings

Brickwork   Although brickwork is naturally associated with loadbearing structures its weathering properties and excellent appearance and range of colour make it a very suitable cladding to other materials such as structural concrete. When used non-structurally in this way bricks need not be bonded and various straight-jointed patterns may be employed. The brickwork should be tied to the background structure with twisted wire ties or metal cramps at 900 mm centres horizontally and vertically arranged in a diagonal pattern. The weight of the brickwork should be taken at each storey level by metal angles or by projections from the structure, below which a compressible joint should be provided (figure 3.26) for reasons given on page 92. Similar support should be provided over openings unless the brickwork is reinforced to act as a lintel as shown in Part 1, figure 5.17.

As in normal cavity walling the space behind the brickwork allows free draining of any water percolating the facing and, if wide enough, the application of thermal insulation to the outer face of the inner leaf.

Methods of support   The desirability of providing adequate allowance for site adjustment of the fixings for ties and supports and of reducing the amount of labour in forming holes and chases to accommodate them, especially in reinforced concrete backgrounds, has resulted

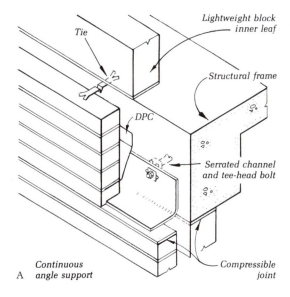

A Continuous
   angle support

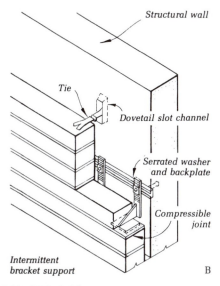

Intermittent
bracket support   B

Figure 3.26 Brick cladding

fixed vertically they and the bolt heads are serrated to increase the loadbearing capacity of the component (A).

Support may be provided by (i) continuous angles or (ii) brackets positioned at each perpend of the base course (A, B). The latter are easier to set out and adjust on site than the former. Where bricks are laid on a continuous angle fixed directly against the wall face they need to be recessed in order to accommodate the necessary thickness of the angle and maintain a regular joint width (A). This can be avoided by the use of a lighter angle section welded to backplates similar to that of the perpend bracket shown in B or by the use of perpend brackets themselves.

BS 8200: *Code of Practice for Design of Non-loadbearing External Vertical Enclosures of Buildings* recommends that metal fixings to external facings or claddings on high rise buildings should be of stainless steel, phosphor bronze or aluminium bronze alloy.

The use of continuous angle supports closes the cavity and necessitates a damp-proof course and weep holes wherever they occur. Angles and brackets usually provide for the support of half-brick facing up to storey height, but fixings can be designed for two-storey heights. The brickwork above each support level is tied back to the structural wall, or to an inner leaf if on a framed structure, by strip ties linked to channels in the former case or built in the joints in the latter as shown in figure 3.26.

Corbelling Corbel courses can be supported from concrete walls or beams by continuous angle support at the base with restraint being given above by ties back to the wall. The frequency and form of the ties will depend upon the depth of bed of the corbel bricks, the steepness of the corbelling, and whether or not the bricks are specials with holes or grooves formed in them. Although it is relatively simple to hold back a few courses of corbelling in this way the amount and, sometimes, the complexity of the steel supporting system required to hold back very deep corbelling puts in question the suitability of this particular use of brick as a cladding.

Natural stone The choice of stone for cladding will be influenced by such factors as the design of the building, its situation and the aesthetic and technical considerations relevant to each particular case. Natural stone may be broadly classified as follows in order of hardness and durability and approximate density:

Igneous rocks, such as granite	2560–3200 kg/m^3
Metamorphic rocks, which include marbles, slate, quartzite	2630–3040 kg/m^3
Sedimentary rocks, such as limestone and sandstones	1950–2750 kg/m^3

in the development of methods of support and tying-back which fulfil these requirements.

In these methods the fixing of ties and supports is by means of steel channels cast-in or bolted to the structural background into which the tie components fit and to which the support components are bolted by means of tee-head bolts (figure 3.26). Channels taking support components are normally fixed horizontally and provide for horizontal adjustment, the components being designed to provide vertical adjustment (B). Where such channels must be

Table 3.4 Thickness of natural stone external cladding slabs

Stone	Slab size (mm)	Minimum thickness (mm)	Traditional mortar for bedding and jointing*
Limestones			
Ancaster Freestone	610 × 450	25	Lime and stone dust
	1220 × 610	38	
	1525 × 760	50	
Ancaster Weatherbed	900 × 450	25	Lime and stone dust
	1525 × 610	38	
	1830 × 900	50	
Doulting Freestone	760 × 610	50	2:5:7 cement, lime, stone dust
Hornton	1070 × 450	38	Spot bedding: with 1:3 cement/sand; jointing:lime with 10% cement
Painswick	760 × 450	75	2:5:7 cement, lime, stone dust
Portland	760 × 760	50	2:5:7 cement, lime, stone dust
	1525 × 900	75	
	1830 × 1220	100	
St Adhelm (Box Ground)	760 × 450	50	ditto
Sandstones			
Auchinlea Freestone	610 × 450	100	1:1:6 cement, lime and sand
Berristall	1220 × 610	50	
Bolton Wood	900 × 610	50	
Crosland Hill (York Stone)	1830 × 300	50	1:3 cement and sand
	2440 × 685	75	
Darley Dale	1525 × 900	75	1:1:4 cement, lime, stone dust
Dunhouse	760 × 610	75	
Pennant	900 × 900	50	1:5 cement and stone dust
Woodkirk (York Stone)	2440 × 900	75	Hydraulic lime and stone dust
	3050 × 1525	100	
White Mansfield	610 × 450	25	1:3 cement and sand or stone dust
	610 × 610	38	
	900 × 760	50	
Granites			
Corrennie	1525 × 760	38	Lime or cement and sand
Kemnay	2440 × 1220	38	ditto
Rubislaw	900 × 610	25	
Shap	1220 × 610	50	
Slates			
Sawn	1525 × 760	38	Up to 1830 × 760 mm available in some slates but 1220 × 610 mm
	1220 × 610	25	recommended as average size slab
	1220 × 610	19	Only in string or apron courses up to 530–610 mm high
	900 × 450	13	Only for small areas of small slabs bedded solid to wall
Natural Riven	610 × 610	25	
	450 × 300	13	ditto
Marbles	Irrespective	19	For cladding only up to first floor level**
	of	25	For cladding rising above first floor level**
	slab size	38	Many Local Authority surveyors require this as a minimum

Note: The maximum size against each stone indicates that recommended by the quarry for cladding slabs, or the maximum size available from the quarry. The mortar, where indicated, is that recommended by the quarry.

Methods of establishing the suitable thickness of a cladding slab by reference to the load to be carried are given in BRE Information Paper 7/98: *External Cladding: How to Determine the Thickness of Natural Stone Panels*. These can sometimes result in smaller thicknesses than those given in this table.

* Selection of appropriate proprietary bedding compounds may also apply.

** See also BS 8298: *Code of Practice for Design and Installation of Natural Stone Cladding and Lining*, for thickness related to type of stone and height of slabs above ground.

Igneous and metamorphic rocks are extremely hard, durable and water resistant and are capable of taking a high polish. They are of high density and strength which permits them to be used in thicknesses of between 13 mm and 50 mm (see table 3.4). Due to the strength of these stones comparatively thin slabs may be fixed with light cramps and wires without risk of the edges splitting. Sedimentary rocks, formed by the redisposition of older rocks by the action of air or water, are generally softer, less durable and more absorbent and exhibit a highly laminar structure. Such stones should be laid with their natural bed at right-angles to the face of the wall.[36] This principle applies whether the stone is bonded into a brick or block background or is used as a non-bonded cladding. Consistent with this, certain minimum thicknesses are desirable according to the nature of the stone and table 3.4 indicates the thicknesses recommended by the quarries for a number of typical building stones suitable for external claddings. This list, of course, is not comprehensive nor extensive, and does not imply that other good building stones are not suitable for this purpose. It will be seen that some sedimentary rocks may be used as thin as 38 to 50 mm, but this thickness does not afford much material at the edges to give a secure anchorage for cramps. A minimum thickness of 75 mm or even 100 mm is recommended for many stones. Thin slabs are not always the most economic, depending upon the site fixing problems, and furthermore, due to various causes during the life of the cladding, failures can occur in practice.[37] BS 8298, referred to in the notes to table 3.4, includes a requirement to establish the strength of slabs thinner than those previously recommended. Slabs should be about 0.28 to 0.37 m² face area and limited to 55 to 68 kg in weight if they are to be manhandled. Slabs as heavy as this would need three operatives to handle since the Construction (Design and Management) Regulations stipulate 20 kg as the maximum to be lifted by one person.

New sawing techniques have now made it possible to cut stone much thinner than the smallest thickness referred to here, but the slabs need reinforcing. Methods used are described on page 112.

Care should be taken that limestone and sandstone are not used in juxtaposition in such a way that the soluble salts formed by the decomposition of the limestone are washed on to the sandstone since this may produce rapid decay of the sandstone. Similar decay may occur where cast stone and sandstone are placed in juxtaposition and brick can be damaged in the same way.

Metal anchors   Natural stone claddings are fixed to the background structure by means of metal anchors which are designed to hold the slabs back to the wall and keep the faces in correct alignment and, in the case of some, to provide support for the stone. Although these fixings may assist in relieving stones below of the weight of those above, the weight of the facing should usually be brought back to the structure at 2.4 to 3.0 m intervals by angles bolted to the structure, by suitable projections or by metal corbels (figures 3.27 and 3.28). All metal anchors should be of non-ferrous metal such as stainless steel or bronze, since galvanising and bitumen painting of iron or steel give temporary protection only and may result in subsequent staining and spalling of the stone (see page 107 for reference to BS 8200 on this).

Figure 3.27 shows various types of metal anchors and supports for sedimentary stones. These are usually in strip form and reasonably substantial when supporting the weight of a thick slab. The ends are accommodated in mortices or grooves cut in the edges of the slabs. The thinner igneous and metamorphic slabs are invariably secured with wire cramps and dowels as shown in figure 3.28, which necessitates less labour on the hard dense stone, drilling and surface sinking only being required.

The ends of all forms of anchors and corbels must be fixed firmly to the background and this is usually done by bedding in mortar in mortices cut, or in the case of concrete cast, in the background material. Forming of individual mortices in concrete can be avoided by casting in dovetail section pressed metal channels to form horizontal or vertical slots which accommodate the dovetail-shaped ends of various types of cramps as shown in figures 3.27 and 3.28.[38]

Fixing   As indicated above sedimentary stone slabs are supported by strip metal anchorages which vary in form according to their function and position relative to the joints. A typical distribution is shown in figure 3.27. The cramps in the horizontal joints would be type 2, which, being solidly built in to the structural wall, could provide a degree of support, or type 3, which would fulfil only a restraining function. Those in the vertical joints would be type 4, 6, 7 or 8, type 7 being capable of providing some support. Over the openings support is given by an angle rag-bolted to the structure or by metal corbel brackets, such as those shown in figure 3.28 set in grooves at each vertical joint. Either of these would be used to provide a positive bearing for the weight of the cladding at the intervals given above, which is likely to be at storey height. Both the continuous angle and the metal corbel can be welded to backplates as described under brick cladding and secured to a cast-in channel with the same resultant advantages.

Fixing methods for igneous and metamorphic stone slabs, which are often much thinner than sedimentary slabs, are shown in figure 3.28. As indicated earlier wire cramps and dowels are mainly used for restraining the

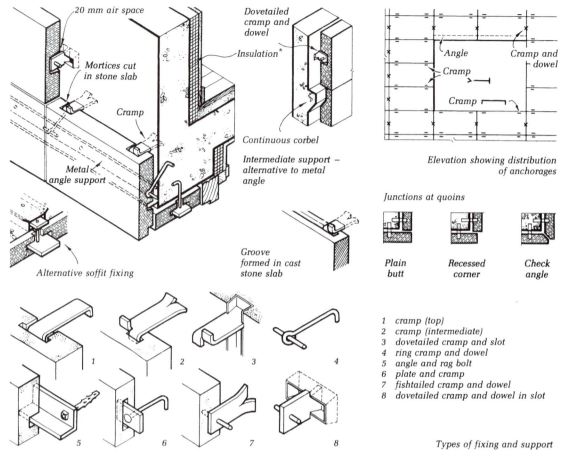

20 mm air space

Mortices cut in stone slab

Cramp

Metal angle support

Alternative soffit fixing

Dovetailed cramp and dowel

Insulation*

Continuous corbel

Intermediate support – alternative to metal angle

Groove formed in cast stone slab

Angle

Cramp

Cramp and dowel

Cramp

Cramp

Elevation showing distribution of anchorages

Junctions at quoins

Plain butt

Recessed corner

Check angle

1 cramp (top)
2 cramp (intermediate)
3 dovetailed cramp and slot
4 ring cramp and dowel
5 angle and rag bolt
6 plate and cramp
7 fishtailed cramp and dowel
8 dovetailed cramp and dowel in slot

Types of fixing and support

* Also under floor finish and in ceiling to satisfy current energy conservation requirement.

Figure 3.27 Stone cladding – sedimentary stones

slabs as these necessitate only drilling into these hard stones and the various types used are shown. 'S'-hooks are used to align the edges of adjacent slabs when fixing is by wire cramps into one edge only of two adjacent edges. The mortar dabs position the 'unfixed' edge relative to the background and the 'S'-hook prevents it falling outwards. Positive bearing is given by metal corbel brackets set into grooves in the backs of adjacent slabs as for sedimentary stones.

Methods of support and tying-back for these hard stones other than those illustrated in figure 3.28 have been developed for the same reasons as for brick cladding and can be used with advantage, especially where large areas of facing are involved.

In these methods the slabs are supported individually by support and restraint brackets incorporating dowels which engage in the edges of the stone. Dowels engaging in the vertical edges provide the easiest method of erection, the support brackets being at the bottom and the restraint brackets at the top: the former are of heavier construction and are secured by tee-head bolts to cast-in or bolted on steel channels (figure 3.29). If the dowels of the support brackets are twisted to engage in the horizontal edges two brackets will be required to each slab, which permits larger, heavier slabs to be carried. When relatively small slabs are used or the slabs are laid horizontally, in each case requiring brackets at close vertical intervals, a more economic method may be used in which support and restraining brackets are fixed to vertical steel channel members secured to the supporting wall behind the vertical joints or, when the supports are in the horizontal joints, within the width of the slab. By this means fixings to the main structure are limited to those required by the vertical channel members.

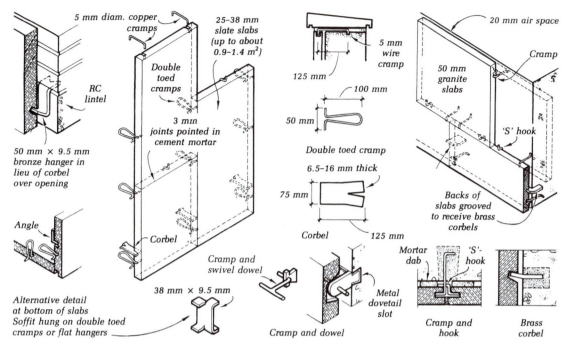

Figure 3.28 Stone cladding – igneous and metamorphic stones

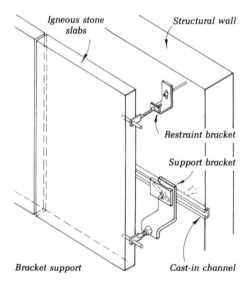

Figure 3.29 Stone cladding – igneous and metamorphic stones

Setting, jointing and pointing Sedimentary stones are bedded in mortar and suitable mortars must be used.[39] The width of the joints should not be less than 5 mm, metal strips being used as screeds. The finish to the face of the mortar joints may be carried out as the work proceeds in the same mortar used for bedding. Alternatively, joints may be raked back 19 mm and pointed with specially prepared mortar of selected colour and equal strength to the bedding mortar. Slabs should be set with a space behind not less than 20 mm wide. This isolates them from the background and avoids salts in the bricks, mortar or concrete being taken up in solution by rain passing through the joints or slabs and then being deposited in the stone when the rainwater evaporates through the cladding, causing staining and decay of the stone. It also allows for differential movement of the cladding and structural background. Ventilation openings should be provided at the top and bottom of the clad area to allow some air flow to evaporate off any moisture entering this space, especially when thermal insulation is positioned on the outer face of the structural background. However, ventilated space behind over-cladding must be limited in area by cavity barriers to prevent fire spread by the potential flue effect of the void. See page 333 and Building Regulations, Approved Document B3, section 10: *Concealed Spaces (Cavities)*.

Unless precautions are taken cladding slabs on very high concrete buildings can become loose at the fastenings due to the cumulative effect of 'creep' in the great height of concrete structure. To overcome this compressible joints at every floor, or not more than 6 m apart, are now generally required, together with the air space behind the slabs referred to above.

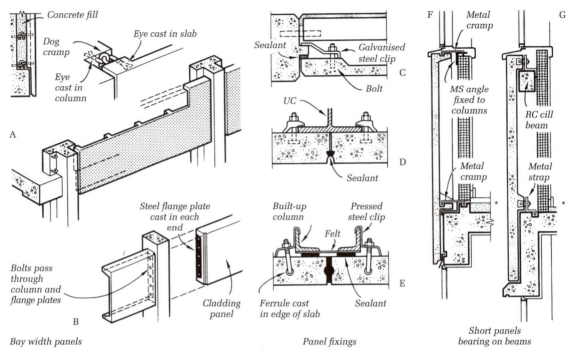

Labels in figure:
Concrete fill
Dog cramp
Eye cast in slab
Eye cast in column
A
Bolts pass through column and flange plates
Steel flange plate cast in each end
Cladding panel
B
Bay width panels

Sealant
Galvanised steel clip
Bolt
C
UC
D
Sealant
Built-up column
Pressed steel clip
Felt
Ferrule cast in edge of slab
Sealant
E
Panel fixings

F
Metal cramp
MS angle fixed to columns
Metal cramp
G
RC cill beam
Metal strap
*
*
Short panels bearing on beams

Floor and ceiling to incorporate insulation to prevent thermal bridging through the structural floor.

Figure 3.30 Precast concrete cladding – horizontal

Very thin stone slabs   As stated earlier the development of new sawing techniques has made it possible to cut granites and marbles, and some limestones, much thinner than necessary for the methods of fixing just described. With some types of stone this can be down to 4.5 mm.

Small size panels could be used for facing work (see section 3.11.1), but for the larger sizes required for cladding the panels must be reinforced, which is done by applying a backing of suitable material by means of epoxy resin adhesive. The backing may be galvanised or stainless expanded steel or fibreglass mesh, neither of which increases by very much the overall thickness of the composite panel, or a honeycomb of aluminium giving an overall thickness of about 24 mm when backing a 5 mm thickness of stone.

Depending on the type of stone and its thickness panels can be as large as 3000 mm by 1500 mm and are fixed mechanically by various forms of patented metal fixings, which vary with the manufacturer of the panels. These panels can be secured with adhesives recommended by the panel manufacturer. Face fixing with screws can also be used. The thinner panels may be incorporated in curtain walling, fixed in rebates by retaining beads or by being bonded direct to the face of the curtain wall framing by the structural silicone glazing techniques described on page 125.

Precast concrete   Precast concrete, including cast stone which is essentially concrete made with crushed aggregate and white cement, may be used as a cladding in a similar manner to natural stone. The slabs are lightly reinforced and may have fixings cast into the back. Since concrete slabs can rarely be produced to fine dimensional tolerances allowance must be made for this in the joints.

Self-supporting cladding panels may be designed to span horizontally between columns (figure 3.30). The weight is usually reduced to a minimum by casting the body of the panel down to 40 mm, rigidity being achieved by forming ribs at the edges, and at intermediate positions on larger panels, in which the main reinforcement is concentrated. Fixing is by projecting stirrups and in situ concrete filling or by metal plates cast in the ends of the panels.[40]

Concrete panels may be finished in a variety of colours and textures by careful choice of aggregate which may be exposed by washing or spraying the surface before hardening or by scrubbing or wire brushing. Polishing or acid spraying may be carried out after manufacture. Formwork

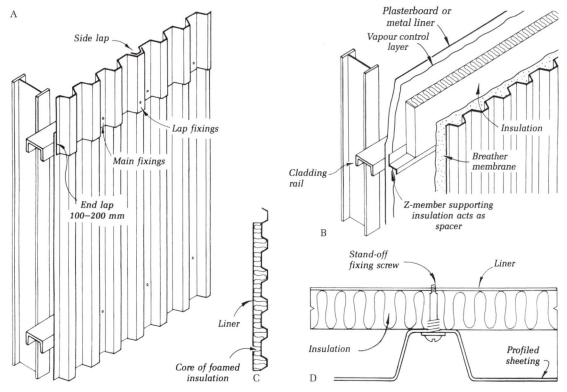

Figure 3.31 Profiled metal sheeting

may be lined with patterned timber, plastic or rubber sheet to provide surface texture (see page 366 and also *MBS: Finishes*, chapter 5).

Metal   Steel, stainless steel, copper, bronze and aluminium may be used in sheet form for claddings to walls and structural frames in the following ways:

Profiled metal sheeting   This application of metal sheeting uses sheets pressed or rolled into various profiles to give the necessary stiffness to span between cladding rails spaced from two to three metres apart, depending on the depth of profile. Aluminium and steel are generally used for this purpose and these can be coated with various materials to give protection against the weather and provide a coloured finish; some of these coatings permit profiling after the coating process.

The sheeting is supported by cladding rails consisting of hot- or cold-formed steel angles or channels which are bolted to the structural frame and to which the sheeting is fixed usually by self-drilling/self-tapping screws through the troughs (figure 3.31 A) although various types of hook

bolts may also be used (see figure 3.36). Some manufacturers have developed secret fixings incorporating concealed clips and interlocking edge joints which avoid fixing through the face of the sheeting. If aluminium sheeting is used on steel rails, bimetallic–electrolytic corrosion must be avoided by the use of isolating tape and plastic washers. With screw or hook bolt fixing weather exclusion is achieved by one corrugation side laps, except in severe exposures when more may be necessary. End laps of 100 to 200 mm are normal according to severity of exposure. Where possible end laps should be avoided by making use of full height sheets, the normal lengths of which are from 10 to 13 m. If end laps are essential they should be located as near as possible to the top of the façade. The end joints between the sheeting and windows and door heads may be made with Z-flashings (see figure 3.36 D).

Thermal insulation can be incorporated with these claddings by backing the sheeting on site by rigid board or by insulation carried either on supports fixed to the face of the cladding rails (figure 3.31 B) or on the rails themselves, in each case being held in position by liners of metal or plasterboard on the inner face.

A vapour control layer is required behind the liner on the warm side of the insulation, with a breather membrane on the outside, to protect the insulation from a reduction in its efficiency due to saturation by condensation, which can occur on the inner face of the sheeting due to radiation to the night sky. The increasing thickness of insulation to attain the U-values required by the Building Regulations aggravates this problem since it further isolates the sheeting from the building's heat so that it becomes comparatively much colder.

The Z-supports shown in B form thermal bridges, the effect of which could be reduced by the use of two-part rails incorporating a thermal break of polyurethane-resin or other suitable insulant.

In order to avoid site placing of the insulation, profiled sheet may be obtained with rigid board insulation bonded to the back or with a lining bonded to it by a core of foamed vapour resistant insulation, the insulation in both forms being kept back from the edges of the sheet to provide for normal side lap joints (C). Where insulation thus lies between sheeting and rails with no supporting member to act as a spacer as in B, spacers must be provided at the fixing points in order to prevent compression of the insulation. Alternatively, specially designed 'stand-off' screws may be used (D).

Curving of this type of profiled sheet along the length of the corrugations can be carried out to a relatively small radius with corrugations 60 mm deep. If such a sheet is erected between adjacent structural columns with the corrugations horizontal, it will act as an arch spanning from column to column to form a strong and stiff cladding, eliminating the need for cladding rails. Single skin or double skins with insulation between may be used.

Metal panelling   This can take two forms, the first based on simple flat panels usually of steel or aluminium, the second based on what are termed composite metal panels. These consist of a sandwich of two skins of steel or aluminium sheet which, in the most basic form, are formed into trays and joined edge to edge to form a box either with an insulating core laminated to them or injected as a foam to bond them together and space them apart, or with metal Z-spacers at intervals with insulation placed between.

Flat metal panels   These will vary in thickness according to size and shape: a very large panel, for example around 7 m by 900 mm, would be in the region of 5 mm thick. This is necessary to impart the stiffness needed against wind pressure and to achieve flatness of surface. In order to keep the thickness to a minimum the edges of the panel may be turned up to form a tray to give added stiffness (figures 3.32 and 3.33) or angles can be welded to the back for the same purpose. Stiffness and flatness can also be achieved by bonding two thin skins together by a plastic core to form a sandwich panel with an overall thickness of 4 to 10 mm, (these are not composite metal panels in the current sense of this term since the core is not thick enough to provide a useful level of insulation). The panel skins may be lightly profiled (figure 3.33 A) or microprofiled, that is profiled to a very shallow depth around 1 mm by stamping

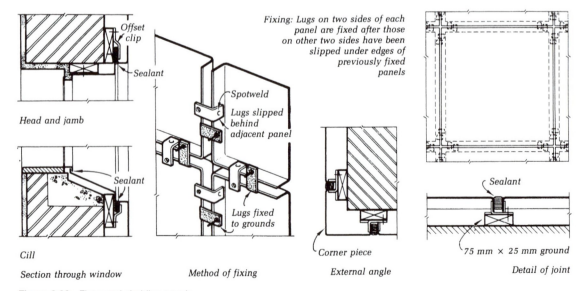

Head and jamb

Cill

Section through window

Method of fixing

Corner piece

External angle

75 mm × 25 mm ground

Detail of joint

Offset clip

Sealant

Sealant

Fixing: Lugs on two sides of each panel are fixed after those on other two sides have been slipped under edges of previously fixed panels

Spotweld

Lugs slipped behind adjacent panel

Lugs fixed to grounds

Sealant

Figure 3.32   Flat metal cladding panels

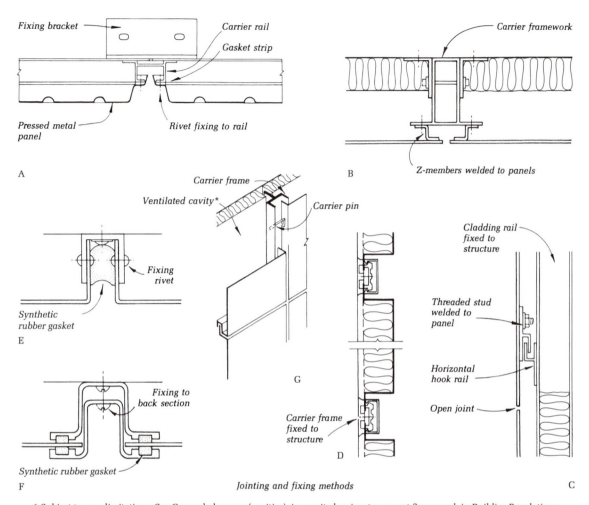

A

Fixing bracket — Carrier rail — Gasket strip — Pressed metal panel — Rivet fixing to rail

B — Carrier framework — Z-members welded to panels

E — Fixing rivet — Synthetic rubber gasket

F — Fixing to back section — Synthetic rubber gasket

G — Carrier frame — Ventilated cavity* — Carrier pin — Carrier frame fixed to structure

D

C — Cladding rail fixed to structure — Threaded stud welded to panel — Horizontal hook rail — Open joint

Jointing and fixing methods

* Subject to area limitations. See Concealed spaces (cavities), i.e. cavity barriers to prevent fire spread, in Building Regulations, Approved Document B3, section 10.[41]

Figure 3.33 Flat metal cladding panels

or pressing, to give additional stiffness and to conceal any curving or unevenness in the surface of the panel.

Panels may be face fixed (A) or be fixed by means of threaded studs welded to the backs of the panels which are bolted to steel brackets fixed to the structure, by bolting through edge members to supporting brackets (B) or by hooks or lugs welded or bolted to the backs of the panels which engage with metal supports fixed to the structure (C) or with adjacent panels (figure 3.33). Tray panels with open slots in the return edges may be hung on dowels fixed to the supporting framework as shown in G, the panels being secured by secret screwing the top edges to brackets on the vertical channel. All of these give concealed fixing.

The panels may be butt jointed with sealant, open jointed with an open drained joint in the form of a channel member behind (A), or open jointed with the general construction of the cladding on the rainscreen principle (C, G).[41] If the panels are formed into trays deep joints may be formed and sealed with gasket or sealant (E).

Alternatively, the panels may be fitted into a metal sub-frame fixed to the wall or structural frame. This framework, which is normally prefabricated, may be made up of simple top-hat sections to which the panels are secured by matching sections and which produce a recessed joint (F) or it may be made up of sections similar to the exterior face or 'carrier' part of a normal curtain wall section (D).

If flat or tray panels with the edge returns bent out are used the panels may then be fixed in the manner of normal glazing as shown. Panelling with any form of fixing or sealing which produces closed joints should have the space behind drained and ventilated if the structure of the wall permits water vapour to pass through from the interior of the building.

Insulation may be applied to the face of the structure behind the panels as in other forms of cladding or it may be bonded to the back face of flat or tray panels.

Composite metal panels These are referred to at the beginning of this section and may also be called 'flat composite metal panels'.

The core of these panels, as already indicated, usually comprises an insulating material which in bonding the two skins together gives stiffness to the panel and also provides thermal insulation. However, on some existing construction the core may be found to contain a honeycomb of metal or paper or be of light balsa wood set end grain to the skins all of which provide stiffness but not insulation. These, together with the Z-spacers used in some panels, introduce thermal bridges that may limit their application to unheated buildings such as garages and warehouses. The skins themselves may be further stiffened by light profiling. By these means the thickness of the skins may be kept down to 1 mm or less and the spanning ability of the panels between supports increased.

The edge treatment of these panels varies according to the form of assembly and fixing to the structure. The panels may be (i) mounted on to a normal curtain wall framing using cover strips or structural gaskets for fixing, (ii) mounted on a carrier sub-frame fabricated from sections similar in form to the front part of a curtain wall section and fixed back to the structure (figure 3.34 A), the panels being secured to the carrier as in the first method, or (iii) mounted as an edge to edge panel assembly on metal cladding rails or direct on to structural members. In this method the panels may be fixed direct through the face or through the joints (B) or by secret fixings of various forms (C). Flush or recessed joints may be formed by the shaping of the panel edges (B), by the incorporation of specially formed edge extrusions (D) or by incorporation of carrier sub-frame members in the edges (figure 3.44 D).

Steel and aluminium may be finished with various types of coloured coatings such as vitreous enamel, baked polyester powder and other plastic coatings, and aluminium may be anodised (see BS 4842: *Specification for Liquid Organic Coatings for Application to Aluminium Alloy Extrusions, Sheet and Pre-formed Sections . . .*).

Composite panels may undergo dimensional changes due to differential expansion of the internal and external surfaces, especially when the core is an insulating material, or by the expansion of air trapped within hermetically sealed panels. Where panels have sealed edges

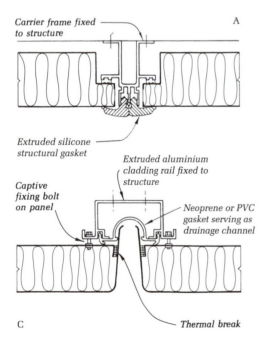

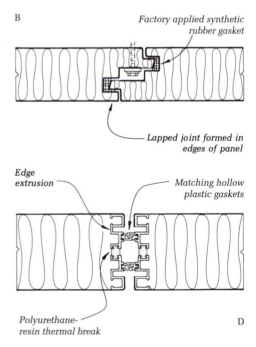

Figure 3.34 Composite metal cladding panels

the difference in dimensional change between inner and outer surfaces will result in bowing or bulging of the surface which has expanded most or contracted least. Hermetically sealed panels will bulge if the air inside them expands, or the surfaces will become concave if the external air pressure exceeds that of the air within the panel. Separation of the outer from the inner skin permits unrestricted movement of the individual skins and so prevents distortion (figures 3.34 C, D and 3.43 F).

Fibre cement Since the introduction of restrictions on the use of asbestos fibre alternative natural and synthetic fibres have been adopted as reinforcement to cement in the production of components for the building industry.[42] The material may be obtained with factory applied coatings or with integrally incorporated colours, but the latter are limited in the range of colours obtainable.

Fibre cement may be used as a cladding in the following ways:

Slate and tile hanging, siding Fibre cement slates are rectangular and are centre nailed. Fibre cement tiles are square, laid diamond pattern, and are double nailed at the centre. In addition a loose rivet is used to hold the slates or tiles together in threes (figure 3.35 A). Siding is basically flat sheeting used in strip sizes 600 mm to 1.2 m long.

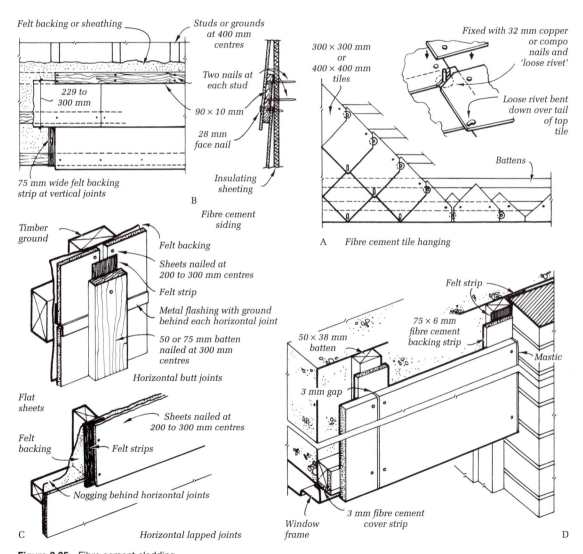

Figure 3.35 Fibre cement cladding

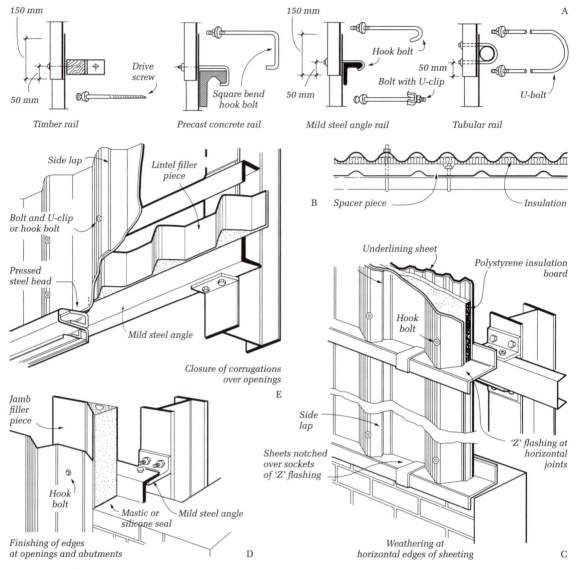

Figure 3.36 Fibre cement cladding

Fixing is to battens with a 25 mm or 38 mm lap at the head and vertical joints are protected by felt backing strips (B).

Flat sheeting This may be fixed to grounds either with lapped horizontal joints with a felt strip behind the vertical joints, or with horizontal butt joints, Z-flashings and vertical cover strips (C, D).

Profiled sheets These have been used for many years as cladding to industrial and temporary buildings. The sheets are lap jointed at the sides and the upper and lower edges

may be either lapped or butt jointed. A variety of metal fixings is available to attach the cladding to steel angle framing, steel tubing, concrete rails, or timber rails or inserts at spacings from 900 to 1800 mm depending on the depth of profile. These are illustrated in figure 3.36 A. Fixing is usually through the crown of the corrugation although when the rails are thin cold-formed sections it is through the trough. Bolt heads are normally sealed against the weather by a plastic washer and cap. Where the upper and lower edges are lapped, it is necessary to shoulder the diagonally opposite panels in a similar manner to interlocking tiling to avoid excessive thickness at the junction of four sheets. Horizontal butt joints are rendered waterproof by the use of a Z-flashing and additional fixings (C).

A comprehensive range of accessories, such as flashings, filler pieces (D, E) and internal and external angles, is available, and in many instances the cladding may accommodate integral window frames. Underlining sheets may be used to produce a double skin cladding incorporating board type insulation as in C or a quilt insulation using spacer pieces between inner and outer skins to prevent compression of the quilt (B).

Figure 3.36 shows the application of one type of sheet, which is typical of others.

Glass reinforced cement Claddings may be formed from GRC, that is glass fibre reinforced cement which, although fibre reinforced, is not included in the category of 'fibre cement' products. It is produced by adding alkali resisting glass fibre to Portland cement, silica sand and water. The material is weather resistant, non-combustible and fire resistant. Compared to reinforced concrete it has a high strength/weight ratio, develops early high impact resistance, is easy to form into a variety of shapes and can be worked with simple tools.

GRC claddings are typically thinner and lighter than those of normal reinforced concrete, which are usually 50 mm or more in thickness compared to 10 to 15 mm for GRC products. They are formed either by spraying into moulds or by casting, extruding or pressing the material in premixed form. The method used depends primarily on the number of units required and complexity of shape. Panels are normally limited to 4 m in length and to an area of 7 to 8 m^2.

Single skin panels may be stiffened by turning the edges to form a tray, by profiling in curved or bent form or by means of ribs (figure 3.37 B, C), although the latter may give rise to pattern staining or 'ghosting' on the panel face. Increased rigidity may also be attained by sandwich construction of two skins of GRC with an insulating core between, but problems of bowing have been encountered where this has been adopted due to the differential thermal movement between the outer and inner skin (see page 116).

Fixing and jointing Fixing devices, which must be non-ferrous, such as threaded studs or sockets, should be cast in thickened areas of GRC extending to at least 12 times the width or diameter of the anchorage or bolt (figure 3.37 A, B). Four fixings per panel are usually used of which one is fixed and the others designed to allow for movement, which is achieved by employing oversized or elongated holes, spring or rubber bushes and large diameter low friction washers (A). It is particularly important to align all

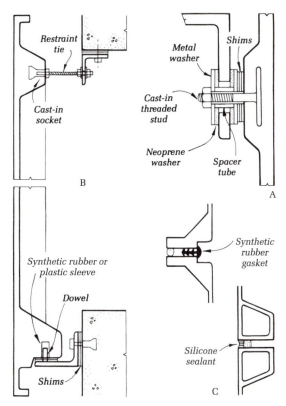

Figure 3.37 Glass reinforced cement cladding

fixings by packing out with shims wherever necessary so that a panel is not drawn out of line at one fixing point thus setting up stresses leading to subsequent cracking of the panel after completion of the work.

The weight of the panels, especially large storey height panels, is best taken at the base and other fixings used only to retain the panels in the vertical position (B).

Joints between adjacent panels may be formed with gaskets, sealants (C) or cover strips, or as open drained joints. The latter are not often used because of the difficulty in forming the complex edge shape required to accommodate the baffle strip in the joint. If gaskets are to be used successfully care must be taken to produce smooth edges in the panel to receive them.[43]

As with other forms of cladding, GRC panels may be prefixed to light steel or aluminium frames for subsequent fixing to the structure.

Thermal insulation may be bonded to the back of panels of either flat or tray form or may form the core of a sandwich panel, although this can result in the bowing of the panel referred to above. Alternatively, the insulation may be applied to the face of the structure itself.[44]

Glass reinforced plastic GRP was originally developed in the UK during the second world war as an alternative to moulded plywood in aircraft. Since then it has had many applications, not least as a composite material for cladding panels to buildings. These panels can be produced in large areas to single or even double storey heights. GRP comprises chopped strands or fibres of glass, bonded together with a plastic resin of polyester or vinylester (see *MBS: Materials*, chapter 13).

Some characteristics of GRP compare with reinforced concrete in that the two base materials compensate for the weakness of the other. Glass fibres are strong in tension but weak in compression, and plastic resins the reverse. GRP is light in weight and easily mouldable into a variety of shapes by 'laying up' by hand, although some mechanisation is possible for mass production. With this process, sheets of the glass fibres are placed in a smooth faced mould and brush coated with a plastic resin containing a setting agent. Variations known as fibre reinforced plastics (FRP) can be made from fibres of glass, carbon or aramid, with epoxy, vinylester or polyester polymers. Powdered fillers can be added to provide body and colour. Without filler the material is translucent.

In view of the low elastic modulus of GRP and the thinness of the material used, units must be suitably profiled to curved or bent shapes or be made of sandwich or stressed skin construction with a core of expanded or foamed plastic in order to obtain adequate rigidity, although the latter method gives rise to the problem of bowing to which reference has already been made and to the possibility of delamination of core from outer skin. The incorporation of ribs to stiffen the panel, especially when it is wide and shallow in profile, is also possible by laminating timber, metal or foamed polyurethane sections into the panel or by forming them in GRP itself (figures 3.38 and 3.39 A, C). The effect of surface patterning, which may be apparent if these ribs lie behind areas of flat panelling, can be overcome by texturing the surface of the panel or reduced by the introduction of areas of woven glass reinforcement in front of the ribs.

Panels of single skin construction are generally stiffened round the edges by return flanges (figure 3.38). Long narrow units are relatively stiffer than wide ones because of the reduction in effective span, but they result in a greater number of joints to be formed during assembly and greater risk of failure at these points. Where necessary for structural reasons flanges to units and fixing points may have metal stiffening members incorporated during laying-up (figure 3.38).

Fixing and jointing Panels may be attached directly to the structure by bolting through the face so that the fixings are exposed, by back fixing to metal lugs bonded into the panel (figure 3.39 B, C) or by bolting through the flanges to lugs or brackets fixed to the structure as in figure 3.38. Fixing may be at top and bottom only as shown, in which case the adjacent flanges are bolted together.

Thermal movement must be provided for in the fixing of GRP panels, in particular in view of the relatively high thermal expansion of this material. Stress build-up can be avoided by fixing rigidly at one point only with all other fixings designed to permit movement as described for GRC claddings.

By forming the panels with flat edges rather than returned flanges they can be mounted into normal curtain wall framing by glazing techniques and as with GRC the panels can be prefixed to a metal carrier frame for subsequent fixing to the structure.

Jointing techniques may be used as for other claddings except that the open drained joint is rarely adopted because of the same edge problems as with GRC. Figures 3.38 and 3.39 C show examples of panel to panel gasket joints in both of which the gaskets are in compression and do not depend on an accurate and precise edge finish to the panels in order to effect a weather sealed joint.

Thermal insulation may be provided in the form of sandwich construction, by incorporating quilt or rigid board insulation in the panel assembly (figure 3.38) or by attaching it to the face of the structure.[45]

Where window units are incorporated in the panels they can be retained in position by filler strip type structural gaskets (figure 3.38; see also Part 1, figure 2.4 A).

Curtain walling The term is here taken to mean a system of cladding comprising a frame or grid of members fixed to the face of a structure, usually at each floor level, and an infilling of panels, glazed or solid, as may be required to perform the functions of both window and wall. Structural glazing, that is suspended glass sheets (see page 130) is, in principle, a curtain wall system, but is not usually included within this category.

The curtain wall must fulfil the same functional requirements as any other system of external walling and it will be considered here under these headings. The main problem in the design of curtain walls lies in the framework which holds the panels and this is normally of metal or timber. Three methods of framing may be adopted in the construction of metal curtain walling based on (i) the patent glazing principle and using patent glazing sections,[46] (ii) metal window fabrication technique, (iii) extruded aluminium box mullions and transoms (figure 3.40 A), many forms of which are developments of the normal patent glazing bar. Patent glazing, with its carefully arranged drainage channels and weep holes, and the metal window technique of 'factory cladding' were both used long before the term 'curtain wall' was introduced.

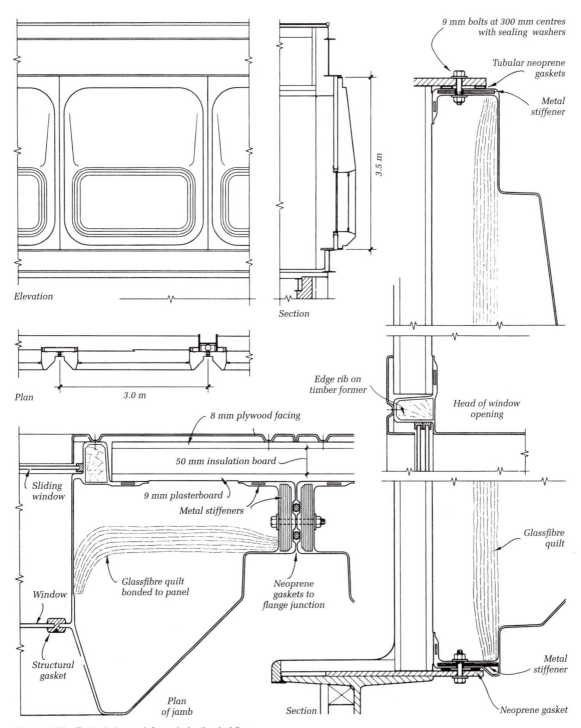

9 mm bolts at 300 mm centres
with sealing washers

Tubular neoprene
gaskets

Metal
stiffener

3.5 m

Elevation

Section

Plan

3.0 m

8 mm plywood facing

Edge rib on
timber former

Head of window
opening

50 mm insulation board

Sliding
window

9 mm plasterboard

Metal stiffeners

Glassfibre
quilt

Glassfibre quilt
bonded to panel

Neoprene
gaskets to
flange junction

Window

Structural
gasket

Metal
stiffener

Plan
of jamb

Section

Neoprene gasket

Figure 3.38   Typical glass reinforced plastic cladding

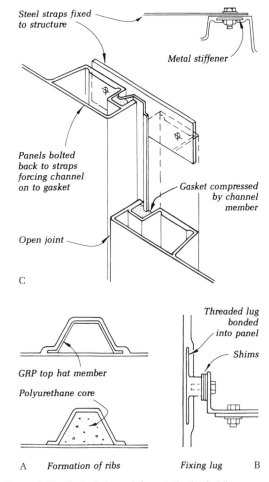

Figure 3.39 Typical glass reinforced plastic cladding

In the majority of proprietary systems of curtain walling, the main members are vertical and are fixed to and span between the floor slabs or beams. Horizontal transoms are fitted between these verticals (B). A method of fixing to the structure in which the principal members are horizontal, thus permitting the employment of normal weathered jointing technique, uses 'stub' columns to support the framing (C). These form a projection on the inside of the building unless absorbed in the cavity between the outer panel and a back-up wall, should this be required by fire regulations.

The members may also be prefabricated into large frames which are hoisted into position on site and fixed back to the structure. It is possible to incorporate into these frames before transport to site the infilling panels, windows and other glazing required in the curtain wall.

Strength and stability Curtain walling, like other claddings, carries only its own weight between supports, but it must be capable of resisting wind forces and of transmitting them to the structure. Wind loads increase in severity as the height above ground and the degree of exposure increase and the members of the framing and their fixings must be designed accordingly. Fixings should be of stainless steel or non-ferrous metal (see page 107) and so designed that should one fail the remaining fixings are capable of taking all the loading on the walling. This provides a margin of safety and prevents progressive failure of a number of fixings.

Fixing devices must be capable of adjustment in any direction to provide for inaccuracies in the structural surfaces to which the framing is attached. Cast-in anchor channels are commonly used in concrete frames to provide the horizontal adjustment. Fixing to steel frames is to plates welded to the steelwork at the required fixing points. Bolt holes should be slotted and packing pieces or shims used to provide for movement and adjustment. Plastic washers should be interposed between adjacent surfaces to allow adequate tension in the bolts combined with sufficient reduction in friction to permit differential movement (figure 3.41 A, B, D, E).

Movement in curtain walling Thermal movement of all parts will occur. In order to maintain dimensional stability this movement must be limited or be allowed to take place freely. Differential movement is likely to occur between (i) the framing and the structure, (ii) the vertical and horizontal members of the framing, (iii) the framing and the infilling panels, or (iv) the inner and outer surfaces of composite panels.

Differential movement between framing and structure is due to the fact that the curtain walling is more exposed to varying external influences than the structure which it protects. The latter, in addition, may achieve a fairly constant temperature due to its higher thermal capacity and to internal heating and air conditioning. Movement between the framing and the structure and between the members of the framing system is provided for in the detailed design of joints and fixing devices.

Methods of attachment designed to allow for differential movement of the structure and framing are shown in figure 3.41 A to D. To prevent the accumulation of thermal movement [47] in mullions extending over several storeys, each section is jointed to the next in a manner that allows each to move without influencing the other by providing space for expansion between the mullion ends and by placing spigots between the mullion sections. Mullions are fixed on one side only of the joint, so permitting the other to slide freely over it (figure 3.42 A, C). This provides also for differential movement due to elastic

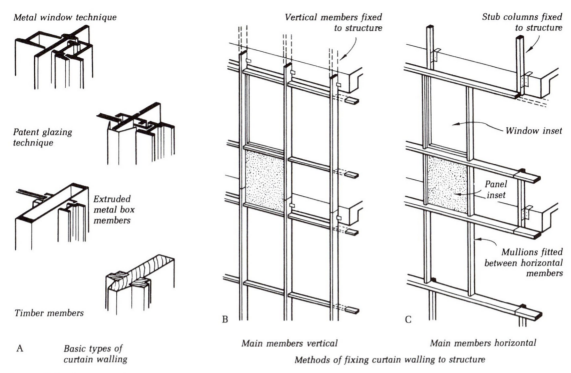

Metal window technique

Patent glazing technique

Extruded metal box members

Timber members

A   Basic types of curtain walling

Vertical members fixed to structure

B

Main members vertical

Stub columns fixed to structure

Window inset

Panel inset

Mullions fitted between horizontal members

C

Main members horizontal

Methods of fixing curtain walling to structure

Figure 3.40 Curtain walling

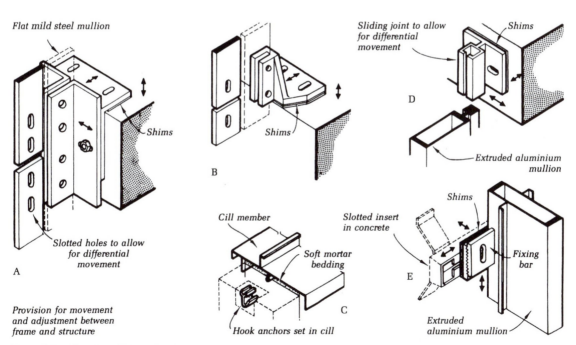

Flat mild steel mullion

Shims

Slotted holes to allow for differential movement

A

Provision for movement and adjustment between frame and structure

Shims

B

Cill member

Soft mortar bedding

Hook anchors set in cill

C

Sliding joint to allow for differential movement

Shims

D

Extruded aluminium mullion

Slotted insert in concrete

Shims

Fixing bar

E

Extruded aluminium mullion

Figure 3.41 Curtain walling – movement

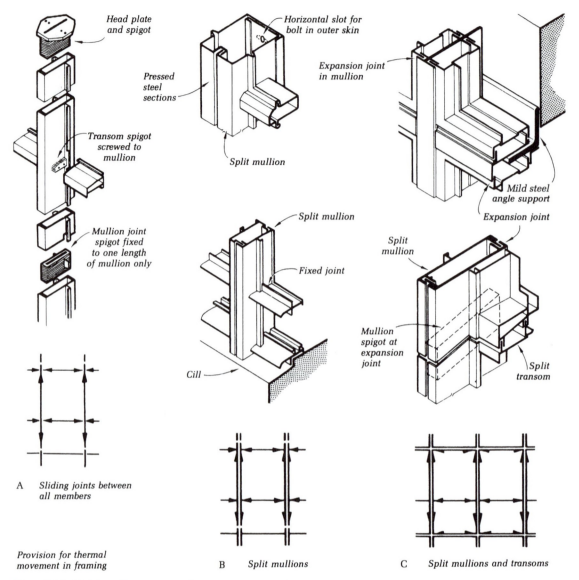

Head plate and spigot

Pressed steel sections

Transom spigot screwed to mullion

Mullion joint spigot fixed to one length of mullion only

A Sliding joints between all members

Provision for thermal movement in framing

Horizontal slot for bolt in outer skin

Split mullion

Split mullion

Fixed joint

Cill

B Split mullions

Expansion joint in mullion

Mild steel angle support

Expansion joint

Split mullion

Mullion spigot at expansion joint

Split transom

C Split mullions and transoms

Figure 3.42 Curtain walling – movement

deformation of the structural frame and for the creep which occurs in a concrete frame. Movement between mullions and horizontal transoms is allowed for by sufficient tolerance at the joints or by the use of split members. A typical spigotted joint in which the space for expansion would be filled with a neoprene gasket or sealant is shown in figure 3.42 A. Details of typical split member construction are shown at B and C. In B the mullions only are split to allow for horizontal movement. C shows two examples of the split mullion and transom technique which permits

erection of the walling in the large pre-assembled units referred to on page 122.

Differential movement between infilling panels and framing is accommodated by sufficient tolerance to permit free movement at the edges of the panels, which necessitates either flexible weatherproofing at the joints or drainage of the joints.

Glazed panels must be detailed with care. Contraction of the frame rather than expansion of the glass is likely to be the cause of breakage. Clear glass should have an edge clearance of 3 mm for widths up to 760 mm and 5 mm for widths over 760 mm. Opaque and heat absorbing glass should have 6 mm clearance, irrespective of its dimensions. Only two setting blocks (dense plastic or hardwood) should be used to obtain the correct bottom clearance and up to 3 mm face clearance should be allowed between glass and frame and glass and bead. Internal stresses may be set up by differential expansion due to differences in temperature at the edges and centre of a glass panel (particularly in coloured, opaque or heat-resisting glass) caused by shading of the edges by frame or beads. This can lead to breakage of the glass. To prevent this it is recommended that the maximum edge cover should be not more than 10 mm. In this connection it should be noted that edges of the glass must be smooth and free from shelling, chipping or grazing. This is a particular problem with wired glass, which should not be used in coloured form and only in transparent or translucent forms if well ventilated at the rear to assist in cooling (see figure 3.44 E). The temperature difference between the edges and centre of a glass panel will be less if the frame is dark in colour since this will attain higher temperatures than lighter or polished materials of greater reflectivity. The technique of bonding the glass to the frame by silicone sealant, as described below, reduces the possibility of such thermal breakage by exposing the whole area of glass to the sun.

Weather resistance   The problem of providing weather tightness in curtain walling is aggravated by the use of impervious infilling materials such as glass, metal or plastic. Unlike natural materials such as stone or brick, which are partially absorbent, these materials are unable to take up moisture. Under wind pressure a large volume of water running down the face of the wall will, therefore, attempt to enter through the joints, since these are the only potential points of entry. Such impervious facings, therefore, require joints which, while providing sufficient tolerance for movement, still remain weatherproof. Modern jointing methods have evolved from the basic techniques of the 'rebate and cover strip' and the 'drained patent glazing bar' (see figure 3.43 A, B, D). The former relies on the principle of breaking down wind pressure by a joint cover and preventing rain penetration by bedding the panel in a suitable mastic composition. The latter relies on adequate drainage, through the interstices of the glazing bar, of a limited amount of water which may enter through the capping or fixing clip.

The methods of weatherproofing joints in curtain walling need to be flexible both in effect and in application. There are three categories of non-rigid jointing: sealants, gaskets[48] and metal cover strips or beads. A comprehensive joint design may incorporate more than one of these features.

Sealants   Oil-based sealants require to be protected from dust, sunlight and air as far as possible (figure 3.43 B, C) and the joints should allow for replacement of the mastic without dismantling the cladding. Absorbent panels should be sealed to prevent absorption of the oil base. Other types of sealant are chemically inert and are not affected by acids, alkalis, vegetable oils or ultra-violet light. They may be used between absorbent surfaces without staining, since they do not contain oil.

Silicone sealants have strong adhesive properties and their use produces a strong seal with high shear and tensile strength; they are resistant to ultra-violet rays as well as to the weather. Their use permits glass (and other materials) to be bonded directly to the face of curtain wall framing without a cover strip of any form, resulting in a glass to glass joint to produce a completely flush façade (figure 3.44 A). The butt joints may be flush or recessed and are sealed with silicone. Most types of glass can be bonded in this way either as single or double glazing; the use of reflective, solar control or screen printed glasses hides the bond area round the edges of the panels. The panels may be bonded on all four edges in which case the adhesion of the sealant is the sole means of retaining the glass on the frame. Where some mechanical restraint is required the glass is bonded on two opposite edges only, usually the vertical edges with positive support and restraint given by the horizontal transom members. Many proprietary forms of silicone bonded glazing systems incorporate thin fins projecting into the joints to provide a degree of positive support as shown in figure 3.44 B, D and E. As much work as possible is carried out under controlled conditions in the factory where, if required, the panels can be assembled into larger units on a light carrier frame and fixed on site to the main framing (figure 3.44 D, E). Alternatively, the panels may be bonded directly to the parts of split mullions and transoms for fixing direct to the structure (C). The reduction or elimination of exposed metal framing reduces the thermal bridge effect and the exposure of the whole surface of the glass to the sun reduces the possibility of breakage by thermal movement (see page 122).

Metal and thin stone panels, such as granite or marble (see page 112), may also be fixed in this manner using the same edge details as for glass, although some degree of mechanical support is usually provided for stone panels, by such means as concealed metal dowels, for example, or by the use of two-side bonding. Any applied coatings on framing or panels must be such that they adhere to the base material at the bond point as well as adhering to the sealant.

Gaskets   These are specially shaped solid or hollow synthetic rubber strips which grip the panel and establish a seal by their close fit between the panel and the adjacent

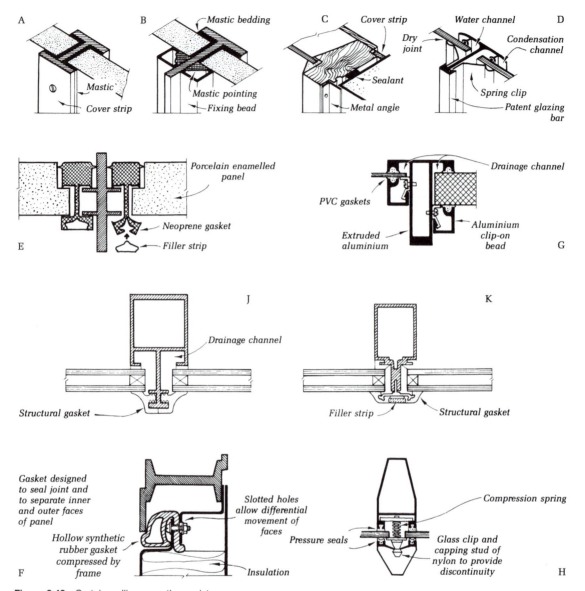

Figure 3.43 Curtain walling – weather resistance

framing, which is usually specially shaped to accommodate a particular section. As explained in Part 1, chapter 2, gaskets may be forced into contact with adjacent surfaces by the use of filler strips inserted into shaped grooves with special tools (figure 3.43 E), by compressing and deforming a hollow section (F) or by compressing a solid gasket by tightening a cover moulding against it (G, H).

Some curtain wall systems use gaskets to form a face-sealed joint, based on the principle of the cover strip, in which the supporting frame and its junctions with glazing and panelling are completely protected from the weather

(figure 3.43 J, K). The gaskets for each panel are mitred in the factory to form moulded corners and when inserted on site in the front face of the mullions and transoms press tightly against each other to form a weathertight junction; in some systems the junction between the two gaskets is formed not by contact but by a filler strip inserted between the two as shown in K. These applications of gaskets reduce the thermal bridge effect by eliminating exposed metal framing.

Suitable materials are synthetic rubbers such as neoprene, ethylene propylene diene monomer (EPDM) and

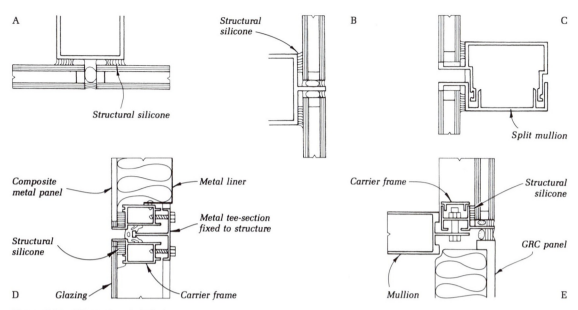

Figure 3.44 Silicone bonded glazing

butyl (cured). Plastics such as PVC should only be used where they are protected from direct exposure and weathering or are easily renewable. Silicone is now widely used as it has proved to be physically stable and durable and can be obtained in various colours. However, the use of silicone for structural gaskets, as in face-sealed joints, where exposed to strong wind pressures on high buildings, may be limited since they may not have adequate strength. Where silicone, neoprene and EPDM gaskets are to surround a panel, the corners are usually preformed in the factory as indicated above. PVC can be fused together by cutting with a hot knife across the mitre.

Cover strips In using metal cover strips for jointing the panel is held against a flange by a sprung clip bead or capping (figures 3.43 G and 3.45 E). Any water entering through the joint between the panel and retaining clip is drained downwards through the front of the mullion. Structural gaskets can be used to form the same type of joint (figure 3.43 J, K).

Thermal insulation and condensation The framing system and the spandrel panels must achieve a satisfactory level of overall thermal insulation. The thermal insulation may be provided separately from the curtain wall in which case it will be supported on the structure, or it may be incorporated in the curtain wall itself in the panel construction.

A wide variety of composite panels is available with differing face materials and forms of insulation (see, for example, under *Metal*, *GRC* and *GRP* claddings). Where glass is used for the spandrel panels double glazing is now commonly adopted, as for the window glazing. Whether single or double glazing is used insulation material may be bonded directly to the inner face of the glass or be contained in a thin gauge aluminium tray, the edges of which are bonded to the glass, in which case the tray acts as a vapour control layer. This technique can, of course, be applied to metal infilling panels (figure 3.44 D).

The framing system itself also permits heat loss, since the members are in direct contact with inside and outside air. This leads to a flow of heat across the members which act as 'thermal bridges' and results in condensation on the inner surfaces. This thermal bridge effect can be reduced by the introduction of 'thermal breaks' in the form of either (i) low conductivity materials incorporated in the section of the framing members (figure 3.45 A) or (ii) physical discontinuity between inner and outer parts of the framing members using plastic connectors (figure 3.43 H and 3.45 B) or plastic spacers at the junctions of mullions and transoms, the only points of contact in this case being the fixing screws at intervals securing the outer to inner part of the section. As indicated earlier this effect can also be reduced by the use of silicone bonding of panels to framing and by the use of gaskets to form face-sealed joints.

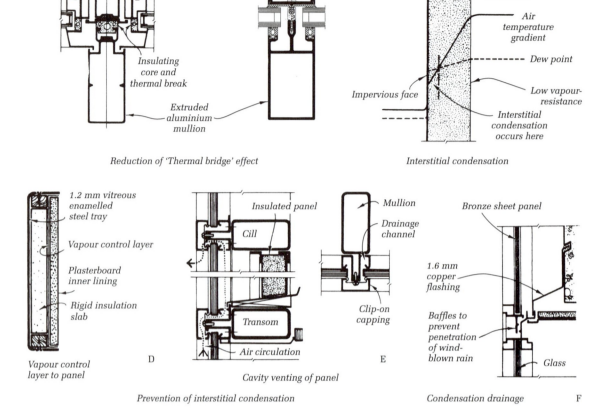

Figure 3.45 Curtain walling – thermal insulation

Interstitial condensation The problem of interstitial condensation needs particular attention in curtain walling systems which are infilled with well insulated panels having impervious external facing materials. In well heated buildings the internal relative humidity is generally higher than that of the external atmosphere. This results in a vapour pressure differential between inside and outside. Since most insulating materials are porous, it is possible for moisture-laden internal air to penetrate the interstices, where condensation will occur if the temperature gradient through the thickness of the panel is steep enough to fall below the dewpoint temperature of the air in the panel. In homogenous porous materials such as brick, the condensate will evaporate to the atmosphere, but where an impervious external skin prevents evaporation the trapped moisture can do damage to the back of such materials as stove enamelled steel, or painted glass (see figure 3.45 C). Furthermore, insulating materials become less effective when their conductivity is increased by dampness.

A vapour control layer of suitable material placed on the room side of an insulated panel will help to reduce the passage of moisture-laden air from the room into the insulation. This may be formed of well lapped and sealed aluminium foil. The metal inner skin of a composite panel will itself be the vapour control layer. It is, however, difficult to maintain a vapour seal at the joints between panel and framing elements, and some residual vapour will penetrate into the panel unless careful perimeter and lap sealing is achieved or hermetically sealed panels are used (figure 3.45 D).

Ventilated cavities Where insulation is separated from the outer facing, either by use of double skin construction (figure 3.45 E) or the use of a back-up wall (figure 3.27) providing insulation as part of the building fabric, condensation on the inner face of an impervious external panel can be reduced if the cavity is vented as indicated in figure 3.45 E. Any slight condensate which may form can drain away through the ventilation holes. Such a ventilated system as this may be made to minimise the entry of water through the joints as in normal rainscreen construction provided there are adequate air seals at all points between the interior of the building and the curtain wall cavity and provided the air inlets are not less than 6 mm wide to avoid blockage by water during rainy periods and are sufficient in area to permit the air pressure in the cavity rapidly to equalise with that of the external air. As in rainscreen construction the cavity must be compartmented to prevent high pressure air being dissipated into areas of low pressure within the cavity: this is automatically achieved if there are no unsealed joints between transoms and mullions round the panels. (See also the note to figure 3.33.)

In high buildings particularly, on the faces of which upward wind currents can develop, any drainage openings should be detailed to prevent rain being blown up into the cavity as shown at (F).

Sound insulation Sound insulation against airborne noise can only be achieved by providing walls of adequate mass or by discontinuity in the construction. The panels generally available for use in curtain wall systems are usually very light and discontinuity is difficult to achieve. The back-up wall required by fire regulations can contribute mass to a panel wall and, if discontinuity can be achieved between this and the external facing element, reasonable sound reduction can be expected.

The problem of structure-borne sounds is increased by the use of a framing system which has elements common to a number of rooms. In theory this can be alleviated by introducing discontinuity at the junctions of framing members and by cushioning the anchorages, although few proprietary systems incorporate such devices.

Fire resistance In addition to the previous references to the flue effect that an air space can have behind claddings and other attached walling systems, further reference is made in chapter 9 to the problem of fire resistance in curtain walling.

Materials of construction Aluminium, used for framing elements, is favoured by many manufacturers for panel infilling because of its lightness, moderate cost and ease of forming. Due to the high coefficient of thermal expansion (see table 4.1, page 188) large thermal movements will occur which must be allowed for in detailing. Steel anchorages should be given a coat of protective paint or be isolated by plastic or fibre washers from the aluminium. This is to prevent the effect of electrolytic action between the dissimilar metals and to provide an intermediary for differential movement.

Mild steel, either hot-rolled or cold-formed, may be used for framing elements but is mostly used in sheet form for panels. Despite the disadvantages of potential corrosion, mild steel is relatively inexpensive and strong. Unlike carbon steel, stainless steel is corrosion resistant,

although having a higher coefficient of expansion. The material is expensive and narrow strips are cheaper than wider and squarer sheets, so that it is cheaper to build up facings to panels from a number of interlocking strips. The material may be used for framing, or more usually for pressed cover sections to aluminium cores.

Bronze is expensive but has excellent weathering qualities. It is used for panels and pressed cover sections. As with steel, contact between bronze and aluminium must be avoided.

Where building regulations permit, timber may be used for framing and panelling. The material is cost effective and has the advantages of being non-corrosive, easily shaped and of low thermal transmittance. Careful detailing is necessary to allow for moisture movement by including sufficient tolerances between panels and framing. Flexibility is achieved by loose tongueing and interlocking sections (figure 3.47 D).

If correctly used glass is an excellent panel material since it is comparatively inexpensive, weather resistant, corrosion resistant and aesthetically acceptable. It may be used for this purpose either as single or double glazing, although double glazing is now virtually standard installation practice for all new and refurbishment work in the UK. Coloured wired glass should not be used since it is susceptible to thermal breakage. Transparent or translucent glass placed in front of a coloured surface with a cavity between should be removable to permit dust and condensation staining to be cleaned off the rear surface.

Glass block walling Glass blocks may be used for non-loadbearing external cladding and for infill panels. These are built with the same types of block and in the same way as the partitions described in Part 1, section 5.8.2, with mortar joints and the same provision for movement at head and jambs but, because of the greater lateral pressures caused by wind, it is necessary to make provision for increased lateral restraint.

Panel sizes are limited to 14 m^2 or, if intermediate steel stiffeners are incorporated, up to 23 m^2, with a maximum width of 8 m and height of 6 m. Panel reinforcement of two linked 6 mm steel wires is placed in every second bed course.

Increased edge restraint is provided by perforated steel strip anchor straps fixed by bolts or self-tapping screws to the jambs and head, either directly to the structure or to a metal frame fixed prior to block laying. These are placed in the same horizontal joints as the wire reinforcement and at the same centres at the head, penetrating at least one and a half blocks into the blockwork. Expansion joints not more than 4–6 m apart should be placed in long lengths of wall.

Curved walls may also be built up in the same way. It is considered advisable to separate a continuous run of curved wall from any flat area of wall at their junction by an expansion joint or support and at the junctions of any converse curves in the wall.

These walls, expecially those of large extent, are usually designed and detailed by the specialist suppliers of glass blocks.

Structural glazing   This is a method of using glass sheets to produce a façade of glass uninterrupted by framing members, as does silicone bonding. Unlike the latter, however, it does not make use of the normal curtain wall framework to support the glass, the self-weight of which, together with the wind loading, is carried by itself. The glazing may be suspended from the building structure or be supported at the base.

In the first method each pane is suspended at the four corners (figure 3.46 A) and, if sufficiently large, at midpoints along the edges, by bolts engaging with brackets fixed directly to the structural frame of the building (B) or, alternatively, to internal vertical glass fins set at right angles to the glazing or to a subsidiary supporting structure of light metal framing behind the glazing, through both of which the façade loads are transferred to the main structure. The supporting brackets may incorporate all four fixings at the corner junctions in one component and will vary in form, as will a subsidiary supporting structure, according to the architect's design. The bolt heads are countersunk into the glass to give a flush external face and

resilient bushes and washers are used to prevent the contact of glass with metal and to permit movements (C). When the glazing is supported at the base lateral restraint will be provided by means similar to those just described, the fixings in this case being designed to permit differential movement between façade and structure.

Toughened glass must be used, 10 mm or more in thickness, in either single or double glazing. The maximum size of sheet used depends on the size of the available toughening oven. Most types of toughened glass may be used and the edges are butt jointed and flush sealed with silicone sealant.[49]

A full description of methods of structural glazing will be found in *MBS: External Components*, chapter 6.

Overcladding   This is the term used for the process of applying a facing or cladding to an existing building needing renovation to provide, for example, improved weather resistance and appearance to its deteriorating façades and, sometimes, improved thermal insulation to the walls. It has developed over recent years and makes use of many of the forms of claddings described in this chapter. Before any overcladding is designed or installed certain aspects require careful consideration, such as the possibility of the spread of fire through the cavity behind the overcladding and into the building through the windows and the consequent need of cavity barriers, the possibility of deterioration continuing after the walls have been overclad and the ability of the fabric to provide secure fixing for the overcladding.[50]

3.11.3 Infilling panels

Most materials used in the forms of cladding described in the previous section may be employed in the construction of infilling panels. The technical problems of weather exclusion, thermal and sound insulation, condensation and fire resistance associated with infilling panels are similar to those encountered in curtain walling.

Brick and block infilling panels are considered under 'Panel walls' on page 91 and glass block panels on page 129. Lightweight infilling panels present the same problems of edge jointing as heavy panels but are even more liable to thermal or moisture movement and need careful design at the junction with the structure, particularly at head and jambs.

Metal-framed infilling panels   Complete storey-height infillings, sometimes referred to as 'window walls', are constructed in a similar manner to curtain walling systems, the vertical framing elements being so fixed at the base and head as to allow for thermal movement (figure 3.42 A). Methods of providing suitable fixing at jambs to allow

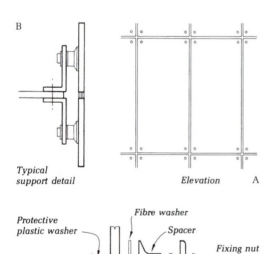

Typical support detail

Elevation   A

Protective plastic washer

Fibre washer

Spacer

Fixing nut

Toughened glass

Washer

C

Detail of bolt fixing

B

Figure 3.46   Structural glazing

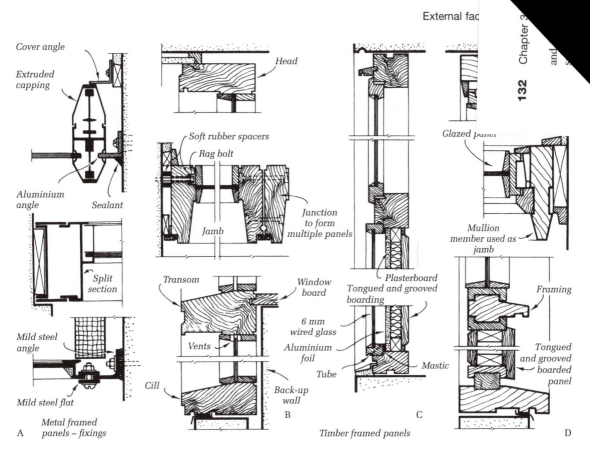

Cover angle

Extruded capping

Aluminium angle

Sealant

Split section

Mild steel angle

Mild steel flat

Metal framed panels – fixings

A

Head

Soft rubber spacers

Rag bolt

Junction to form multiple panels

Jamb

Transom

Vents

Cill

Window board

6 mm wired glass

Aluminium foil

Tube

Back-up wall

Plasterboard
Tongued and grooved boarding

Mastic

B

C

Timber framed panels

Glazed panel

Mullion member used as jamb

Framing

Tongued and grooved boarded panel

D

Figure 3.47 Infilling panels

for movement are illustrated in figure 3.47 A. Spandrel height panels may be screwed or clipped to metal angle framing fixed to the floor slabs or between columns.

Concrete infilling panels This term is usually applied to the forms of horizontal cladding which span between columns and which are described on page 112.

Timber-framed infilling panels It has been common practice for many years to infill between brick walls with timber-framed walling based on traditional stud walling technique, the timber being cut and assembled on site.

Trends towards prefabrication have brought about the introduction of pre-assembled frames which can be erected as units and finished on site or are already finished before erection. It is common practice to construct the panel with a frame of rebated sections, similar to a large timber window frame with all necessary mullions and transoms, and to infill this with glazing, openable sashes and fixed panels in the manner of curtain wall technique. The junctions of head, cill and jamb with the structure are

usually designed to allow sufficient erection tolerances. Typical details of this type of infilling panel are shown in figure 3.47 B and C.

Proprietary curtain wall systems provide a further method of timber infill panel walling. These consist of various sections forming the main frame into which are fitted various types of panels. A typical example is shown in D.

In all methods of installing panels or frames (including window frames between structural elements) it is an advantage to form a rebate to reduce the risk of wind blown rain entering the joint directly. This may be provided by shaping the basic structural material such as brick or concrete, or by forming the rebate with a facing element as shown in some of the illustrations in figure 3.10.

3.11.4 Façade engineering

With increasing efforts to produce a satisfactory internal environment by passive measures rather than by mechanical means the building façade is being made a vehicle of such measures. These are designed to reduce solar glare heating, to enhance the lighting of the interior and sometimes to provide controlled natural ventilation, and make use of external louvres, either fixed or tiltable, venetian and roller blinds, sometimes controlled by photocells, and external fixed horizontal light shelves reflecting daylight through the façade glazing onto the ceilings and into the rooms inside. Blinds can be accommodated either outside or inside a curtain wall or within a space between two glazed skins of the wall sufficiently wide to take them. These measures work together with passive measures adopted internally, such as stack induced natural ventilation and adequate cross-ventilation paths and the cooling effect of the thermal mass of exposed floor soffits, to assist in reducing high air temperatures in summer. Taken altogether these can permit a reduction in or the elimination of mechanical ventilation and air conditioning.

Photovoltaic panels can also be incorporated in the façade to contribute to the energy requirements of the building.

A double-skinned façade with a wide enough cavity can act as a buffer zone between external and internal environments. With controlled dampers at top and bottom the space forms a thermal flue in summer when the dampers are opened to let external air rise through it to remove heat gains; in winter, with the dampers closed, it forms a thermal buffer. The cavity can accommodate any required louvres, shutters or blinds and, if of sufficient width, open mesh maintenance walkways at each floor level. The skins can be of curtain walling or structural glazing or a combination of both.

The design of façades on these lines can be a complex process involving computer modeling, and a new discipline has emerged dealing with this, now termed 'Façade engineering', with its own experts who can give advice to architects at the design stage.

3.12 Movement control

The nature of movements in buildings generally is referred to in the next chapter in section 4.9, where references are given to more detailed discussions of the subject elsewhere. Only the practical steps necessary to prevent damage to walls through movement will be discussed at this point.

3.12.1 Settlement movement

The methods used to minimise differential settlement have been described in the previous chapter. In masonry walls, provided the mortar is weak, slight movements can be taken up in the joints without damage to the wall and in normal circumstances no special provision is made. Monolithic concrete walls are more sensitive to move-

ment, and when these are not reinforced, particularly with no-fines concrete, reinforced foundations may be required. Tall buildings may require a cellular raft to provide a foundation of sufficient stiffness.

3.12.2 Moisture movement

Most walling materials absorb water and in doing so they expand; on drying out they contract. This is termed moisture movement, and it is usually reversible except in concrete, mortars and plasters. These have an initial drying shrinkage on first drying out after setting which exceeds any future reversible movement. The magnitude of moisture movement varies with the material and in some it is so great as to necessitate special measures in construction. Table 3.5 gives a broad grouping of materials relative to their moisture movement.[51]

Control of moisture movement Special precautions should be taken with walls built of sand-lime or concrete bricks or concrete blocks and these are primarily the use of weak mortars and the provision of vertical movement joints to control cracking. These are referred to under *Blockwork* in Part 1, chapter 5, where the advantage of a weak mortar is described and the need to break long walls by control joints into panels with lengths not exceeding one-and-a-half times to twice the height is stressed (see figure 3.48 A).

Table 3.5 Degrees of moisture movement

Materials having very small moisture movement	Well-fired bricks and clay goods.* Igneous rocks. Most limestones. Calcium sulphate plaster
Materials with small moisture movement	Some concrete and sand-lime bricks. Some sandstones
Materials with considerable moisture movement calling for precautions in design and use	Well-proportioned ballast concretes. Cement and lime mortars and renderings. Some concrete and sand-lime bricks. Lightweight concrete products. Some sandstones
Slab and sheet materials with large moisture movement calling for special technique of treatment at joints and surrounds	Wood-cement materials. Fibrous slabs and wallboards. Fibre cement sheeting. Plywoods and timber generally

* *Note*: Fired clay products may show an initial expansion on wetting that is irreversible. Walls built with 'kiln-fresh' bricks may expand twice as much as similar walls built with bricks that had been exposed to the weather for a fortnight. A slow moisture expansion, greater than that shown by the standard tests, may also take place over a period of years.

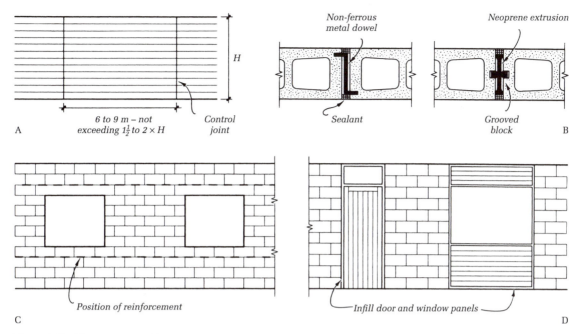

A   6 to 9 m – not   Control
exceeding 1½ to 2 × H   joint

Non-ferrous
metal dowel   Neoprene extrusion

Sealant   Grooved
block   B

C   Position of reinforcement

Infill door and window panels   D

Figure 3.48   Movement control in concrete masonry

Control joints are essential in long imperforate walls and should be provided at intervals of 6 m. Such joints may be vertical or may follow the bonding lines in toothed fashion and where in external walls must be made weatherproof by suitable resilient filling. Where stability of the panels requires them non-ferrous metal dowels are placed across the joint in every other course with provision made to ensure that sliding can take place as the panels on each side move, such as covering them with greased paper. Alternatively, grooved blocks with a neoprene extrusion may be used (figure 3.48 B).

The stress concentration which occurs in the narrow sections of wall above and below window openings can be distributed through the wall by steel reinforcement in bond beams in the lintel and cill courses (see Part 1, figure 5.30) or by masonry reinforcement in the bed joints immediately above and below these courses respectively (figure 3.48 C). Such reinforcement must not be carried across any control joints in the walling, nor should any applied finishes on the wall faces. As suggested in Part 1 these narrow sections of wall can be avoided by designing doors and windows within storey-height infill panels as shown in D, the vertical joints on each side constituting control joints between the infill panel and the imperforate blockwork panels on each side.

Shrinkage in structural concrete is normally restrained by reinforcement (see page 88). When shrinkage or contraction joints are considered desirable in a concrete wall

to control random cracking and the joint must be water resistant, a PVC water bar or stop is incorporated in a similar manner to that shown in expansion joint B in figure 4.37. At a contraction joint, however, the reinforcement would carry across the joint line and no gap would be formed in construction, although precautions would be taken to prevent bonding of the concrete on each side of the joint. Those materials with large moisture movements are generally most suited to internal use and some slab and sheet materials in this category are commonly used in the construction of partitions. Methods of providing for movement in these are suggested in Part 1, table 5.3 on page 112.

3.12.3 Thermal movement

The cause and magnitude of thermal movement is described in the next chapter (section 4.9.3).

Brick walls up to about 30 m in length which are free to expand can usually safely accommodate thermal movements in the normal range of temperatures. When a wall exceeds this length, or where movement is restrained in some way such as shown in figure 4.36 F, the effect of stresses set up by thermal movement should be examined. Vertical expansion joints should be provided to break the wall into sections about 12 m in length and to provide space for an unrestrained movement in each section of about 16 mm in 12 m. Such joints in the external walls of buildings should never be more than 15 m apart in order to avoid cracking due to thermal contraction and they must be made weatherproof by the use of non-extruding resilient filling. Joints in boundary walls need not necessarily be filled in.

Expansion joints in concrete block walls should be about 6 m apart. Thermal movement in partitions is usually quite small. Only in the case of long partitions near boiler rooms or cold stores need consideration be given to this type of movement, although a joint must be formed in a partition when the latter crosses an expansion joint carried through the floors of a building. Methods of detailing this are shown in figure 4.37 F and G. In the former the wall on one side of the joint is built up first and one wing of the strip fixed to it by screws inserted and driven through the slots formed by the cutout lugs in the opposite wing. The latter is then secured by the lugs which are bedded in as the wall is built up.

Expansion joints not more than 15 m apart or where there is an abrupt change in plan should normally be provided in dense concrete walls. These may be formed as shown in figure 4.37 A, B, or with crimped copper strips *K* as in H and J. Applied finishes may require special treatment as at C, D.

Reference is made on page 84 to the use of vertical expansion joints to limit differential thermal movement between the inner and outer leaves of large areas of cavity walling, and methods of dealing with thermal movement in curtain walling are described on page 122–3.

Unless the roof slab is arranged to slide on the head of a wall a joint should be provided in the wall where one is provided in the roof slab.

In positions where a sliding joint is required as in figure 4.36 F, this can be detailed as shown in figure 4.37 E.

Notes

1 See also the Brick Development Association Handbook to BS 5628 (DG 10) by Haseltine and Moore.

2 See page 136.

3 For a more comprehensive graph for brickwork and similar graphs for various types of blockwork see BS 5628-1: *Code of Practice for the Use of Masonry. Structural Use of Unreinforced Masonry.*

4 Much of this data was resourced from *Principles of Modern Building*, volume 1, 3rd edition, The Stationery Office.

5 See also BS ENs 772-3 and 772-7: *Methods of Test for Masonry Units.*

6 Approved Document A, section 2C, which applies only to residential buildings up to three storeys and to certain other small buildings.

7 A column is defined as an isolated member of which the width does not exceed four times its thickness. The term 'pier' is defined as a thickened section of a wall placed at intervals along its length (see figure 3.5 F, G, H).

8 Defined as 'support which will restrict movement in the direction of the thickness of the wall, or, in relation to a column, movement in the direction of its thickness or width'.

9 In buildings up to six storeys, above which they should be calculated.

10 The Building Regulations limit the use of this type of anchor to houses of not more than two storeys.

11 Compare with propped cantilever frame, page 145.

12 See also *Brick Diaphragm Walls in Tall Single-storey Buildings* (DG 11) by W G Curtin, G Shaw, J K Beck and W A Bray, Brick Development Association.

13 See also *Design of Brick Fin Walls in Tall Single-storey Buildings* by W G Curtin, G Shaw, J K Beck and W A Bray, Brick Development Association.

14 See *MBS: Materials*, chapter 8. Methods of thermal insulation described in Part 1, section 5.3.3, for solid brick walls may be used for concrete walls, or permanent shuttering of suitable slab materials, such as mineral wool batts or expanded polystyrene, may be used in the construction of the wall.

15 For cellular foundations see page 54.

16 For examples of the use of no-fines concrete see BRE Reports 153, 156 160, 191 and 275.

17 See page 176.

18 See *Overturning*, page 75.

19 See page 176 for reference to the omission of reinforcement from concrete cross walls.

20 See *MBS: Introduction to Building*, chapter 5, section 5.8, figure 5.16, with reference to shear panels.

21 See page 74.

22 See pages 77 and 78 (figure 3.4).

23 For a more detailed consideration see *Brick Development Association* publication (DG18) *Brick Cladding to Steel Framed Buildings* by Bradshaw, Buckton and Tonge.

24 See *Fire Precautions in New Hospitals (Firecode)*, NHS Estates, TSO, and for definition of limited combustibility see note 20 on page 341.

25 For the general requirements regarding these walls see chapter 9, section 9.4.1.

26 The rights and obligations of adjoining owners in connection with any building work carried out on a party wall are laid down in the Party Wall etc. Act, 1996, TSO.

27 It should be noted that the Building Regulations AD A, section 2C, limit any difference in height of the ground levels on either side of a normal wall to four times the thickness of the wall.

28 This type of wall is quite different from the diaphragm load-bearing wall described on page 85.

29 This material goes under the name of *Voltex* and is produced by CETCO Europe Limited of Birkenhead.

30 This material is marketed as *Flexiseal* and is produced by RIW Ltd of Bracknell.

31 Natural Rock Asphalt and some asphalt mixtures have been tested satisfactorily to 130 kN/m².

32 For further consideration of the main methods of preventing entry of water into basements see BS 8102: *Code of Practice for Protection of Structures Against Water from the Ground.*

33 For other causes see BRE Digest 245, *Rising Damp in Walls: Diagnosis and Treatment.*

34 These tubes are marketed by Hydrotek-Wallguard Ltd of Chipping Ongar.

35 BRE Digest 245: *Rising Damp in Walls: Diagnosis and Treatment*. BS 6576: *Code of Practice for Diagnosis of Rising Damp in Walls of Buildings and Installation of Chemical dpc's*.

36 See also *MBS: Materials*, page 81.

37 See BRE Information Papers IP 6/97 *External Cladding using Thin Stone*, IP 17/98 *Use of Lightweight Veneer Stone Cladding Panels*, and IP 10/01 *Lightweight Veneer Stone Cladding Panels*.

38 See AD A, section 3, on fixing and support of claddings.

39 BS 5628-3: *Code of Practice for Use of Masonry* gives guidance on suitable mortars. See also table 3.4.

40 For recommendations on fixing see BS 8297: *Code of Practice on the Design and Installation of Non-Loadbearing Precast Concrete Cladding*. See also AD A, section 3.

41 See also *Advanced Construction Technology, Part 8.6: Rainscreen Cladding* by R Chudley and R Greeno, Pearson.

42 These include polyvinyl alcohol (PVA), polyacrylic nitı. (PAN) and natural cellulose.

43 See *MBS: Introduction to Building*, chapter 2, section 2.5.3.

44 See *MBS: Materials*, chapter 10.

45 See also BRE Digest 161: *Reinforced Plastics Cladding Panels*.

46 See *MBS: External Components*, section 5.13, on patent glazing.

47 See table 4.1, page 188.

48 See Part 1, *Joints*, chapter 2 and *MBS: Materials*, chapter 16.

49 Fittings for structural glazing are marketed under the trade name of *Planar*, by Pilkington Group Ltd.

50 See BRE Report 93, *Overcladding: External Walls of Large Panel System Dwellings*.

51 For percentage movements in individual materials see *MBS: Materials*, table 1.6.

4 Multi-storey structures

The frame and the loadbearing wall are first considered in relation to high-rise construction, followed by a consideration of the factors relevant to the choice of an appropriate structure. Framed structures generally are examined after which construction with steel frames in hot-rolled and cold-formed steel sections is described, then frames in in situ cast and precast concrete followed by in situ and precast walls. Various methods of prestressing concrete are described, the chapter concluding with a consideration of methods used to control movements in buildings as a whole to prevent damage.

The frame and the loadbearing wall High-rise construction is used for many building types. For some the framed structure is the most suitable form, for others the loadbearing wall. The advantages and disadvantages of these two forms have been described in Part 1.

In a framed structure, the wall being relieved of its loadbearing function, it is possible to fulfil the enclosing functions by forms of construction more suited to the purpose than heavy loadbearing walls, and to provide a structure lighter in weight and often more quickly erected. Such a system is widely used but the point made in Part 1 (page 70, *Loadbearing masonry*) must be stressed here: that in certain circumstances loadbearing wall construction, fulfilling as it does both the loadbearing and enclosing/dividing functions, will prove less expensive in terms of overall cost than a framed structure. This is particularly so in case of individual small-scale and domestic buildings, although in the case of large programmes involving one basic type of small-scale building, permitting bulk ordering of materials and the use of a prefabricated system to reduce site labour, a framed structure may prove more economical than one based on the loadbearing wall. In the case of taller buildings with suitable plan forms, such as

some types of flat blocks, the wall used as a loadbearing element can again be a more economical structure than a frame. This has been more fully discussed in terms of brick and plain concrete walls on pages 76 and 87. See also page 176.

The structural frame of a building is generally a small proportion only of the total building cost and varies on average from 10 to 40 per cent of the total cost.[1] Small economies in the structure will, therefore, usually have a slight effect only on the total cost. Nevertheless, since the cost of the structural frame is often the largest single item in the total cost, it should be kept to a minimum, although it must always be borne in mind that the cheapest structure may not necessarily produce the cheapest building. For example, reduction in the cost of the structure at the expense of increased depth of beams will lead to an increase in the cost of finishings, particularly external claddings, due to the increased overall height of the building. As a further example, in some proprietary forms of prefabricated steel building frames beams of rather wide span are closely spaced and the resulting light load carried by the wide span beams produces beams which are slightly wasteful in steel (see page 138). The extra cost is, however, more than balanced by the economies of standardisation and economy in roof cladding.

The frame, therefore, must not be considered in isolation but in relation to the other elements of the building and as a cost viable comparison with traditional construction practice. A cost comparative analysis applied to different ways of constructing the walls to a simple two-storey office building can be used to emphasise this, and at the same time show the economic advantages of loadbearing wall construction over framed construction for small buildings. An external loadbearing cavity wall can be used with intervals for normal window openings, compared with a structural frame of the same overall dimensions, with solid and glazed infilling panels. Although the cost of the frame will be little more than half that of the loadbearing walls, the frame cannot enclose space. The cost of the infilling panels must be added to the frame costs to make it comparable in functional terms with the wall construction. The total cost of the frame and the infilling panels will be rather more than one-and-a-half times the total cost of the loadbearing walls together with the windows.

However, frame construction could be justified if fully glazed infill panels are required, but if the glazed area could be reduced by replacement with cavity wall panel construction the final result would be quite illogical since the cavity wall itself is capable of carrying the floor and roof loads without the frame. The justification for a framed structure in circumstances such as these rests on the advantages given by a frame in other directions, such as the large areas of glazing which are considered necessary in some building types.

4.1 Choice of appropriate structure

Multi-storey wall construction may be carried out in masonry or in plain or no-fines concrete and the use of these is described in chapter 3. The use of reinforced concrete for this type of construction is described later in this chapter. For multi-storey framed structures both steel and reinforced concrete are used. Although a structural frame must be considered in relation to the other elements of the building it is, nevertheless, necessary to consider it separately in broad terms at the outset of a scheme, as in this chapter, in order to establish a basis from which to begin.

There is no simple formula by means of which the most appropriate structure for a particular building may be selected and because there are so many variable factors affecting choice and economy these can be discussed only in general terms.

4.1.1 Site

The available site, its cost, the nature of the subsoil and planning restrictions on the height of buildings in the particular area must all be considered. When the cost of the land is low and a large area is available, the automatic choice of a multi-storey structure is illogical. Apart from increased construction costs, such things as the extra cost of external walling due to the increased ratio of external wall to floor area in a tall building, with its repercussions on the heating of the building in terms of greater heat loss,[2] make the serious consideration of a low structure essential. But where the cost of land is high, as in central city areas,

a multi-storey structure must invariably be used in order to reduce to a minimum the required area of site to economise on its cost.

The height of a tall structure may be limited by economics or by local planning limitations. Construction costs increase with building height and additional storeys lead to increased column, foundation and wind-bracing costs, so that for each particular project there is an economical height limit at which the structure may be stopped, even though it may be lower than that permitted by local regulations. The economical height may be determined by assessing the return on investment, this being calculated with regard to the area of the building, land costs, the number of floors and the cost per unit volume or area, as well as many other factors. This, of course, necessitates the use of some form of cost analysis and cost planning. Above the economical limit there are diminishing returns on the money invested in the building. If the maximum height should be limited by local planning limitations rather than by the cost of construction, it may be necessary to place the maximum number of storeys within the permitted height and this will have a profound influence upon the choice of a type of floor to give the thinnest possible construction. In commercial buildings the rental value of any extra floors gained in this way must be capitalised before comparing the cost of a selected floor with that of alternative floor systems.

Significance of the soil   The type of soil on which the structure is to rest will influence its design, above as well as below ground. For example, in the case of soil with a low bearing capacity and high compressibility, on which differential settlement might occur, a low structure imposing minimum point loads on the soil through simple foundations would be most suitable, but if the location of the site made it necessary to build high over a small area, expensive foundations such as piles or caissons taken to a firm stratum might be required. In this case it might prove cheaper to increase the span of beams in the frame in order to restrict the number of columns and thus the number of expensive foundations. The relationship between soil, foundations and superstructure is considered in more detail in chapter 2.

4.1.2 Type and use of building

An analysis of the problem on the basis of the site considerations – a balancing of costs and use requirements as they arise from site conditions – will usually give in broad terms an indication of the most suitable kind of structure, and this will lead to a final choice from a few alternative systems within the broad lines suggested. This choice will be influenced by the nature and use of the building and the architect's conception of the finished building.

Plan forms requiring a repeating unit both horizontally and vertically, and in which room widths may be multiples of the structural unit, lead logically to the use of a simple skeleton frame. The intensity and nature of the superimposed loading must be considered so that a suitable type of floor structure can be selected and the most economic span chosen for the particular floor type. When a square structural grid can be adopted, or at least one in which the larger side of each bay does not exceed one-and-a-third times the length of the shorter side, the use of a two-way reinforced floor slab with four-edge support becomes economically possible, as it can result in a thinner and lighter slab when load and span conditions are suitable. Certain building types, such as flats, may be constructed more economically in loadbearing wall construction than as a frame. The standard of natural lighting required from outer walls will influence the height from floor to ceiling according to the depth of the rooms on plan.

These factors will involve the consideration of column centres, wall spacings, storey heights and infillings and claddings and of the implications of providing the building services, especially heating, air-conditioning ducts and electrical trunking. Should the solution suggested by these considerations be at variance with that suggested by site considerations, then a review of both must be made in order to arrive at a final satisfactory solution.

4.1.3 Span and spacing of beams

Span The column spacing in a framed structure, which determines the span of the beams, is perhaps the most important factor influencing the cost of the structural framework. Generally, the cheapest frame will be that with the closest column spacing and, therefore, the lowest dead weight.[3] Apart from frames with exceptionally tall columns, that is in those of normal storey height, a column costs less than a beam. Although in practice the selection of the most economic structure may not always be possible, because of the space requirements of the building, it is desirable to bear in mind this economic factor as a 'yardstick' by means of which can be assessed the cost consequences of adopting a particular solution.

Spacing The spacing of the beams is dependent upon a number of factors:

- The economic span of floor slabs.
- The economic loading of beams.
- The economic loading of columns.

In the case of a building not requiring wide span beams the economic span of the floor slab will be a decisive factor in fixing the spacing of the main floor beams. For multi-storey buildings such as flats, offices and other types

with light and medium loading this will be from 3 to 4.5 m, to give a floor slab in the region of 125 to 150 mm thick. Greater spacing will necessitate the use of a deeper floor of tee, hollow beam or other type of construction. For a detailed consideration of floor construction reference should be made to chapter 5.

In establishing the spacing of beams in a structural frame another important influence, apart from the requirements of planning, is the relationship between load and span of beams. Since both the bending moment and, especially, the deflection of a uniformly loaded beam are more rapidly increased by an increase in span than by an increase in the load per metre run, it is generally uneconomical to carry light loads over long spans. This can be seen from the expressions for bending moment and deflection:

$$\text{BM} = \frac{wl^2}{8} \text{ and } d = \text{constant } (c) \times \frac{wl^4}{EI}$$

The bending moment increases with the square of the span l, and the deflection with fourth power, but both only directly with the load per metre run, w. If the span of a beam is doubled the deflection will be sixteen times as great but if the load per metre run is doubled the deflection will be only twice as great. In other words, an increase in span necessitates a much greater increase in the 'I' value of the beam in order to keep deflection within acceptable limits, compared with that necessitated by a proportionate increase in the load per metre run. This being so, when beam spans must be large it will usually be cheaper to space the beams further apart, particularly when floor loading is light or medium, even though the load per metre run on the beams is thereby increased. The savings in cost due to the smaller number of beams required will be greater than the increase in cost due to the wider floor slabs and the increase in size of the remaining beams and columns due to the greater load they carry. Note that E is known as *Young's modulus of elasticity* (N/mm^2) and I the *second moment of area* (cm^4). See also Part 1, section 3.3.2, *The beam*.

Heavy loading on wide span beams also has economic repercussions on the columns of a frame. Variations in loading cause only relatively small variations in column weights since the decisive factor in most columns is stiffness against buckling, so that where beam spans are great an increase in the unit load on the beams due to spacing them at wider intervals has the effect of imposing greater loads on fewer columns without necessitating a great increase in their weight. This is particularly beneficial where the column heights are great and the greater tendency to buckle results in large column sizes.

It will be appreciated from what has already been said that for a given span of floor slab, which will establish the load per metre run on the supporting beams, there will be an economic span for the beams. For types of buildings with light to medium floor loads and normal concrete floors used at their economic span this will be from 4.25 to 6 m.

When the column centres have been tentatively established, studies of typical bays are made for comparative purposes in terms of construction and costs. An analysis will be made of the various possible framings in structural steel or reinforced concrete or in both, some of which will prove to be more economic than the others. Wherever possible the frame layout should be based on a regular 'grid', a term derived from the pattern of beams at each floor level, this pattern being determined by the setting out of the columns of the frame. The advantages of a regular grid are given in Part 1, page 120.

Transference of loads The imposed loads on building frames are transmitted through successive units or stages to the foundations, that is, for example, from floor slab to beam and from beam to column, and the total cost of the structure is made up of the cost of each successive stage. Generally, to achieve maximum economy the number of stages through which the load is transferred should be kept to a minimum, although since they are interpendent and modifications to one will affect the others, the introduction of an extra stage may in some circumstances be justified. For example, the introduction of secondary beams may result in a saving of so much dead weight in the floor slabs as to make it economically advantageous.

As a general principle loads should not be transferred horizontally if it is possible to carry them axially downwards from the point of application. Thus setbacks on section resulting in columns bearing on beams should be avoided where possible, as this necessitates the use of heavy or deep beams and increases the cost of the structural frame, as does the introduction of wide span beams on lower floors, carrying a number of columns in a closely spaced grid above (see figure 4.4 B, D).

4.1.4 Wind pressure

Another factor to be considered in the selection of the most appropriate structure is the need for the structure to be able to resist wind pressure. This is a most important consideration in the case of very tall buildings as far as both foundations and superstructure are concerned. The recommendations of BS 6399-2: *Loading for Buildings. Code of Practice for Wind Loads* concerning wind pressure on buildings as a whole and upon claddings are described in chapter 3. They apply to framed structures as well as to loadbearing wall structures.

The means adopted in wall structures to provide stiffness against wind forces are described in the section on

reinforced concrete wall construction on page 176 and in the case of masonry structures, on pages 76 and 90.

Action of frame under wind pressure A structural frame may be considered as acting under pressure of wind either as a cantilever in which the floors remain plane or as a portal frame in which the floors bend. In analysing the structure in terms of wind resistance the height-to-width ratio is extremely important in deciding in which of these ways the structure is likely to react. A very tall, slender building will act more closely to a cantilever while a short, squat structure will act as a portal frame. In practice most buildings act partially in both ways.

In whichever way a particular structure may behave it will move under wind pressure, in the same way that a cantilever deflects under load, and in the case of tall buildings particularly, where the movement could be large, this must be limited if it is not to result in cracks in finishes and discomfort to occupants of upper floors. In large structures with heavy, stiff claddings, wind forces are not likely to produce much lateral deflection, but in framed structures with light curtain walls the basic structure required by dead and imposed loads may have to be stiffened in order to limit lateral movement, or side-sway. A maximum permissible deflection at the top may be fixed and the bending moments produced by this may be calculated and the structure designed accordingly. In the United States a standard for maximum deflection of 0.002 times the height of the building has been established as a limit for buildings having a rigid outer skin, and 0.0001 times the height for buildings with curtain walls which provide little rigidity to assist in reducing the deflection. Fluid viscous dampers can be applied to tall buildings to limit deflection due to wind action.

Wind bracing The necessary stiffness or windbracing can be obtained in a number of other ways, used either separately or in combination:

- By the use of deep beams or girders producing very stiff joints with the columns (steel and reinforced concrete).
- By the use of brackets or gussets to produce stiff beam to column joints (steel).
- By the use of diagonal bracing in vertical panels of the frame (steel and precast concrete).
- By constructing solid walls called shear walls in suitable positions or by using stair wells or lift shafts, if these are constructed in monolithic reinforced concrete, running the full height of the building so that they act as stiff, vertical members to which wind loading is transmitted by the floors. Shear walls may be introduced in panels of the frame to fulfil the same stiffening function as diagonal braces (steel and reinforced concrete).

The method of windbracing to be adopted must be considered at the beginning of the design stage because it will have a bearing upon floor thicknesses and floor types, beam depths and column sizes and the introduction or not of stiffening walls.

In multi-storey buildings any tendency to uplift at the foundations will usually be more than counterbalanced by the total weight of the building, which will keep the resultant pressure within the middle third of the base. In extreme cases, however, when buildings are very tall and slender, and especially if they are of considerable length, it may be necessary to resist uplift by means of tension piles or ground anchors incorporated in the foundations.

4.2 Framed structures

4.2.1 Choice of material

The materials most suitable for the construction of multi-storey structural frames are steel and reinforced concrete. Timber, apart from its use in domestic type buildings up to four or five storeys high,[4] and the aluminium alloys, by reason of their particular characteristics, are not suitable for this purpose. A comparison of the characteristics of these four structural materials is given in chapter 8.

The principal factors influencing the choice between steel and concrete are:

● the availability of materials and labour
● cost
● speed of construction
● possibility or otherwise of standardising the sizes of the structural members
● size and nature of site
● fire resistance required
● flexibility of design.

With these in mind the two materials may be compared in order better to understand the consequences of the choice of one or the other.

Site considerations   Both steel and cement are made under factory conditions and both are subject to British Standards. The strength of steel is controlled and established during manufacture, but that of concrete is often dependent upon strict supervision on a building site. The exception will be with ready mixed truck deliveries, but this will still be subject to the potential for contamination during site transportation and placement. On a large contract effective supervision and the use of carefully controlled procedures are likely to lead to better quality concrete than on a small contract. Site conditions, quality of labour and doubt regarding the possibility of obtaining

first class supervision for a particular project might lead to the choice of steel.

Assembly of steel frames   The fabrication of the individual members of a steel frame is carried out 'off-site' in the steel supplier's premises and assembly only is necessary on the site itself. Steel assembly is carried out with bolted and welded connections by skilled labour, quickly and accurately within small tolerances. Completion at an early stage in the building programme permits the laying of floors and roofs and the rapid sealing-in of the whole structure to permit other trades to follow on continuously without interruption through adverse weather. The fixing of lower floors can, of course, be commenced before the frame has been completed up to roof level and while the upper members are being fixed in position. The obstruction of floor space by the vertical props required for in situ cast concrete floors may be avoided by the use of telescopic shutter supports off the beams or by the use of precast concrete or steel deck floors.

Construction of concrete frames   With in situ cast concrete structures all work, except possibly the cutting and bending of reinforcement, is carried out on the site, the erection of formwork and placing of reinforcement by skilled and unskilled labour and usually the making and placing of concrete by semi-skilled or unskilled labour. The proportion of site work to prefabricated work is greater in the case of in situ concrete and building operations are complicated by formwork and centering, which has to be in position for a considerable time to allow the concrete to strengthen before the frame can take up loads and subsequent lifts be commenced. Much site space is needed for the storage and mixing of materials, although this has largely been resolved by specifying quality guaranteed depot supplied 'truck-mixed' concrete. The type of formwork used and the nature and size of the building will, of course, have an effect upon the actual method of work and the speed at which it progresses.

Cost and speed of erection   With a repetitive skeleton frame and the employment of a suitable contractor, reinforced concrete is likely to be somewhat less expensive but, if in situ, generally longer in construction than steel, although construction can be rapid with a disciplined design approach, careful and detailed planning of the site work, good site organisation and properly trained operatives and good supervision.

To some extent cost and speed of erection are dependent upon the availability of materials and labour. When materials are scarce prices rise and delivery periods become extended. When they are abundant prices and delivery periods fall.

On most sites construction time to permit early occupation of the completed building is a most important consideration and is sometimes the decisive factor in the choice of the framing material. In recent years the introduction of automated methods in the fabricating workshop, the use of hollow fire-protection techniques and steel deck floor construction, together with modern organisational methods have so improved the completion times of steel framed buildings that the overall cost, taking into account early occupation, has been cheaper than if constructed in concrete. However, it can be shown that the principles of this 'fast track'[5] building, as it is termed (see Part 1, section 2.2.2), involving the design and management considerations referred to here, can be applied to concrete construction with similar overall economic results.

The use of precast concrete framing members for multi-storey structures helps to overcome some of the disadvantages of in situ work and reference should be made to page 162 where this is discussed.

Design considerations In situ reinforced concrete often allows greater flexibility in design because of its monolithic nature and because it is not confined to standard sections. There is, however, an economic limit set to this flexibility by the increased costs of formwork for excessive changes of section and, generally, reinforced concrete can be used most economically when the structure is designed to permit the maximum re-use of a minimum amount of formwork. This fact is not only of vital importance to the structural engineer but also to the architect designing in reinforced concrete, for it imposes on him a strict discipline in terms of form and detail of which he must be aware during the design period of any project.

When the plan for some reason or other must be irregular, reinforced concrete may prove to be cost effective and quicker in construction since a large number of complicated connections might be necessary in a steel frame.

Sometimes it is desirable to extend a framed building. When the frame is of steel the problem is relatively simple, the new members being bolted to the existing after they have been exposed by the removal of any fire-resisting casing. With reinforced concrete it is necessary to expose sufficient of the reinforcement in the existing members to obtain a satisfactory bond with the new work or, alternatively, sufficient to permit the welding on of extension bars to provide the necessary bond.

In choosing between steel and concrete it must be borne in mind that each material should be used in the way which best exploits its particular potentialities. Solutions which are favourable to steel are not necessarily so to concrete, and if comparisons are to be realistic it is important that they should be made on the basis of the cost of a building as a whole based on the structural solution most favourable to each material.

Fire resistance Reinforced concrete itself, with normal cover to the reinforcement, provides a substantial degree of fire resistance which can be varied by variations in cover thickness and other means. Steel, because of its behaviour under the action of fire (see page 317), requires a protective cover varying in nature and thickness according to the standard of fire resistance required. The material commonly used for this purpose in the past, but not so much today, was concrete, poured round the steel members within a timber formwork. Although support for this formwork can be obtained from the structural frame itself the process slows down the progress of work. Alternatives are sprayed vermiculite and hollow protection in the form of various combinations of renderings or plaster on metal lathing or in the form of fibre or gypsum plasterboard or in the form of intumescent paint (see pages 335 and 337) on a matrix of expanded metal. These methods have the advantage of reducing the weight of the casing and speeding up the progress of work, but are not all necessarily cheaper than, nor even as cheap as the concrete casing. An alternative to protective casing is to cool the steel by water in the event of fire by techniques which have now been developed and are described on page 337 together with other methods of protecting steelwork.

As pointed out in section 9.5.3 if a fire safety engineering approach is adopted a protective covering to steelwork may, in some circumstances, not be required.

4.3 The steel frame

Steel framed structures are designed with the connections between the separate members treated as either non-rigid or fully rigid joints. The significance of these two forms of connection is explained in Part 1, pages 43 and 44. In reality the stiffness, that is the degree of continuity between the members, of either type of connection lies between these two extremes according to the manner in which the connection is formed (see pages 150 and 157). Research is continually in progress in this area to provide design data which will lead to a more rational design approach and to economies in the use of steel.

As shown in the small-scale examples in Part 1, chapter 6, the frame consists of horizontal beams in both directions and vertical columns called stanchions, usually all of standard rolled sections of various sizes, joined together by welding or by bolts with cleats, to form a stiff, stable skeleton capable of transmitting its own dead weight and all other loads to the foundations, to which the feet of the stanchions are bolted. All the steel members are commonly encased with concrete or other suitable material to which reference has already been made, the thickness depending of the degree of fire resistance required. The beams carry one of the many types of in situ, precast concrete or steel floors described in chapter 5.

4.3.1 Basic elements of construction

The multi-storey steel frame is built up of hot-rolled mild steel sections[6] standardised in shape and dimensions. These are described and illustrated in Part 1, section 6.4.1. There are beam, channel, angle and tee sections and of these the beam section is the basis of the frame for both beams and stanchions. Where necessary these can be strengthened by the addition of plates welded to the flanges to create *compound sections*, although this is rarely necessary since both Universal beam (UB) and column (UC) sections are produced in related 'families' of different weights (mass per metre), the use of which eliminates a great deal of plating and compounding for heavily loaded beams, and the need for flange and seating plates to stanchions (see figure 4.1 A).

A form of deep beam cut from a standard beam section is known as a *castellated beam* and is shown in figure 4.1 B. The web of a Universal beam is cut along its length on a castellated line and the two halves are then placed 'point' to 'point' and welded at the junctions to produce an open web beam of the same weight per metre run as the original section but with a depth 50 per cent greater than that of the original and, therefore, having a greater moment of inertia and resistance to deflection.

In recent years it has become economically possible to make circular cuts in the web rather than the zig-zag cuts for a castellated beam. In the same way the two halves of the cut beam are welded together, forming circular holes in what is called a *cellular beam* (figure 4.1 C).

The use of these beams, or girders, permits services to be accommodated within a structural zone rather than under a solid web beam, thus reducing the overall depth of the floor construction and, therefore, of the floor to floor height. The beams are, also, particularly suited to long span use where the loading is light, that is where stiffness rather than loadbearing capacity is the critical consideration.

In a cellular beam the spacing and diameter of the openings, or cells, can be varied within limits, thus varying the width of web between. As well as giving flexibility in relation to services passing through the beam, this makes it possible to provide sufficient web width for the connection of secondary beams without need to infill cells for this purpose. At supports and other points of high shear, instead of filling in the adjacent cells with plates, ring stiffeners round their perimeter may be used thus preserving the cells for services.

When used compositely with a concrete floor (see page 154) cellular beams may be fabricated from different beam sections to give a top compression flange (which acts with the floor) smaller than the bottom tension flange.

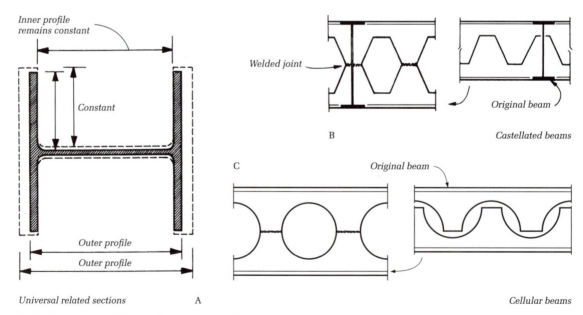

Inner profile remains constant

Welded joint

Constant

B

Original beam

C

Castellated beams

Original beam

Outer profile

Outer profile

Universal related sections   A   *Cellular beams*

Figure 4.1   Structural steel sections and built-up beams

A range of rectangular hollow steel sections is available which, because of their efficient shape, are particularly useful for columns and other compression members.

High-tensile steel can be used for the sections instead of mild steel. It is more costly but its high yield point permits greater working stresses, usually 50 per cent more than for mild steel, resulting in a saving in weight which can offset the extra cost of the material. Deflection criteria may limit the development of the maximum stresses since its elastic modulus is approximately the same as that of mild steel, although methods have been devised to overcome this (see page 155). Even so an ultimate saving in weight of 8 to 10 per cent is possible. Although not widely used for complete frames, it is useful where a local reduction in size of beam or stanchion is essential and this in turn leads to a reduction in the amount of casing and to a saving in height and floor space.

The number of standard and non-standard sections now available to the designer is large and necessarily so if design is not to be limited. Nevertheless, the least expensive frame is not necessarily that which contains least material, although weight can be a useful broad guide, and it is usually more economical to standardise the size of sections to some extent on any particular project rather than to relate the size of every member exactly to the calculations. Any saving effected by using the minimum size in every case may be outweighed by the cost of excessive 'firring' and packing out in order to obtain some degree of uniformity in the finished overall beam and stanchion sizes, and by the cost of making more complicated connections. It must also be borne in mind that the material is cheaper than labour, that standard sections held in stock are cheaper and more readily available than others, and that it is more economical to buy a large number of members of the same size than to buy a few lengths of a number of different size sections. A regular grid layout for the frame, as indicated in Part 1, chapter 6, assists such a standardisation by equalising the loading on different parts.

4.3.2 Frame layout

Planning requirements permitting, economy in the steel frame is obtained by the adoption of a regular and reasonably close spacing of stanchions. Short span beams are more economical per unit of length than long span beams (see chart, figure 4.2).

For normal conditions of loading the cheapest beam will be the deepest available section giving the required strength. Although this may be in excess of the strength required it can still be cheaper than a shallower but heavier section. Take as an example a Universal beam spanning 6 m carrying a total distributed load of 110 kN. The shallowest beam suitable is a 254×146 weighing

43 kg per metre run which will carry 111 kN, but a 356×127 beam will carry 126 kN over the same span yet weighs only 39 kg per metre run, resulting in a saving of 24 kg of steel on the beam.[7] As indicated earlier such savings must be balanced against increased costs which may arise in other parts of the building due to increased overall floor heights. The savings on the frame are likely to be greater than the increases in other directions in the case of large buildings of the warehouse class, where the area of external walls and internal walls and partitions is small relative to the total floor area.

As a preliminary rough guide an economic depth of from one-fifteenth to one-twentieth of the span should be allowed for beams. A deflection not exceeding 1/360 of the span is normally considered satisfactory. Stanchions are usually Universal column sections.

The simplest form of frame is a skeleton 'cage' made up of Universal beams and Universal columns with the members standardised as far as possible and the weight kept to a minimum, having regard to all other relevant factors (figure 4.2 A). Of the beams at each floor level those in one direction will be 'floor' beams carrying the floor slabs and those at right-angles to them will be 'tie' beams which are necessary to provide lateral stability to the skeleton frame as explained in Part 1. As a steel frame is made up of one-way spanning stages a rectangular rather than a square grid layout is generally the most economic (see 1 to 4, figure 4.2).

Advantages of the cantilever By projecting the ends of the main beams beyond the outer stanchions as cantilevers the negative moments set up at the supports serve to reduce the positive moment in the centre spans, which can then be longer without an increase in the size of the beams, or, alternatively, a reduction in the size and weight of the beams can be made.[8] The optimum projection for the cantilever bays, producing approximately equal positive and negative bending moments, is one-fifth to one-quarter of the overall length (figure 4.2 B). The external walls are freed of stanchions and services can be freely run round the building, but stanchions some distance inside the walls are required and this may be inconvenient in respect of the planning of some types of building. The double beams or double stanchions that must be used in order to achieve the continuity of the cantilever beams at the supports result in more complicated construction (see page 151).

Where the spacing of the floor beams is great enough to necessitate secondary beams (see figure 4.2, note to frame layout 4) these may span over the tops of the floor, or primary main beams so that they are continuous (figure 4.3 A) and thus result in a reduction of the positive moments in each span with the accompanying advantages already referred to in respect of the cantilever projections (see also

A

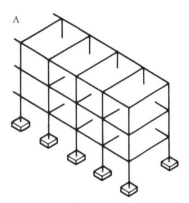

Simple 'cage' frame

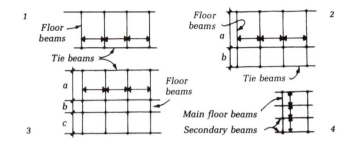

B

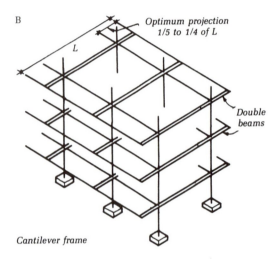

Cantilever frame

Layout	Member	Spans	
		Practicable range (m)	Economic range (m)
1	Floor beams	3.65 to 15.25	4.25 to 6.00
	Slab	2.40 to 7.30	3.00 to 4.25
2	Floor beams	3.00 to 15.25 for economy 'a' should not be more than $1\frac{1}{5} \times$ 'b'	4.25 to 5.50
	Slab	As for 1 above	
3	Floor beams	a.c. 3 to 15.25 b. 1.8 to 15.25 for economy 'b' should be from $\frac{1}{8}$ to $\frac{2}{3}$ of $(a+c)$	4.25 to 5.50 2.40 to 3.00
	Slab	As for 1 above	
4	If spacing of floor beams is greater than 5.50 m secondary beams may be used to keep slab span within the economic limits		

C

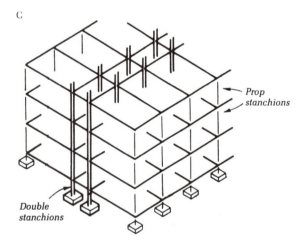

Propped cantilever frame

Figure 4.2 Steel frames

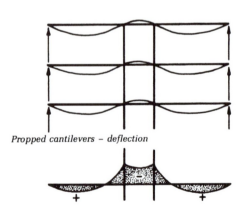

Propped cantilevers – deflection

Propped cantilevers – bending moments

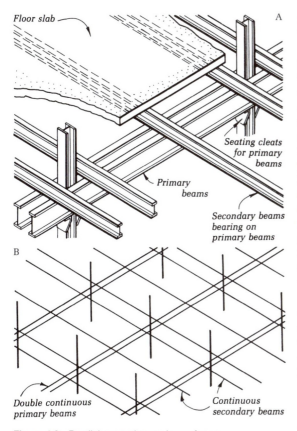

Figure 4.3 Parallel or continuous beam frame

Part 1, figure 3.30 C, E). If the frame layout is also multi-bay in the direction of the primary beams (figure 4.3 B) these may be paired in parallel on each side of the stanchion and similar advantages are gained at the inner bays of these beams, whether or not cantilever end projections are employed. The reduction in depth possible in both sets of beams is such that the overall depth of the floor structure is, for spans up to 15 m, comparable with that in a conventional frame, but within it is a further space for running services between the secondary beams in addition to that between the primary beams.

If the cantilever projections in a cantilever frame such as in figure 4.2 B are made long in proportion to the centre span then the floor space will be uninterrupted except for the central stanchions. This may possibly be in the line of corridor walls as in C, but span for span this is more extravagant in material because of the very high cantilever moments. In addition, wind stresses in the stanchions are high because of the small distance between the stanchions, and this method is usually justified only when some form of good lateral bracing is incorporated,

such as lift shafts, gable walls or suitably placed cross walls, in order to resist the wind load.

The outer end of long cantilevers may be simply supported by props or struts in the plane of the walls to give what is known as propped cantilever construction as shown in figure 4.2 C. In this, although the centre span is small the negative bending moments over the supporting stanchions and the deflection of the cantilevers are reduced as shown on the diagrams in C, even though 'hogging' of the centre span may still occur (see Part 1, page 44). The props are designed with free or hinged ends at the floor levels. Their function is to provide end support to the cantilevers, to transmit to the main structure the wind load at each floor level and to transmit their proportion of the vertical load to the foundations. They can be comparatively small in cross section as they provide no rigidity to the frame against lateral wind pressure and, as a consequence of this, they will transmit no turning force to the foundations due to wind pressure.

If it is not desired to cantilever out at all floors, it may still sometimes be necessary to set in the outer stanchions at the lower floors either for architectural reasons or in order to simplify the problems of foundations adjacent to a building line. This may necessitate heavy compound or plate cantilever girders, particularly if headroom requirements prevent the use of the most economic depth (figure 4.4 A).

Certain types of buildings require wide span beams throughout; others, based generally on a close, regular grid, require a variation of that grid to permit the provision of large, unobstructed floors at certain levels such as ballrooms and restaurant areas in hotels, or the provision of wide shopfronts. These will require stronger beams in the form of some type of girder (B).

4.3.3 Girders

Spans greater than the economic limits suggested earlier require the use of the deeper, heavier beam sections. In the past, when the loads were heavy, it was the practice to increase the flange area by the addition of plates to the top and bottom flanges of a standard beam section, forming a 'compound girder'. The wide range of Universal sections now available and the ease with which sections may now be automatically welded up has made the practice rare.

For heavier loads and for spans in the region of 15 to 36 m, where a section larger than that obtainable from standard beam sections would be required, it is necessary to use plate girders or trussed girders built up of plates or smaller standard sections. The weight of steel in relation to the cube of the building is increased in such cases, resulting in an increase in cost. This may be minimised if ample depth can be allowed for the girders. The increased depth

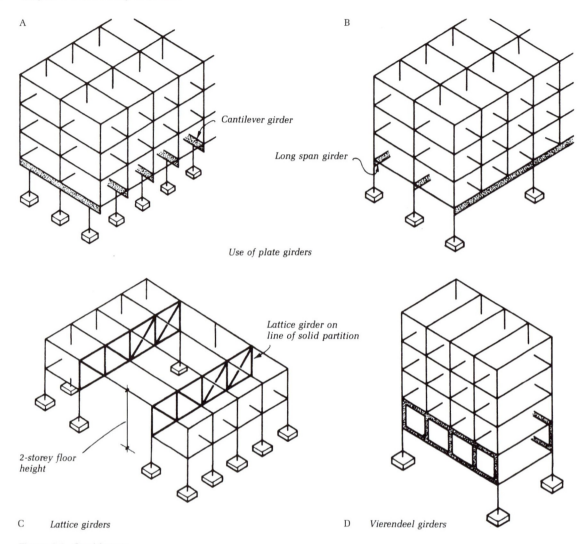

A

Cantilever girder

Long span girder

Use of plate girders

B

Lattice girder on
line of solid partition

2-storey floor
height

C   *Lattice girders*

D   *Vierendeel girders*

Figure 4.4   Steel frames

of lever arm results in a decrease in flange stresses, and this in a considerable reduction in the weight of the flanges, at the expense of a slight increase only in the weight of the thin web of a plate girder or the slender braces of a trussed girder.

Plate girders   A plate girder consists of a web plate welded to flange plates to form a large I-section, the thickness of the flanges being varied according to variations in stress by curtailment of the flange plates as the stresses reduce. When necessary the webs are strengthened against buckling by stiffeners cut to fit between the flanges and welded to them and to the web (figure 4.4 A). Stiffeners are usually placed at the ends and under any point loads

and then at equal distances between at a spacing not greater than one-and-a-half times the depth of the web. For very large spans and heavy loads it may be necessary to increase the thickness of the web plates towards the bearing where the shear increases, in the same way that the flange plates are increased where the bending stresses increase.

An alternative to a flat flange plate is a tee split from a Universal beam section, between two of which the web plate may be welded (B).

When extra wide flanges are necessary, or to give lateral stiffness in long beams, the girder is made with two webs to form a 'box girder'. These are not so liable to twist or bend laterally as a single-web girder.

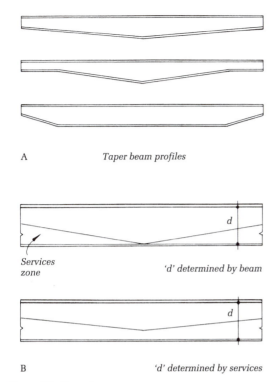

A   *Taper beam profiles*

Services zone   *'d' determined by beam*

B   *'d' determined by services*

Figure 4.5   Taper beams

The economic depth for plate and box girders may be taken as one-twentieth of the span. When a depth shallower than the economic limit is used, careful consideration must be given to deflection to ensure that this is not excessive. The width of the flanges is normally from one-fortieth to one-fiftieth of the span. If wider than this adequate flange stiffening is necessary. For spans much greater than 18 to 21 m plate girders are very heavy and become uneconomic.

Plate girders, because they are welded up from steel plate cut to size and shape, are not limited to parallel flanged form. It is possible, therefore, to taper the web with its maximum depth at the centre of span where the bending stresses are at a maximum, resulting in a more efficient use of material (figure 4.5 A). The advent of automatic plate cutting and welding has reduced the cost of this process and made this type of plate girder or beam economically viable in situations where the triangular spaces at each end between beam and ceiling can be used as services zones (B), thus making the depth of the beam at the centre overlap the services zones and so reducing the overall depth of the floor (see page 209). Beams tapered at the supports to half the depth at the centre generally give a good balance between the cost of the beams and the cost

of other elements favourably affected by the reduction of floor depth, such as external claddings.[9]

Trussed or lattice girders   Where wide spans must be covered and sufficient depth is available, the use of a deep, triangulated beam is likely to be less expensive than a plate girder. These are known as trussed, lattice or framed girders and usually, in a multi-storey frame where height is likely to be limited, can most easily be accommodated in a storey height, for example, on the line of an internal dividing wall (figure 4.4 C) or, in the case of a multi-storey industrial building, exposed within a service floor sandwiched between the production floors. A trussed girder is usually made up of angle and Tee sections of various sizes for all members, connected by welding. The top and bottom flanges, or booms, are normally formed by a Tee section or a tee split from a beam section. When the span is great or the loading is very heavy it may be necessary to fabricate the girder from channel and beam sections as shown in figure 4.6 C, or even to fabricate the booms and struts as plate girders in order to attain adequate strength, especially in compression members liable to buckling. Trussed girders are triangulated in various ways, but for building work the arrangement shown, producing an 'N' girder, is most common. The struts or compression members are vertical and the sloping ties, lying in the plane of diagonal tension, are in tension.

The economic depth of trussed girders is from one-sixth to one-tenth of the span.

Lightweight lattice girders prefabricated in standard depths and lengths from hot- and cold-rolled sections are widely used for roof and floor framing in structures where the loading is relatively light (see chapters 5 and 8 and Part 1, section 3.3.4).

Space frames   These are described on page 272. Although more commonly applied to roof construction, the space frame can with advantage be applied to multi-storey frame construction. A light, rigid member with considerable stiffness results from forming a three-dimensional frame, and triangular section frames can show substantial savings over conventional girders – some have quoted savings as much as 20 per cent. Used for floor framing, wide spans can economically be covered giving an open floor through which all services can be freely run.

Vierendeel girders   A Vierendeel girder may be used as an alternative to a floor height trussed girder when the diagonal ties of the latter would cross door or window openings, or for some other reason would be inconvenient (figure 4.6 D). The Vierendeel girder has no diagonal members, the shear normally carried by these members being transferred to the bearing by the stiffness of the

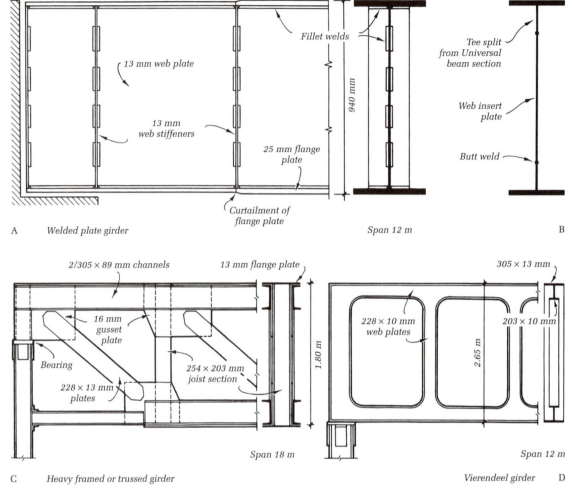

A   *Welded plate girder*   *Span 12 m*   B

C   *Heavy framed or trussed girder*   *Vierendeel girder*   D

Figure 4.6   Steel girders

chords and vertical members and by the rigid joints connecting them, as shown in figure 4.6 D. In such circumstances the chords would lie in the planes of the floors and the vertical members would appear as columns. These girders are expensive in steel and more costly than a trussed girder.

    A form of girder known as a *stub girder* is in principle a relatively shallow form of Vierendeel girder which makes use of composite construction (figure 4.7). In this the floor slab acts as the compression chord and, usually, a Universal column section as the tension chord. Shear is transferred by short lengths of beam sections (the stubs) welded to the lower tension chord and linked to the floor slab by shear studs (see page 155). The openings between the stubs permit the secondary or floor beams to pass through, spanning as continuous beams over the lower

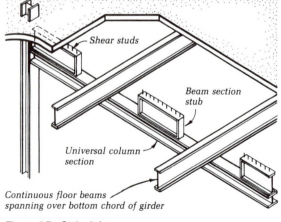

Figure 4.7   Stub girder

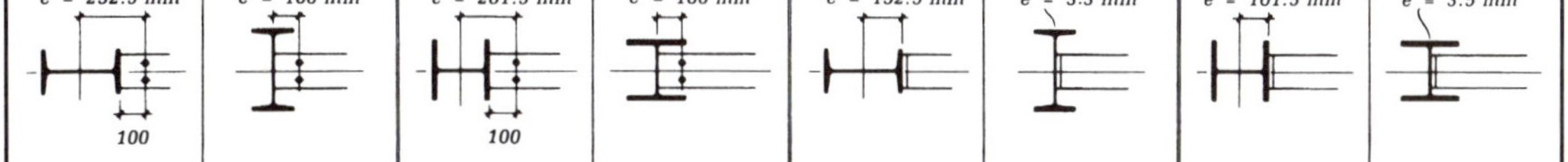

Bolted connections				Direct welded connections			
305 × 165 × 46 kg UB		203 × 203 × 46 kg UC		305 × 165 × 46 kg UB		203 × 203 × 46 kg UC	
$e = 252.5$ mm	$e = 100$ mm	$e = 201.5$ mm	$e = 100$ mm	$e = 152.5$ mm	$e = 3.3$ mm	$e = 101.5$ mm	$e = 3.5$ mm
$Z = 646.4$ cm³	$Z = 108$ cm³	$Z = 449.2$ cm³	$Z = 151.5$ cm³	$Z = 646.4$ cm³	$Z = 108$ cm³	$Z = 449.2$ cm³	$Z = 151.5$ cm³
Relative stress in stanchion $f = M/Z = We/Z$ (for constant load W)							
$\dfrac{W \times 25.25}{646.4} = 0.04$	$\dfrac{W \times 10}{108} = 0.09$	$\dfrac{W \times 20.15}{449.2} = 0.05$	$\dfrac{W \times 10}{151.5} = 0.07$	$\dfrac{W \times 15.25}{646.4} = 0.024$	$\dfrac{W \times 0.33}{108} = 0.003$	$\dfrac{W \times 10.15}{449.2} = 0.02$	$\dfrac{W \times 0.35}{151.5} = 0.002$

Note: Z = section or elastic modulus. See also end note 4 to Part 1, chapter 3.

Figure 4.8 Stanchion/beam relationship

chord, and the spaces on each side of these permit the passage of underfloor services.

4.3.4 Stanchions

Universal and compound stanchions The Universal beam and Universal column sections, used alone or plated to produce compound sections if necessary, are commonly used for the stanchions.

Stanchions are usually positioned with the web in line with the main beams which they support so that the beam is connected to the stanchion flange. This is because smaller bending stresses due to the eccentric bearing will be set up in the stanchion than if the beam is connected to the web. In the case of cleated joints, regulations require a minimum eccentricity (e) of 100 mm to be assumed at the beam connection and when the beam is connected to the stanchion web rather than the flange, greater bending stresses always result, in spite of the smaller total lever arm, due to the small value of the minimum section modulus (see figure 4.8). With column sections the difference is less marked and with direct welded joints a connection to the web will, in fact, produce less bending stresses in the stanchion because of the extremely small eccentricity at the joint.

As most stanchions fail by buckling in the direction of the least dimension, and since strength in this respect depends upon the value of l/r (effective length or height divided by the least radius of gyration), those sections with the widest flanges are most efficient for use as stanchions since they give closer values for the radius of gyration in the direction of both principal axes. This is shown clearly on the graph in figure 4.9.[10] In shorter stanchions, where buckling is less critical, it can be seen that strength is to some extent proportional to weight, but with increase in

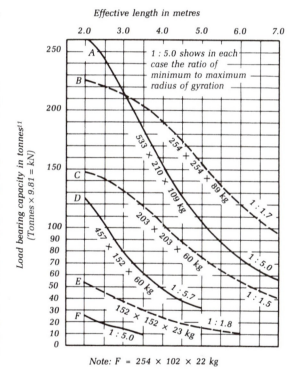

Note: F = 254 × 102 × 22 kg

Figure 4.9 Effect of wide stanchion flanges

height, strength reduces more rapidly in sections having a high ratio of minimum to maximum radius of gyration. It will be seen that, with columns of equal dimensions in both directions, it is possible to obtain a section differing little in its resistance to buckling about both axes.

r columns Solid and hollow
, the same r value in all directions
r space than any other equally strong
ial load, but they present difficulties in
ons to them and in a multi-storey frame
ity suffers. In addition, the solid steel col-
and expensive.

of efficiency, that is in terms of strength to
weight , o, as distinct from size, the solid circular column is less efficient than a Universal column section with
a wide flange. For example, a 100 mm diameter solid steel
column, weighing 61.7 kg per metre run, has an r value of
25 mm while a $152 \times 152 \times 37$ kg column section has a
greater minimum r value of 38.7 mm. Where the effective
length of a stanchion can be reduced by adequate restraint
in the direction of least r, the efficiency of the Universal
section will be further enhanced.

The caps and bases of solid circular columns are of
thick steel plate bored to fit over the ends of the column,
which are turned to produce a small shoulder. The plates
are 'shrunk' on to the column by being heated and then
forced on to the turned ends where they cool and shrink
thus tightly gripping the column (figure 4.10 A).

A hollow circular column has a greater r value than a
solid circular column of the same diameter, that for the
former being approximately $0.35 \times$ mean diameter and for
the latter $0.25 \times$ overall diameter. The tube section, with its
smaller area of material and load-carrying capacity but
greater r value, has advantages, therefore, in cases where
buckling rather than direct stress will be the critical factor.

Cap and base plates and connecting cleats are normally welded to hollow circular columns (B). Tubular
columns are often applied to two-storey frames where the
total height of column can be in one length to preserve
continuity.

Hollow rectangular stanchions A hollow square section permits connections to be made to it more easily than
to circular columns, and has an equal r value about both
normal axes slightly greater than that for a hollow circular
column of the same dimension, being $0.41 \times$ mean
breadth. As with hollow circular columns this section is
particularly useful for long, slender but lightly loaded
stanchions and will show an economy in material over the
normal Universal section. They are selected from the standard range of rectangular hollow sections to which connecting cleats have normally been welded. In recent years,
however, a drilling method known as 'Flowdrilling' has
been devised in which the steel of the stanchion wall melts
as the drilling proceeds and is pushed through the hole
being drilled for approximately 10 to 12 mm to form a
short tube which can later be tapped to permit bolt fixing
of the cleats.

Lattice or braced stanchions Stanchions can be formed
from pairs of beam or channel sections braced together by
diagonal or horizontal plates. These are called respectively
laced and battened stanchions, and are used in cases where
very tall but relatively lightly loaded stanchions are necessary, requiring a maximum r value in both directions, or in
cases where the load is too great for even the largest rolled
section (see figure 4.10 D, E). These sections can easily be
arranged to give equal r values about each axis by adjusting the distance between the pair of basic members. For
heavy beam sections the distance between webs will be
about three-quarters of the depth of the beam, and for
channel sections placed back to back the distance between
webs will be about half to two-thirds of the depth of the
channel according to the particular section used. Limits on
the spacing of lacing bars and battens are defined in BS
5950-1: *Structural use of Steelwork in Building. Code of
Practice for Design. Rolled and Welded Sections.*

4.3.5 Connections

Connections between the members of a normal steel frame
are made by means of welding and bolts with angle cleats
and plates.

The subject of connections is introduced in Part 1, chapter 6, where types of bolts, the nature of welding and the
basic types of connections used are described and reference
is made to their detailing relative to fabrication. These will
now be considered further and others will be described.
Welded connections are discussed in more detail under
Welded construction in section 4.3.8 of this volume.

Iron and steel castings New alloys have been developed in this field which, compared with the brittle nature
of normal cast iron, are ductile and weldable and it is now
possible for complex joints and junctions to be cast in a
single operation rather than being machined or fabricated
from several parts. Castings of high quality and excellent
finish can be produced, making them suitable for components which will be exposed to view.

Stanchion bases For reasons given in Part 1, section
6.4.2, the foot of a stanchion must be expanded by means
of a base plate which will act as an inverted cantilever
beam and is designed accordingly.

There are two types of base: (i) the slab or bloom base
and (ii) the gusseted base.

Slab or bloom base This consists of a base plate thick
enough to resist the moments caused by the bearing pressure. When the area of base required is such that its projection beyond the stanchion is small (Part 1, figures 6.2
and 6.3) these moments will be small and the thickness of

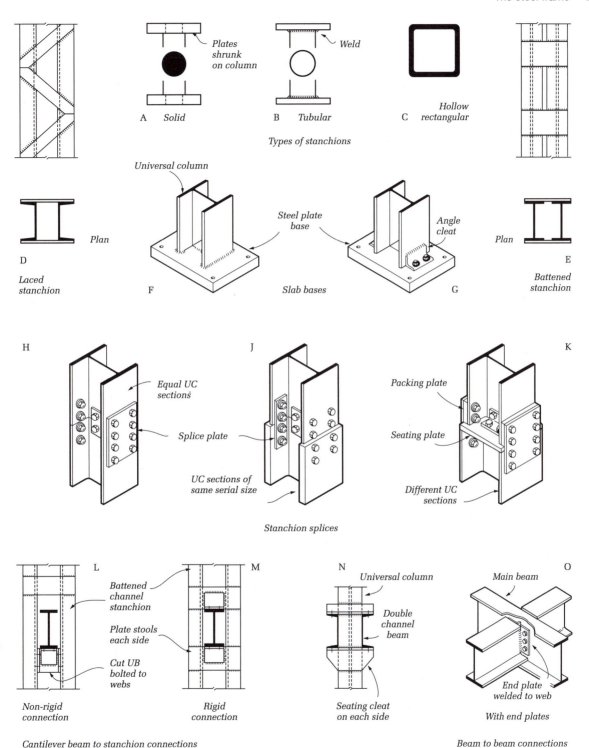

Figure 4.10 Steel frames

the base plate will be relatively small. As its area increases so must its thickness. It is fixed to the foot of the stanchion by welding (figure 4.10 F) or by a pair of angle cleats when circumstances make it advisable to send stanchion and slab separately to the site (G). These fastenings are required mainly to secure the slab to the stanchion.

This base requires less fabrication than the gusseted base (see below) and there is a saving in depth of the cover necessary to give a clear floor compared with that required when there are large gusset plates, but it sometimes requires more material, particularly in the case of large bases in which the slab may be very thick.

The underside of base plates and slabs need not be machined when the base will be grouted to a concrete foundation, provided the underside is true and parallel with the upper face. A bloom base is always used for solid circular columns.

Gusseted base This consists of a base plate stiffened by gusset plates which act as ribs (figure 4.16 E). The size of the base plate as that of a bloom base depends on the safe bearing pressure that the concrete foundation can resist. The breadth of the base generally varies from two to three times the breadth of the column, and the height of the gusset plates from one-and-a-half to three times the breadth of the column. Gusseted bases are commonly used when the stanchion transfers high bending moments to the foundation. The holding down bolts for all stanchion bases must be well tied down to the foundation block (figure 4.16 F).

Base location Figure 4.11 shows the fabrication of a temporary bolt 'box' or template. This is used to provide for accurate positioning of the holding down bolts in a wet concrete foundation. Large square steel washers ensure

that the bolts are well anchored. The template and the expanded polystyrene around the bolts are removed when the concrete has hardened, thus permitting marginal bolt movement to locate with the holes in the stanchion base. The stanchion is positioned to the bolts and steel packing under the base allows for fine adjustment with the retaining nuts to ensure that the stanchion is vertically aligned. On alignment, the underside of the base and the bolts can be grouted in place.

Stanchion caps A cap must be provided to the stanchion if the beams rest on top of it. This is illustrated in Part 1, figure 6.2 and described on page 122 of that volume.

Stanchion splices Joints in the length of stanchions, generally termed splices, are necessary for transport, assembly and fabrication purposes because it is difficult and expensive to handle lengths much greater than about 10.50 m (see figure 4.10 H). Stanchions should be fabricated in lengths as long as possible and the limit of 10.50 m for an individual length results in stanchion lengths of two to three storeys in height, depending on the floor-to-floor height of the structure. The joints should be made as near as possible to the beam level, but in order that splice plates shall not obstruct beam connections they are usually placed about 300 to 450 mm above floor level.

Any change in the size of section is made at these points (J) and where the whole of the end of the upper length does not bear on the lower length a horizontal seating plate of the same overall size as the lower stanchion must be placed between the two lengths (K). Packing plates will be necessary between the upper smaller section and the splice plates. It will be seen by reference to figures 4.1 A and 4.10 J that the use of Universal related sections

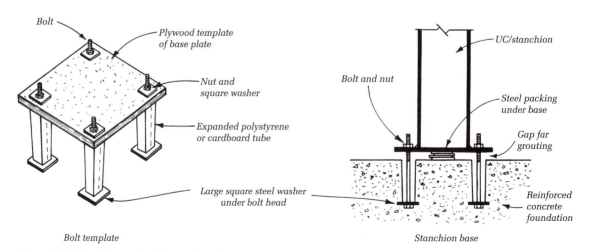

Bolt template

Stanchion base

Figure 4.11 Steel frame stanchion base location

of the same serial size reduces the need for seating plates. When the ends of each adjacent length of stanchion are machined and bear directly one on the other, transmission of load will be by direct compression so that the splice plates serve only to secure the two lengths. If the ends are not machined and are not in full contact the splice plates and bolts must be sufficient to transmit the entire load.

Beam to stanchion connections   Direct compression connections are only possible where the beam rests on top of the stanchion (Part 1, figure 6.2). This is normally not possible at lower beams where loads have to be transmitted by shear connections as shown in Part 1, figures 6.3 and 6.4.

Bearings for cantilever beams passing through a double member stanchion can be provided by stools formed from pieces of beam section fixed between the stanchion members on which the beams rest and to which they are fixed by stool cleats as shown in figure 4.10 L. Greater rigidity at the connection may be provided by plate stools connected to the flanges of the stanchion members and secured to the flanges of the beam with angle cleats (M). The top plates and cleats are fixed after the erection of the beam. Where the beam itself is a double member, projecting seating cleats are fixed to the flanges of the stanchion to take the beam members (N).

The maximum stress permitted in stanchions depends on the ratio of effective length to radius of gyration. The former varies with the number and disposition of the beams connected to the stanchion, as this affects the degree of fixity and restraint given to the stanchion. The effective lengths for standard conditions of restraint are given in BS 5950-1.

Beam-to-beam connections   The direct compression connection, in which one beam bears directly on the top flange of the other, is the most economical, but in multistorey frames this is generally not suitable because of the considerable depth of the pair of beams one on top of the other (but see page 145). Shear connections are therefore usual with web connected to web. End plates are welded to the web of the secondary joists in the fabrication workshop and the connection is completed on the site by bolting to the main beam (figure 4.10 O).

When the main beam is much deeper than the secondary beam, for example in the case of a deep beam section or a plate girder, erection is facilitated by using deep seating stools made up of plates and angles or of a piece of cut beam section. These have the added advantage of forming web stiffeners at the point of application of the load, which is desirable in deep beams. Where the flanges of both beams are to be level, that of the secondary beam must be notched to clear the flange of the main beam.

Tension joints are generally avoided but may have to be used when the secondary beam must pass under the main beam. The connection depends entirely upon the nuts of the bolts by which the secondary beam is hung from the other. Direct connection of flange to flange should be avoided whenever possible, the better method being to hang from the flange of the main beam by means of angles and gusset plates.

4.3.6 Wind bracing

Adequate lateral resistance against wind pressure may be provided in a number of ways. These may take the form of:

- rigid frames
- diagonal bracing (X or K)
- shear walls.

Rigid frames   In many cases where the beams are reasonably deep and satisfactory connections to the stanchions can be made the structure may be stiff enough as a whole to resist excessive lateral deflection. Connections can be made by means of top and bottom flange cleats of sufficient size and bolts in sufficient numbers. The cleats must be stiff enough to transmit moments without excessive distortion and often would become uneconomically thick, especially when the stanchion is narrow and necessitates the use of long cleats to accommodate the required number of bolts. The thickness of the cleats can be reduced by the use of plate or gusset stiffeners.

A better method is to weld a steel fixing plate on to the end of the beam, making use of high tensile steel friction bolts for fixing (see Part 1, figure 6.4); when greater stiffness is required the plate and bolts may be extended beyond the beam. Since the columns in such a frame carry bending moments they will be larger than with other methods. Where wind loading is not excessive, that is in buildings which are not too tall, this method can produce an economic solution. Very tall structures with relatively small base areas, however, require other methods.

Diagonal bracing   In this method the principle of triangulation is applied to certain frame panels up the height of the building to produce stiff vertical elements which resist the wind pressure transmitted from the external walls by the floors of the structure acting as rigid diaphragms. The remainder of the frame is then required to carry only the gravity load (see figure 4.12 A).

Shear walls   These are stiff walls capable of resisting the lateral wind loading. In principle they act in the same way as the braced frames in diagonal bracing, wind pressure being transmitted to them by the floors (figure 4.12 B).

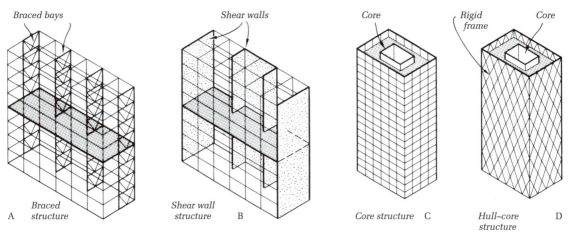

Figure 4.12 Wind bracing

They can, in fact, be formed within the panels of the frame itself. Even lightweight wall panels and 102.5 mm brick panels with door openings in them increase to a considerable degree the stiffness of the structural frame.

Lift shafts and stair wells in tall buildings are commonly formed of reinforced concrete, producing a stiff annular core or cores for the full height of the structure, and these may be used to provide the necessary resistance to lateral wind pressure. The floors transmit the wind pressure from the external walls to the core (C).

A development of this for very tall buildings, known as the 'hull–core' system, utilises a rigid or braced framework for the perimeter structure to form stiff walls interacting with each other to produce an external annular structure, called the hull. This acts together with the internal core through the floors to produce a very stiff total structure to resist wind forces (D). In some circumstances, especially if the plan form approaches a square, the perimeter structure may be made rigid enough to provide in itself all the resistance to the wind forces, the pressure on a windward wall being transferred through the floors wholly to the two flank walls of the annular perimeter structure. This relieves the internal core of these forces and avoids the need to provide heavy shear walls at this point. Such a structure is sometimes referred to as 'pierced tube' construction.

4.3.7 Composite construction

Although it is convenient to consider structural steel and reinforced concrete framed construction as two separate entities, there are many instances where the two forms of construction are compatible and complementary. Not least, the use of concrete as fire protection to the more vulnerable structural steel frame.

Due to the potential for failure of steel elements under load at temperatures above 500 °C, structural steel members in multi-storey frames must be adequately protected from the effect of fire, and this is fully discussed in chapter 9. Solid concrete casing is common and, while it is used primarily for the fire protection of the steel,[12] BS 5950-1 permits allowance to be made for the stiffening effect of the casing on beams and stanchions, resulting in higher permissible stresses for the steel provided the casing gives a minimum cover of 50 mm to the steel section and is reinforced with steel fabric.

Composite stanchions   BS 5950-1 also permits some of the stanchion load to be carried by the concrete, but limits the thickness of the casing which may be considered for this purpose to 75 mm from the overall dimensions of the steel section. These allowances do not apply to sections greater than 1000 mm × 500 mm nor to box sections. In a multi-storey steel-frame building, having solid concrete casing to the members, a saving in steel can, therefore, be effected by taking the casing into account. This results in smaller overall sections and, therefore, a saving in the concrete casing content. In the case of stanchions, the more lightly loaded a stanchion is, the greater will be the benefits deriving from this procedure.

Composite stanchions can also be formed with the concrete placed internally, relative to the steel. Multi-storey stanchions with two opposing channels bolted together have been used, the channels being filled with concrete, designed to carry 28 per cent of the load, and showing a saving of 50 per cent of steel compared with normal stanchions.

BS 5950-1 contains conditions in which the frictional resistance between the top flange of a steel beam and the floor slab it supports may be considered sufficient to provide adequate lateral restraint to the beam. As this restraint is more often than not available in practice, solid concrete beam casing is not essential for this purpose and lightweight protection such as vermiculite spray and hollow casings of various types can be used with economic advantage. In some circumstances hollow casings to the stanchions may show economies, but in others it may be more advantageous to use concrete casings to carry some of the stanchion load.

Composite beams Concrete floor slabs may be used to act together with the steel supporting beams. A concrete slab adequately bonded to a steel beam will act in the same way as in a reinforced concrete T-beam. This is particularly useful when the steel is encased, as the lever arm of the composite section will be greater by about half the thickness of the slab. The concrete casing by itself does not, however, provide sufficient bond between the steel and concrete to transmit the shear stresses, and this is essential if composite action is to be achieved. The greatest shear stress is at the neutral axis, which is always near the top of the combined section, so that a satisfactory bond can be obtained by welding shear connectors to the top flange of the beam. That most commonly used is the shear stud (figure 4.13 A) fixed by an electric spot welding gun. These may be used with both in situ cast and precast slabs. Alternatively, with precast slabs, high strength friction grip bolts may be used to produce sufficient friction between beam and slab to transmit the shear stresses (B). The problem of tension arising in the top of the member and compression in the bottom over the supports in continuous composite beams may be dealt with as follows:

- The concrete may be reinforced longitudinally with rods.
- The beam may be strengthened by welding on top flange plates, ignoring the concrete entirely, as well as plates on the bottom flange to strengthen that.

High tensile steel beams Concrete is also used to stiffen high-tensile steel beams in order to permit them to be stressed to their working limit. In normal building structures these beams cannot always be stressed to their limit because the amount of deflection might be unacceptable, high-tensile steel having an elastic modulus approximately equal to that of mild steel. In this system the beam is deflected to the same extent that it would be under full working load and the concrete is cast as a casing round the tension flange. After this has gained strength, the deflecting force is removed, and as the beam reverts to its original unstressed state, it compresses the concrete. Under

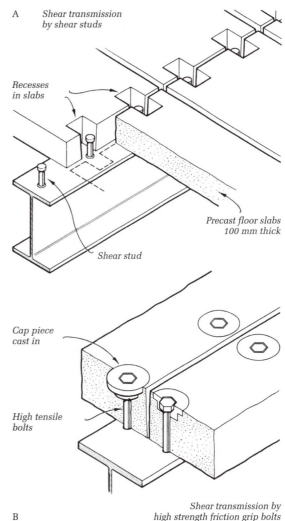

A *Shear transmission by shear studs*

Recesses in slabs

Precast floor slabs 100 mm thick

Shear stud

Cap piece cast in

High tensile bolts

Shear transmission by high strength friction grip bolts

B

Figure 4.13 Composite beams

working load the tensile stresses in the bottom flange are resisted by the compressive stresses in the concrete casing and this has the effect of reducing the deflection of the beam, thus permitting the high-tensile steel beam to be stressed to its limit so that its high strength may be fully utilised.[13,14]

Prestressed steel beams Prestressing may be applied to mild steel beams in order to allow the use of reduced cross sections resulting in savings in weight and cost. The post-tensioning cables may be passed through a box tension flange with anchor plates at the ends (figure 4.14 D) or they may be external to the beam at the sides (E) or below the tension flange (A, B, C). In the latter examples it will

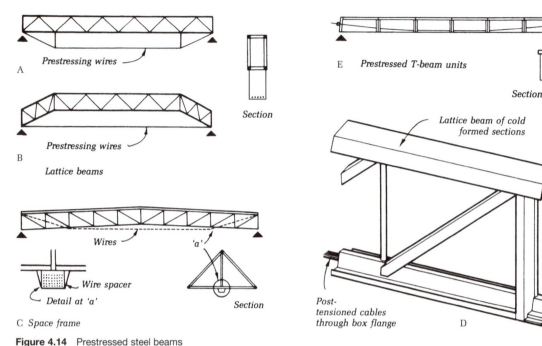

A *Prestressing wires*

Section

B *Prestressing wires*

Lattice beams

C *Space frame*

Wires '*a*'

Wire spacer

Detail at 'a' *Section*

E *Prestressed T-beam units*

Section

Lattice beam of cold formed sections

Post-tensioned cables through box flange D

Figure 4.14 Prestressed steel beams

be seen that the beam may be bent up so that the cable is straight or the cable may be strutted off the tension flange to form a trussed beam. Depth/span ratios in the region of one-thirtieth to one-thirty-fifth can be attained.

4.3.8 Welded construction

Substantial savings in the weight of steel arise when joints are welded and allowances made in the design of the frame for the effects of rigid connections. These savings can be as high as 20 to 25 per cent of the weight of a comparable frame in bolted construction. The economies are due to (i) the elimination of bolt holes; (ii) simpler connections with no bolt heads; (iii) the possibility of developing fully rigid end connections leading to the use of lighter members; and (iv) in the case of built-up members, such as plate girders, the use of single plates, welded directly one to the other without the addition of angles. Although not usual with multi-storey frames, where the frame is exposed the appearance is enhanced by the cleaner and neater joints, and less maintenance costs are involved in re-painting, because of the smaller sections and the absence of bolted connections.

The use of welding permits the better use of hollow and cold-rolled steel sections, both of which, when used in appropriate circumstances, tend towards further reductions in weight. Flexibility in design arises from the fact that built-up sections may easily be fabricated, thus freeing the

designer from the restrictions of the standard rolled sections. In addition, the material in such members may be proportioned to the moments it is called upon to resist in the different parts, thus producing maximum structural efficiency; greatest advantage may be taken of this in single-storey rather than in multi-storey frames.

Workshop fabrication of the component parts, as indicated in Part 1, is now invariably carried out by means of welded connections, whether or not the site connections are bolted or welded. As noted, site welded connections have several advantages, not least a saving in weight over bolted cleats. However, when a steel frame building comes to the end of its economic life, bolted connections are more easily removed. Undamaged sections will have greater opportunity for re-sale and re-use.

Methods of welding Two methods are normally employed in structural work: (i) oxy-acetylene welding, (ii) metal arc welding. In the former, the flame from an oxy-acetylene torch heats the two surfaces to be joined at their point of contact, and molten metal from a steel filler rod held in the flame at the same time fuses into the two surfaces. This process is generally limited to the welding of thinner cold formed steel sections and for site cutting of all sections when alterations are required. In the second process, an electricity generator or mains transformer is used to produce a high amperage current that passes through the metal filler rod, which acts as an electrode, and

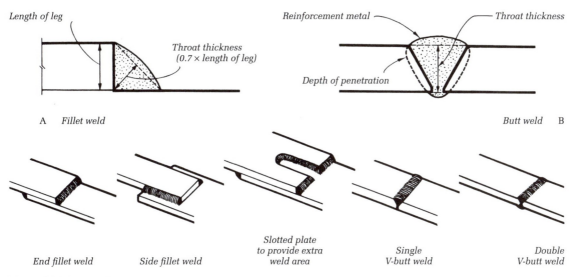

Length of leg

Throat thickness (0.7 × length of leg)

Reinforcement metal

Throat thickness

Depth of penetration

A   *Fillet weld*

Butt weld   B

End fillet weld   *Side fillet weld*   *Slotted plate to provide extra weld area*   *Single V-butt weld*   *Double V-butt weld*

Figure 4.15 Types of welds

the material to be joined. An arc is produced, which heats and melts the surfaces and the end of the filler rod.

Types of welds   There are two types of weld: a butt weld and a fillet weld. The fillet weld is that in which the two surfaces joined are basically at right-angles one to the other (figure 4.15 A), the butt weld is that in which the two surfaces to be joined butt against each other (B). There are varieties of each of these and some are shown in figure 4.15. The strength of a weld is based upon its area, made up of the effective length and the throat thickness.[15]

Types of welded connections   Varying degrees of rigidity may be produced in welded joints according to their design, but unless especially designed to produce full fixity, a welded joint will generally possess no more rigidity than a normal bolted connection. Welded joints may be in the form of direct or indirect connections. A direct connection is one in which the members are in direct contact one with the other and are joined to each other directly by welds without the use of cleats or plates. In the case of beams, particularly, this necessitates accurate cutting to length and complicates the assembly process, so that more often than not an indirect plate connection is used such as with fin or end plates (see Part 1, figure 6.3 E), which overcomes these disadvantages. Stanchion bases and caps, and stanchion splices, may be formed with direct or indirect connections (figure 4.10 F, G); figure 4.16 A, B, C show splices formed by direct connections. Temporary angle cleats are bolted to the webs to enable the lower and upper stanchion lengths to be held in position while the necessary site welds are executed. Splices are sometimes

formed as shown in D. The welded base (E) is a gusseted form, as referred to on page 152. When used to transmit bending moments to the foundation the holding-down bolts must be well secured to the base slab as shown in F and and figure 4.11.

4.3.9 Simple and fully rigid framing

BS 5950-1 provides for simple and continuous design methods for structural steelwork. The majority of steel frames have been designed on what is known as the simple design method, in which all the beams are assumed to be simply supported at the connections with the stanchions, in disregard of the obvious and proved partial fixity existing at the ends of beams with bolted connections. Continuity is produced by direct welded, that is fully rigid, connections and in this method the interaction of beams and stanchions is calculated for all conditions of loading resulting in a considerable saving in steel. It is not, however, so readily applied to multi-storey as to single-storey frames, because of the large number of joints.

Where a fully rigid connection is made between a beam and stanchion it may be necessary to increase the shear strength of the stanchion web. This may be accomplished (i) by welding in horizontal stiffeners at the level of the beam flanges, (ii) by the further addition of web plates, or (iii) by means of a triangulated system of stiffeners, which may be designed to act either in conjunction with the stanchion web, or entirely alone (figure 4.16 G).

The use of gussets or brackets at the junction of the beam and stanchion, or deepening the end of the beam (H, I) has the effect of reducing the shear stress at the

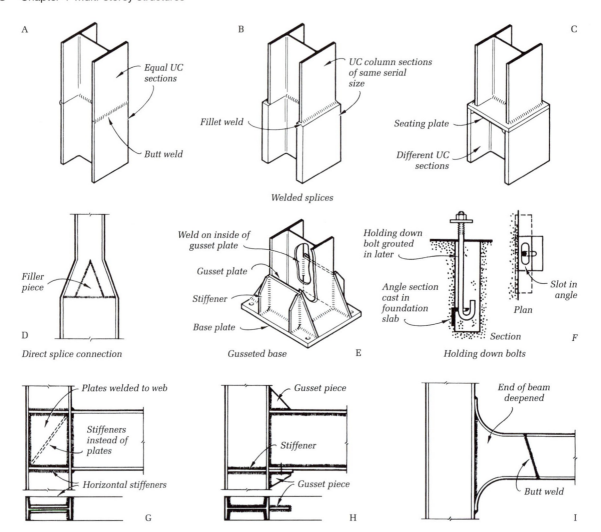

A

Equal UC
sections

Fillet weld

Butt weld

B

UC column sections
of same serial
size

Fillet weld

Welded splices

C

Seating plate

Different UC
sections

Filler
piece

Direct splice connection

D

Weld on inside of
gusset plate

Gusset plate

Stiffener

Base plate

Gusseted base

E

Holding down
bolt grouted
in later

Angle section
cast in
foundation
slab

Holding down bolts

Plan

Section

Slot in
angle

F

Plates welded to web

Stiffeners
instead of
plates

Horizontal stiffeners

G

Gusset piece

Stiffener

Gusset piece

H

End of beam
deepened

Butt weld

I

Figure 4.16 Welded connections

connection and can eliminate the need to strengthen the stanchion web. It also increases restraint at the joint.

4.3.10 Cold-formed steel sections

These are made from steel strip cold formed to shape in a rolling mill, press-brake or swivel bender, those formed by rolling being known as cold-rolled sections, the others as pressed-steel sections.

Pressed-steel sections These are largely used for flooring and roofing units and wall panels, and for metal trim such as skirtings and sub-frames, the lengths of which are limited by the maximum width of the press-brake or bending machine (figure 4.17).

The press-brake is a machine which has a long horizontal former rising and falling with a pressure of 150 tonnes. This former presses the steel strip into a suitably shaped horizontal bed to form folds in sequence as shown in figure 4.17 A, by means of which simple sections are shaped, usually up to 3 m long, although some machines produce lengths up to 7 m. The swivel bending machine folds the strip by means of clamps (B), in lengths up to 2 m, the folds or bends being made in sequence as before. These methods can produce sections economically in small quantities. The press-brake can handle steel strip in thicknesses up to 20 mm but the swivel bender only up to 3 mm.

Cold-rolled sections For structural members of greater length, cold-rolled sections are used. These are formed

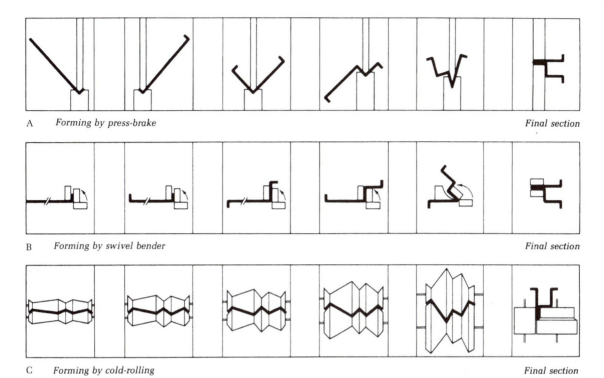

A *Forming by press-brake*   *Final section*

B *Forming by swivel bender*   *Final section*

C *Forming by cold-rolling*   *Final section*

Figure 4.17   Cold-forming methods

into the required shape by passing metal strip between six to fifteen progressive sets of forming spindles or rollers, each pair of which adds successively to the shaping of the strip, the final pair producing the finished section (C). The basic sections rolled are plain angles and channels, lipped channels, and zeds (see figure 4.18). Outwardly lipped channels are commonly called top-hat sections. The length is limited only by considerations of transport. The maximum width of strip which can be formed is rather more than 1 m, in thicknesses from 0.3 to 8.0 mm.

There is virtually no limit to the shape that can be rolled so that the designer can choose a shape best fitted for any particular purpose although, as the manufacture of the rollers for sections outside the basic range is expensive, the economic advantage of choosing such a section must be considered carefully if the quantity required is small.[16] Where possible the section should be made to fulfil more than one function by shaping it, for example, to avoid the casings often necessary with hot-rolled sections and to permit the direct attachment of claddings and windows. The strength of individual members may readily be varied by varying the gauge of the steel while maintaining the same overall sizes. These possibilities can lead to increased efficiency in the structure and to substantial economies in the weight of steel used.

Application of cold-rolled sections   As indicated in Part 1, chapter 6, cold-rolled sections are most efficiently used with structures of moderate loads and span, in which circumstances they can be less expensive than hot-rolled members. Erection of the structure is often quicker and easier because of its light weight and rigidity. Cold-rolled sections are used for roof trusses, lattice beams and rigid frames, where they are probably used to greatest advantage, but they can be used for two-storey frames throughout, such as steel-framed housing, and for three-storey frames in which the bottom storey stanchion lengths are in hot-rolled sections in order to keep the overall size the same as the cold-rolled lengths above. As an alternative to the use of hot-rolled sections, a heavier section with the same overall size can be obtained by spot welding cold-rolled reinforcing sections on the inside.

Channels placed back to back and box sections are suited to axial load, the latter having considerable torsional strength. There appears to be little economic advantage in using cold-rolled sections for columns as far as cost of material and fabrication are concerned, but there are considerable savings in erection costs, which may be 10 to 15 per cent cheaper than with hot-rolled sections. When fire-resisting casing is not required external columns can, as mentioned above, be shaped to act as window mullions

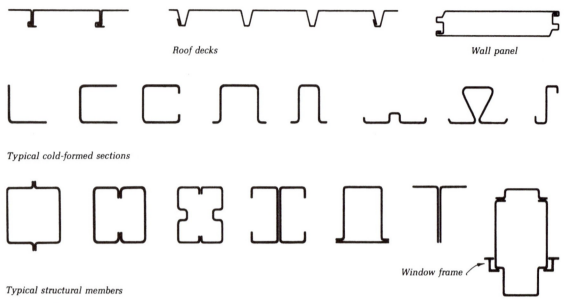

Roof decks   *Wall panel*

Typical cold-formed sections

Typical structural members

Window frame

Figure 4.18   Cold-formed steel sections

to accommodate windows by direct fixing as shown in figure 4.18. I-section beams formed from channels placed back to back are only likely to be economic for light loads over short spans, where the smaller hot-rolled sections might not be used to their limits. The wider choice of section possible with cold-rolled sections gives them an advantage in these circumstances. Greater savings over hot-rolled work are found by using cold-rolled lattice beams over intermediate spans with light loadings although, as indicated above, the most economic field is in roof structures where the advantages of the low dead/live load ratio and ease of handling and erection show most clearly (see chapter 8, page 283).

Connections   These are made by various types of welds, self-tapping screws, bolts and cold rivets (see Part 1, section 6.4.3) and sections can be formed to push-fit into each other, thus avoiding the use of gusset plates. For example, top-hat flange sections can be used for beams into which the bracing members fit, so that the node connection is direct. Details of construction in cold-formed steel are given in Part 1, chapter 6.

Structural considerations are similar to those in the design of light alloy structures. The design of thin wall structures requires special consideration, due to the possibility of local instability and, as in aluminium sections, lips to the edges may be provided to give increased stiffness to the section.[17]

Protection   The need for protection from corrosion is important because there is not so much margin for wastage of metal as in hot-rolled sections. Phosphating followed by paint dipping and stoving is usually adopted for internal work and where the structure is to be exposed to the weather, hot-dip galvanising can be used. Mild-steel strip is widely used in the production of cold-rolled sections but high-tensile steel with rust-inhibiting qualities is likely to be used to an increasing extent because of its structural advantages.

4.4 The reinforced concrete structure

Reinforced concrete, because of its particular characteristics, can be formed into walls as well as into beams and columns to form a skeleton frame, and floor slabs can be designed without projecting beams to carry them. A reinforced concrete structure may, therefore, consist solely of walls carrying slab floors, slab floors and columns only or a combination of columns, beams and loadbearing walls, each being used to fulfil most satisfactorily the functions required at various points (see figure 4.23 A, D). Staircases and lift shafts often must be enclosed in solid walls, and it is logical to make these of reinforced concrete capable of both enclosing the areas and carrying the floor loads, rather than to surround the areas with beams and columns and then enclose with non-loadbearing panel infillings. Such enclosures, being monolithic in form, result in very broad annular columns running right through the building which can be used to provide resistance to wind pressure on the structure.

Greater flexibility in planning and design is possible with reinforced concrete than with steel[18] but little can be salvaged when a reinforced concrete framed building comes to the end of its economic life.

At the beginning of its structural life reinforced concrete is fluid or plastic in character and this gives rise to two important factors concerning the nature of the structure for which it is used:

● The ease with which a monolithic structure may be obtained, producing a rigid form of construction with the economies inherent in this form.
● The ease with which almost any desired shape may be formed either for economic, structural or aesthetic reasons. For example, the material may be disposed in accordance with the distribution of stresses in the structural members, placing most where the stresses will be at a maximum and reducing it where they will be at a minimum.

These two factors together, monolithic continuity giving particular distributions of stresses, and variation in the disposition of material according to the stress distribution, produce characteristic concrete forms which are most obvious, as far as building structures are concerned, in single-storey structures. Nevertheless, these characteristic forms do find a place in multi-storey frames, and it should be made clear at this point that the designer's freedom to cast concrete in almost any shape is limited by the cost of the formwork or shuttering into which the concrete must be poured.

Shuttering costs This forms a large proportion of the total cost of a reinforced concrete structure as can be seen from the following approximate percentage break-down:

Concrete	40%	Materials	28%
		Labour	12%
Shuttering, including erection and stripping	32%	Materials	12%
		Labour	20%
Reinforcement	28%	Materials	20%
		Labour	8%

Shuttering costs for beams alone may be as high as 40 per cent of the total cost of the beam. It will be seen that the percentage labour content in shuttering is far greater than that in steel fixing or concreting, so that economies in shuttering will have a significant effect in reducing the cost of the concrete work. Such economies are the outcome of simple structural forms repeated a number of times, making the construction of the shutters a simple matter and enabling them to be used repeatedly to the

maximum extent. In addition, the use of prefabricated formwork panels, or flying forms, results in a considerable reduction in the labour content in the cost of shuttering. Complicated shapes, particularly if curved, appearing only once in a structure lead to high shuttering costs.

In situ cast structures Up to comparatively recent times multi-storey reinforced concrete frames have always been erected as in situ cast structures for which all the constituent concrete materials have been brought to the site, mixed and placed in formwork erected in the position the concrete will finally occupy in the completed structure. Such frames are invariably of monolithic construction by which full continuity throughout columns, beams and slabs is attained.

Advantages and disadvantages The advantages of monolithic or fully continuous construction are:

● reduced deflections in the members
● reduced bending moments distributed more uniformly throughout the structure than in discontinuous structures. The reduced moments result in lighter members. The greater uniformity in distribution will, in members of uniform section sized to the maximum bending moment, have the effect of involving less waste of material at the points which are less highly stressed
● in the case of beams there is a less rapid increase in dead weight with increase in span because, due to the stress distribution, a great deal of the extra material is required over the supports which will take its weight directly.

Against these advantages must be placed the following disadvantages:

● the adverse effect of differential foundation settlement, which has been described on page 41
● adverse effect of temperature movement. Movement due to temperature changes has a similar effect upon a continuous structure to that of foundation movement, and close attention is necessary at the design stage to the maximum possible movement and to the use of expansion joints at appropriate points to prevent accumulative movement throughout the whole structure.

Cross-wall construction in in situ cast concrete is known as *box frame* construction. This, together with in situ cast external wall construction is dealt with in this chapter rather than in that on 'Walls and piers' (chapter 3) because of its total monolithic character enabling the walls and floor slabs to act together. For ease of reference reinforced concrete floors have been considered in a separate chapter, although they do, in fact, form an integral part of most reinforced concrete structures.

Precast structures A precast concrete component may be defined as a component cast in a position other than that which it will finally occupy in the completed structure and which, after removal from the forms and maturing, requires to be placed and fixed in position.

The technique of precasting concrete for structural purposes was originally applied to the manufacture of floor and roof slabs, but the process has now developed to such an extent that whole building structures can be erected from factory produced precast components involving columns, beams, floor and roof slabs, wall panels and cladding.

Advantages and disadvantages In terms of site work the great advantage of the precast structure is that the speed and simplicity of assembly compares favourably with that for a steel frame and this, allied to the relative cheapness of concrete, makes it an extremely valuable method of construction. In addition to the saving of time and labour on the site factory production makes possible a closer control of the concrete than is often possible on the site, particularly in the case of small projects, and leads to a saving in materials and an improvement in quality. Formwork and its support is greatly reduced, the site is less obstructed and, in cases where the concrete is to be exposed, the production of satisfactory surface finishes is facilitated. The difficulties arising from the shrinkage of fresh concrete are eliminated because all maturing takes place before the components are built into the structure.

The principal disadvantage is that the continuity and rigidity of structure attained in an in situ cast structure are more difficult to achieve in the precast form and account must be taken of this at the design stage.

Size of components As with normal in situ work, precast work should be designed to produce the maximum repetitive use of a minimum amount of shuttering. Individual components should be simple in form and they should be as large as methods of transport and assembly will permit in order to reduce the number of joints in the structure. As far as transport is concerned the limit on the size of a factory cast component is in the region of 18 to 21 m by 2.4 m overall. For multi-storey work a crane is invariably employed for lifting and placing purposes and the size and weight of component should be related to the capacity of the crane likely to be used, since it will most economically be employed when hoisting at its maximum capacity. Individual joints can as easily be made in the case of heavy as in the case of light components when the units are supported by crane, but with smaller, lighter units more joints must be made and the crane must make a greater number of lifts at greater cost.

Precast concrete itself tends to be more expensive than in situ concrete because of factory overheads and transport costs, but against this must be placed, in terms of the structure as a whole, the savings in time and labour on the site, so that the costs of precast and in situ structures are generally about the same, unless the units are precast a considerable distance away from the site making transport costs high.

Site casting Precasting can, or course, be carried out on the site and some large contracting firms do, in fact, carry out much of their casting work in this way, although other comparable firms make a practice of carrying out all such work in a factory, even though this may involve transport over long distances.

Site casting does reduce the amount of handling and avoids transport costs but a large amount of site space is required for the casting beds although methods such as *Tilt-up* construction minimise this (see Part 1, page 15). Provided that the quality of control and supervision usually available in the factory and similar means of efficient vibrating and cleaning of shutters are available on the site, good results are possible. The need for wet concrete vibrating equipment can in fact be eliminated by the use of self-compacting concrete which incorporates recently developed plasticizers and fine sand to facilitate its flow and consolidation in the formwork by means of its own weight. When structural components must be of such a size as to prohibit transport from a factory there is no alternative to site casting. Precasting of facing material or of components requiring a high degree of surface finish is probably best carried out under factory conditions.

4.4.1 Choice of structure

The choice of a particular reinforced concrete structural system as most suitable in any given case will depend largely upon the nature and purpose of the building. For example, the structure of a building to accommodate heavy, evenly distributed loads might most economically be developed as flat slab construction, whereas one in which considerable concentrated loads caused by machinery would occur could most economically be formed with a beam and column system in which the various elements could be designed more easily with regard to the local loading at any point. In the case of flats requiring a high degree of fire resistance and a measure of sound insulation in the separating walls, a box-frame structure is suitable and can be economic whereas for an office block, requiring large areas which can freely be divided up in different ways by non-loadbearing partitions, this would be unacceptable.

Linked with these considerations will be that of resistance to wind. A box-frame structure provides ample resistance in a transverse direction but will require stiffening longitudinally by lift or stair enclosures or by solid walls.

The frames to heavily loaded structures may have beams sufficiently deep to provide the necessary rigidity at the joints with the columns without any other form of bracing. In other cases, where solid walls must be provided on plan for functional reasons, it may be economical to transfer the wind loads to these entirely by the floor slabs so that the columns may be relieved of lateral pressure. When a precast concrete frame appears to be suitable the decision must be made whether to use diagonal bracing within the frame or solid walls at suitable points to provide wind bracing.

As pointed out in chapter 2, the choice of structure is closely linked with the soil conditions on the site and the economic design of the foundations, and the consideration of all three must take place at the same time.

Economic considerations The most economic structure is not always that in which the amounts of steel, concrete and shuttering are all kept to a minimum, and the most economical beam is not necessarily one in which the 'economic percentage' of steel is provided, that is the amount of reinforcement which permits the safe working stresses in the steel and concrete to develop at the same time. In fact, it is normally only in slabs and sometimes in rectangular beams that this is possible. For economy of shuttering it may be desirable to maintain a constant depth for a continuous beam, although the bending moments vary considerably at different points, rather than attempt to reduce the amount of concrete. In order to standardise shuttering in this way to reduce its cost, considerable variations in the concrete mix and steel content, as a means of standardising the size of beams and columns, can often be justified. In some circumstances it may be necessary to increase the steel content in order to restrict the depth of the beam for reasons of headroom, or in order to keep floor to floor heights to a minimum, and the extra cost may well be counterbalanced by savings in other directions.

4.4.2 The materials of reinforced concrete

Concrete itself has considerable compressive strength. The crushing strength of normal concrete will depend on its composition, but is generally in the range of 7.5 to 40 N/mm^2. Concrete is weak in tension, its strength in this respect being only one fourteenth to one eighth of its compressive strength. BS EN 206-1: *Concrete. Specification, Performance, Production and Conformity* in conjunction with BS 8500-1 and 2: *Concrete. Complementary British Standard(s) to BS EN 206-1* give the characteristic compressive strengths for various concrete grades. It may be noted here that concretes with much higher compressive strengths up to 150 N/mm^2 have been developed in recent years and these show benefits especially in the construc-

tion of high-rise buildings. In structural members in which both compressive and tensile stresses occur under load, the full compressive strength of the material cannot, therefore, fully be developed, and in order to overcome this deficiency a material, strong in tension, is introduced in the tensile zones to reinforce the concrete at those points. Steel is used for this purpose at the present time, but fibre-reinforced polymers are being considered as alternatives. Polypropylene and steel fibres can be used in the concrete mix as secondary reinforcement to counter the effect of drying and shrinkage cracking.

Reinforcement Steel is used, either as mild steel or high-tensile steel bars or wires, because it can easily be produced in forms suitable for the purpose, and it possesses to a large degree characteristics which are essential in any material to be used for reinforcement. Assuming adequate tensile strength these are:

- a surface which will satisfactorily bond with the concrete so that when the steel is stressed it will act together with and not pull away from the concrete
- a coefficient of lineal expansion much the same as that of concrete, so that under changes of temperature undesirable stresses will not be set up
- a relatively small elongation under stress to avoid excessive deflection.

Steel reinforcement is obtainable in a variety of forms of bars and fabric.

Bars Circular section rolled mild steel bars or rods are mostly commonly used for all forms of reinforced work. High-tensile steel rods can also be used. Due to the higher working stress, less steel will be required to provide the same strength as mild steel, but in developing its higher strength it stretches more than mild steel and the cracks in the concrete around it will be larger. This might be undesirable in circumstances where corrosive fumes, for example, could attack the steel. The problem of the formation of large cracks is closely linked with that of bond between the steel and concrete. The better the bond the less risk there is of large, concentrated cracks developing, and as a means of increasing bond and limiting cracking to fine, well distributed cracks, *deformed bars* can be used. The greater bond stress obtainable makes it possible to stress the steel to higher limits and thus develop its strength to a maximum. Greatest advantage is obtained when a large number of small diameter bars are used rather than a few larger bars because the surface area in contact with the concrete is thereby increased. The use of these smaller bars eliminates the necessity of end hooks (see figure 4.20), thus economising in steel and simplifying prefabrication.

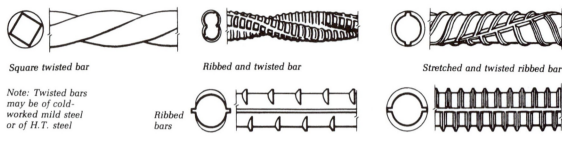

Square twisted bar   **Ribbed and twisted bar**   **Stretched and twisted ribbed bar**

Note: Twisted bars may be of cold-worked mild steel or of H.T. steel   *Ribbed bars*

Figure 4.19 Deformed bars

Deformed bars are produced in a number of ways. Firstly, as high-tensile steel bars rolled with projecting ribs or corrugations along the length or, secondly, from mild steel bars which are cold-worked to increase the ultimate tensile strength and raise or eliminate the yield-point of the steel, the amount of increase depending upon the nature of the basic steel and the amount of cold-working. Both stretching and twisting are used as methods of cold-working and may be applied to circular ribbed bars or to square bars which become deformed by the twisting process and thus afford better bond. Some examples are shown in figure 4.19, with standard hooks and bends shown in figure 4.20.

The importance of eliminating the yield-point lies in the fact that when mild or high-tensile steels are used for reinforcement, both of which have yield-points, the bond between reinforcement and concrete begins to break down when this stress is reached, so that in practice design is based on this value. The use of cold-worked mild steel or high-yield-point steel makes it possible to work to much higher stresses, particularly if deformed bars are used. Within certain limits such steel used as reinforcement can be stressed to its ultimate strength.

Fabric The use of this form of reinforcement is an economic way of reinforcing large areas such as floor and roof slabs. It is produced in two main forms: as a mesh of wire or rods electrically welded at the points of crossing or as expanded steel sheets.

Mesh fabric is manufactured either from hard (cold) drawn steel wire or from small cold twisted steel bars, both of which, due to drawing or cold-working, have

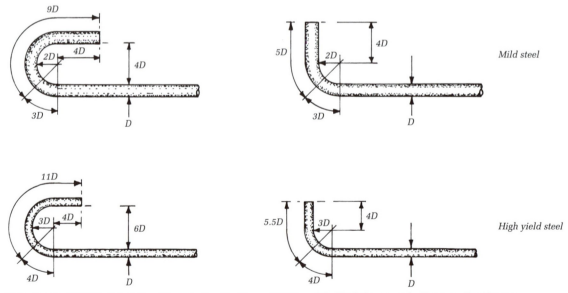

Mild steel

High yield steel

Note: See also BS 8666: *Scheduling, Dimensioning, Bending and Cutting of Steel Reinforcement for Concrete. Specification.*

Figure 4.20 Typical reinforcement hooks and bends

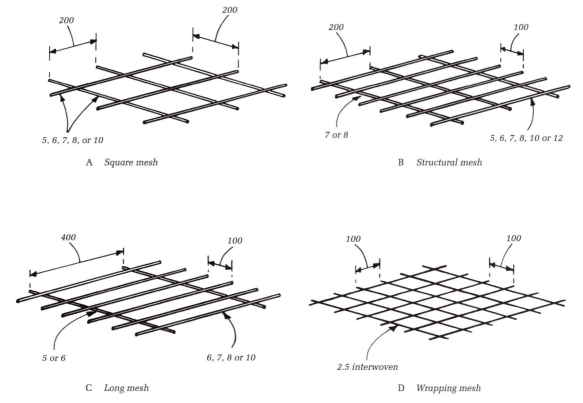

Note: Dimensions in millimetres. See also BS 4482: *Steel Wire for the Reinforcement of Concrete Products. Specification* and BS 4483: *Steel Fabric for the Reinforcement of Concrete. Specification.*

Figure 4.21 Typical fabric reinforcement

greater strength than mild steel, that of the wire being considerably greater. It is supplied as square or rectangular mesh as shown in figure 4.21, in pieces 4.8 m long × 2.4 m wide or in rolls up to 72 m long. Its use avoids the necessity of tying together separate bars.

Expanded metal fabric of steel sheets slit and stretched to form a diamond-shaped mesh is supplied in pieces and for reinforcing purposes has a mesh of 75 mm or more.

Aggregates Various materials are employed as aggregate, the selection depending upon the purpose for which the concrete is used. They may be divided into (i) heavy, (ii) lightweight aggregates.

Heavy aggregates These include the natural sands and gravels and crushed stones covered by BS EN 12620: *Aggregates for Concrete*, and crushed brick. These are normally used where strength and durability are required, although many lightweight aggregates are also used for structural concrete.

Lightweight aggregates These were for many years used for reinforced concrete floor, roof and wall slabs, and are now used also for general reinforced concrete construction. Satisfactory materials are foamed slag, expanded or sintered clay or shale, and sintered fly ash. Expanded slate and imported pumice are also satisfactory. These aggregates can be used with the addition of sand to provide a satisfactory grading to give the necessary strength and impermeability, the resulting concretes having maximum densities from 1400 to 2000 kg/m^3 with characteristic cube crushing strengths from 15 to 60 N/mm^2. The modulus of elasticity of these concretes is less than that of gravel concrete so that deflection in beams and slabs tends to be higher.[19] Depths of structural components, therefore, need to be greater. Multi-storey structures have been constructed in lightweight reinforced concrete for many years, making use of both in situ and precast concrete.[20]

Fire resistance When a high degree of fire resistance is required, the type of aggregate used is important as this largely affects the behaviour of concrete under the action of fire. The classification of aggregates in respect of fire resistance is given on page 318 in terms of spalling and non-spalling types. BS 8110-2: *Structural Use of Concrete. Code of Practice for Special Circumstances* permits a reduction in the size of reinforced concrete members made with lightweight aggregates. For example, for fire resistance periods of four hours and two hours, columns may be reduced from 450 mm and 300 mm to 360 mm and 240 mm minimum overall size respectively.

Size of aggregate   This should be as large as possible consistent with ease of placing round reinforcement. For heavily reinforced concrete members the nominal maximum size is usually taken to be the lesser of the following:

- 6 mm less than the minimum lateral distance between main reinforcement bars, or
- 6 mm less than the minimum concrete cover to reinforcement (see below).

Where the reinforcement is widely spaced as in solid floor slabs, the nominal aggregate size is usually taken as the lesser of:

- 40 mm
- the largest reinforcement bar diameter, or
- the concrete cover to reinforcement (see below).

As a general rule, the maximum size of aggregate should never be greater than one quarter of the minimum thickness of the concrete member. However, for practical reasons such as placement and consolidation of concrete, the maximum nominal size of aggregates is usually limited to 14 or even 10 mm. For general purposes, 20 mm is usually satisfactory and, for relatively thin beams, ribs and structural topping to preformed block floors, 10 mm is normal.

The following is a guide to the minimum concrete cover to reinforcement in super-structural situations:

- ends of reinforcement, 25 mm or 2 × bar diameter (take greater)
- longitudinal reinforcement in columns, 40 mm
- binding wire in columns, 25 mm
- columns with a minimum dimension less than 190 mm with reinforcement less than 12 mm diameter, 25 mm
- longitudinal and transverse reinforcement in beams, 25 mm
- binding wire in beams, 20 mm
- longitudinal and transverse reinforcement in slabs, 20 mm (may reduce to 15 mm if the nominal size of aggregate is less than 15 mm).

Note that adequate concrete cover to steel reinforcing bars is achieved by applying proprietary plastic 'clips' or 'rings' between the formwork/shuttering and the steel bars. See also Building Regulations, Approved Document B: *Fire Safety* to ascertain the minimum cover for fire protection.

4.5 The in situ cast concrete frame

For small span structures a rectangular grid layout similar to that for a steel frame, with one-way spanning floor slabs, can be satisfactory, but with large spans or heavy loading a square grid with two-way spanning slabs is more economical because of the resulting reduction in thickness and dead weight of the slab. Codes of Practice restrict the thickness of floor slabs to a fraction of the span[21] as a precaution against excessive deflection so that there is a limit to the possible reduction in dead weight of slab for any given span. Up to the point at which deflection ceases to be the factor governing slab thickness, no advantage, therefore, is gained by using a two-way spanning slab in place of a one-way span, because the thickness of slab and, consequently, its dead weight, must be the same in both, and there will be little difference in the amount of steel required. After this point has been reached advantage can be taken of the economies resulting from the use of a two-way spanning slab on a square grid.

One-way spanning slabs   In the case of one-way spanning slab construction the transverse, or tie-beams, necessary in a steel frame to provide lateral rigidity to the frame are not essential to an in situ cast reinforced concrete frame, since each floor is cast as the frame rises and can provide rigidity to the frame. Where such transverse beams are omitted, lateral stiffness against wind pressure must, of course, be provided by the floor slab, which should be made strong enough to fulfil this function. Figure 4.22 shows ways of framing in this manner. A shows the floor beams running parallel with the main external walls resulting in a flat ceiling for the length of the structure, an advantage in certain types of buildings, such as offices, where movable partitioning is likely to be changed in position from time to time. With no transverse beam projections such partitioning can be standardised to the floor to ceiling height and be freely placed in any position. The supporting columns may vary in position along each beam relative to those carrying the other beams, although, unless essential for planning reasons, this would not be done because of the variations caused in beam shuttering and possible variations in foundation loading. When the width of the building necessitates two lines of internal columns these may be placed the width of a corridor apart and the floor slab between them thickened to form a stiff longitudinal beam (B and figure 5.9). This will act with the columns as a rigid inner structure to resist the wind pressure transferred to it through the outer spans of floor.

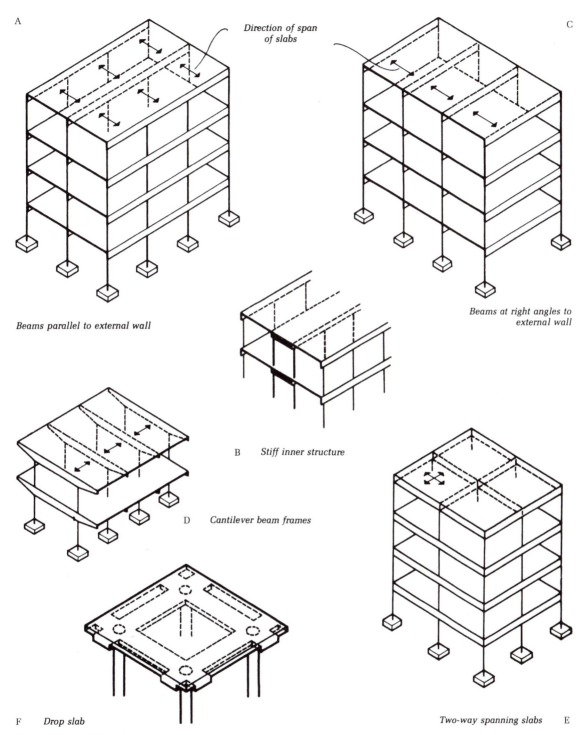

A

Direction of span
of slabs

C

Beams parallel to external wall

Beams at right angles to
external wall

B   Stiff inner structure

D   Cantilever beam frames

F   Drop slab

Two-way spanning slabs   E

Figure 4.22   Reinforced concrete frames

Figure 4.22 C shows the floor beams running at right-angles to the main external walls which are free of beams, thus permitting lightweight infilling panels to run from floor to slab soffit on elevation. If the beams are made sufficiently deep, internal columns may be omitted giving wide, unobstructed floor areas where these are necessary. D shows cantilever beam frames, the advantages and disadvantages of which are, in principle, the same as those constructed in steel (see page 143). The longitudinal beams necessary with steel can, however, be omitted as shown, provided the floor slabs give adequate lateral stiffness. The propped cantilever principle (see page 145) may be applied to a reinforced concrete structure and, in some cases, it may be economic to omit the cantilever beams and design the floor slab itself to cantilever over longitudinal beams. The soffit can be sloped up to a shallow outer edge beam which will be supported by the outer 'props'.

Two-way spanning slabs As already mentioned at the beginning of this section, the rectangular grid layout can be economical for small spans and lightly loaded structures, but when larger spans and heavy loads are involved the square grid with two-way spanning slabs shown in E is likely to be cheaper. Although normal beams carrying a simple solid slab of this type will show economies over one-way spanning slab construction, in the case of wide-span grids certain variations will result in greater economies by further reducing the dead weight of the floor slab. The normal slab and deep beams may be replaced by a drop slab in which the beams are replaced by a thickening of the slab to form wide, shallow bands over the lines of the columns as shown in F. The effect of widening the beam to a band is to shorten the span of the slab, with a consequent reduction in its thickness, dead weight and amount of reinforcement.

The floor slab, which is an integral part of the structure and has a significant effect upon the economics of the building as a whole, may be constructed in various ways. Types of floors are discussed in chapter 5.

Adaptability of concrete Concrete in solution is a very adaptable material. It permits considerable design flexibility and latitude in the form of structural members for structural or other reasons such as planning and lighting, and readily permits the interaction of beams and floor slabs to produce more economical members. Some examples of this adaptability are shown in figure 4.23. The use of beam and column framing with loadbearing concrete walls in the same structure is logical when the latter can fulfil an enclosing, as well as structural, function (A) and is often adopted to provide or assist in the resistance to wind pressure (D). B and C show ways in which members may be formed to fulfil dual functions, to accommodate services or to distribute material according to the stress distribution in a member.

T- and L-beams Where the floor slab is cast monolithic with a beam, as is usual, part of the slab may be utilised to act with the beam to form a flanged beam (see figure 4.24). The necessary tension reinforcement in a beam can be accommodated in a relatively thin rib of concrete which needs to be only wide enough to accommodate the steel and to provide for shear stresses. By combining such a rib with the floor slab the necessary resistance in compression may be obtained with a minimum depth of beam and without the use of compression reinforcement. Where the slab extends on both sides of the rib the beam is termed a T-beam; where it is on one side only of the rib it is termed an L-beam. The width of slab which may be assumed to act as the flange of a T- or L-beam is laid down in BS 8110-1: *Structural Use of Concrete. Code of Practice for Design and Construction.* Beams are normally continuous over supports and in the case of T- or L-beams the reversal of stresses at the point of support presents a problem since the rib, in compression at those points, is generally insufficient in area to resist the compressive stresses. The problem may be solved by any of the following methods:

- providing compression reinforcement
- deepening the rib by means of a haunch, which increases the area
- widening the rib for its full depth
- widening the bottom of the rib only to provide a lower flange.

For most building frames compression reinforcement is invariably used and is generally no more expensive than the provision of haunches, which is the most common alternative. However, in heavily loaded frames, where shear stresses are likely to be high at the supports, haunches may be preferable. The flaring or widening of the rib and the provision of a lower flange are rarely used as they complicate the shuttering (see figure 4.24).

4.5.1 Layout of reinforcement

In detailing the reinforcement in a member the arrangement of the bars should be as simple as possible with sufficient space left between the bars for each to be surrounded by concrete. The minimum distance between bars must be greater than the maximum size of aggregate used. The space needed between the bars, together with the thickness of external cover, is often a governing factor in determining the size of a member. BS 8110-1 gives recommendations on the spacing of bars.

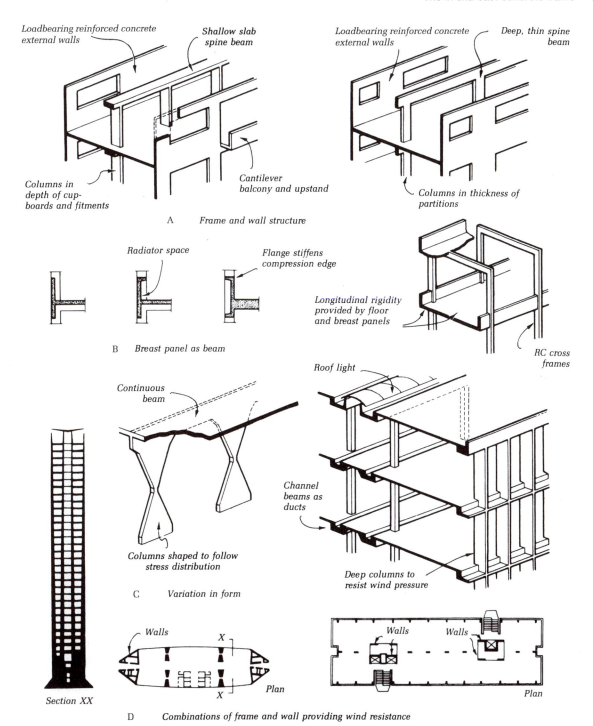

Loadbearing reinforced concrete external walls

Shallow slab spine beam

Columns in depth of cup-boards and fitments

Cantilever balcony and upstand

Loadbearing reinforced concrete external walls

Deep, thin spine beam

Columns in thickness of partitions

A   Frame and wall structure

Radiator space

Flange stiffens compression edge

B   Breast panel as beam

Longitudinal rigidity provided by floor and breast panels

RC cross frames

Roof light

Continuous beam

Channel beams as ducts

Columns shaped to follow stress distribution

Deep columns to resist wind pressure

C   Variation in form

Section XX

Walls

X

X

Plan

Walls

Walls

Plan

D   Combinations of frame and wall providing wind resistance

Figure 4.23   Adaptability of reinforced concrete

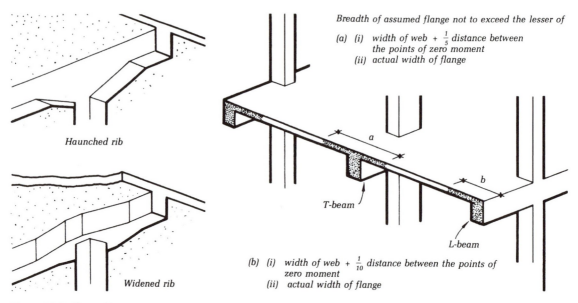

Breadth of assumed flange not to exceed the lesser of

(a) (i) width of web + $\frac{1}{5}$ distance between the points of zero moment
(ii) actual width of flange

(b) (i) width of web + $\frac{1}{10}$ distance between the points of zero moment
(ii) actual width of flange

Haunched rib

Widened rib

T-beam

L-beam

Figure 4.24 Flanged beams

The minimum number of different bar sizes should be used, and the use of the largest size consistent with good design will reduce the number of bars to be bent and placed.

Bars must be extended beyond any section sufficiently far to enable the required grip to be developed or be hooked. Hooked bar ends occurring in a tensile zone may cause cracking, and to avoid this the hooks should be omitted and the bars made longer to allow for the loss of the hooks. Alternatively the bars should be bent into a compression zone.

Tension bars continuous round a re-entrant angle, as in a cranked slab, should have a radius large enough to reduce the outward pressure of the steel to that which the concrete can resist in shear and tension. If this is not possible the tension bars should be linked by stirrups to compression bars, or better, the bars should be separate and should extend beyond the intersection sufficient for bond (figure 4.25).

Links must be provided to avoid the possibility of buckling of compression reinforcement and the bursting out of the concrete. To make the reinforcement stiff during concreting, and to hold stirrups in position, bars in the corners of beams are provided and the stirrups are continued round the tension side. At all points of intersection the bars must be wired together to prevent displacement during concreting.

Some typical details of reinforcement layout are shown in figure 4.25.

4.6 The precast concrete frame

The precast concrete structural frame has developed as a result of attempts to combine the advantages of the steel frame with the economy of the concrete frame. A precast frame will generally be less expensive than a steel frame encased in concrete and comparable in cost with an uncased steel frame. In multi-storey buildings a steel frame must be encased with some appropriate fire resisting material and in many cases concrete is used to obtain sufficient protection. The amount of concrete necessary is usually almost as much as that required for a comparable frame constructed of reinforced concrete, and the concrete frame would show a saving in cost over that of the encased steel frame. In addition to this, the process of casing the steel does away with much of the saving in time associated with steel construction.

The in situ concrete frame involves a considerable time lag between the pouring of the concrete and the removal of all shuttering and temporary supports, resulting in a delay in the re-use of shuttering and in the obstruction of working areas for long periods. Shuttering can be complicated in the case of slabs and beams. By applying the technique of precasting the disadvantages of in situ work can be avoided and benefit derived from some advantages linked with the steel frame.

The frame may be (i) partially or (ii) wholly precast. In the first method the horizontal members only are precast,

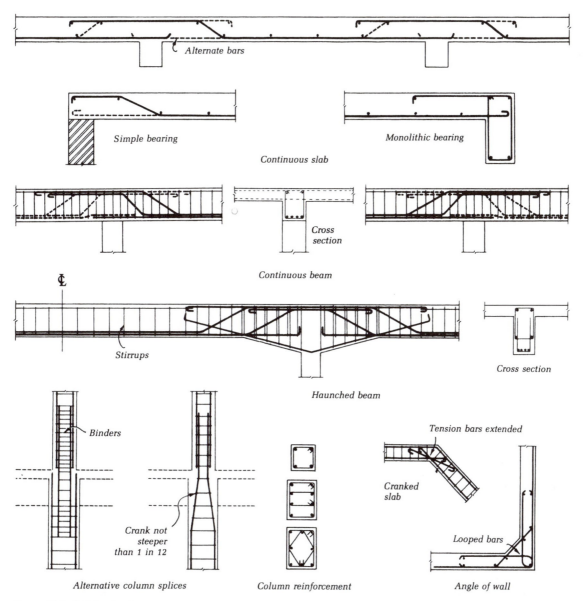

Alternate bars

Simple bearing

Monolithic bearing

Continuous slab

Cross section

Continuous beam

Stirrups

Cross section

Haunched beam

Binders

Tension bars extended

Cranked slab

Crank not steeper than 1 in 12

Looped bars

Alternative column splices

Column reinforcement

Angle of wall

Figure 4.25 Reinforcement

the columns being cast in situ with continuity simply achieved in the normal way as each section of the column is cast. The factors in favour of casting the columns in situ are, firstly, the simplicity of achieving continuity by this means and, secondly, the fact that solely from the point of view of shuttering there is little in favour of pre-casting. This is because column shuttering is simple in form, takes up little space, requires relatively little labour to erect and strip and involves a negligible wastage in re-use. In contrast, in the case of beams and slabs, bending stresses are

set up in the shuttering while the concrete is wet, necessitating heavy forms and a considerable amount of propping. This is avoided when the shuttering is supported by the ground or a production bench and the formwork can be lighter and cheaper. Floor areas are obstructed by props for considerable periods while the concrete attains sufficient strength to support its own self-weight. Considerable wastage in horizontal shuttering occurs in stripping and re-erecting,[22] and this work takes longer than in the case of columns. These disadvantages are avoided when the horizontal members are precast, so that the arguments in favour of this are considerable.

Nevertheless, arguments in favour of in situ cast columns based only on shuttering disregard questions of quality of concrete and of time and labour spent on the site and there is now a wide use of wholly precast frames, especially since the development of multistorey columns precast up to four storeys in height, which minimise site construction time considerably.

4.6.1 Methods of fabrication

Precast frames can be fabricated in a number of different ways:

- from individual beams and columns, the columns sometimes being cast in more than one storey height
- from 'frames' composed of column sections and beam lengths forming a single cast unit
- from precast units acting as permanent structural shuttering to cast in situ concrete to form a composite structure.

Precast beams and columns Details of this method are shown in figure 4.26. As in the case of a steel stanchion a precast concrete column must be connected by some means to the foundation slab, and methods of accomplishing this are shown in A.

Column joints, usually of such a nature as to ensure continuity, must be made at each floor level in the case of storey height columns or at less frequent intervals if multistorey height columns are used. The attainment of satisfactory continuity at the column connections necessitates either the exposure of relatively long lengths of reinforcing rod at each joint in order to obtain sufficient bond length or the use of some form of connecting plate. The section of column left open for the jointing of the rods must then be boxed in and concreted solid in situ. While such a joint is being made and until the in situ cast concrete has gained sufficient strength, the upper section of column must be adequately supported and held fixed in position. To facilitate this a number of methods of jointing have been evolved to permit the self-weight of the upper section to be transferred to the lower section while the joint is being made so that bracing only is required to hold it in a vertical position (B).

Beam to column connections With multi-storey height columns provision must be made for the connection of the beams at the intermediate levels and this is accomplished by providing either haunches or projecting steel sections as a seating for the beams (C, D). The whole or partial omission of the concrete at the appropriate points accord-

ing to the number and disposition of beams to be connected has been adopted in the past. The exposed reinforcing rods are stiffened against bending under the weight of the upper column lengths during erection by welding mild steel rod bracing to the main rods (E). Alternatively, at such points the reinforcing rods are replaced by a length of Universal section or by boxed channel sections welded edge to edge (F).

Connections between beams and columns can be made in various ways according to the degree of rigidity and continuity required. F shows a method in which the beam ends are boxed or rebated so that the in situ concrete filling makes the whole joint monolithic. Continuity rods are placed between the beam ends to transfer the negative bending moment over the column. In the case of three-way connections or corner columns the continuity rods for the centre or corner beams are usually cast in the column and bent up out of the way until the beams are in position. When the columns can be haunched to provide the beam seating, a simple connection can be made using a mild-steel dowel to provide a positive beam fixing within the small beam rebate which is filled with in situ concrete (C). Site welded steel bearings and continuity bars passing through holes in the column are shown in D. The bars bond with the in situ topping to the beams. Alternatively, bolted connections can be made, using projecting steel brackets cast into the columns or bolted to cast-in steel connectors, to provide a seating for the beam ends. To these the beams are bolted through steel bearing plates cast in a rebate on the underside of the beam ends to ensure a full bearing of beam on the column support.

Beams may be quite independent of the floor or roof slab which they support or, more generally, they may be designed to act integrally with the slab in the form of T- or L-beams. This necessitates some form of shear connector and in situ concrete to enable beam and slab to act together. The connectors are usually in the form of bent steel rod stirrups projecting from the top of the beam (figure 4.26 D, G). In situ concrete is required at the beam position to integrate the beam and the precast floor or roof elements which are usually notched or troughed at the ends to receive it (see figure 5.10 A). When the slab elements run parallel with the beam, sufficient width of in situ concrete must be allowed to provide the flange to the beam (figure 4.26 G). As shown in this illustration precast soffit elements can be incorporated to avoid the use of normal shuttering.

Lateral rigidity A structure of precast beam and column units can be given lateral rigidity by making certain parts of in situ construction to which the precast portions are securely tied. Less rigidity is then required at the joints between the precast units and construction is simpler. The

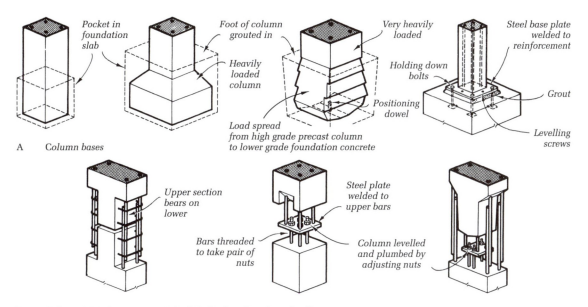

A   Column bases

- Pocket in foundation slab
- Foot of column grouted in
- Heavily loaded column
- Very heavily loaded
- Load spread from high grade precast column to lower grade foundation concrete
- Positioning dowel
- Steel base plate welded to reinforcement
- Holding down bolts
- Grout
- Levelling screws

B   Column joints (space concreted solid after bars have been fixed)

- Upper section bears on lower
- Bars threaded to take pair of nuts
- Steel plate welded to upper bars
- Column levelled and plumbed by adjusting nuts

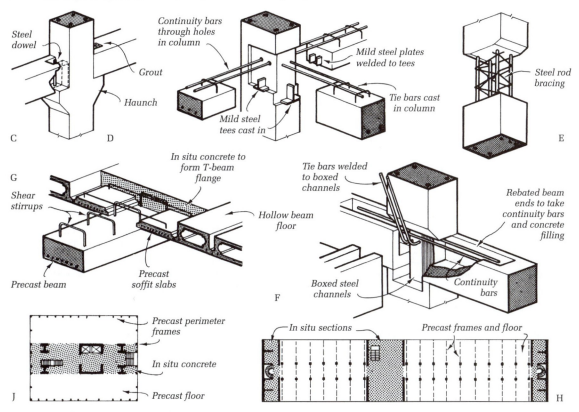

- Steel dowel
- Grout
- Haunch
- Continuity bars through holes in column
- Mild steel tees cast in
- Mild steel plates welded to tees
- Tie bars cast in column
- Steel rod bracing

C   D   E

G

- Shear stirrups
- Precast beam
- In situ concrete to form T-beam flange
- Hollow beam floor
- Precast soffit slabs

- Tie bars welded to boxed channels
- Rebated beam ends to take continuity bars and concrete filling
- Boxed steel channels
- Continuity bars

F

J
- Precast perimeter frames
- In situ concrete
- Precast floor

- In situ sections
- Precast frames and floor

H

Figure 4.26   Precast concrete structures

in situ work can be in the form of end and intermediate bays of the building constructed with cross walls running the width of the structure, as shown in H. To these the precast portions between may be tied at each floor level by tensioned cables passing through ducts in the floor slabs, the floors thus forming wind girders spanning between the in situ blocks. In building types with a central core of services, stairs and lifts, the whole core can be of in situ cast reinforced concrete and precast concrete members can be used on the outside walls as shown in J. These can be tied back to the in situ core by cables as described above, or by bolting the precast floor slabs to the precast perimeter frames and the core.

Lateral rigidity may also be obtained by the use of diagonal braces placed in the vertical plane at various points in the structure, similar to the arrangement in a steel frame.

Precast frame units   These can be formed in various ways but each type consists essentially of a pair or more of columns linked by a beam and so formed that beam and column connections do not occur at the same point, thus overcoming the difficulties of assembly which arise when they do coincide. A pair of columns linked by a beam is easier to brace temporarily while the joints are being made than separate columns. Such units are, therefore, easier and quicker to erect than separate beams and columns. These units are suited to a layout in which there are no lateral beams, the floor and roof slabs spanning directly between lines of support running parallel to each other.

Figure 4.27 shows types of such units. In A the column joint is located at the top of the beam and in B at the points of contraflexure in the columns where the bending moment is at a minimum. When the perimeter columns of a building are closely spaced to act as window mullions, frame units can be formed of two columns, a head beam and a cill beam. The method of linking the units varies according to the treatment of the elevation. In C the head and cill beams will be hidden by cladding and the columns will be exposed. The head beam, that is the floor beam, can project on each side and meet its neighbour at the centre of the adjacent bay without detriment to appearance. The cill beam acts as a brace to the frame and may or may not provide support to cladding. When the whole of the frame is to be exposed on elevation, care is needed in the arrangement of the joints. The head and cill beams are kept within the line of the columns and when the units have been erected in alternate bays they are joined by separate beams at cill and head which are bolted to them (D).

These illustrations show storey height frames with a pair of columns, but they can be constructed three or four columns wide and with only a top beam, provided the beam is substantial enough to withstand the hoisting stresses and the weight is within the capacity of the lifting crane. Two- or three-storey height frames can also be used. These multiple frames reduce construction time by the reduction in the number of joints to be made.

Column connections are illustrated in figure 4.27 A, B, C, details 1, 2, 3. They may also be made by means of high tensile steel bars passed through holes formed through the height of the columns. These are connected at the joints by screwed couplers and tightened by a torque-controlled spanner. Connections between beam ends are made by coupling plates or rebated ends and dowels. Rectangular frames may be formed of half-columns at each side and half-beams at head and foot. The half-columns are channel shaped so that when erected a void is formed between them which is filled with in situ concrete, which can be reinforced if necessary. The half-beams are bolted together.

Composite structure   Reference is made above to the use of precast soffit elements as permanent shuttering to portions of in situ cast work required to form T- and L-beams at the junctions of precast beams and slabs. The use of reinforced precast concrete units as permanent shuttering, designed to act with in situ concrete to form a composite structure as illustrated in figure 4.27 E, is a means of obtaining the continuity and rigidity inherent in in situ cast work without the use of normal formwork. It also reduces the amount of precast work which factory overheads and transport costs tend to make more expensive than in situ work.

For economy the units should be shallow, but for ease of handling their thickness should not be less than one-fortieth of the length. In order to obtain units of reasonable length, therefore, the section should be of such a shape as to give stiffness to a thin member (see figure 4.27 E) or be stiffened in some other way. In the case of beams the precast element should extend no higher than the soffit of the floor slab, so that in some instances the precast unit may be only a shallow strip carrying the tensile reinforcement similar to the prestressed element shown. To stiffen this during transport, stirrups and any top reinforcement for the beam may be introduced. Increased stiffness is given if diagonal bars are used to form the reinforcement into a lattice girder similar to the elements of this type used for floor slab construction shown in figure 5.2 G.

Satisfactory bond between the precast units and the in situ cast concrete is essential in order to transfer shear stress. A roughened surface on the precast unit is usually adequate, but this does assume good site supervision and workmanship in forming the junction between the two. A definite mechanical bond is preferable, achieved by means of projecting wire stirrups or castellations. When composite beams or slabs are continuous the negative moments over the supports must be resisted by reinforcement placed in the in situ cast concrete.

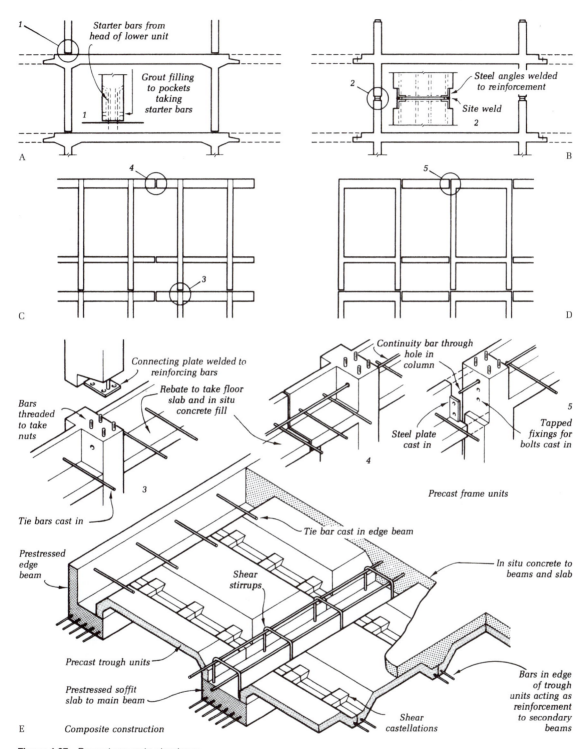

1 Starter bars from head of lower unit

Grout filling to pockets taking starter bars

A

Steel angles welded to reinforcement

Site weld

2

B

4

3

C

5

D

Connecting plate welded to reinforcing bars

Rebate to take floor slab and in situ concrete fill

Bars threaded to take nuts

Tie bars cast in

3

Continuity bar through hole in column

Steel plate cast in

4

5

Tapped fixings for bolts cast in

Precast frame units

Tie bar cast in edge beam

Prestressed edge beam

Shear stirrups

In situ concrete to beams and slab

Precast trough units

Prestressed soffit slab to main beam

Shear castellations

Bars in edge of trough units acting as reinforcement to secondary beams

E   Composite construction

Figure 4.27   Precast concrete structures

Application of prestressing   Figure 4.27 E shows a floor for heavy loads which incorporates a wide shallow beam in which the tensile zone is a precast prestressed slab, the compression zone being of in situ cast concrete. The side shuttering of the beam is formed by the edges of the precast trough units forming the lower part of the floor slab. These trough units are very thin but their form, which provides the shuttering for main and secondary beams, makes them stiff enough to carry the live loads during the casting of the in situ concrete.

The combination of precast and in situ cast concrete is particularly economical when allied to prestressing in the range of 6 to 9 m spans, for which normal prestressed concrete is not generally economical. In a normal prestressed concrete beam the concrete throughout is of high quality, but in a composite beam the lower precast and prestressed section only need be of high quality concrete, thus effecting an economy due to the smaller volume of high quality prestressed concrete to be manufactured and transported to the site.

Precast units and in situ concrete may be combined in columns as well as in beams and slabs. By using a precast concrete casing with an in situ cast core, time and labour can be saved by the elimination of normal shuttering and a good finish is obtained when the surface of the columns is to be exposed. It also permits the construction of the next floor to proceed more quickly while still maintaining full monolithic junctions with the floors and beams above and below. Fibre reinforced plastic hollow tubes can also be used to confine the concrete.

4.7 The reinforced concrete wall

4.7.1 In situ cast external wall

The reinforced concrete loadbearing wall used as the enclosing wall to a building is the alternative to its use as a dividing element in the concrete box frame described below. The wall areas over openings act as beams and those areas between openings as columns, thus no projections occur internally (see figure 4.23 A). These openings may be wide, since with normal cill heights there is ample depth of wall between window head and cill above to act as a deep, thin beam and the wide, narrow window is a characteristic of this form of construction. Alternatively, the whole height of the wall may be regarded as a beam pierced by any necessary openings for windows.

Sufficient width of wall must, of course, be left between openings to act as columns taking all the vertical loads. The problems of appearance and thermal insulation are the same as with the plain concrete wall, but the danger of cracking due to possible unequal settlement is reduced because reinforcement is present to resist any tensile stresses set up.

4.7.2 The concrete box frame

This is a form of cross-wall construction in which the walls are of normal dense concrete and, with the floors, form box-like cells as shown in figure 4.28. As in the case of brick or block cross-wall construction, it is suited to those building types in which separating walls occur at regular intervals and are required to have a high degree of fire resistance and sound insulation. The most common building type in this category for which it is suitable is the multi-storey flat or maisonette block. The advantages listed on page 89 in respect of brick and block cross-wall construction apply also to the box frame.

In concrete walls of normal domestic scale, about 2.4 m high and 100 mm thick (subject to greater thicknesses as deemed necessary for thermal and sound insulation, and fire resistance), failure is almost wholly related to the strength of the concrete and very little to the slenderness of the wall. Reinforcement, therefore, may be nominal in amount or may be omitted altogether provided that the concrete is sufficiently strong to resist the stresses set up under load. For multi-storey blocks in the region of ten or eleven storeys high the mix would be designed to give a strength of around 15 N/mm^2 at 28 days, although for the two lowest storeys a stronger mix might be necessary as well as the inclusion of reinforcement.

Cracking due to the shrinkage of concrete is normally overcome by the inclusion of shrinkage reinforcement. Such cracking generally occurs only if the shrinkage is resisted by some restraint, such as that offered by changes in the plane of a wall or by a previously poured lift of concrete which has been permitted to take up its shrinkage before the next lift is poured on to it. Provided that concreting can proceed without undue delay and that the walls are in simple, straight lengths, shrinkage reinforcement in the walls may safely be omitted.

Lateral rigidity   Although the junctions of walls and floors in a box frame are monolithic, if the walls are not reinforced the structure can only provide rigidity in the length of the building to the extent of the precompression set up in the walls by the floor loads and self-weight of the walls, as explained in the case of normal cross-wall construction (see page 91). Additional stability must normally be given by staircase and lift shafts of reinforced concrete, or by the inclusion of longitudinal walls at certain points in the plan. The box-walls themselves provide rigidity in the transverse direction.

Many box frames have been constructed with the end, or gable walls similar in form to the internal cross walls. The solid external concrete wall suffers certain disadvantages (page 87 and Part 1, page 68) and to it must be applied thermal insulation and, generally, some external facing for

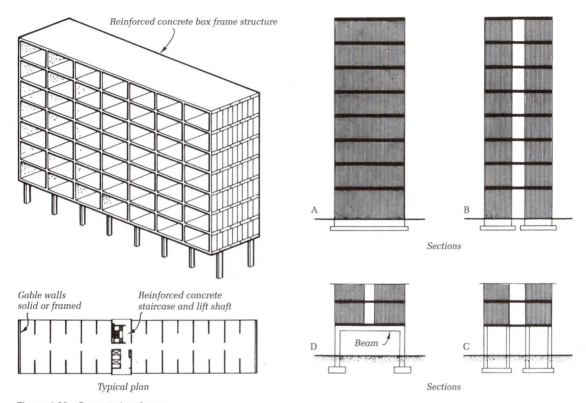

Reinforced concrete box frame structure

Gable walls solid or framed

Reinforced concrete staircase and lift shaft

Typical plan

A

B

Sections

D

Beam

C

Sections

Figure 4.28 Concrete box frame

the sake of appearance and to ensure weather resistance. As both can be applied as satisfactorily to a frame as to a solid wall and since the latter, used as a gable wall, is more expensive than the frame, it appears logical to use a frame in this position.[23] With regard to thermal insulation this should be provided at the ends of the box walls as heat losses at these points can be high, leading to condensation on the internal faces of the walls adjacent to the exposed ends. Depending upon the relationship of infilling panels and wall ends the insulation may be applied to the inner faces or end of the wall as shown in figure 3.10.

Relative costs For low-rise blocks the box frame is unlikely to be as economical as either a reinforced concrete frame or brick cross-wall construction. In high-rise blocks, however, it is more cost effective than the frame when plan requirements permit the walls to be spaced 5.00 m or more apart but less economic when the walls must be placed closer together. It should be noted that the optimum spacing of the walls is 5.00 to 5.50 m with the cost rising a little more rapidly with a decrease than with an increase in the spacing. It should also be noted that cost comparisons should take into account the effect of other

elements that have a bearing upon the total cost of the building as a whole. For example, the cost of infilling panels of bricks or blocks at the separating wall positions, to provide the necessary degree of fire resistance and sound insulation, must be added to the cost of the reinforced concrete frame to make it comparable with the box frame.

In its simplest and most economic form all the box walls run in a straight, unbroken line from back to front of the building and are supported directly by a strip foundation (figure 4.28 A). Although effectively a strip, the foundation is more likely to be a reinforced beam supported on piled foundations, particularly if the structure is several storeys high. Walls may be pierced by openings or be in completely separate sections on the same line, or staggered relative to each other provided that each section is in the same position throughout the height of the building (B). If the upper floors are to be supported on columns at ground level the necessity of beams and the disposition of the columns will depend upon the arrangement of the walls above. Straight, unbroken box walls can act as deep beams spanning between the supporting columns with any necessary reinforcement placed in the tension and shear zones. If the walls are broken extra columns must be introduced

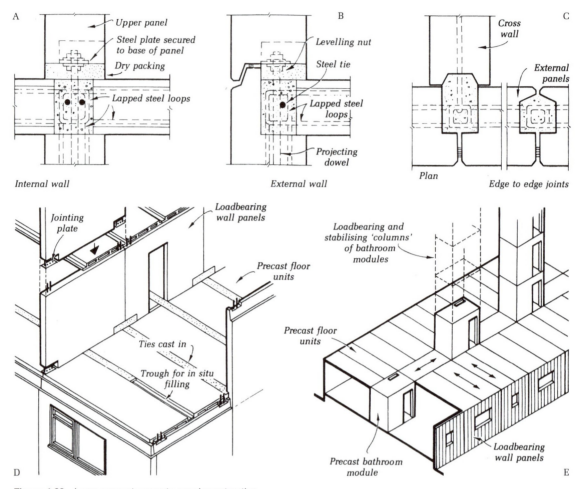

Figure 4.29 Large precast concrete panel construction

to enable each wall section to act as a beam (C) or, alternatively, a separate beam must be introduced to pick up the sections and transfer the loads to the columns (D).

4.7.3 Large precast panel structure

In this form of construction the loadbearing elements are large panels not less than storey-height, used with precast floor and roof units (figure 4.29 D). Window openings may be cast in the external panels which are usually finished with an exposed aggregate or tooled or profiled surface; thermal insulation is applied to the internal face. Internal panels can be made smooth enough to make plastering unnecessary.

The method is most suitable for residential buildings since the dense concrete panels can provide, as well as the strength for loadbearing, the degree of fire resistance and sound insulation required at the separating walls. Cellular, cross-wall and spine-wall plan forms may be used. The advantage of the cellular plan is its inherent stability and the fact that all walls may be loadbearing so that the floor panels may be two-way spanning.

Types of panels External wall panels are commonly solid and may be finished on the external face in various ways by means of exposed aggregate, ribbing or other forms of profiling. As with any form of concrete cladding or facing adequate cover to its reinforcement must be provided and any tooled, profiled or applied aggregate finish must not reduce this cover. Alternatively, and with regard to thermal insulation requirements, panels may have a smooth outer face with fixings cast in to take various forms of facing or overcladding. Dry linings incorporating insulation can be applied to the internal face.

Internal loadbearing wall panels are solid or cored and between 125 and 225 mm thick with nominal reinforcement. Adequate sound insulation can be achieved with a thickness of 190 mm if plastered on both sides and rather thicker if not.

Floor panels may be of solid or cored construction. The former may be reinforced as two-way spanning slabs and they also provide better air-borne sound insulation. The latter are lighter in weight, but can span in one direction only and are commonly prestressed.

Casting panels Horizontal casting is used for complicated panels which present some difficulty in casting, such as those with openings in them and those which are to have an integral or applied surface finish. When cast these are preferably removed from the moulds by means of pivoting mould beds or by vacuum pads, in order to avoid damage. The former method avoids the need for reinforcement to resist lifting stresses.

Vertical casting is preferable for wall and floor panels required to have a fair face both sides since this has the advantage of eliminating face trowelling. The moulds can be arranged in batteries with ten or more compartments, the division plates being of thick steel or concrete panels or of ply facing on both sides of a steel frame. A production system of using two concrete panels, initially cast horizontally with a very smooth, true face, as the mould faces to reproduce a run of similar units was developed during the 1960s by the Building Research Establishment. The first panel is cast between the initial pair and has a true face on each side. After curing the three panels are spaced apart to provide the mould for two further panels, the five then being used to produce four more, and so on.

Structural connections In situ concrete is commonly used to form the structural joints between panels. The method used to form the horizontal joint between the wall panels is shown in figure 4.29 A, B. A threaded bar or dowel projecting from each end of the lower panel provides, by means of nut and washer, temporary support and a means of levelling the upper panel while a top nut and washer, when tightened down, enables it to act as a vertical tie between upper and lower panels. The space at the ends of the floor slabs is filled with in situ concrete and after this has set and the upper panel has been levelled the gap above it is dry packed with cement mortar. It is most important that this dry packing extends the full thickness of the slabs and is thoroughly compacted, and that when it has set the lower nuts are run down to ensure contact between the upper slab and the packing, thus off-setting the initial shrinkage of the mortar and ensuring a full transfer of load only through the packing and concrete filling. Vertical edge joints may be made as in C, the ends of the

panels being rebated to take a concrete filling, or recessed to hide the joint as in the right-hand detail.

Methods of weatherproofing external joints are illustrated in B and C, but reference should be made to Part 1, chapter 2, where this subject is discussed in detail and alternative methods are shown.

Disproportionate collapse The Building Regulations require buildings to 'be constructed so that in the event of an accident the building will not suffer collapse to an extent disproportionate to the cause'.[24] The application of this requirement in terms of storey height (basements included) varies depending on the building type and its occupancy. With precast panel structures, in particular, precautions against collapse due to accidental removal of loadbearing members must be taken by providing adequate horizontal and vertical ties between the members of the structure. Steel loops or stirrups are cast-in the edges of wall and floor slabs to tie together the component parts of the structure and continuous steel reinforcement acting as a peripheral tie is required at each floor and roof level (figure 4.29 B) to which the loops in both directions at each level are bonded. Additional ties in the direction of the span may be encased in the floor or roof panels (D) and transverse ties may be placed in the in situ concrete in the joints between the transverse wall panels (A).

Where strength requirements necessitate more than one vertical tie dowel at each end of a panel as described above they can be accommodated by a larger steel plate welded to the reinforcing bars projecting in rebates in the bottom corners of each panel (D). This method is similar to that used for precast columns shown in figures 4.26 and 4.27.

Lateral rigidity With cross and spine wall plans, overall rigidity can be provided by in situ cast lift and stairwells, but this has the disadvantage of mixing precast and in situ work on the site. Fully precast construction uses bathrooms precast as reinforced concrete boxes or modules complete with floor and ceiling and lift and stairwells precast in storey or half-storey heights which, when erected on each other, form structural 'columns' running the full height of the building. A number of these vertical units along the centre of the block form a structural spine to the remainder of the structure which is fabricated from large precast floor panels and storey height loadbearing wall panels (E).

For low-rise buildings the precast multiple T-beam slabs described for floors on page 202 may be used vertically for walls. The ribs are placed externally and the slab acts as the structural outer leaf of a cavity wall, the roof and floors being carried on corbels cast on the back of the slab. For tall single-storey buildings such as those referred to under diaphragm and fin walls (page 85) these slabs, with the base sunk into pocket foundations, may be made to act as propped cantilevers comparable to the masonry fin wall.

4.8 Prestressed concrete

Prestressing is the process of imparting to a structural member a compressive stress in those zones which under working loads would normally be subject to tensile stresses. It is, in fact, a process of precompressing by means of which the tensile stresses produced by the applied load are counteracted by the compressive stresses set up before the application of the load. This can very simply be seen in the process of removing a row of books from a bookshelf. The row of books in itself has no tensile strength and unless supported by a shelf would fall apart, but by applying pressure by a hand at each end the row may be made to act as a beam and be lifted off the shelf. The pressure of the hands sets up a compressive stress which overcomes the tensile stress which the weight of the books would set up and which would cause the books to part from each other.

Although particularly advantageous in concrete work for reasons given below, prestressing may also be used in brick and stone masonry and in steel and timber construction.

Advantages of prestressing   Normal reinforced concrete is not able to benefit fully from the high-quality concrete and high-tensile steel now available because of the low straining capacity of concrete in tension, which results in cracks appearing in the concrete around the reinforcement at loads well below the normal design load. While generally not dangerous these cracks in practice are usually limited to about 0.25 mm in width, thus limiting the stresses which may be applied to the reinforcement, so that neither the qualities of modern high-strength concrete nor of high-strength steel may be fully developed.[25] In a prestressed member, however, the concrete is at all times under compression so that there is a complete absence of cracks. In the event of an overload, provided that this is within the elastic limit, the cracks formed will close again after removal of the load without harm to the structure. The high compressive strength obtainable in present day concrete can, therefore, be fully used, while at the same time the high-tensile qualities of modern steel may also be fully utilised because the steel is not used as normal reinforcement to take the tensile stresses which the concrete is unable to resist, but, as will be seen later, is used solely as a means of producing the compressive stress in the concrete.[26]

Reduction in the depth of beams and slabs, thus producing higher stresses, is therefore possible without giving rise to crack formation, and depth/span ratios of 1:20 for beams and 1:40 for slabs are common, although for beams

much smaller ratios are possible depending upon loading conditions. The applied compressive stress, in addition to cancelling out the tensile stresses due to bending, considerably reduces those tensile stresses caused by shear so that the webs of prestressed beams can be much thinner than in normal reinforced concrete beams, resulting in I- and box-sections as typical prestressed concrete forms. The smaller sections thus possible produce considerable savings in steel and concrete. They result in dead weights of up to 50 per cent less than with normal reinforced concrete, and because the high grade steel used can be stressed to its limit, a saving in the weight of steel of one-tenth to one-fifth can be shown over that of normal reinforcement. Although high-tensile steel is more expensive than mild steel there is a saving in cost because of the small amount required.

The decrease in the dead to live load ratio considerably reduces costs over medium and long spans, increases maximum spans, and makes prestressed concrete much more suitable for wide span members carrying light loads than normal reinforced concrete. The lightness of prestressed work, due to reduced depth and web thickness and to the reduced amount of steel, and the longer spans economically possible, results in lower column and foundation costs.

Applications of prestressing   In its application to building work prestressed concrete is mostly used for beam and slab members in precast construction. When applied to complete monolithic structures prestressing presents complications in design and construction. For wide spans, freely supported beams are generally considered preferable to continuous beams in order to avoid similar complications (see page 185). For spans below 6 m normal reinforced concrete construction is generally less expensive than prestressed concrete. Between 6 and 9 m prestressed work may or may not prove more economical according to the particular application, having regard to such factors as the reduction in size and numbers of columns and foundations likely to result from the use of prestressed work. In this range the composite form of construction described on page 174 is likely to be the most cost viable. For spans greater than 9 m prestressed work will usually show economic advantages over reinforced concrete, especially when the imposed loading is light, as in roof construction. For multi-storey frames as conventionally planned, prestressed concrete is not, in most cases, likely to be cheaper than reinforced concrete.

Columns, being compression members, are normally not prestressed. However, in tall columns particularly, where bending stresses may be high due to wind pressure or an eccentric load such as applied by a travelling crane, prestressing can usefully be applied. Tensile stresses in walls, even loadbearing walls, are normally not such as to justify prestressing but in tall retaining walls where bending stresses may be high, prestressing can be economical. As indicated in chapter 3, when applied to retaining walls prestressing, by preventing crack formation, has advantages in terms of the water resistance of the wall (see page 102).

4.8.1 Principles of prestressing

The prestress, or precompression may be induced in a beam entirely without the use of steel by means of external jacks, in the same manner that hand pressure is applied to a row of books, provided that sufficiently solid abutments are available as in the case of a bridge (see figure 4.30 A). The principles of prestressing can usefully be considered on the basis of this method.

The pressure will be of a uniform intensity over the whole section if applied on the neutral axis of the beam, and if of equal intensity to the tensile stresses induced by the imposed load will cancel them out. As will be seen, this results in a final compressive stress in the upper fibres, assuming a beam of uniform cross section, of twice that set up by the imposed load. This precompression of the compression zone is neither necessary nor does it make maximum use of the concrete in carrying its load, since, in terms of the final compressive stress, which must not exceed the maximum permissible strength of the concrete, that induced by the imposed load is only one half of this

maximum. For greatest efficiency it is necessary to apply the prestress in the tensile zone only.

The distribution of the prestress across the section depends upon the point of application of the pressure (B), and its intensity is calculated in the same manner as are the bending stresses caused in a column or wall by an eccentric load.[27] This will be clearly appreciated if the beam is visualised as a 'horizontal' column. It will be seen that by applying the pressure at some point within the lower third, compressive stresses are induced in the bottom portion and smaller tensile stresses in the top portion of the beam. By the selection of an appropriate pressure, which is kept to a minimum to economise in steel, and point of application, the stresses across the section may be so apportioned that when acting together with those set up by the dead load of the beam the resulting stress at the top is zero while the stress at the bottom represents the maximum permissible compressive stress of the concrete (C), thus making maximum use of the strength of the concrete. Since the forces due to prestressing and the dead load act simultaneously the upper fibres of the concrete are not, in fact, subjected to the tensile stresses set up by the prestressing, nor the lower fibres to the excess compression indicated.

When the live load is applied additional compressive stresses are set up in the top and additional tensile stresses in the bottom fibres, and these forces, acting together with

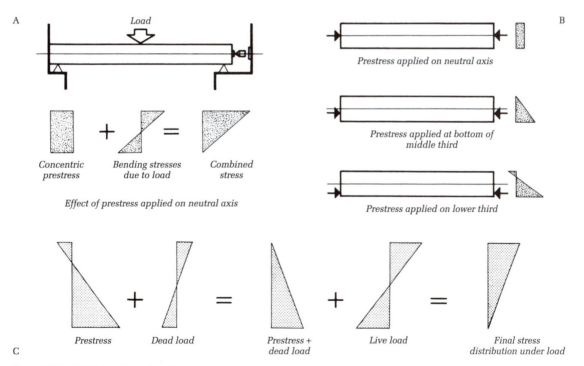

A

Load

Concentric prestress + *Bending stresses due to load* = *Combined stress*

Effect of prestress applied on neutral axis

B

Prestress applied on neutral axis

Prestress applied at bottom of middle third

Prestress applied on lower third

C

Prestress + *Dead load* = *Prestress + dead load* + *Live load* = *Final stress distribution under load*

Figure 4.30 Principles of prestressing

the residual forces from the combination of dead and pre-stressing loads, result in a compressive stress in the top fibres and a smaller compressive, or a zero, stress in the bottom. Greatest economy is obtained if the maximum compressive stress in the bottom fibres, due to dead and prestressing loads, is equal to the maximum stresses set up by the live load, thus producing zero stress at the bottom and a maximum permissible concrete stress at the top when the beam is under load.

4.8.2 Methods of prestressing

The method of applying the precompression by means of jacks, which presupposes sufficiently strong abutments, is of limited use and rarely practicable for normal building works. The alternative method used consists in principle of stretching, or tensioning, high-tensile steel bars or wires which are then anchored to the concrete member. On release of the tension on the steel a compressive force is applied to the concrete as the steel seeks to contract to its original, unstretched, length. The anchorage may be by means of bond between steel and concrete or by external mechanical means at the end of the member, and these two methods form the main difference between the two systems of pre-stressing known as *pre-tensioning* and *post-tensioning*.

Pre-tensioning   In this system high-tensile steel wires are tensioned before the concrete is cast round them, and then, when the concrete has attained sufficient strength, the wires are released and, in seeking to regain their original length but being bonded to the concrete, induce in the concrete the required compressive force. Based on Hooke's law that within the elastic limit stress is proportional to strain, the amount of elongation required in the steel wires (both in pre- and post-tensioning) to produce a particular compressive force in the concrete can be easily calculated (see page 185 for the effect of certain stress losses).

As strong abutments are required between which to stretch the wires pre-tensioning is invariably applied to precast units and is usually carried out in a factory, although a prestressing bed set up on a site as a means of avoiding factory overheads might prove more economical for a very large contract. Factory production is generally preferable since the need for close quality control of the concrete preparation and its placing and of the stressing of the steel is more likely to be satisfied under factory conditions than on the site.

Although pre-tensioning can be applied to individual members formed and stressed in their own moulds, the most usual method is that known as the 'long line' system in which the wires are stretched within continuous moulds between anchorages 120 m or more apart. The wires pass through templates at each end which position them

correctly and the ends are gripped in anchor plates (figure 4.31 A). Spacers are placed at various intervals along the mould according to the required lengths of units. The anchor plates are then jacked away the calculated distance to stretch the wires, the concrete is poured and after it has hardened sufficiently the wires are released and are cut between each unit. This practice is very common for repetitive shaped units such as lintels.

At the extreme ends of pre-tensioned members the bond between steel and concrete is not fully developed, and for a short length, varying from 80 to 120 times the diameter of the wire according to the quality of the concrete and the roughness of the surface, the wires contract considerably in their length with a consequent loss of stress in the wires, the stress at the cut end being zero. At the same time this contraction is accompanied by a lateral swelling which forms a cone-like anchor (B). The length in which this occurs is termed the transfer length and requires reinforcement for shear in the form of stirrups. The lateral swelling of the released wires tends, of course, to occur throughout their length, thus further increasing the bond between wires and concrete.

Several small diameter wires are used so that the greatest surface area is obtained to increase the bond, and the usual diameters lie between 2 and 5 mm. These wires have ultimate tensile strengths ranging from 1540 N/mm^2 for the larger diameters to 2310 N/mm^2 for the smaller. It is essential that the wires be thoroughly degreased and allowed to rust slightly in order to produce a satisfactory surface. Careful control of the concrete mix and vibration are used to produce high quality concrete, and some form of curing is normally applied to accelerate the hardening.

Post-tensioning   In this system the concrete is cast and permitted to harden before the steel is stressed. The steel, which is usually in the form of high-tensile steel cable or bar, if placed before concreting, is prevented from bonding with the concrete either by being sheathed with polythene or tarred paper or by being coated with bitumen. Alternatively, the prestressing steel can be introduced after the concrete has set by casting in an extractable rubber core, inflatable tubes or duct-tubes at the appropriate positions, which are removed before the steel is inserted. It is also possible to place the wires in recesses outside the concrete and protect them. The cable or bar is anchored at one end of the concrete unit and stressed by jacking against the other end to which it is then also anchored (figure 4.31 C). The steel is subsequently grouted under pressure through holes at the ends of the unit to protect it from rust and to provide bond as an additional safeguard.

There are a number of methods of anchoring and jacking the prestressing steel, some of which are illustrated in figure 4.32.

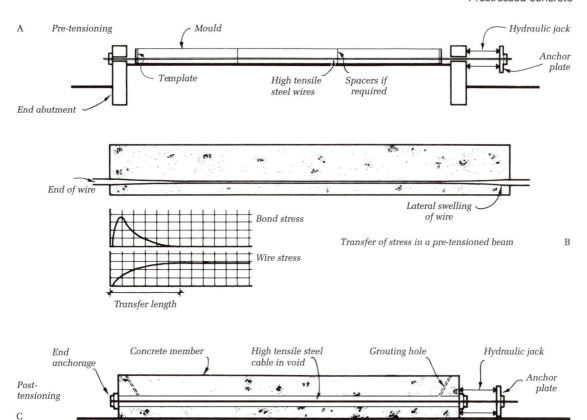

Figure 4.31 Methods of prestressing

The Freyssinet system This uses numerous cables of multi-strand wires positioned in a spiral flexible tube sheathing, and an anchorage device cast into the end of the concrete member consisting of a concrete cylinder with a central conical hole and a conical concrete plug grooved on the outside to take the cable wires which are laid between the cone and the cylinder (figure 4.32 A). The special double-acting Freyssinet jack incorporates a stressing piston and a wedging piston (B). The wires, led through grooves spaced round the head of the jack, are wedged to the stressing piston which is operated until the required extension of the wires is obtained. Then the wedging piston is used to force the plug into the concrete cylinder to anchor the wires. The wires are then released, the cable grouted through the conical plug, the wires cut flush and the face of the anchorage protected with a pat of mortar.

The Magnel–Blaton system differs from the Freyssinet system in the form of anchorage used and in the manner of stressing the wires (C). The wires are stressed in pairs by a normal single-acting jack bearing on the anchorage and are secured by steel wedges to grooved steel plates, each

of which anchors eight wires. These plates are arranged in layers, the number depending upon the size of the cable, and bear on a steel distribution plate. The wires in the cable are held about 5 mm apart throughout their length by spacer grilles.

The Gifford–Udall system and the *PSC system* both stress the wires one at a time. They are anchored individually, in the former system by means of a pair of conical half-wedges driven into a steel barrel accommodated in an anchor plate, in the latter, by a single-piece split sleeve driven into the tapered hole of an anchor block (D). The prestressing wires in these post-tensioned systems are usually from 5 to 7 mm in diameter.

Secondary reinforcement is usually required in the concrete immediately behind the anchorages, and vertical stirrups at the ends of the beam to distribute the local loading from the anchorage of the cables.

The Lee–McCall system (otherwise known as *Macalloy*) uses alloy steel rods instead of cables. The rods are from 13 to 40 mm in diameter and are anchored, after stressing

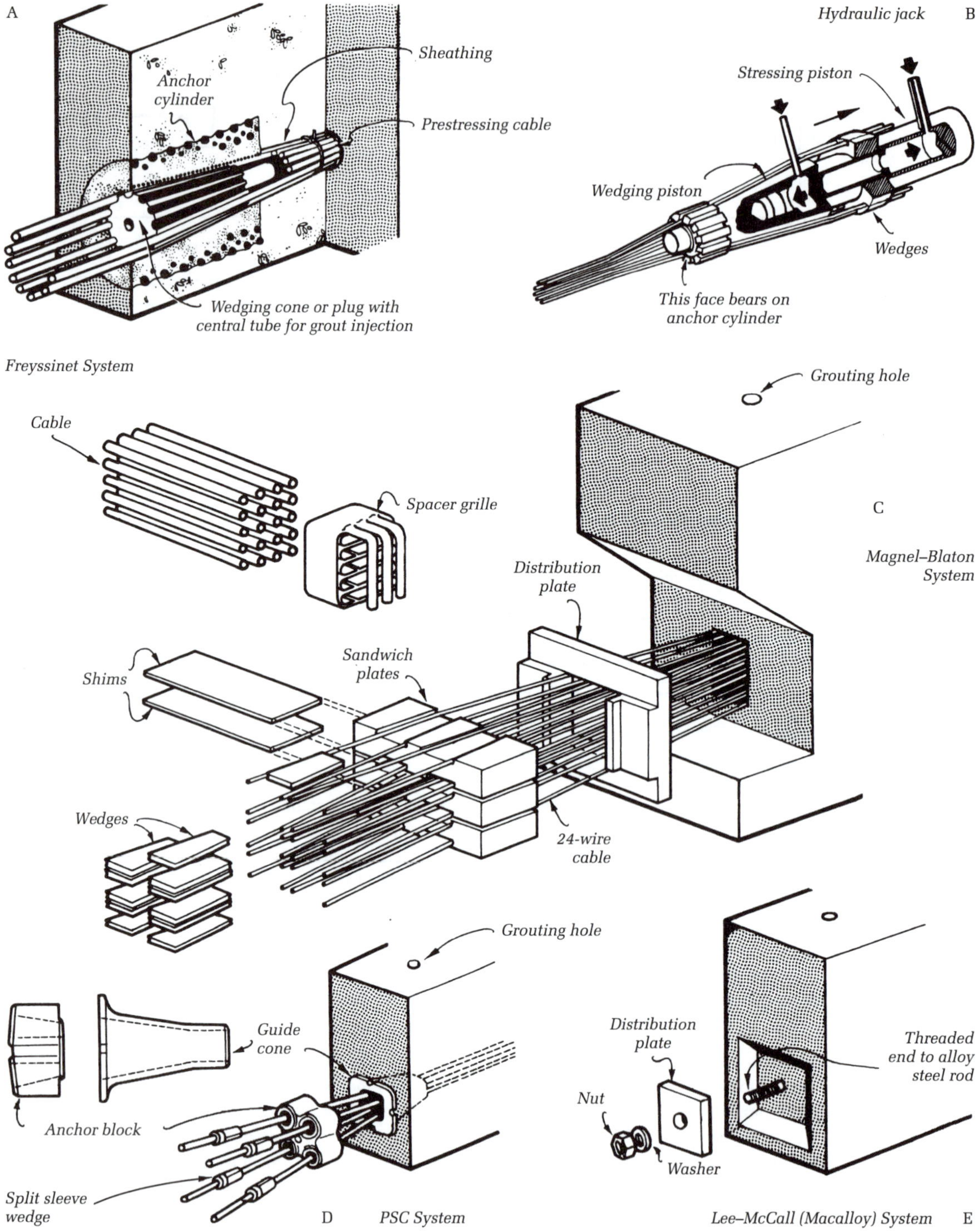

A

Sheathing

Anchor cylinder

Prestressing cable

Wedging cone or plug with central tube for grout injection

Hydraulic jack　B

Stressing piston

Wedging piston

Wedges

This face bears on anchor cylinder

Freyssinet System

Cable

Spacer grille

Grouting hole

C

Magnel–Blaton System

Distribution plate

Shims

Sandwich plates

Wedges

24-wire cable

Grouting hole

Guide cone

Distribution plate

Nut

Anchor block

Threaded end to alloy steel rod

Split sleeve wedge

D　PSC System

Washer

Lee–McCall (Macalloy) System　E

Figure 4.32 Post-tensioning systems

by jack, by means of a special nut screwed on to the threaded end of the rod, the thread of nut and rod being so designed that the load is transferred by degrees to the nut in such a way that stress concentrations are largely eliminated (E). It is possible with this system to re-stress the rods at any time before grouting in, so that the loss of prestress due to shrinkage and creep in the concrete, which occurs in the early life of a prestressed member, may be wholly restored if desired.

Post prestressing After prestressing a concrete member a gradual reduction in the prestressing force commences and continues for a considerable period. This is due to the shrinkage of the concrete, the creep of the concrete and the creep of the steel. The creep of a material is the increase in strain, i.e. lengthening or shortening, which continues to take place after the stress on the material has become constant, so that it will be evident that the creep in the concrete of a prestressed member which is under compression from the stressing wires, will result after a time in the shortening of the beam, whilst the creep of the steel in tension will result in a lengthening of the wires, which, together with shrinkage of the concrete, leads to a loss in the initial prestress. In determining the initial prestress an allowance must be made for these losses, together with those due to elastic shortening of the concrete as it is stressed and, in the case of post-tensioning, to anchorage slip.

Distribution of prestress The distribution of prestress over a section is discussed on page 181 in terms of the point at which maximum stresses are set up by the external loads, that is, the point of maximum bending moment. At other sections the dead and live load moments will be less and the stresses due to prestressing will be excessive, in large beams particularly. A reduction in the moment of resistance of the section at these points can be made by varying the eccentricity of the prestressing wires.

This can be accomplished in two ways: (i) by using straight cables and varying the section of the beam as shown in figure 4.33 A, B, or by raising the centre of a beam of constant cross section, as in C, or (ii) by curving the wires upwards from their lowest point as in D. As the shear forces tend to increase as the bending moment decreases, reduction of the section as in B may not always be desirable and, in any case, since variations in section along the length of a beam increases formwork costs, a constant cross section is preferable except for very large spans. The curving upwards of the prestressing cable gives a vertical component which helps to resist the shear forces in the beam and enables high shear loads to be taken.

Application of pre- and post-tensioning Pre-tensioning is most suitable for the production of large numbers of

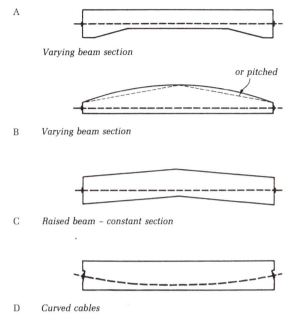

A

Varying beam section

or pitched

B   *Varying beam section*

C   *Raised beam – constant section*

D   *Curved cables*

Figure 4.33 Prestressed concrete beams

similar units, particularly if they are of a cross section too small satisfactorily to accommodate the relatively large post-tensioning cables. In pre-tensioning the wires must be straight so that shear resistance from curved-up wires is not obtainable. Generally speaking, the method is not suited to prestressing on the site. Beams range generally from 4.5 to 23 m in length, the maximum length depending upon transport and handling. Beams up to 30 m or more can be made as 'specials'.

Post-tensioning is invariably used for prestressing on the site and for large members. In most cases it is not economical for members less than 9 m long because the cost of the anchorages relative to the length is high, while the cost of jacking is the same as for a long beam. It may be cheaper to use reinforced concrete for a large number of small units if, for some reason, pre-tensioning is not suitable. The general range of spans is 15 m upwards. Post-tensioning has the advantage, particularly with members carrying heavy shear loads, that the cable can be curved upwards to provide added shear resistance.

When continuity over supports is essential the necessary stressing in the zones of negative moment can be effected in a number of ways, illustrated in figure 4.34. In in situ cast members either the cable must be undulating, which gives rise to friction losses (A) or the member itself must be 'undulating' in form (B) to vary the point of maximum prestress. Precast members can be prestressed in the normal way and continuity be provided over the supports

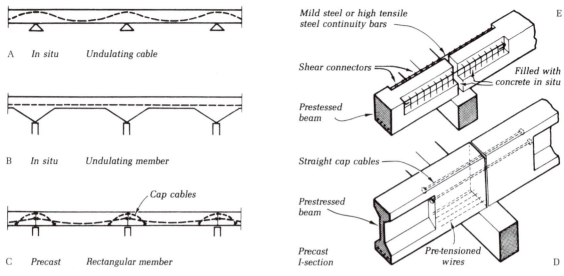

A   *In situ*   *Undulating cable*

B   *In situ*   *Undulating member*

C   *Precast*   *Rectangular member*

Cap cables

Mild steel or high tensile steel continuity bars

Shear connectors

Prestessed beam

Filled with concrete in situ

Straight cap cables

Prestressed beam

Precast I-section

Pre-tensioned wires

E

D

Figure 4.34   Prestressed concrete beams – continuity over support

by cap cables, curved in the case of rectangular sections or straight in the case of I-sections (C, D), or by continuity bars set in in situ filling in rebates in the beams ends (E).

By the use of post-tensioning it is possible to build up a beam from a number of precast concrete units or segments placed end to end like the row of books mentioned earlier. These units can be produced on the site, but being small are usually manufactured in a factory to benefit from the advantages of factory production. Holes for the cables can be formed through the units by light steel tubing or duct-tube, and the units are assembled by being placed end to end with stiff mortar in the joints or with an epoxy structural adhesive applied to the adjacent faces to form glued joints, the whole being post-tensioned (figure 4.35 A, C and figure 8.29). Alternatively, when the members are wide, or in slab form, the cables may be placed in gaps between the precast units as shown at B. These methods reduce site work and avoid expensive formwork. Assembly may be on the site or in the factory; if assembled on the site the cost of transport of a large beam can be eliminated while gaining the advantage of factory production.

4.9 Movement control

All buildings move to some extent after construction. Within limits this movement can be accommodated by the fabric of the building without damage to structure or finishes. When greater movement is anticipated provision must be made for it to take place freely without damage to the building.

Apart from overturning forces and possible overstressing of materials movement is caused by settlement, changes in moisture content and changes in temperature. Space does not permit a full discussion of the nature and effects of these movements. Reference should be made to other sources which deal with these in detail.[28]

4.9.1 Settlement movement

The problems with settlement, particularly differential settlement, are discussed in chapter 2 and the means by which this can be minimised are described. Proper foundation design, relative to the nature of the structure, will usually keep these movements within acceptable limits, but in some cases it will be less expensive to provide for movement in the structure. This is done on subsidence sites as described on page 64.

Settlement joints are often provided at the junction of parts of a building which vary considerably in height or in loading (figure 2.16 A), or, where floors run through, provision for movement between the parts is made by a flexible or hinged bay at the junctions (figure 4.36 A). Claddings and windows in such bays must be designed to permit free movement.

4.9.2 Moisture movement

The magnitude of moisture movement in a structural frame is likely to be small. It is primarily in walls and claddings that provision must be made for this and reference should be made to page 132 where this is discussed.

A   *I-section built-up beam with internal prestressing cables*

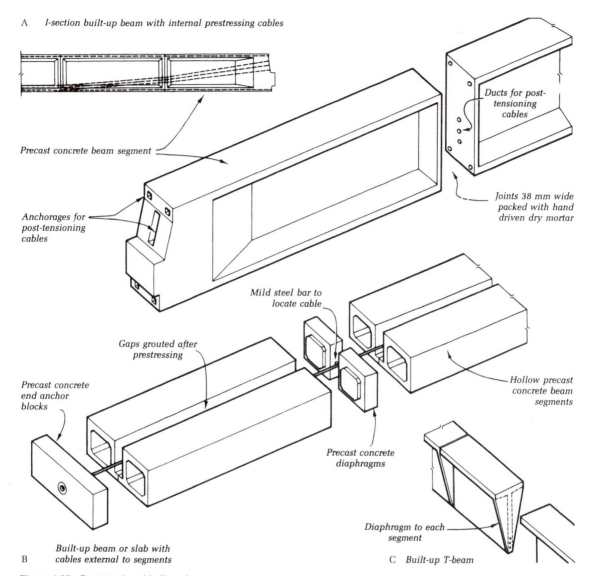

Precast concrete beam segment

Ducts for post-tensioning cables

Anchorages for post-tensioning cables

Joints 38 mm wide packed with hand driven dry mortar

Mild steel bar to locate cable

Gaps grouted after prestressing

Precast concrete end anchor blocks

Hollow precast concrete beam segments

Precast concrete diaphragms

Diaphragm to each segment

B   *Built-up beam or slab with cables external to segments*

C   *Built-up T-beam*

Figure 4.35   Post-tensioned built-up beams

4.9.3 Thermal movement

This is caused by variations in the temperature of the structure and its parts. The magnitude of the movement will depend upon (i) the variation in temperature, (ii) the coefficient of thermal expansion of the materials of which the building is constructed, and (iii) the length of the structure or its part (figure 4.36 B). In Britain the seasonal variation between a cold winter night and a hot summer day may be as great as 50 °C. Some indication of the coefficient of thermal expansion of some common building materials is given in table 4.1[29] together with the approx-imate increase in a length of 30 m for a 28 °C rise in temperature. Note that considerable variations in the coefficient values can occur between samples of any material.

The roof of a building, being exposed fully to radiation from the sun during the day and radiation to the cold sky at night, will be most seriously affected. In buildings with a simple rectangular plan, up to about 30 m in length, thermal movement will usually be small and can take place freely in any direction. Precautions against the effect of such movements on walls and partitions in the form of longitudinal and diagonal cracks (figure 4.36 C) must, however, be taken[30] (see page 311).

Table 4.1 Thermal movement of building materials

Material	Coefficient of thermal expansion per °C × 10⁻⁶	Approximate increase in a length of 30 m for a 20 °C rise in temperature (mm)
Concretes	10–14	11.76
Mild steel	12	10.00
Aluminium alloys	24	20.00
Brickwork	8	6.70
Limestones	3–4	3.36
Sandstones	12	10.00
Granite	8–10	8.40
Slates	9–11	9.24
Glass	9–11	9.24
Fibre cement	8	6.70
Wood:		
along grain	4–6	5.00
across grain	30–70	58.80
Plastics		
(glass reinforced)	20–35	29.40

Expansion joints When the length of the building exceeds much more than 30 m expansion joints are usually provided, placed at intervals not greater than 30 m or at other suitable positions not exceeding this distance apart. These subdivide the length of the building so that the amount of movement within each section is limited, and they provide space for expansion so that damage to the structure is avoided (D). The actual spacing and widths of the joints can be calculated from the coefficients of expansion and a selected rise of temperature.

When the form of the building is such that expansion at some points is restrained, movement will be greater at the unrestrained areas. This can be limited by the provision of expansion joints at the points of restraint as shown in E and F. The design of the end joints in F must permit a sliding action. Expansion joints should similarly be placed where sudden changes occur in plan (G), and where floors and roofs are weakened by large openings in the structure.

Expansion joints should not be limited to the roof slab. They should also be formed in the external walls extending some distance down and inwards to enable the stresses set up by expansion to be distributed. It is common practice to carry expansion joints through the whole of the structure from top to bottom, particularly in monolithic reinforced concrete buildings. In a framed building the simplest method is probably to use double columns and beams as shown in figure 4.37 L, but when double members are undesirable the joint may be formed in the centre of a bay with cantilevered floor slabs or beams, or on the line of a column with sliding bearings to the floor structure as shown in L. In suitable steel-framed structures joints are sometimes provided in the roof and top storey only, reliance being placed on the flexibility of the upper stanchions.

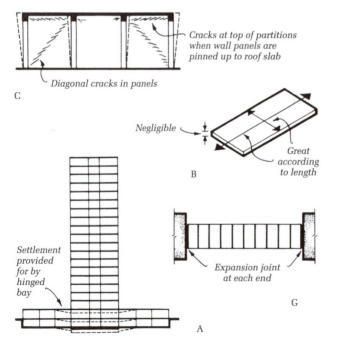

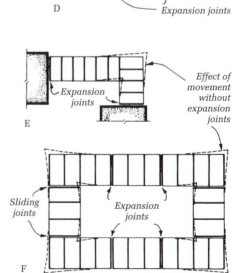

Figure 4.36 Movement control

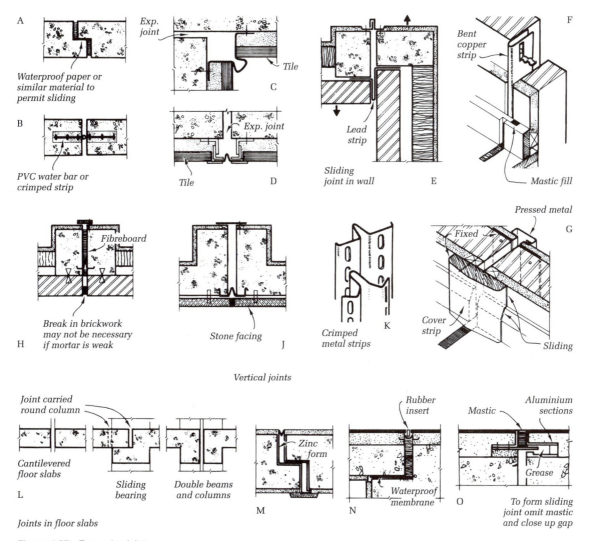

Figure 4.37 Expansion joints

Typical details of expansion joints are shown in figure 4.37. As it is often difficult to remove the shutter board forming a narrow gap between two in situ cast concrete members bitumen-impregnated soft fibreboard is frequently used for this purpose and left permanently in position (H). With wider gaps the board may be withdrawn and the space filled with non-extruding resilient filling (N), or left open (J) as circumstances require. Waterproof resilient filling or crimped 0.6 mm copper strip (K) is used to form a flexible weather-resisting barrier in joints in external walls and between columns (A and J).

Expansion joints in floors should be detailed in the same way to prevent water used for cleaning passing to the ceiling below (M, N). Alternatively a grease seal may be

used (O). In positions where a sliding joint is required (figure 4.36 F), this can be detailed for floor slabs as shown in figure 4.37 O, for walls as E and for roof slabs in principle as figure 8.50 D.

For further consideration of expansion joints in walls and roofs, reference should be made to pages 133 and 310 respectively.[31]

Notes

1 The higher proportions would be for buildings in which the structure predominates, in which services are few, little internal sub-division is required and the area of enclosing element to floor area is small, as in warehouses, or may not exist, as in car parks.

2 See Part 1, section 2.6.1, *Planning*.

3 Although weight is not the only factor in minimising cost: see page 140, section 4.3.2, *Frame layout*.

4 See Part 1, chapter 6.

5 The 'fast track' construction process requires efficient planning of site operations to run concurrently rather than consecutively, thereby consolidating the project programme and further enhancing progress by undertaking work activities immediately behind design detailing.

6 See page 158 for a description of cold-formed steel sections and their uses, and Part 1, section 6.4.3.

7 Data based on grade 43 steel, now graded S 275 to BS EN 10025: *Hot Rolled Products of Structural Steels*. Grade 50 steel is now graded S 355.

8 See Part 1, page 43, *Beam action*.

9 For a more extensive consideration of the application of this type of beam see article on 'Taper beams for long office spans' in the *Architects' Journal*, 1 October 1986.

10 See also Part 1, section 3.3.3.

11 Approximate figures for guidance only, based on grade 50 steel, now graded S 355 to BS EN 10025.

12 Where the structural steelwork may be adversely affected by moisture from the adjoining earth BS 5950-2 recommends that it shall be solidly encased with concrete at least 100 mm thick.

13 This type of beam is produced under the proprietary name of *Preflex*.

14 See also BS EN 1994-1-1, Eurocode 4: *Design of Composite Steel and Concrete Structures*.

15 General requirements and permissible stresses for welded construction are laid down in BS 5950-1. For details of methods and forms of joints see also BS EN 1011-1 and 2: *Welding. Recommendations for Welding of Metallic Materials*.

16 The method is only economic for the production of large quantities. Up to 1500 m may need to be run off before the cost of setting up the machine is covered, depending on the complexity of the section to be rolled.

17 The design of cold formed steel structures is covered by BS 5950-1: Part 5 *Code of Practice for Design of Cold Formed Thin Gauge Sections*. BS EN 10162: *Cold Rolled Steel Sections* gives the properties of a range of sections.

18 In view of the wide range of structural options which reinforced concrete permits, the British Cement Association provide access to concrete design codes and standards. These include guidance at the initial conceptual design stage towards the most economic solution, by means of comparative charts and graphs giving economic span ranges, economic depths of slabs and beams related to span and load, column sizes related to ultimate load and similar information for typical forms of construction in in situ, precast and prestressed concrete.

19 See *MBS: Materials*, table 8.21.

20 See 'Guide to the structural use of lightweight aggregate concrete', published by the Institute of Structural Engineers.

21 See table 5.1.

22 Comparable figures for the re-use of plywood shuttering are 4 to 6 times for in situ work and 20 times for precast work. For large precast contracts the use of steel forms can result in more than 100 times re-use.

23 The cost of mounting and demounting the shuttering for the gable walls is about six times the cost of that for the internal walls.

24 Regulation A3. See AD A, section 5, with specific reference to BS 8110-1.

25 The use of special types of reinforcement in normal reinforced concrete to give increased bond and to limit cracking to fine, well distributed cracks is discussed on page 163.

26 Concrete with a 28-day crushing strength of 30 N/mm^2 is generally used, although for pretensioned work 40 N/mm^2 concrete is sometimes used. Steel wire with ultimate tensile strengths up to 2310 N/mm^2 and high-tensile bars with ultimate strengths up to 1110 N/mm^2 are used.

27 This is explained in Part 1, page 38.

28 See BRE Digests 227 and 228, *Estimation of Thermal and Moisture Movements and Stresses*, and *Design for Movement in Buildings*, by S J Alexander and R M Lawson, CIRIA.

29 See BRE Digest 228 for a more comprehensive list.

30 See BRE Digest 361, *Why do Buildings Crack?* for a consideration of this.

31 For a comprehensive treatment of the subject of expansion and construction joints, see *Design and Construction of Joints in Concrete Structures*, by M N Bussell and R Cather, CIRIA.

5 Floor structures

This chapter is concerned with upper floor construction in concrete and steel appropriate to wide spans and heavy loading, together with some special types of floor construction. Factors influencing the choice of floor type are outlined followed by descriptions of the different forms of floor construction. Methods of providing for the distribution of services in floors are then discussed, the chapter concluding with a brief reference to the control of movement due to setlement and thermal changes.

The broad categories of floor structures have been discussed in Part 1, chapter 8. An indication of the economic application of these relative to span and superimposed loading is given in table 8.1 of that volume and the types of floor construction suitable for short spans and for longer spans with light loading are described. In this chapter those types of floor appropriate to wide span and heavily loaded buildings will be considered, together with some special types of construction. As pointed out in Part 1, in large-scale and multi-storey buildings the floors are normally main structural elements closely related to the general structure of the building and as such they must be considered at the design stage.

5.1 Upper floors

5.1.1 Choice of floor

The main factors influencing the choice of floor type, together with some indication of the manner in which each may be relevant, are discussed below. Reference should also be made to *Choice of appropriate structure* on page 137 and to pages 162 and 166.

Nature of the building structure With a steel frame, providing in itself all the necessary lateral rigidity, a pre-cast concrete floor or a steel deck floor laid quickly on the steel beams but contributing little to the rigidity of the frame could be suitable. With an in situ reinforced concrete frame an in situ concrete floor cast in with the frame and designed to provide lateral rigidity would be logical and could permit the omission of the beams parallel to the floor span.

In a reinforced concrete frame with beams on all four sides of the floor panels the latter may be designed as two-way reinforced slabs to minimise thickness and weight. With a steel frame, for which a rectangular grid layout is generally more economical than a square layout, one-way spanning floor construction will usually prove more suitable, particularly over short spans.

The height of the structural frame will also have a bearing upon the choice of floor type. The weight-saving economies of hollow-slab floor construction are of little significance relative to the advantages of solid floor slabs where the structure is less than five storeys in height. For heights of five storeys and above, because of the greater effect of the columns on the cost of the frame, there is economic advantage in reducing the dead-weight of the floors by the use of hollow-slab forms.

Loading Flat slab construction is economical for heavy uniformly distributed loading, which with normal beam and slab construction might require very deep beams. Because of the smaller overall thickness the total floor to floor height is reduced, and this in turn will effect overall economies by a reduced height of structural frame and reduced area of external cladding. Where heavy concentrated loads must be carried a diagonal beam floor might be selected because such loads are dispersed throughout all the members of the grid, thus avoiding concentrations of high stress and resulting in a reduction in beam depths.

Span This may sometimes be fixed by plan requirements, but as pointed out in Part 1 it may often be reduced to fall within the economic range of a cheaper type of floor system (see Part 1, table 8.1).

With increase in the span (and in the load) there will normally be an increase in the thickness and weight of a solid floor slab. Any increase in the weight of the floor will impose a greater load upon the structural frame and the foundations, with a consequent increase in the cost of these elements. The desirability of reducing the dead weight of the floor to a minimum, for reasons of structural economy, has resulted in the large variety of floor systems now available. Each system is most economical over a limited range of spans and loading, having regard solely to the relationship between weight of floor, the distance it will span and the load it will carry.

Other factors, however, are also involved. For example, the weight of floor beams and floor ribs may be reduced by an increase in their depth.[1] But such an increase may increase the total height of the structure, the area of external cladding and the cubic contents of the building, while the total floor area remains the same. Thus possible economies effected in the structural frame and foundations by the reduction in floor weight must be related to the extra cost of the greater height and the greater area of external facing and a balance struck between the two.

Degree of fire resistance required This is frequently a determining factor in the choice of floor. Many buildings, because of their high fire load, must be divided into fire-tight compartments by walls and floors. These must have a degree of fire resistance sufficient to withstand the complete burnout of the contents of any compartment and prevent the spread of fire to other parts of the building. The considerations involved in respect of forms of construction and the nature of the materials used with regard to their action under fire are discussed in chapter 9.

Provision of services The large number of services and extensive equipment required in many types of building necessitate early consideration of the means of housing them both vertically and horizontally. Horizontal runs of small diameter pipes and cables can often conveniently be accommodated in the topping or screed to the floor but if the services are extensive this can result in a screed 75 to 150 mm thick, adding to the depth and weight of the floor. Alternatively, they can be run in a raised floor above the floor slab. A further alternative is to run the services below the floor slab and beams and to conceal them by a sus-

pended ceiling, although this is not possible if the floor soffit must be exposed to provide thermal mass for environmental reasons (see reference to this in section 3.11.4). They may often economically be run freely in the actual floor depth by the choice of an appropriate type of floor construction. These methods of accommodating services are described in section 5.2.8.

In highly serviced buildings such as laboratories, hospitals and industrial processing plants, the number of services and the amount of room required for their distribution may necessitate a considerable depth of space below the floor. This depth may be used to advantage in the adoption of a deep, light support structure to the floor. When plant is involved, the depth required may be so great as to result in what is, in fact, a service floor sandwiched between the production or processing floors.

Whether service pipes, ducts and cables are housed within the floor structure, within voids above or below the structure, or separately in a service floor plant room, the building designer must provide for this by allocating sufficient space. Depending very much on the building's function, it is not unusual for the services to accommodate the space equivalent of one floor in six in modern commercial or office buildings. In financial terms, the cost of installing building services as a proportion of the total cost of construction can exceed 70 per cent of the capital allocation, particularly for buildings of the type mentioned above.

Degree of sound insulation required Weight of structure is important in connection with insulation against airborne sound. The greater the weight the greater the insulation provided. The degree of insulation provided by a boarded timber floor, as pointed out in Part 1, may be acceptable in the first floor of a house but is inadequate in most other buildings. Many types of light concrete floors can provide insulation against airborne sound sufficient for some buildings, but for others their inherent insulation value must be increased by the provision of a suspended ceiling underneath or a floating floor on top. A solid concrete floor of sufficient thickness and weight can give a reasonable degree of insulation against airborne sound but has little effect on impact sound. Apart from the increase they make in airborne insulation, floating floors are widely used to reduce the transmission of impact sound. For a detailed consideration of sound insulation see *MBS: Environment and Services*, chapter 6, and *MBS: Introduction to Building*, chapter 7.

Cost, speed of erection, adaptability Consideration of cost must enter into the choice of the floor, but not solely in terms of the floor itself. As with all other building elements the effect of each type of floor upon the remainder of the building must be examined and the economic consequences assessed and compared. One aspect of this has already been considered briefly under *Span* on this page. Speed of erection may be an over-riding factor in some circumstances and in others adaptability to non-rectangular panels, which, due to site or other limitations, are sometimes unavoidable.

5.1.2 Strength and stability

Reference is made in Part 1, page 181 to the need for adequate strength and stiffness in floors in carrying their dead load and superimposed loadings. In their design these loadings must be calculated or assumed on the basis of values laid down in regulations or codes of practice, and maximum permissible deflections must be established.

Floor loading The dead load is usually based on the weights of materials specified in BS 648: *Schedule of Weights of Building Materials.* BS 6399-1: *Loading for Buildings. Code of Practice for Dead and Imposed Loads* requires the dead load of any partitions not definitely located in the design of the building to be allowed for as an additional uniformly distributed load per square metre of floor of not less than one-third of the weight per metre run of the partitions with a minimum of 1 kN/m^2.

The superimposed loads to be assumed in the design of a floor vary with the different uses to which the parts of the building may be put. Values for these for buildings of different occupancies are given in tables in BS 6399-1 in the form of (i) uniformly distributed loads per square metre of floor area ranging from 1.5 kN/m^2 for individual dwellings to 7.5 kN/m^2 for boiler rooms and motor rooms, with higher values for some areas in industrial occupancies, and (ii) concentrated loads assumed to act at a point except in the case of punching and crushing loads when they are assumed to act over the actual area of application.

A floor slab must be designed to carry whichever of these loads produces the greater stresses, the concentrated loads being considered to be applied in those positions which result in the maximum stresses or, where deflection is the primary design criterion, which result in maximum deflections. Where the slab is capable of effective lateral distribution of its load consideration of the concentrated load is not required.

Floor beams, unless designed as composite beams (see pages 155 and 174), are designed as individual elements supporting the loads transferred to them by the floor slabs bearing on them, with the exception of beams and ribs spaced at centres not greater than 1.5 m which are designed as parts of a ribbed floor slab.

In the case of a single span of a beam supporting not less than 50 m^2 of floor at one general level, the imposed

Table 5.1 Limitation of deflection

Reinforced concrete beams and slabs:	
maximum deflection 1/250	BS 8110-1
Solid slabs	
The ratio for two-way spanning slabs is to be based on the shorter span.	
Steel beams:	
maximum deflection 1/360	BS 5950-1
Timber beams and joists:	
maximum deflection 1/333	BS 5268-2
Deflection should be limited to 14 mm for long span domestic floor joists	

load may, in the design of the beam, be reduced by 5 per cent and by a further 5 per cent for each 50 m^2 supported in excess of this, subject to a maximum reduction of 25 per cent.

Deflection In order to minimise the cracking of plaster and other ceiling finishes the deflection of floor slabs and beams must be limited in accordance with the nature of the finishes. In Codes of practice this is covered by laying down a fraction of the span as the maximum permitted deflection in the case of steel and timber, and in the case of reinforced concrete by laying down such values of the ratio of span to depth[2] as will limit deflection to a given fraction of the span. These are shown in table 5.1.

Where the floor structure is used to transmit wind forces to strong points (see page 139) it must be stiff enough to fulfil this function and buckling must be taken into account in its design.

5.2 Upper floor construction

As indicated in Part 1 upper floors may be constructed of timber, reinforced concrete or steel. The choice of a particular type will depend largely upon the factors already discussed.

The advantages and limitations of these different types of floor are discussed in Part 1 (pages 172–3) where the construction of timber floors is described in detail, together with that of a simple in situ cast solid concrete slab floor. Concrete floor construction will now be considered further and types of steel floor and special floor construction will be described.

5.2.1 Reinforced concrete floors

These fall into two broad categories, in situ cast and precast, in each of which there is a wide variety of types. Many types of precast floor involve the use of in situ topping of concrete acting structurally with the precast components to form a composite construction.

In situ cast floors These floors, being a wet form of construction as indicated in Part 1, require temporary support until the concrete is strong enough to bear its loads. The length of time during which the support must be left in position depends upon the type of cement used in the concrete, the air temperature and the weather when the work is in progress. Table 10.1 on page 359 gives a guide to the number of days after pouring at which the formwork may be removed or 'struck'.

Solid concrete floor slab This type is commonly used when the slab is to act as a membrane supported on columns without beams, as in the flat slab and plate floors which are described later, or where a high degree of lateral rigidity is required to be provided by the floor. In buildings up to four storeys in height in its simplest form it may often prove more economic than hollow block construction.[3] It gives maximum freedom in design on plan and section since it can easily be made to cover irregular plan shapes and can easily be varied in thickness at different points according to variations in load or span. It is a heavy floor but has good resistance to fire and sound.

The simplest form of solid floor is the one-way spanning slab described in Part 1, which, as explained there, is economic only over small spans up to 4.60 m.

For large spans or heavy loadings a two-way spanning slab should be used in which the reinforcement is designed to act in both directions, the proportion of load taken by each set of reinforcement depending upon the ratio of long to short side of the floor panel. The most economic application is to a square grid. The distribution bars in the upper diagram of Part 1, figure 8.18 would, for a two-way spanning slab, be replaced by main reinforcing bars. The economies of this type of floor arise from the reduced thickness and weight of the slab which result from two-way spanning. In practice, however, the minimum thickness of a slab is limited to a given fraction of the span in order to avoid excessive deflection (see page 193). In many cases it is also limited by the requirements of fire resistance. There is, therefore, an advantage in using a two-way spanning slab only when the thickness of the slab ceases to be governed by these considerations.

Normal in situ solid floor slabs are not generally prestressed, this being applied most economically to rib or beam members although when large spans are required advantage may be derived from the thinner and, therefore, lighter slabs made possible by the use of post-tensioning.

A beamless floor consists of a reinforced slab resting directly on reinforced concrete columns with which it is monolithic. There are no projecting main or secondary beams, the slab acting as an elastic diaphragm bearing on point supports. The columns may or may not have flared heads. BS 8110-1 covers the design of both types under the term 'flat slab', but when designed without column heads the construction is commonly called a 'plate' floor and for the sake of distinction that term will be used here.[4]

Plate floor The slab, or plate, is reinforced at the bottom in each direction over the whole of its area with concentrations of reinforcement along the lines of the column grid. These form wide 'column bands' within the plate thickness. A mat of mesh or rods in each direction is formed in the top of the plate over the columns. It is most economic when the grid repeats uniformly, although this need not be the same in each direction. However, because of the large area of intersection of the wide column bands it is possible, within certain limits, to displace columns from the regular grid. This gives a flexibility which is useful in terms of planning (see figure 5.1 A). Any such column layout must, of course, be the same throughout the height of a multi-storey building. A regular spacing of columns of up to approximately 5.50 m in each direction produces the most economical grid.

The system is most efficient for light and medium loadings as in flats and offices. With domestic loading and a grid spacing up to 4.5 m, the plate thickness would be from 125 to 150 mm. For 3.50 kN/m^2 loading over a span of about 5 m a plate thickness of approximately 200 mm would be necessary. For maximum economy the thickness of the slab must be kept to a minimum consistent with deflection requirements and those of shear resistance at the columns. A minimum of three bays in each direction is, therefore, desirable, together with a half-bay projection of the plate beyond the external columns in order to provide restraint on the outer panels of the plate. Nevertheless, it is possible to use it over two bays only and also possible to reduce the plate projection beyond the external columns to twice the plate thickness, or even to eliminate this projection altogether. But the latter necessitates an uneconomic amount of reinforcement in the slab.

Wallings approaching the weight of normal cavity panel walls with a light-weight inner skin must be carried on the outer column bands between or close to the external columns. Lightweight panels, or curtain walling requiring only lateral support from the floors, may be carried by the plate projecting up to half the grid beyond the external columns.

Practical advantages arising from the use of the plate floor are simplification of shuttering and reinforcement, reduction in dead weight compared with beams and slab and a flat soffit throughout facilitating the use of standard height partitions, particularly useful in office blocks. In many cases also an overall reduction in the total height of

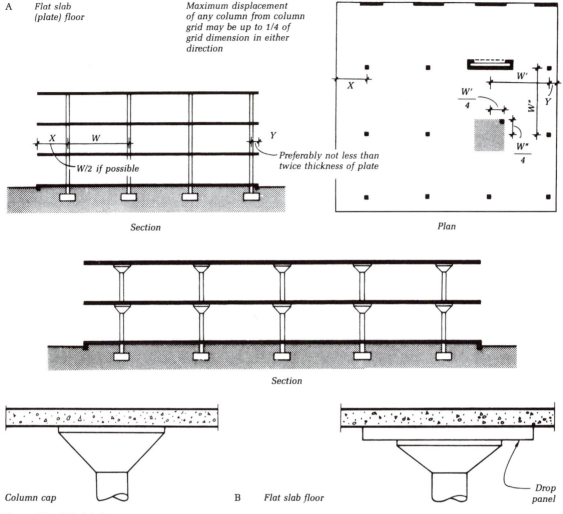

A   *Flat slab (plate) floor*

Maximum displacement of any column from column grid may be up to 1/4 of grid dimension in either direction

Section

W/2 if possible

Preferably not less than twice thickness of plate

Plan

Section

Column cap

B   *Flat slab floor*

Drop panel

Figure 5.1   Flat slab floors

a building results, with consequent economies in carcassing and finishings. When designed within the limits of maximum efficiency, savings of from 15 to 20 per cent are possible on the cost of structure alone over a normal slab and beam system.

For wider spacings of columns and heavier loadings necessitating greater depths of slab it is possible to reduce the dead weight of the panel between the column bands by using concrete hollow blocks similar to those used for normal hollow block floors (see figure 5.2 D). The open ends of the blocks are sealed with sheet steel strip or other suitable material to form ribs 300 mm apart which take the reinforcement spanning in two directions between the 'column bands'. Alternatively, normal rectangular grid construction may be used (see page 198).

Flat slab floor   This is the other form of in situ concrete floor without projecting beams and is termed coloquially 'mushroom construction' because of the expanded or flared column heads which are part of its design (see figure 5.1 B).

As in the plate floor, the slab is reinforced in both directions with 'column bands' running on the lines of the column grid, the slab acting as an elastic diaphragm supported directly on the heads of the columns.

The system is designed for heavy evenly distributed superimposed loads and is economical for loadings of 4.5 kN/m^2 or more and in cases where there is little solid partitioning on, or large openings in, the slab. It is, therefore, suitable for such building types as warehouses and others with heavy imposed loadings and large imperforate, undivided areas of floor.

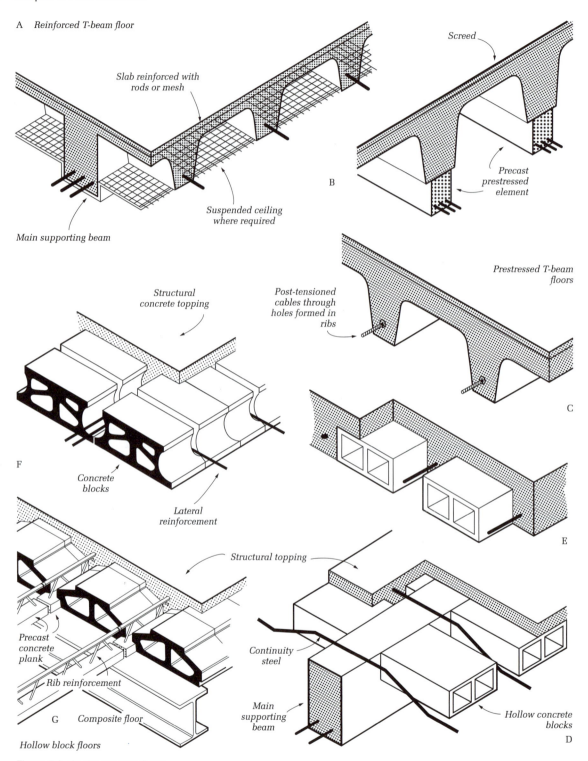

A   *Reinforced T-beam floor*

*Slab reinforced with
rods or mesh*

Screed

*Precast
prestressed
element*

B

*Suspended ceiling
where required*

Main supporting beam

*Structural
concrete topping*

*Post-tensioned
cables through
holes formed in
ribs*

*Prestressed T-beam
floors*

C

F

*Concrete
blocks*

*Lateral
reinforcement*

E

Structural topping

*Precast
concrete
plank*

*Continuity
steel*

Rib reinforcement

*Main
supporting
beam*

*Hollow concrete
blocks*

G   *Composite floor*

D

Hollow block floors

Figure 5.2   In situ concrete floors

For maximum efficiency the columns should be on a regular grid of about 6 to 7.50 m in approximately square bays. In order to provide adequate resistance to the compression stresses in the bottom of the slab over the points of support, and to increase the resistance to shear and punching stresses at these points, the heads of the columns are expanded to give the typical 'mushroom' cap. This will be square or circular depending upon the shape of the column. In some circumstances it is necessary to thicken the slab over this cap as shown in B to form what is termed a drop panel. The main advantages over normal beam and slab floors, as with plate floors, are the reduction in floor to floor height due to the elimination of beams and simpler formwork; advantages which must be balanced against its relatively great weight. For grid spacings of 6 to 7.50 m and for loadings of 4.5 kN/m^2 and over, the slab thickness will be from 225 to 300 mm.

As with plate floors, not fewer than three bays in each direction are desirable, with half-bays cantilevering beyond the outer column bands to obtain restraint on the outer panels.

In situ T-beam or ribbed floor This is illustrated in figure 5.2 A and consists of a series of T-beams cast monolithically side by side to produce a relatively thin slab with ribs on the underside. The basis of a T-beam is described on page 168, and by applying the principles to slab construction a floor lighter than a solid slab results.

It is an expensive floor to construct with normal shuttering and proprietary steel forms, hired by the general contractor or used by the makers as subcontractors, are generally employed. These forms produce ribs at 600 mm centres 90 or 100 mm wide at the bottom and slightly wider at the top. The depth of the ribs and the thickness of the top slab, which is reinforced with rods or mesh, are adjusted according to the load and span. Hardwood fillets can be embedded in the bottom of the ribs to provide a fixing for lathing or battens. The underside may then be enclosed with a ceiling and the resulting voids between the ribs can accommodate services and recessed lighting fittings (see figure 5.17 B).

As an alternative to shuttering of proprietary steel forms, permanent shutters of thin glass reinforced cement trough units may be employed. In order to reduce the amount of temporary support required the GRC soffits for the ribs are formed into planks similar to the rib planks in figure 5.2 G; to this can be added the main bar reinforcement for the rib. The edges of the GRC trough units bear on the edges of these planks and, together with the plank soffits, form a smooth permanent finish to the underside of the floor.

This type of floor, being cast monolithic with the main supporting beams, may be used to stiffen the structural frame in buildings up to three or four storeys high. Because of its low dead weight it can economically be used for spans up to 9 m. It can be prestressed by means of precast pretensioned elements bonded to the bottom of the ribs as shown in figure 5.2 B or by means of posttensioned cables run in holes formed at the bottom of the ribs (C). To cover spans up to 14 m for light loads these methods would require a floor depth in the region of 450 mm overall. For spans less than 9 m normal reinforced concrete is cheaper.

In situ hollow block floor Based on the T-beam principle this type of floor is lighter than the simple solid slab floor, and as it provides a flat soffit the applied ceiling required with an in situ T-beam floor is not necessary (figure 5.2 D, E). Like the solid concrete floor, in most forms it requires shuttering over its whole area. On this hollow concrete blocks are laid end to end in parallel rows, about 75 to 100 mm apart, according to the width of rib required. Reinforcement is laid in these spaces and concrete poured between and over the blocks to form a series of T-beams. When shear requirements at the end of spans necessitate an increased concrete section or when a flange to a main T-beam is required the blocks are laid to stop short of the supporting beams to allow for this, as shown in figure 6.14 B.

The thickness of the structural topping is not less than 25 mm for the shallowest of this type of floor and increases for greater depths. This thin slab and the blocks only may be punched with holes for the passage of services.

The main types of this floor are patented and are designed and erected by specialist firms. They are normally designed as one-way reinforced floors and the majority are most suitable for spans up to about 6.70 m. They can be adapted to two-way reinforcement by closing the ends of the blocks and spacing them out to form ribs at right-angles to the normal ribs as described under *Plate floor*, page 194. One type makes use of blocks made in such a way that in addition to the parallel spaces forming the ribs of the T-beams, in this case 450 mm apart, narrower lateral spaces 450 mm apart are formed at right-angles to the ribs as shown in figure 5.2 F. These take lateral reinforcement. The bottom edges of the blocks are lipped, the lips touching when the blocks are laid in position to form a soffit to the ribs, thus reducing shuttering problems. The concrete in the ribs of this type of floor must be well tamped since any possible honeycombing of the concrete at the base of the ribs cannot be observed.

A composite form of the in situ hollow block floor is shown in G. In this a thin precast reinforced concrete plank forms the soffit of an in situ T-beam rib. The widths of the planks can be varied to suit grid requirements. To suit heavy load or wide span requirements the depth of the ribs is increased by the use of deeper filler blocks and the thickening of the structural topping. The reinforcement and the projecting bond steel is in lattice form expanded from cold rolled steel strip or is formed of welded rods. Its particular form provides good bond between the plank and the in situ concrete and stiffens the plank during handling and erection. The top boom can serve as compression reinforcement when required and additional tensile reinforcement can be introduced in the form of high-tensile or cold-twisted steel bars. Like the precast rib and filler floor (figure 5.7) no shuttering is required and temporary support is provided by ledgers and props.

All these floors are monolithic with the supporting beams.

Square and rectangular grid floor   This is a development of the two-way spanning slab in which the two sets of reinforcement are concentrated in the ribs, as shown in figure 5.3. This type of ribbed floor construction is also known as *waffle* or *honeycomb*, because of the regular pattern of soffit recesses. This finish can be left exposed in buildings with a utility function such as car parks, but where a level ceiling is required threaded inserts cast into the underside of the concrete ribs will provide fixings for a suspended ceiling. The combination of a ceiling void and the recesses, particularly if rectangular, will create space for service pipes, cable troughs, ducting and some items of plant such as extract or fans, refrigeration units and filters.

From the structural perspective, the short span of the reinforced concrete ribs results in a considerable reduction in the floor thickness and the total dead weight of the floor. Application is most suited to relatively wide-span floors where the weight of a solid slab would be excessive. The optimum spacing of the ribs is controlled by the minimum practical thickness of the slab they carry. The latter is controlled by requirements for fire resistance, depth required for effective accommodation of reinforcement and the

maximum permitted ratio of span to depth for the slab (see table 5.1). Formwork moulds or pans are made to suit standard rib spacing (centres) of 600, 800 and 900 mm.

Further examples of this construction method are referred to on page 362, with particular reference to the temporary support and the metal or plastic waffle moulds or pans that determine the soffit profile. See also section 8.3.7 and figure 8.41 B. The illustration shows the disposition of pans and steel reinforcement in the ribs. As indicated, depending on the material, pans can be left in place as permanent formwork, but are usually struck and re-used several times.

In situ diagonal beam floor   This type of floor, like the square and rectangular grid floors, is a single layer grid construction (see section 8.1.8). It consists of two intersecting sets of parallel beams equally spaced and set at 45 degrees to the boundary supports as shown in figure 5.4. The beams are basically all of the same depth and cross section and are rigidly connected at their intersections and to continuous edge beams. The support and end restraint provided by the shorter and stiffer corner beams to those of longer span, together with bottom slabs provided in the corners to resist uplift due to the negative bending moments in these areas (A), results in a considerable reduction in the bending moments in the longer beams compared with those in a normal slab and beam floor of the same span. Thus a depth/span ratio of 1:30 is possible. In addition to bottom slabs the short beams across the corners are sometimes, for greater efficiency, made slightly deeper and wider than the other beams.

For greatest efficiency the grid layout should give a subdivision of three panels along the shorter side of the structural bay. Where internal columns are necessary these are preferably located on the intersection of the beams (B). Crossed cantilevers, springing from the heads of the columns within the thickness of the floor, may be necessary if the layout results in beams of excessive length as shown in B. Cantilever beams are also required if the columns are located within a panel instead of at an intersection of beams (C). Ideally, this type of floor should be square on plan, or subdivided into squares, to give maximum efficiency.

The floor slab need not be integral with the beams. Although it is generally of in situ concrete it may be of precast concrete units, of profiled metal or fibre cement sheets with a topping, or of timber joists with boarding, since the beam structure is designed independent of the slab.

This floor is most economic when used to carry heavy superimposed loads, particularly concentrated loads, over spans of 15 m or more.

Prestressing can be applied to this type of floor because of its beam structure. When prestressed the beams are

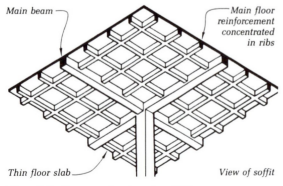

Main beam —
Main floor reinforcement concentrated in ribs

Thin floor slab —
View of soffit

Figure 5.3   Square and rectangular grid floor

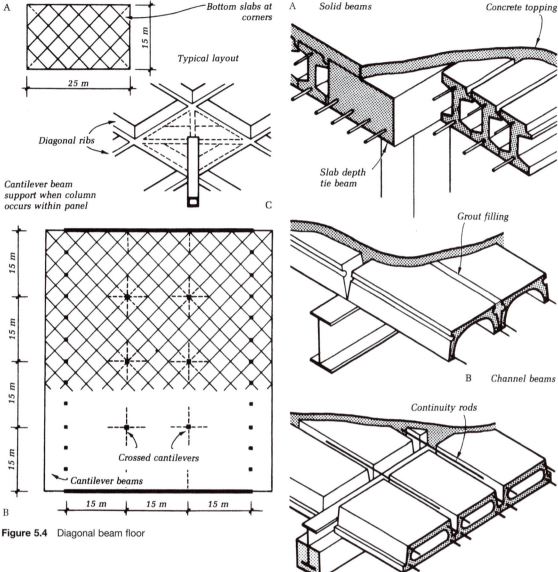

Figure 5.4 Diagonal beam floor

Figure 5.5 Precast beam floors

generally precast in panel lengths, supported on temporary runners and props, and post-tensioned together as shown in the roof example in figure 8.41 C.

Precast floors Most types of in situ concrete floor require shuttering over the whole of their area which must be kept in position for a considerable period until the concrete has gained sufficient strength. The use of precast floors reduces or eliminates shuttering and results in shorter construction times as indicated in Part 1, chapter 8, where the advantages and disadvantages of this form of construction are outlined. The two basic categories of precast beam and precast rib and filler block construction will now be considered in greater detail.

Precast beams These are illustrated in figure 5.5. They provide the simplest form of precast concrete floor and are widely used because of the speed with which they can be assembled and used as a working platform. The floor consists of a number of beams placed side by side and, to a very large extent, spanning between supports independent of each other. The cross section varies between different manufacturers, some having a closed soffit, others an open soffit requiring an applied ceiling. All are designed to produce a more or less hollow floor to reduce the dead weight.

Solid beams   In this type some form of I-section is normal (A). The beams are laid with edges touching or, in some cases, overlapping slightly, so that the bottom flanges form a flat soffit. If continuity over supports is required portions of the top flanges must be removed to allow the insertion of reinforcement and in situ concrete.

Hollow and channel beams   The solid units are relatively narrow and heavy, and developments in the form of channel and hollow beam sections of greater unit width, and much the same weight, enable floor areas to be laid more quickly (figure 5.5 B, C and Part 1, figure 8.18).

Units are made in various widths and depths vary according to span and load. Each beam is reinforced in the bottom corners and in some types in the top corners also. The sides are splayed or shaped to form a narrow space between the beams. This is filled with grout to assist the units to act together in some measure, the adjacent faces of the beams being grooved or castellated to provide mechanical bond. The grout is normally taken up to finish flush with the top of the beams, a structural concrete topping being used only when extra strength is required for heavy loads over long spans or when shear studs are used for composite construction. Continuity over supports is obtained by the insertion of steel rods in the joints prior to grouting (C). These should be welded to the steel beam if they are to serve as the anchors to the precast units required in buildings over four storeys high.

In all types of precast beam floors the units are made to the correct lengths between bearings with the ends designed to suit the type of bearing. The use of foamed plastic cores to form the voids in hollow beams, instead of pneumatic tubes, gives greater design freedom at the ends, which can be solid, because the cores are not withdrawn (see figure 5.10 A). Notches, holes and fixings for applied ceiling and floor finishes can be incorporated. The underside of the units can be left smooth or keyed for plaster in types having a flat soffit forming a continuous ceiling.

The beams are delivered to the site, hoisted and placed in position on their supports and the joints grouted up. A minimum of in situ filling is required and no temporary shuttering or supports. Spans up to 6 m or more are possible although the economic limits for most types are 3.50 to 4.80 m.

Steel and concrete plate floor   In order to dispense with downstand beams to achieve a reduced overall depth of floor, steel floor beams may be integrated into the floor structure, exposing only the bottom flange (see figure 5.6). This produces a form of flat slab construction in conjunction with a steel frame and is accomplished by the use of welded built-up sections or Universal column sections which span in the shorter direction, with precast floor units

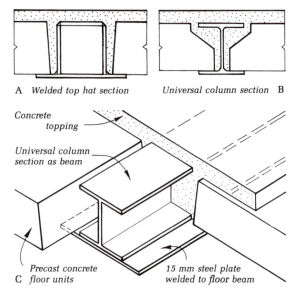

A   *Welded top hat section*   *Universal column section*   B

Concrete topping
Universal column section as beam
C   *Precast concrete floor units*   *15 mm steel plate welded to floor beam*

Figure 5.6   Steel and precast concrete plate floor

spanning in the longer direction together with tie beams within the thickness of the slab. The top hat welded section shown in A is the simplest of the built-up sections which may be used for the beams but is, nevertheless, complicated in its fabrication and the use of a Universal column section provides a simpler solution (B), but necessitates shaped ends to the precast units to allow positioning on the bottom flange. By welding a 15 mm steel plate to the lower flange, projecting 100 mm on each side of the flange, a bearing is provided on which plain ended units may be placed (C).[5] The space between the beam web and the floor units is filled with concrete as in other forms of precast floors; with an increased thickness of top screed and the use of shear studs composite action may be achieved for heavy loadings over spans up to 12 m. Because most of the steel is surrounded by concrete with only the bottom plate exposed a high resistance to fire is achieved. The reduction in overall floor depth and its effect on total building height results in the same economic advantages given by an in situ concrete plate floor which are referred to on page 194. Steel trough units instead of precast concrete may, in a similar manner, be used to produce an in situ composite slab and this construction is described on page 205.

Precast ribs and fillers   This type of floor consists basically of precast reinforced concrete ribs spanning between the main supports and carrying hollow blocks or slabs to fill the spaces between the ribs (figures 5.7 and 5.17 E). In most systems a flush soffit is produced. The in situ topping may be simply a screed (figure 5.7 A) or may be structural,

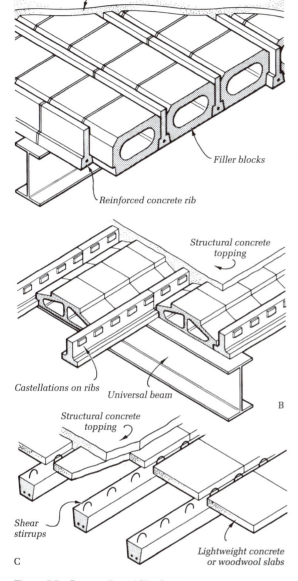

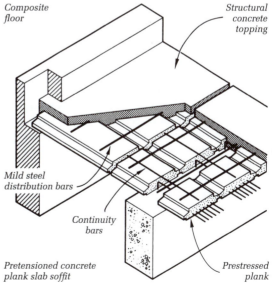

The pots, blocks or slab units are usually rebated at the bottom edges and sit on projecting flanges or lips at the bottom of the ribs so that a flush soffit results (A, B). In those systems designed as a composite construction with a structural topping, the filler blocks are either shallower in depth than the ribs, or are deeply splayed, as shown in B, so that the latter project clear of the fillers and bond with the topping when this is cast. The sides of the ribs are usually castellated to provide mechanical bond. In systems with lightweight slabs resting on the tops of the ribs, bond is obtained by means of steel stirrups or dowels left projecting from the top of each rib (C).

No shuttering is required for this type of floor, although in those systems with structural toppings a few props may be necessary under the ribs until the in situ concrete has matured. Continuity over supports is obtained by means of rods placed between the ribs.

Prestressed floors   Prestressing is now applied to many types of precast floors. This is advantageous where wide spans are involved because it reduces the thickness and dead weight and increases the economic span of the floor.

Pre-tensioning is most commonly adopted so that no stressing is carried out on site and it is applied to all the precast floors so far described – solid, hollow and channel beams and rib and filler types.

The economic advantage of combining prestressed precast elements with in situ concrete has been referred to on page 176. Many types of prestressed floor are based on composite construction of this nature, one of which is illustrated in figure 5.8.

Figure 5.7   Precast rib and filler floors

to act with the ribs (B), depending on the load range for which the system is designed.

The ribs, which are placed at centre spacings ranging from about 250 to 600 mm, are manufactured in the required lengths to suit individual contracts. In the case of heavy floors they may be pierced at intervals to permit the passage of lateral distribution rods.

The fillers may be hollow clay tiles or pots, hollow precast concrete blocks, lightweight concrete slabs or lightweight units of solid expanded polystyrene (EPS).

Figure 5.8   Prestressed floor

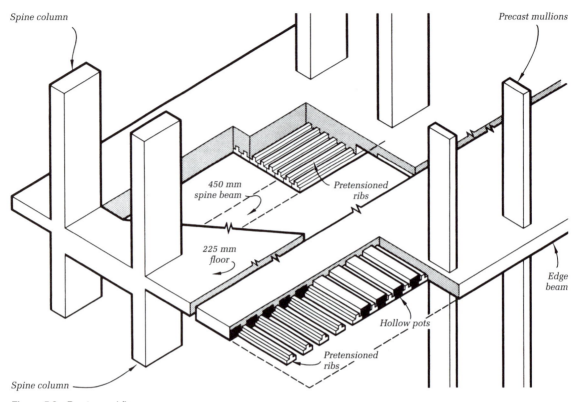

Spine column

Precast mullions

450 mm spine beam

Pretensioned ribs

225 mm floor

Edge beam

Hollow pots

Pretensioned ribs

Spine column

Figure 5.9 Prestressed floors

The prestressed concrete planks are placed close together side by side without filler blocks to form permanent slab shuttering. The edges are grooved to provide a dovetail key for the in situ concrete structural topping in addition to the natural bond between the rough top of the plank and the topping. The planks are produced in widths between 400 and 1200 mm by 50 to 100 mm thick in lengths up to 10 m, and the thickness of the topping is typically 75 mm but varies according to the span and load. Steel distribution rods, and continuity rods over supports, are laid on top of the planks before casting the topping. As with the other system no shuttering is required and only a temporary central prop for spans over 2.40 m.

Prestressed precast T-sections may be used in composite construction either in rib and filler block combination or placed close together to form a flat soffit and filled over with solid in situ concrete (figure 5.9). The latter produces a heavy slab and is useful in circumstances where stiffness for lateral rigidity is required, as, for example, in the spine beam in certain reinforced concrete frames as shown here (see also figure 4.22 B and the relevant text).

Many of these prestressed floors can be used for spans of 9 to 12 m and over. Generally speaking they are

unlikely to be more economical than the normal reinforced types for spans less than 6 m.

Large precast floor panels   It is normal to employ some form of crane on contracts of any size, and their cost effectiveness depends on operating them at maximum capacity. In all forms of precast concrete work regard is paid to this fact. This has led logically to the use, in suitable circumstances, of relatively large precast floor panels. The proprietary hollow beam floor has been developed in the form of precast multiple units, or wide slabs, as shown in figure 5.10 A, which can be as wide as 2.70 m, the width for any particular span being dependent on the lifting capacity of the crane available. Advantages, in addition to the economic use of the crane, are faster assembly, reduction in weight compared with an equivalent area of normal hollow beam floor, reduction in grouting and the simplification of trimming large holes by casting during manufacture.

Prestressed planks are produced in widths up to 1.20 m, and 50 mm thick reinforced planks up to 2.4 m wide (figure 5.10 B). The latter incorporate lattice reinforcement similar to that to the precast rib soffit in figure 5.2 G.

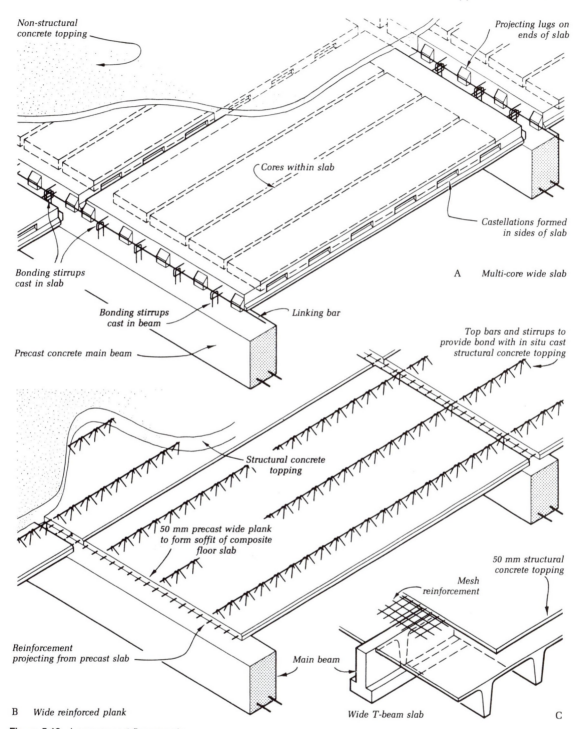

Non-structural concrete topping

Projecting lugs on ends of slab

Cores within slab

Castellations formed in sides of slab

A   Multi-core wide slab

Bonding stirrups cast in slab

Bonding stirrups cast in beam

Linking bar

Precast concrete main beam

Top bars and stirrups to provide bond with in situ cast structural concrete topping

Structural concrete topping

50 mm precast wide plank to form soffit of composite floor slab

50 mm structural concrete topping

Mesh reinforcement

Reinforcement projecting from precast slab

Main beam

B   Wide reinforced plank

Wide T-beam slab   C

Figure 5.10   Large precast floor panels

Wide multiple T-beam or ribbed slabs are also produced, especially for wide spans. These are 2.4 m wide and may be up to 725 mm deep with top slabs 50 or 75 mm thick, the former requiring a reinforced structural topping (figure 5.10 C).

Flat plate floors on precast concrete columns on a grid of about 5.50 m have been precast in large panels. The column bands are cast as panels spanning between column head plates and the whole of the centre panel is cast in one piece. Rebates in the top edges of the panels form channels or strips about 450 mm wide in which in situ concrete is placed to bond with the steel reinforcement left projecting from the edges of each panel. The column head plates need most accurate setting and levelling and it is usually more economical to carry these out as in situ work (see note 18, page 190).

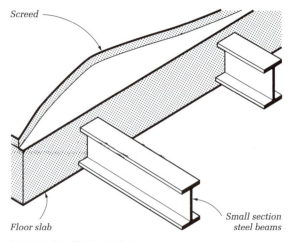

Figure 5.11   Filler joist floor

5.2.2 Filler joist floors

This type of floor is the forerunner of the modern reinforced concrete floor and consists of small rolled steel beams or joists at fairly close centres, surrounded by and carrying a concrete slab as shown in figure 5.11. The steel, or filler, joists are connected direct to the main steel beams or, more commonly, are supported on a continuous steel seating angle bolted to the web of the main beam, every second or third filler only then being connected. Where headroom is not restricted it is preferable to run the fillers over the supporting beams, extending over at least three spans. The maximum spacing of the fillers is related to the superimposed loading on the floor and to the thickness of the concrete unless the concrete is reinforced to span as a slab, or function as an arch, between the filler joists.

The filler joist floor tends to be heavy and is generally used in situations with very heavy loading conditions as in factories, loading platforms and colliery buildings where it can be economical. Compared with a solid reinforced concrete floor it has the advantage that holes may be cut subsequent to its completion anywhere between the filler joists. This can prove useful in certain classes of building involving periodic alterations in equipment and services.

5.2.3 Steel lattice joist floors

Lattice, or open web, joists with flanges of light steel channels, angles and flats, or of cold-formed sections, and with bent rod, angle or tube lacing to form the web, can be used instead of normal timber joists or Universal beams. Continuous timber nailing fillets can be fixed on the flanges to provide fixing for floor and ceiling finishes (figure 5.12 A). The open lattice web gives considerable

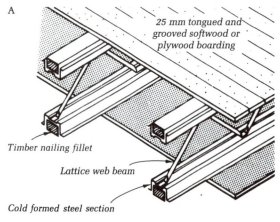

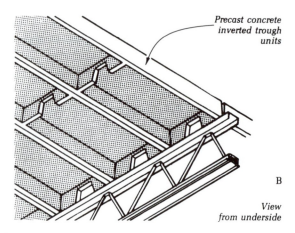

Figure 5.12   Steel lattice joists

freedom in running services through the thickness of the floor.

For a given floor loading and normal joist spacing, greater spans can be covered than with timber joists of the same depth. For example, with normal domestic loading 225 mm by 50 mm timber joists at 400 mm spacing will span approximately 4.90 m whereas a lattice joist of approximately the same depth would span 8.50 to 8.80 m. Over the 4.90 m span a lattice joist of about 175 mm deep could be used. Other forms of composite beam that could be used in this situation are described in Part 1, section 8.4.3.

When a building programme is sufficiently large the use of shallow lattice joists at 400 to 450 mm centres can be competitive for domestic work, having regard to the ease with which services can be run through the floor thickness. But, generally speaking, their most economic use is at 600 mm centres using an increased thickness of flooring. Greater spacing of the joists is possible when light precast concrete trough panels are used to span between them as shown in figure 5.12 B, and a lighter floor is generally obtained than if rolled steel beams are used to carry the slabs.

5.2.4 Steel deck floors

Steel in the form of profiled sheeting working in conjunction with a concrete topping can be used to form a composite floor slab as illustrated in figure 5.13. The steel deck units are fixed to supporting steel beams by welding, self-tapping screws or by cartridge hammer[6] pins and, where composite action with the supporting beams is desired (see page 155), they are fixed by through-deck welding of the shear studs (A).

The deck units are relatively light in weight, can be quickly hoisted in bundles and fixed in position, and when fixed assist in stabilising the steel frame during erection and provide a working platform for materials and other trades before the concrete topping is poured although some propping during construction may be required. The floor is lighter in weight than most forms of in situ and many forms of precast concrete floors.

Figure 5.13 A shows one form of this type of floor which is widely used. The profiled trough units act as permanent shuttering and tensile reinforcement for the in situ concrete topping to form a composite construction. The

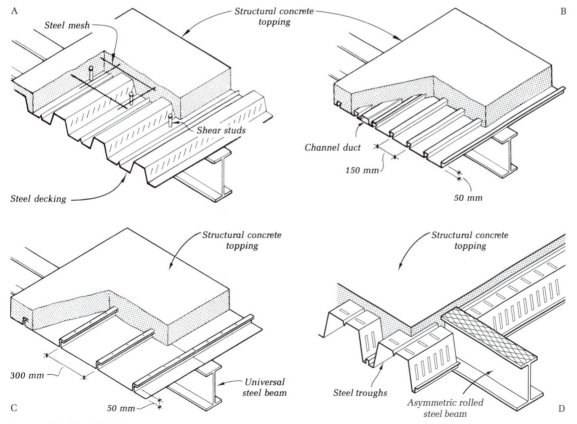

Figure 5.13 Steel deck floors

troughs are indented or perforated to increase the shear bond with the concrete and their depth will vary with the span of the floor to give the necessary structural depth of rib.[7] Steel mesh is incorporated in the concrete which provides distribution steel and the continuity reinforcement required over the supporting beams. Another form uses cold-formed steel decking with folds projecting upwards, which key into the concrete to form the reinforcement to the slab (B, C).

A form of plate floor construction is described on page 200 in which the steel beams are integrated into a floor structure of precast concrete units so that the bottom flange only of the primary beams is exposed on the soffit and it was indicated that profiled steel trough units could be used instead of precast concrete units. This is illustrated in figure 5.13 D, which shows an asymmetrical rolled steel section produced for this purpose instead of the plated steel section in figure 5.6, the wider bottom flange providing the bearing for the trough units. Stop plates are placed at the open ends between the troughs to prevent the concrete filling round the beam passing through when it is poured. The trough units, acting with the in situ concrete form a composite slab structure; composite action between the slab and the supporting beam (see page 205) can be facilitated by a raised pattern on the top flange which provides adequate bond without the use of shear connectors. Both forms of this floor, with only the bottom flanges of the beams being exposed, achieve a fire resistance of 60 minutes.

As indicated on page 114, 60 mm deep trapezoidal profiled steel units similar to that shown in figure 5.13 A can be bent into curves. While not forming deck floors like those described above, these curved units can be used in floor construction by acting as arches springing from the bottom flanges of adjacent supporting steel joists and carrying a filling of concrete to form a horizontal floor surface 75 to 150 mm above the crown. Depending upon the rise of the arch, much of the floor depth may often be within the depth of the supporting beams. Steel mesh is placed in the concrete above crown level to provide distribution steel and the continuity steel over the supporting beams.

5.2.5 Open grating floors

Open flooring, some examples of which are shown in figure 5.14, is used mainly in industrial buildings, particularly for service and operating platforms to machines, where the passage of light and air is required to be maintained.

This type of flooring is made up in steel, aluminium alloy or glass reinforced plastic in panels of varying widths and lengths as required. In the metals it can be formed of parallel strips or flat sheets spaced apart and braced either by similar profile sections (A, B) or by bars intersecting at right-angles at intervals along the length of the panel (C). The junctions of all members are welded or riveted and depths range from 19 to 63 mm. Clear spans are up to 2.4 or 2.7 m.

Open flooring pressed from 16 gauge (1.52 mm) mild steel sheet is also available, produced in sections or planks 225 mm wide and in 1.35 m and 1.80 m lengths. The depth is standard at 38 mm so that variations in loading must be allowed for by variations in the support spacing.

In aluminium alloy this type of floor is also produced as a 150 mm wide ribbed extrusion in depths from 19 to 50 mm, with rectangular or square holes punched in the top plate (D). Clear spans are much the same as for the other types of floors. For all types the supporting structure is formed from various rolled-steel sections to which the floor panels are fixed by means of clips and bolts (A). An insulated clip and stainless steel or cadmium plated bolts are used with the aluminium extrusions to isolate the two metals and avoid possible electrolytic action.

Glass fibre reinforced plastic is formed as a rectangular or square mesh grating in depths of 25 mm or 40 mm similar to C, but with all bars full depth. This flooring provides a high degree of resistance to corrosion.

Perforated cast-iron plates of a similar nature are used externally for walkways and for fire escape stairs. It is a heavier form of construction than the steel panels described above, but is used externally because of the greater resistance of cast iron to corrosion (see figure 7.8).

5.2.6 Spring floors

Gymnasium and dance floors, particularly the latter, should be resilient. This resilience is imparted by the use of a sub-floor of timber bearers, with or without springs or rockers as shown in figure 5.15, on which narrow strip flooring is fixed.

In its most economical form the sub-floor may consist of thin floor battens at 400 to 450 mm centres carried on bearers fixed to the main floor in a staggered arrangement (figure 5.15 A). This gives a bearing at every 900 mm to alternate floor battens and at 450 mm to the intermediate battens. Although not producing as much resilience as a floor incorporating springs it is cheaper than a true spring floor.

Another form without springs, developed in Sweden for gymnasium floors, uses bearers made up of timber boards in 'sandwich' form in which the boards are blocked apart at staggered intervals (B). Resilience is given because the blocks carrying the upper board bear on the middle board at a point mid-way between the lower blocks: not directly on them.

An alternative to these achieves resilience by means of rubber cradles on which the floor battens bear.

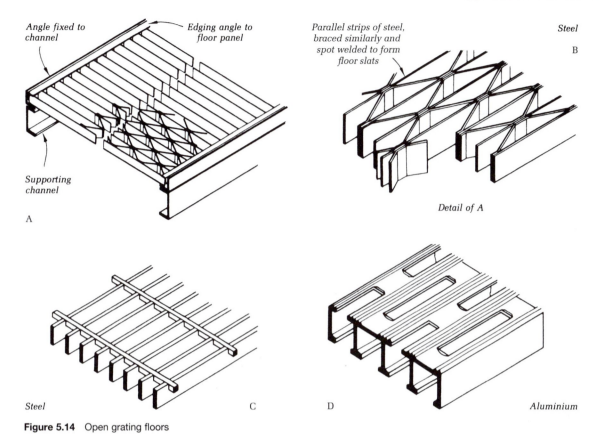

Angle fixed to channel

Edging angle to floor panel

Supporting channel

A

Parallel strips of steel, braced similarly and spot welded to form floor slats

Steel

B

Detail of A

Steel   C

D   Aluminium

Figure 5.14   Open grating floors

Fully resilient floors incorporate special springs or rockers to permit the sub-floor to deflect when in use. In some forms the springs are of a leaf type placed at intervals between the joists and continuous parallel wood bearers to form 'spring joists' on which the strip flooring is laid (C). Other forms incorporate spring and rocker fitments carrying steel beams on which timber bearers are laid (D). Others make use of rocking bars only. Devices can be incorporated to lock the springs when a rigid floor is required.

5.2.7 Raised floors

These are structural sub-floors raised above the normal floor slab to form a cavity to accommodate services; they are also referred to as 'access' or 'platform' floors. They have been developed in response to the increase in cable and ducted services in areas such as laboratories and computer rooms and, because of the flexibility and easy accessibility they provide, have now come to be used to facilitate services provision generally, especially in office

spaces, instead of below-slab accommodation with a suspended ceiling.

Raised floors fall into two broad types according to whether they provide a shallow or deep cavity for the services.

Battened floors   These provide a shallow cavity for cable services usually less than 100 mm in depth formed by timber battens supporting flooring in which there are removable access panels or removable strips (see figure 5.15).

This type of floor can also be constructed with profiled steel trough section sheets laid on the structural floor to form cavities between the troughs.

Pedestal floors   Deep cavities can be obtained with this type of floor so that all normal cable and environmental services can be accommodated. The flooring panels are supported on metal props or pedestals which are adjustable in height to permit levelling of the floor. See MBS *Internal Components*, chapter 4, for a detailed consideration of these floors. Figure 5.16 also shows the principles.

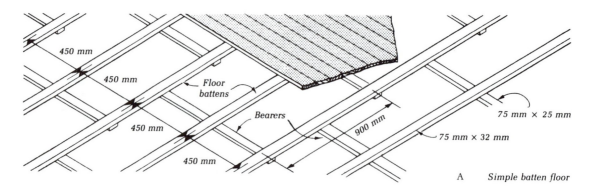

450 mm

450 mm

450 mm

450 mm

Floor battens

Bearers

900 mm

75 mm × 25 mm

75 mm × 32 mm

A   Simple batten floor

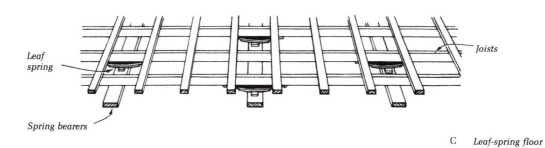

Leaf spring

Joists

Spring bearers

C   Leaf-spring floor

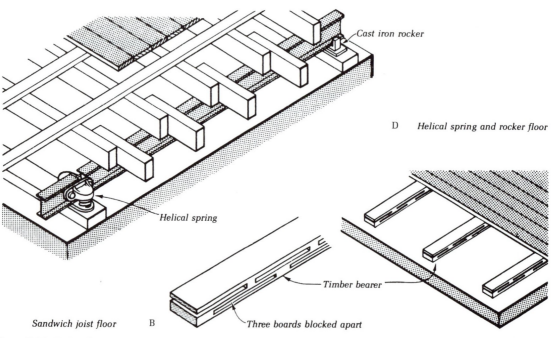

Cast iron rocker

D   Helical spring and rocker floor

Helical spring

Timber bearer

Three boards blocked apart

Sandwich joist floor   B

Figure 5.15 Spring floors

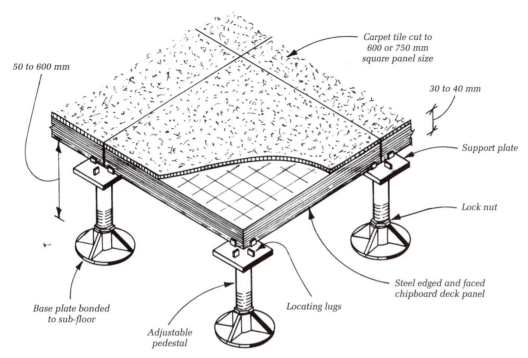

Carpet tile cut to
600 or 750 mm
square panel size

50 to 600 mm

30 to 40 mm

Support plate

Lock nut

Steel edged and faced
chipboard deck panel

Base plate bonded
to sub-floor

Locating lugs

Adjustable
pedestal

Figure 5.16   Raised access or pedestal floor

5.2.8   Provision for services in floors

As indicated at the beginning of this chapter (page 192) services can be accommodated above the structural slab, within its actual thickness, below the slab or below its supporting beams. In both of the latter methods a suspended ceiling is used to conceal the services.

Horizontal distribution   Electrical conduit may be laid on top of the structural slab after it is cast, holes being left at all drops to ceiling points below (figure 5.17 A). The top screeding must be thick enough to bring the finished floor level above that of the conduit. Plastic or metal underfloor ducting for electrical and, especially, telecommunications wiring may be used instead of conduit, in which case the top screed will be thicker. To reduce its weight lightweight concrete or foamed mortar may be used. Circular ducts formed in the floor slab or screed by means of pneumatic cores are described on page 230.

Pipes for water, heating, gas or drainage services may be laid either on top of the slab or fixed to, or suspended from, the soffit. When laid on top of the slab the thickness of the finishing screed may be considerable, especially if some pipes cross each other, so that the layout of the services should be planned early enough to permit the weight of the screed to be taken into consideration in the design

of the floor. As with underfloor ducts lightweight concrete or foamed mortar could be used to reduce the weight of the screed. Where these services are extensive and necessitate a relatively great thickness of screed some form of cavity floor provides an alternative to burying in screed with the advantage of providing flexibility in the running of the services and easy access to them by virtue of the open cavity and top access (see above).

With in situ solid slab floors electrical conduits can be cast in the soffit of the slab by laying screwed conduit on the shuttering with all the necessary outlet, junction and draw-in boxes in position. The pipes for floor and ceiling panel heating may be embedded in the floor slab or screed because they are welded at the connections. The coils of ceiling panels in solid concrete floors are laid on the shuttering and, after testing, the reinforcement is fixed and the concrete cast around and over the pipes so that they are embedded flush with the soffit. A 19 mm screed is applied on top of the pipes and on this the normal blocks are laid. In the areas where no heating panels occur deeper blocks are used (C). Heating panels in precast beam floors are accommodated in recesses formed by the use of shallower but specially strengthened beams, the coils being carried on steel straps and bolt hangers (D).

When the form of construction involves ribs or secondary beams on the underside of the floor the pipes may

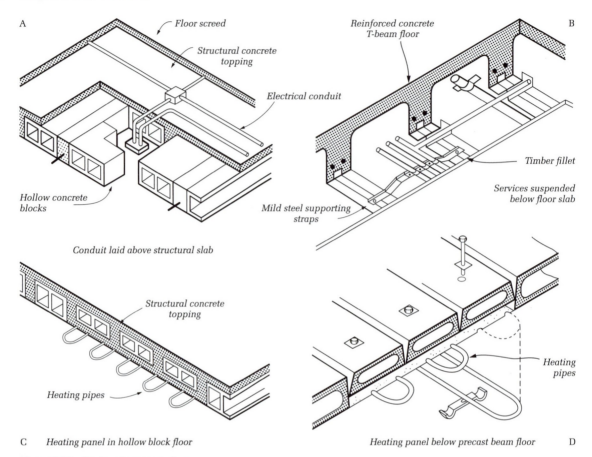

A

Floor screed

Structural concrete topping

Reinforced concrete T-beam floor

B

Electrical conduit

Timber fillet

Services suspended below floor slab

Hollow concrete blocks

Mild steel supporting straps

Conduit laid above structural slab

Structural concrete topping

Heating pipes

Heating pipes

C   *Heating panel in hollow block floor*

Heating panel below precast beam floor   D

Figure 5.17   Services in concrete floors

be suspended from the slab or rib soffit and concealed by a suspended ceiling at the level of the rib or beam soffit as shown in figure 5.17 B. In this situation services may be distributed laterally through circular holes not exceeding one-third of the depth of the floor in diameter formed on the centre of the floor depth. Such holes should not be closer to each other than twice the depth of the floor. In the case of steel deck floors using trough units, such as those in figure 5.13 A and D, pipes may similarly be run in the spaces between the troughs. However, in the latter the web openings in the beam required to permit the passage of pipes will reduce the fire resistance of the beam and this may necessitate fire protection to the underside of the bottom flange.

Where services are required to be run within the slab thickness there are advantages in running them in ducts since they provide easy accessibility for inspection and maintenance and facilitate alterations and additions. Lateral ducts such as these are described on page 227.

Services under floor slab   Where services are run under the floor slab within the depth of the structural floor, openings through steel or concrete beams, used for distribution in lightly serviced buildings, should be kept within the limits given on page 227. Castellated and cellular beams and trussed girders provide within themselves apertures through which services may pass, but even in trussed girders with relatively large spaces between the strut and tie members the flexibility in disposing services is limited and as a consequence services are often placed below the supporting floor beams which further increases the overall floor depth. This has important economic implications (see page 192 under *Span*) especially in wide span multi-storey structures, such as modern commercial buildings in which the structure is required to provide large spaces free of columns, resulting in beam spans of 15 to 18 m which necessitate the use of relatively deep plate or other forms of girders. This type of building is heavily serviced by mechanical and telecommunications systems. For reasons given on page 207, raised floors on top of the structural floor may be used to accommodate many of these services, but they, too, increase the overall floor depth, already relatively large because of the deep beams necessitated by the wide spans required. However, various types of beams and beam systems may be used which facilitate the accommodation of services in this type of building without unduly increasing the depth of the floor structure.

As indicated on page 147, the use of taper beams provides for such long span construction without excessive overall floor depth for the accommodation of services since these can pass through the open spaces below the shallower ends of the beams to a greater or lesser extent within the maximum depth of the beam, depending upon the nature and extent of the services to be accommodated (figure 4.6). This overlapping of the structure and the services zones thus minimises the overall depth of the floor compared with that which would result from the use of parallel flanged girders over the same span.[8] The use of stub girders (figure 4.7) provides openings between the stubs through which services may pass, and the floor with continuous primary and secondary beams described on page 145 (figure 4.3) provides spaces between each set of beams in which services may be distributed.

The advantages of rigid beam to stanchion connections in steel frames have been described in section 4.3.8, one of which is the reduction in depth of the beam, and where services must pass below floor beams the depth of the latter can be kept to a minimum by the use, where suitable, of haunched connections in which a deep steel gusset piece is welded to the underside of the beam end, both of which are welded to a connecting plate which permits a rigid bolted connection to be made to the stanchion. Part 1, figure 6.4 A illustrates this form of connection without a gusset piece – the latter gives greater rigidity.

Vertical distribution The vertical distribution of services will very often necessitate holes through the floors of a building. Small holes up to 150 mm across may usually be formed anywhere in a beam and slab floor, but if there is more than one hole in a floor panel they should not be nearer to each other than one and a half times the width of the larger adjacent hole.

A group of holes more closely spaced than this must be considered a single opening. Holes may also be formed within the width of an individual filler block or section of a hollow beam (figure 5.18 A) and between the ribs of a T-beam floor (figure 5.17 B). In both cases the edges of the hole should be kept clear of the ribs. Apart from such openings concrete floors require to be trimmed. Illustrations of methods commonly adopted are given in figure 5.18.

In in situ solid concrete and hollow block floors openings not exceeding 1 m across can be trimmed by the use

of appropriate reinforcement placed within the thickness of the slab round the opening (B), but for larger openings such trimmers and trimming beams are too shallow and normal beams have to be formed. In flat plate and flat slab floors particular care must be taken in deciding the positions of holes, especially when these lie near to columns and column bands. They should be kept at least one-quarter of the span away from the columns and edges of the slab. This applies also to holes lying near the flanges of T-beams, which must be kept out of the flange area.

Small holes for services are often cut after the concrete floor has been cast but this is uneconomical and the positions of all such openings should be settled before the concrete is poured. Small vertical holes are formed by means of timber boxes, pieces off cardboard rolls, sheet metal bent to a rectangular or cylindrical shape or gas barrel fixed to the shuttering to give an aperture of the required size. Most formers are left in position except those of timber, which are usually removed.

Openings of limited size can be formed in two ways in precast beam floors: one, by the use of special trimming beams carrying reinforced trimmers (C); the other, by the use of cranked steel strap hangers bearing on the beams at each side of the opening, one at each end carrying the ends of the trimmed beams (D).

In the case of rib and filler block floors the necessary blocks are omitted and the shortened ribs carried on special trimmer ribs, the ends of each being supported on mild steel hangers (E). The trimming ribs can be doubled up on each side of the opening to give additional strength. A certain amount of in situ concrete may be necessary to make out openings to sizes not falling within the limits of the precast beams or filler blocks.

Limited openings in the prestressed plank floor can be trimmed within the thickness of the floor by special detailing and the use of cut planks to form trimmers. Holes can be formed in casting, making use of wide planks where necessary. Small holes can be cut on site provided they are near to joints.

5.3 Movement control

5.3.1 Settlement movement

When settlement joints are provided in the structure, as described in the previous chapter, the junctions of the floor slabs must be designed to permit relative movement by rotation. These would be detailed as pivoted joints.

5.3.2 Thermal movement

Typical details of expansion joints in floors are shown in figure 4.37 and reference is made to these on page 188.

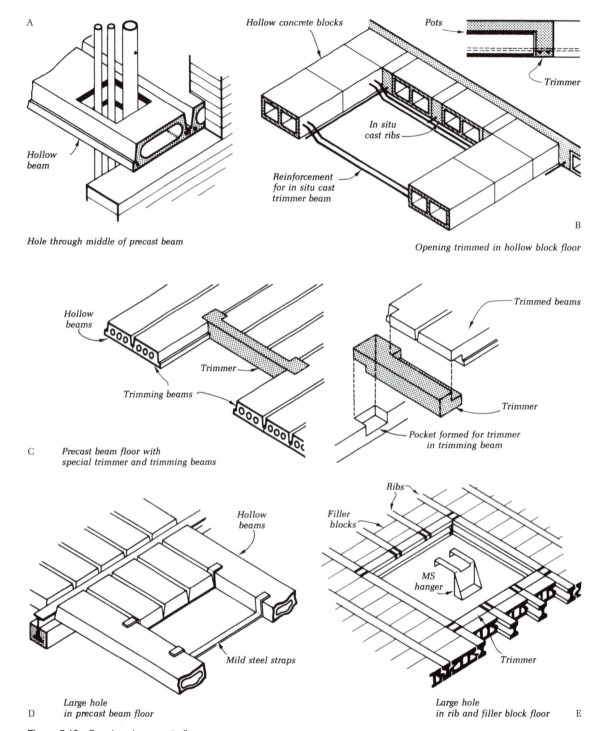

A

Hollow beam

Hole through middle of precast beam

Hollow concrete blocks

Pots

Trimmer

In situ cast ribs

Reinforcement for in situ cast trimmer beam

B

Opening trimmed in hollow block floor

Hollow beams

Trimmer

Trimming beams

Trimmed beams

Trimmer

Pocket formed for trimmer in trimming beam

C   *Precast beam floor with special trimmer and trimming beams*

Hollow beams

Mild steel straps

D   *Large hole in precast beam floor*

Ribs

Filler blocks

MS hanger

Trimmer

Large hole in rib and filler block floor   E

Figure 5.18   Openings in concrete floors

Notes

1 See Part 1, pages 41 and 42.
2 As an approximate guide, under modest load the slab thickness is about 1/24 of the span for simply supported situations. See also table 10.2.1 in *Construction Technology* by R Chudley and R Greeno, Pearson.
3 See section 5.1.1.
4 The term *plate floor* should not be confused with *floorplate* which, as explained in note 2 on page 187 of Part 1, refers to the structural floor of a building irrespective of the form of construction.
5 This type of floor beam is produced under the trademark name of *Slimflor*, by Corus Construction and Industrial. Corus also produce an *Asymmetric Slimflor* rolled steel standard section with top flange narrower than the lower flange (see figure 5.13 D).
6 This consists of a holder to guide the pin axially as it is driven home by a mechanism fired by a cartridge or one fired by a blow from a club hammer.
7 For floor span relative to depth, see BS 5950-4: *Structural Use of Steelwork in Building. Code of Practice for Design of Composite Slabs with Profiled Steel Sheeting*.
8 For a more extensive consideration of the application of this type of beam see article on 'Taper beams for long office spans' in the *Architects' Journal*, 1 October 1986.

6 Chimney shafts, flues and ducts

The large flues and free-standing chimney shafts required by high capacity heating appliances are considered in this chapter. Factors relevant to establishing the necessary size and height of a chimney shaft are discussed followed by descriptions of the construction of reinforced concrete and steel shafts. Flues required for gas-fired appliances are then discussed and described. The chapter concludes with a consideration of the various types of ducts used for distributing services in a building.

The construction of flues and fireplaces to serve heating appliances of a domestic scale has been described in Part 1, chapter 9. The construction of the large flues and chimney shafts necessitated by higher capacity appliances and large heating plants is discussed in this chapter.

6.1 Chimney shafts

A chimney shaft is a free-standing structure containing a flue which by virtue of the heating apparatus it serves is generally larger than a domestic flue.[1] Flues are bounded by flue linings which protect the surrounding structure of the chimney and assist the formation of convection currents by which gases are taken to the outlet of the flue for dispersal to the upper air.

6.1.1 Size of chimney shafts

The height and cross-sectional area of a given flue clearly relate to the basic need to provide a suitable volume of warmed air and gases whose temperature difference to that of the outside air is sufficient to allow convection currents to form within the flue. These currents must be strong enough to overcome the frictional resistance of the flue (and any other airways connected to it through which the

gases must pass) and the inertia of cool air present in the flue in different atmospheric conditions.

The design of large chimneys and flues involves calculations in which the factors of cross-sectional area, height, plant type and size and fuel are related. The size and height of chimneys and flues in relation to plant size and type of installation can be made by reference to simplified tables or graphs.[2]

The temperature differential between the inside and outside of the flue is a critical factor and may be affected by:

- *Atmospheric conditions* High atmospheric temperatures and heavily polluted atmospheres will cause sluggish action and possibly down draughts.
- *Materials of construction* Flue linings of low insulation value will not maintain high enough temperatures within the flue to promote strong convection currents. Flues surrounded by chimneys of high thermal capacity will take a lot of heat from the gases before the surrounding material warms up enough to allow gases to reach high temperatures.
- *Size of flue* Too large a flue is uneconomical in cost and space, likely to encourage condensation and will dissipate the heat of the flue gases too rapidly resulting in sluggish operation. This problem occurs when several boilers are connected to the same stack, some of which are not in use in summer. The remaining boiler(s), often providing hot water only, does not require such a large flue and cannot produce sufficient hot gas for the flue to function. In these circumstances separate flues are preferable, grouped together if possible in a single chimney stack to engender a concentration of flue gases at maximum load sufficient to make the smoke plume rise well into the upper air and disperse the effluent. Undersizing a flue will produce a high efflux velocity of flue gases and spillage of combustion products.

6.1.2 Height of flues

Apart from theoretical factors affecting the design of flues, there are statutory requirements for the minimum heights of flue outlets. The minimum size and height for domestic chimneys is determined in the Building Regulations, Approved Document J: *Combustion Appliances and Fuel Storage Systems*, sections 2, 3 and 4 for solid fuel, gas and oil burning appliances, respectively. For power-generation plant, industrial central heating boilers and other large-scale fuel and waste combustion equipment, the Clean Air Acts of 1956, 1968 and subsequently 1993 (Sections 14 and 15), incorporate requirements limiting the dispersal of dark smoke, grit, dust, gases and fumes from chimneys that must have adequate height and size for the situation. Further guidance is contained in the Chartered Institution of Building Services Engineers publication, *Guide B: Heating, Ventilating, Air Conditioning and Refrigeration*, section 1, appendix 1.A2: *Sizing and Heights of Chimneys and Flues*. The following factors are taken into account in relation to the determination of chimney heights:

Nature and quantity of effluent Conventional fuels (gas, oil and coal) are assessed in terms of their sulphur content, the quantity of sulphur dioxide in their flue gases and the sizes of the furnace installation.

Speed of gases through flue The memorandum in the first Clean Air Act recommends flue gas efflux velocities of up to 15 metres per second irrespective of flue efficiency in enabling the plume of smoke to rise well above the flue terminal before dispersion.[4,5]

Neighbouring sources of air pollution This is affected by the general character of the district. In view of the reduced efficiency of flues in polluted atmospheres, the memorandum gives five classifications of districts ranging from country districts to large cities with varying degrees of pollution.[6]

Presence of other buildings and tall trees These can cause downwash of air around chimneys and consequent premature deposit of effluent on the ground in the vicinity. If the height of the chimney as determined by the previous factors is less than two and a half times the height of any local object some modification such as fan assistance may be necessary to achieve proper dispersion of effluent in the upper air.

6.1.3 Materials and construction of chimney shafts

Generally, the basic parts of a chimney shaft comprise: foundations, structural shell, flue lining (where considered necessary), branch flue connection to furnace(s) and terminal.

Foundations Chimney shafts standing alone are clearly subject to wind loads and foundations must distribute eccentric loading to the subsoil within the allowable bearing pressure. Settlement associated with clay subsoils must be expected since the heat from the base of the chimney will in time be transferred to the subsoil. Due to the concentration of load it is common for chimney foundations to be piled.

Structural shell This part of the shaft must perform as a stable structure cantilevered from the base and having overall stability when subjected to wind pressure from any given direction. The wind load is assessed in accordance with the methods laid down in BS 6399-2, described briefly on page 75. The actual pressure on the shaft varies with its shape and the code gives coefficients to take account of this.

In many cases the structural shell will be designed to give support to lining materials with or without the presence of an air space, and certain openings may be needed for access to air spaces in addition to those required for branch flues.

Chimney shafts are normally constructed of concrete or steel with or without some form of lining. Table 6.1 summarises some types of construction.[7]

Masonry shafts and chimneys Free-standing shafts in brickwork are now rarely built due to their high cost.

Large chimneys may be constructed of masonry within buildings provided precautions are taken to protect the building from damage by heat or through corrosion of any structural steel. Figure 6.1 shows typical methods of constructing such chimneys within buildings.

Reinforced concrete shafts Except for low-rise applications and for aesthetic priorities, these have generally replaced brick shafts as a more economic construction process. Due to the high wind loading experienced in relation to the very tall chimneys of such buildings as power stations in situ reinforced concrete has always been used and the chimneys designed as special structures. Typical reinforced concrete shafts are shown in figure 6.2. Reduction in thickness may be achieved by stepping or tapering the shaft but considerable economy can be achieved by constructing the whole shaft with constant bore and shell thickness thus enabling a rising or slip form shutter to be used.

Chimneys of medium or small size up to about 50 m high are often economically constructed of precast reinforced concrete units in the form of either segmented blocks or horizontal rings and a wide range of single or multiple flue designs are available. Figure 6.3 shows typical blocks used for this purpose.

Table 6.1 Chimney shaft construction

Simple: single-skin construction in one material (principally for gas appliances)

1 Riveted steel plate	4 Fibre cement pipe BS EN 1857	6 Factory made metal chimneys BS EN 1859
2 Welded steel plate	5 Precast concrete flue blocks for building into walls BS EN 1858	
3 Cast iron BS 41		

Composite: double- or multi-skin construction using one or more materials (suitable for oil and coal fired furnaces)

Structure: materials and system of construction		Lining materials		
Precast reinforced concrete, cast in complete plan sections, with main reinforcement threaded and grouted in as building proceeds		1 Precast Moler concrete[3] 2 Precast refractory concrete		
Precast reinforced concrete, panels with the main reinforcement cast into the vertical and horizontal joints between the panels as the work proceeds (Trustack)		1 Moler concrete precast in tubular form with unventilated air space		
Reinforced concrete cast in situ	1 Firebrick with ventilated air space 2 Firebrick with rock wool insulation	3 Moler brick[3] with ventilated air space 4 Acid resisting brick with ventilated air space 5 Acid resisting brick with rock wool insulation	6 Refractory concrete (precast) or in situ with ventilated air space 7 Refractory concrete with rock wool insulation 8 Moler concrete with ventilated air space	
Steel plate,* exposed externally with or without supporting guy ropes * see BS 4076	1 Firebrick 2 Moler brick 3 Acid resisting brick	4 Refractory concrete (in situ, precast, or gunned) 5 Moler concrete (in situ, precast or gunned)	Note: all these linings should be positioned no further than 25 mm from the steel to prevent the build-up of dust, which will cause the lining to bulge	
Steel plate covered with aluminium: air space between the two materials: with or without supporting guy ropes As above but with the air space filled with rock wool insulation		1 Structural steel acts as lining 1 Structural steel acts as lining		

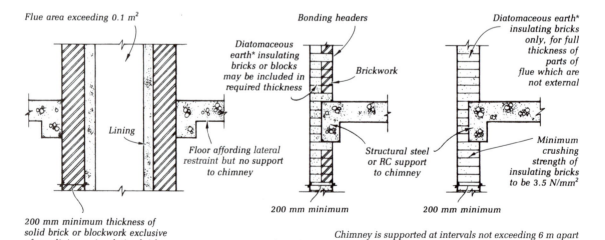

Flue area exceeding 0.1 m²

Bonding headers

Diatomaceous earth insulating bricks or blocks may be included in required thickness*

Brickwork

Diatomaceous earth insulating bricks only, for full thickness of parts of flue which are not external*

Lining

Floor affording lateral restraint but no support to chimney

Structural steel or RC support to chimney

Minimum crushing strength of insulating bricks to be 3.5 N/mm²

200 mm minimum

200 mm minimum

200 mm minimum thickness of solid brick or blockwork exclusive of any lining or insulating brick

Chimney is supported at intervals not exceeding 6 m apart (see BS 5854: Code of Practice for Flues and Flue Structures in Buildings)

* Other lightweight aggregate materials with equivalent heat insulative performance may be considered.[3]

Figure 6.1 Large chimneys

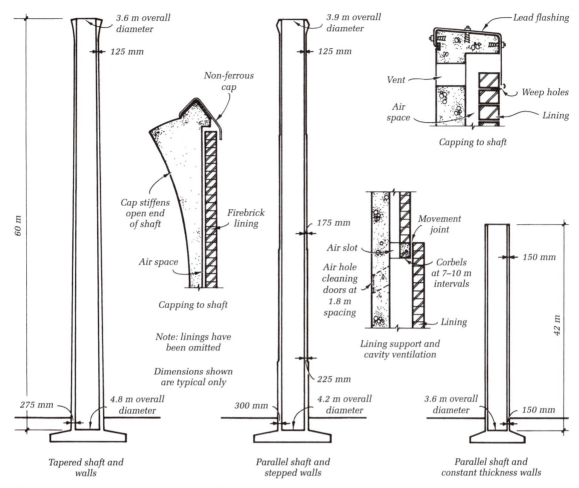

Figure 6.2 Reinforced concrete chimney shafts

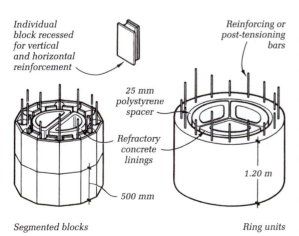

Figure 6.3 Precast concrete chimney shafts

Steel shafts These may be either self-supporting, guyed or stabilised by enclosing framed structures. Steel chimneys can be assembled relatively quickly and may be lined or unlined depending on the fuel being burnt (see *Flue linings* below).

Self-supporting chimneys Self-supporting steel chimneys have been constructed up to 135 m in height but a more normal height is around 75 m. The cantilevered chimney shell is secured to a concrete base by bolts and is usually stiffened by gusset plates at the base (figure 6.4 A). A steel chimney weighs far less than a concrete chimney and resistance to overturning will largely be provided by the mass of the concrete base. Steel chimney shells are either riveted or welded structures, particular care having to be paid to the latter form with regard to the resonant vibrations caused by wind eddies which, if they match the

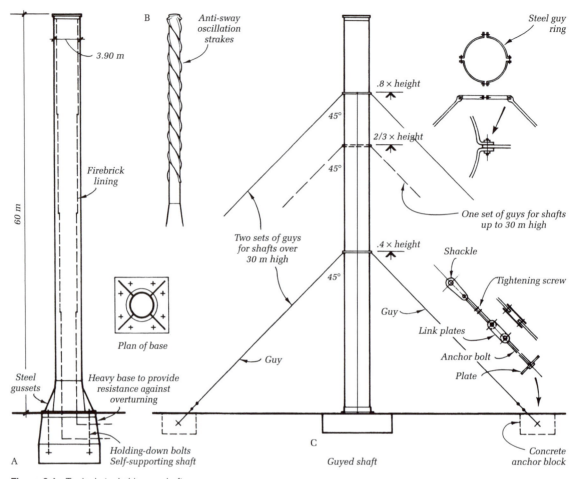

Figure 6.4   Typical steel chimney shafts

natural frequency of the sway of the chimney, may cause the sway to develop to the point of collapse. Unlike reinforced concrete or riveted steel, welded shafts have little structural damping within the structure to resist oscillation and in order to break the wind vortices affecting such chimneys helical strakes are often fitted to the upper part of the stack (figure 6.4 B).[8]

Guyed chimneys   Guyed steel chimneys (figure 6.4 C) are built of 5 mm to 19 mm steel plate depending on their size. Chimneys of up to 30 m are guyed at two thirds of their height and over 30 m at 0.4 and 0.8 of their height. The guys have tightening bolts in the linkage and terminate at anchor plates buried in concrete anchor blocks in the ground. As the guys stabilise the chimney against sway from wind pressure and overturning the concrete foundation only needs to be large enough to resist dead weight.

Where there is insufficient space for guying and a fully cantilevered construction may not be desirable due to site conditions, steel chimneys can be stabilised by erecting around them singly or in groups a framed structure to which the chimney shaft(s) is attached by suitable bracketing (figure 6.5 A).

Steel chimneys are cheaper than either masonry or reinforced concrete but are comparatively short lived. Unless lined or clad it is necessary to protect the metal against corrosion due to condensation caused by the heat loss from such chimneys. For this purpose proprietary metal paints are available and PVC coatings.

Steel construction is also suitable for ejector chimneys which provide induced draught by means of a small high pressure blower discharging into a venturi to produce suction. A typical example of an ejector chimney providing a draught equal to that of a chimney 60 m high by 3 m diameter using natural draught is shown in figure 6.5 B.

The performance of steel chimney shells may be improved both in functional performance and longevity by building them in double-skin form with a steel inner shell

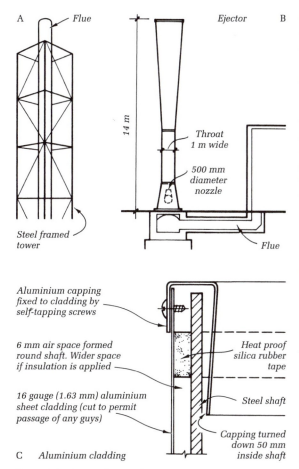

A

Flue

Steel framed tower

14 m

B

Ejector

Throat 1 m wide

500 mm diameter nozzle

Flue

Aluminium capping fixed to cladding by self-tapping screws

6 mm air space formed round shaft. Wider space if insulation is applied

16 gauge (1.63 mm) aluminium sheet cladding (cut to permit passage of any guys)

C   Aluminium cladding

Heat proof silica rubber tape

Steel shaft

Capping turned down 50 mm inside shaft

Figure 6.5   Steel chimney shafts

and an outer loadbearing shell of heavier steel or by cladding them in aluminium and interposing insulation, such as rockwool, between these skins (figure 6.5 C). The performance of such chimneys depends on the workmanship in making the joints gas proof since the steel lining is impervious to vapour and can accommodate high gas velocities and pressures.

Fibre cement chimneys are only suitable for gas appliances (see page 222) and if free standing may be guyed.

Resin bonded reinforced glass fibre chimneys have the advantage of lightness of construction and resistance to acid corrosion, less maintenance and longer life than steel. Guy ropes are normally required or a steel supporting tower. The main disadvantage of such chimneys is the limited working temperature, which is in the region of 235 °C. The chemical composition of the resin bond is critical and it is advisable to fit a temperature indicator alarm and some means of admitting cold air in case of fire.

Flue linings   Flue linings protect the structural shell from the effects of heat and corrosive agents. If suitably chosen,

linings provide thermal insulation around the flue and thus help to maintain the temperature of the flue gases, thereby reducing the production of corrosive agents and promoting strong convection currents by maintaining a high temperature differential between the inside of the flue and the outer air.

Types of linings in common use are: firebricks, refractory concrete, moler bricks and concrete, acid resisting bricks. For small installations linings used include: fibre cement pipes, clayware pipes, flexible metal tubing and proprietary insulating concrete. Special linings of resin bonded glass fibre or plastic may be used under controlled temperature conditions and as specified by the appliance manufacturer.

The properties and characteristics of the main group of lining materials used for larger chimneys are as follows:

Firebricks   These have a 28–32 per cent alumina content and are suitable for temperatures up to 1200 °C. The bricks are moulded in radial form to suit the chimney and they are laid in fire cement or mortar of ground fireclay. Such bricks are not good insulators but are effective anti-corrosion liners, gas tight and of long life. Firebricks must be laid in a manner which allows them to move freely relative to the surrounding chimney shell. Touch headers should be built against the outer shell at 3 m intervals.

Refractory concrete   This is similar to firebrick but can be cast in situ or applied by gun.

Moler bricks   Made from diatomaceous earth in solid form, these bricks are good insulators and effective within the temperature range 150–800 °C. The material may be made to given shapes and due to its low coefficient of expansion may be built tightly against the enclosing chimney shell or if permitted, bonded into masonry structures (figure 6.1), or used as permanent shuttering for concrete shells. The normal thicknesses of moler brick linings are between 76 mm and 114 mm and they are set in mortar made from ground moler bricks and Portland or aluminous cement – depending on the anticipated flue temperature. Moler bricks have low crushing strength but do not need excessive support due to their light weight.

Moler concrete   This is made from diatomaceous earth with aluminous cement and may be cast in situ, gunned or precast. The performance of moler concrete is similar to that of the bricks of this material in the temperature range 150–980 °C.

Acid resisting bricks   These are highly vitrified fire or clay bricks set in acid resisting cement. They produce an impervious lining used when flue gases are likely to be very acidic or at or near their dew point (150 °C or below).

Such bricks are not good insulators and do not stand up to rapid changes of temperature and they should be used in combination with other materials or with an enclosing air space about the lining. Clay acid resisting bricks withstand temperatures of up to 540 °C and those of firebrick up to 1100 °C.

Linings should be taken to the top of brick and concrete shafts to prevent damage to the chimney shell through sudden increase in temperature. They must be protected from the weather by an adequately oversailing capping which also allows for thermal movement (see figure 6.2). Corbel supports are usually spaced at about 7 m for heavy linings such as firebricks to 13 m for lighter linings such as moler bricks. Where air spaces are relied upon to give a shallow temperature gradient through the chimney construction and afford removal of diffused gases, air holes must be provided in the corbels with adjacent cleaning doors (see figure 6.2). In steel chimneys brick linings are usually supported on internal steel angles with a 25 mm gap between firebrick and metal shell which is filled with loam. Alternatively self-supporting moler brick or precast moler concrete liners may be used built against the steel shell, depending on the compressive stress at the base and the amount of movement between the shell and lining anticipated in any particular location.

Branch flues   Where possible branch connections should sweep in to ensure a good gas flow (figure 6.6). Two branches should not enter the stack directly opposite each other unless a splitter plate is provided to divert the gases and avoid turbulence which could cause unstable draught conditions. Generally openings should not occupy more than 25 per cent of the circumference of the chimney unless the structure is supported and stiffened as necessary. In large installations involving a number of boilers branch flues are collected into one horizontal header flue having a cross sectional area of about one and a third times that of the largest or main flue. Such flues are usually made accessible for cleaning and should be kept as short and as well insulated as possible. Brick or concrete construction is mostly used with insulating linings as previously described for flues. Steel or cast iron horizontal headers should be insulated to prevent overheating of the boiler house. Some types of plastic insulation used for this purpose should be separated from the hot metal by an air space for which purpose the flue is wrapped in expanded metal before applying the insulating composition.

Terminals   Due to the cooling effect of the outside air, there is greater danger of corrosion at the terminals of flues. Terminals to large chimneys must be constructed of acid resisting materials such as acid resisting aluminium bronze, chemically pure lead, high alumina cement or

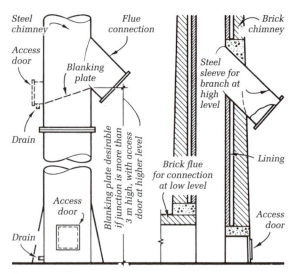

Figure 6.6   Branch flue connections

thick mild steel insulated internally with aluminium. Furthermore the upper external surfaces of chimneys should be protected with surface coatings of acid resistant paint or be constructed of high alumina concrete. Guy ropes used to stabilise steel chimneys should be secured at least 3 m below the terminal to avoid corrosion. Inversions of temperature within the flue can be avoided and effluent velocities from oversized flues can be increased by using a truncated cone with the correctly sized outlet.

Lightning conductors   These should be provided to all high chimneys, reinforcement in reinforced concrete chimneys being connected to an earthing plate at the base and not to the conductor at the top. A lightning conductor provides a low impedance path for the lightning to discharge as directly as possible to earth. Sharp bends must be avoided since the lightning will take a shorter route even if this is through the structure. The conductor consists of an air terminal connected to a conductor tape of copper (about 25 mm × 5 mm section) which is connected either to a copper rod, tube or plate earth or to a suitable water main. The air terminal consists of a copper rod located at the top of the stack, the zone of protection being assumed to be within a cone with its apex at the top of the terminal and a base equal to the height above ground level. The conductor tape should have expansion loops of easy radius at intervals in the length of the stack.[9]

6.2 Gas flues

The subject of chimneys for gas-fired appliances has been introduced in Part 1, chapter 9, in respect of domestic appliances and it is here extended further.

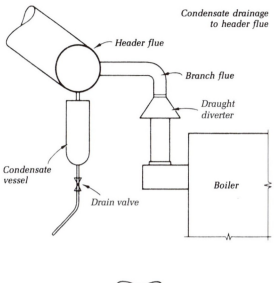

Condensate drainage
to header flue

Header flue

Branch flue

Draught
diverter

Boiler

Condensate
vessel

Drain valve

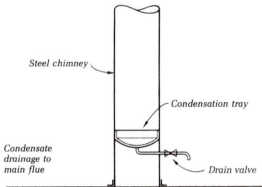

Steel chimney

Condensation tray

Condensate
drainage to
main flue

Drain valve

Figure 6.7   Gas boiler flues – removal of condensate

Compared with the combustion of solid fuels which require 100 per cent excess air to promote adequate natural draught in a flue, gas (and oil) fuels require only 20 to 30 per cent excess air resulting in smaller flues. Gas has a low sulphur content and produces a reasonably low condensate of sulphuric acid on thin walled cold flues such as steel or fibre cement, which in single skin form are only suitable for gas flues.

The burning of gas fuels produces a large quantity of water during combustion and this together with the weak draught resulting from the relatively low temperature of the gases leaving the appliance leads to a considerable accumulation of condensate at the base of the flue. Flues to gas boilers require to be properly drained by incorporating condensation trays or vessels in the base of the chimney or header flue as shown in figure 6.7.

Baffles or draught diverters are often incorporated in the design of gas boilers or should be provided at each boiler flue outlet to allow free removal of the products of

combustion and to divert any down draught away from the combustion chamber. In addition, this 'break' in the flue system enables a quantity of air to mix with and dilute the flue gases thus lessening the risk of condensation within the main flue. Further dilution may be provided by a ventilating grid at the base of the main flue situated in the boiler room and this should be as large as practicable up to the cross-sectional area of the main flue.

6.2.1 Discharge of products of combustion

The Building Regulations, Approved Document J deals with the safe installation of heat producing appliances in buildings including fixed appliances burning gas. In order that a heat producing appliance may function safely, it needs an adequate supply of air and must be capable of discharging the products of combustion to the outside air without allowing noxious fumes to enter the building and without damaging the fabric by heat or fire.

In the Approved Document, section 3 deals with gas burning appliances with a rated input of up to 70 kW (net) such as domestic cooking appliances, balanced flue appliances including boilers, convector heaters, water heaters, etc., decorative log and solid fuel fire effect appliances and individual natural draught, open flued appliances such as boilers, back boilers, etc.

Apart from cooking appliances (which discharge their products of combustion into the air of the room in which they are situated), gas burning appliances should discharge into balanced flues (see page 224) or chimneys of suitable construction or flue pipes. Flue sizes and the positioning of the flue outlets are covered in Part 1, section 9.7.1. Figure 6.8 A on page 222 of this Part indicates the limit set by the Building Regulations on the proximity of a flue outlet to openings into the building.

Appliances other than those covered by Approved Document J (except flueless water heaters) should be installed in accordance with BS 5440-1 and 2: *Installation and Maintenance of Flues . . . etc.* Flueless heaters should be installed in accordance with BS 5546: *Specification for Installation of Hot Water Supplies for Domestic Purposes, Using Gas-fired Appliances of Rated Input not Exceeding 70 kw.*

6.2.2 Materials and construction of gas flues

The word 'flue' means a passage for conveying the discharge of an appliance to the external air and includes any part of the passage in an appliance ventilation duct which serves the purpose of a flue.

In discussing suitable materials and methods of constructing flues it is important to note the difference between a 'chimney', which means any part of the structure

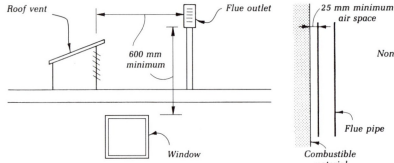

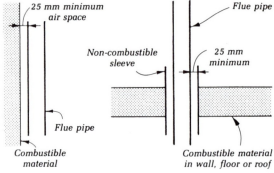

A   *Proximity of flue outlet to openings into building*

B   *Separation of flue pipes from combustible material*

Figure 6.8 Gas flues

of a building forming the wall of a flue, and a 'flue pipe', which means a pipe forming a flue but does not include a pipe built as a lining into a chimney or an appliance ventilation duct.

Gas burning appliances may discharge into a flue in a masonry chimney (see Part 1, page 204) or into a flue pipe. In the former the flue must be lined or the chimney must be constructed of flue blocks conforming to BS EN 1806: *Chimneys. Clay/ceramic Flue Blocks for Single Wall Chimneys. Requirements and Test Methods.*

Flue pipes These may be formed of the same materials as used for those serving solid fuel appliances (see Part 1, page 202) with the addition of sheet metal and fibre cement.

Fibre cement pipes may be used either singly or in groups surrounded by suitable non-combustible material.

There is considerable saving in the space required for grouped flues of fibre cement over normal brickwork and they may be used up to a limit of 30 m, the weight of the flue pipes being separately supported at each floor level. Individual pipes must be securely fixed, with the sockets upwards, and each joint must be filled with fire resistant rope and pointed with fire resistant compound. Suitable terminals are required as indicated earlier.

Fibre cement is not resistant to attack by corrosive flue gas condensates. If condensation is anticipated internal protection should be applied in the form of bitumastic paint or brush-on acid resistant cement finish. Alternatively, fibre cement pipes with vinyl acetate protection already applied may be used. The height above the appliance at which condensation may occur depends on a number of variable factors. Table 6.2, which was included in

Table 6.2 Length of certain flues to gas fired appliances

Situation of flue	Type of appliance	Maximum length of flue in metres	
		If flue is circular or square, or is rectangular and has the major dimension not exceeding three times the minor dimension	*If flue is rectangular and has the major dimension exceeding three times the minor dimension*
(a) Flue formed by a chimney or flue pipe which is internally situated (that is to say, otherwise than as (b) below)	Gas fire	21	12
	Heater installed in drying cabinet or airing cupboard; or instantaneous water heater	12	(not permitted)
	Air heater or continuously burning water heater	6	(not permitted)
(b) Flue formed by a chimney having one or more external walls; or by a flue pipe which is situated externally or within a duct having one or more external walls	Gas fire	11	6
	Heater installed in drying cabinet or airing cupboard; or instantaneous water heater	6	(not permitted)

earlier editions of the Building Regulations, Approved Document, gives some indication of the maximum advisable heights for certain types of appliance and situations of flues serving one appliance only. When a very tall flue is being considered the local gas supplier should be consulted in case internal coating of the flue or other precautions may be advisable. Fibre cement pipes cool very easily and long flues on the outside of a building invariably result in condensation trouble and may even fail to function as a flue. In such circumstances an insulated flue such as described below would be better.

Metal pipes may be used for flues, but preferably only in positions where the flue can be seen and easily replaced if necessary, because of the liability of corrosion. Protected steel or cast-iron pipes and sheet metal with welded or folded seams are suitable.

Insulated metal chimneys Increased use is being made of factory made chimney systems, suitable for gas, oil or solid fuel installations. These systems comprise flue units and components and fittings for both internal and external situations and are generally restricted to heights of between 12 and 15 metres although special support systems can be designed for heights in excess of this (see Part 1, page 201 and figure 9.14).

The flue units are normally formed of a lining (such as stainless steel) and an outer casing of corrosion resistant material with about 25 mm of non-combustible high temperature insulating fill between the two concentric tubes. Most systems may be installed in traditionally constructed dwellings with timber floors and roofs provided account is taken of their self weight. The principal components are the chimney sections, support plates (or brackets externally), fire stops, special flue pipe connectors and terminals. These factory made chimney systems claim to save up to 50 per cent of the cost of a traditional brick chimney and are, of course, much quicker to erect.

Gas flue pipes must not be nearer than 25 mm to any combustible material (if the pipe is double-walled this distance is measured from the inner pipe). Where passing through a wall, floor or roof it must be separated from any combustible material by a non-combustible sleeve sufficiently large to provide an air space round the flue pipe of not less than 25 mm (see figure 6.8 B).

Table 6.3 summarises the application of materials which are suitable for flues to gas appliances.[10]

Table 6.3 Materials for flues to gas appliances

Material	Condition	Suitability	Protection against condensation
Brick Linings:	Corrosion resistant lining	All appliances	Inside face may be lined with acid resistant tiles embedded in an acid resistant jointing material
Clay flue linings and flue terminals			Acid resistant lining may be introduced consisting of suitable clay flue lining, earthenware pipe or fibre cement flue with protective coating of acid resisting material. See below under 'fibre cement'
Vitrified clay pipes, fittings and joints			
			Inside face may be rendered with acid resistant cement
Precast concrete	Protected	All appliances	May be wholly of acid resistant cement
			May be composite, with inside wall made of acid resistant material
			Joints should be made with an acid resistant joint material
Fibre cement	Protected	All appliances	Inside face may be coated by manufacturer or on site with an acid resistant compound. Suitable coatings have been prepared from: (a) vinyl acetate polymer, (b) a rubber derivative base compound
Earthenware pipes	Glazed or unglazed but of low porosity	All appliances	
Metal	Protected or corrosion resistant	All appliances – generally used for connection to chimney or flue pipe	Mild steel acid resistant vitreous enamelled BS 6999 Stainless steel BS EN 10088-1* Protected cast iron BS 41
Factory-made insulated chimneys		All appliances	

* Stainless steel with insulation is available.

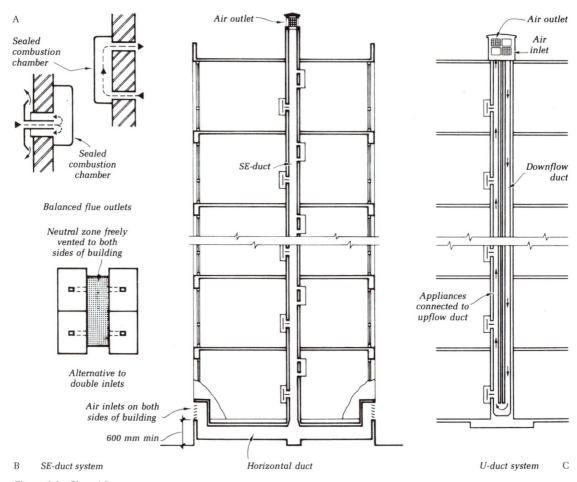

A

Sealed combustion chamber

Sealed combustion chamber

Balanced flue outlets

Neutral zone freely vented to both sides of building

Alternative to double inlets

Air inlets on both sides of building

600 mm min

B   SE-duct system

Air outlet

SE-duct

Horizontal duct

Air outlet

Air inlet

Downflow duct

Appliances connected to upflow duct

U-duct system   C

Figure 6.9   Shared flues

6.2.3 Balanced flues

Certain gas appliances are designed to operate with 'balanced flues'. A 'balanced flue' or 'room-sealed' appliance is so designed that the combustion chamber is enclosed and sealed from contact with the air in the room. Air for combustion passes into the casing from outside the building and the products of combustion are discharged directly through the wall at an adjacent position. The air pressure is the same on both inlet and outlet and the warm gases are discharged without being affected by wind pressure or gusts (see figure 6.9 A). The outlet terminal must be at least 300 mm from any opening into the building which is above or partly above the terminal. Such appliances may also be connected to common flue systems or appliance ventilation ducts using natural or mechanical ventilation (figure 6.9 B and C).

6.2.4 Shared flue systems

The use of shared or common flues reduced considerably the amount of space required for gas flues in multi-storey buildings and facilitates the positional planning of gas fired appliances away from outside walls resulting in functional planning and economic advantages.

One system links up conventional appliances by short branch flues to a main flue. Others use a main flue or duct serving room-sealed appliances.

SE-duct and U-duct systems   These two systems were developed by the South Eastern Gas Board under the Gas Council (now British Gas). They incorporate common ventilation ducts into which room-sealed appliances can be discharged: the SE-duct system and the U-duct system. The former consists of a main duct with air inlets at the

Table 6.4 SE-ducts: required sizes

Rating of appliance (kW)	Number of storeys												
	3/4	5/6	7/8	9/10	11/12	13/14	15/16	17/18	19/20	21/22	23/25	26/28	29/30
Continuous appliances with kW ratings of:	Duct type: 1–200 mm × 300 mm; 2–230 mm × 400 mm; 3–330 mm × 480 mm; 4–380 mm × 560 mm												
2.93	1	1	1	1	1	1	1	1	1	1	2	2	2
4.93	1	1	1	1	1	1	1	1	2	2	2	3	3
5.86	1	1	1	1	1	1	2	2	2	3	3	3	3
8.79	1	1	1	1	2	2	3	3	3	3	3	3	4
11.71	1	1	1	2	2	3	3	3	3	3	4	4	4
14.66	1	1	2	2	3	3	3	3	4	4	4	*	*
17.59	1	1	2	3	3	3	3	4	4	4	*	*	*
20.52	1	2	3	3	3	3	4	4	*	*	*	*	*
23.45	1	2	3	3	3	4	4	*	*	*	*	*	*
26.38	1	2	3	3	4	4	*	*	*	*	*	*	*
29.31	1	3	3	3	4	4	*	*	*	*	*	*	*
Instantaneous water heaters with kW rating of: 28.58	1	1	1	1	2	2	2	3	3	3	3	3	3

* In these situations special ducts may be necessary.

Table 6.5 SE-ducts: required sizes

Rating of appliance (kW)	Number of storeys									
	3/4	5/6	7/8	9/11	12/14	15/16	17/19	20/22	23/29	30
Instantaneous water heaters (28.58 kW) in combination with continuous appliances with ratings of:	Duct type: 1–200 mm × 300 mm; 2–230 mm × 400 mm; 3–330 mm × 480 mm; 4–380 mm × 560 mm									
2.93	1	1	2	2	3	3	3	3	4	4
4.39	1	1	2	2	3	3	3	3	4	4
5.86	1	1	2	2	3	3	3	4	4	*
8.79	1	1	3	3	3	3	4	4	*	*
11.72	1	2	3	3	3	4	4	*	*	*
14.66	1	2	3	3	4	4	*	*	*	*
17.59	1	2	3	3	4	4	*	*	*	*
20.52	2	2	3	3	4	*	*	*	*	*
23.45	2	3	3	4	*	*	*	*	*	*
26.38	2	3	3	4	*	*	*	*	*	*
29.31	2	3	4	4	*	*	*	*	*	*

* In these situations special ducts may be necessary.

base drawing air from the perimeter of the building at ground level and terminating in a specially designed terminal above roof level (see figure 6.9 B) whilst the latter consists of a twin duct in the form of an elongated 'U' with both air inlet and outlet at roof level, thus avoiding the necessity to build horizontal ducts beneath the building (figure 6.9 C). In both systems the air pressure at the inlet and the outlet of the appliances must be equal to allow the residual heat from the flue gases to act as the motive power to move gases up the flue. A number of precast concrete units for which the Gas Council secured patent rights are available comprising duct blocks, storey height duct units, terminals and intakes. Tables 6.4 and 6.5 show the required sizes of SE-ducts in blocks of flats relative to the appliance rating and height of building.

Branched flue system Branched flues may be used where conventional open-flue appliances are installed in tall buildings with a regular floor plan and aspect on successive floors, e.g. blocks of flats and offices of uniform design, floor on floor. The principle is to provide a subsidiary or branch flue for every heating appliance, each connecting to one common or main flue. This main flue extends the full height of the installation, terminating in free air above the top of the building. It is limited to ten consecutive floors. The principle of operation being to provide the necessary draught in the branch flue and to evacuate the gases through the main flue. A branched flue system saves considerable flue space on the upper floors of tall buildings (figure 6.10). The system, also called the shunt system, may be constructed from precast units of refractory concrete formed with two internal apertures for branch and main flues.

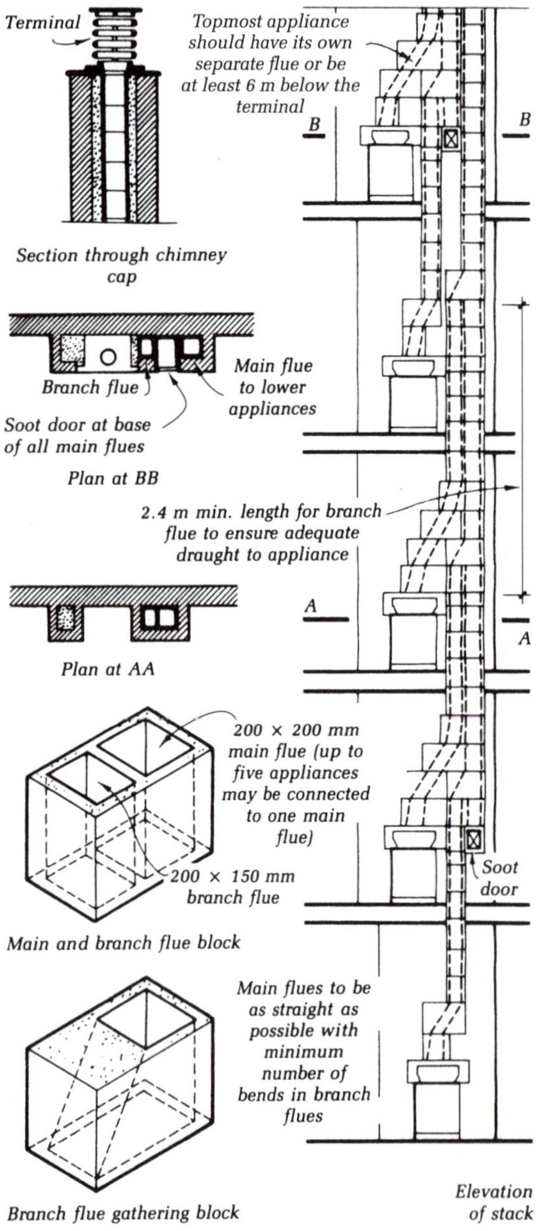

6.2.5 Fan diluted flues

Conventional flue systems operate on the natural draught principle and because the pressure difference is so small the flue has to have relatively large dimensions to keep the flow resistance low. However, if an external source of power is introduced, such as a fan, the area of the flue can be considerably reduced (see ejector chimneys on page 218). In fact, the only limits on area reduction are the noise level, air velocity, and the power consumption which must be kept to an economical level. This is the principle of the mechanical extraction system which, apart from the reduced flue area, has the added advantage of being able to draw combustion products downwards from the appliance and horizontally under floors where necessary. It is, therefore, useful where natural draught flues are not practical. A development in this field is the fan diluted flue (figure 6.11). This system has been specifically developed to solve the problems of ground floor shop premises in mixed tower blocks of offices, shops and flats. The system operates with fresh air being drawn in through a duct by fan, mixed with the products of combustion, and finally being discharged to the atmosphere with a carbon dioxide content of not more than one per cent.

Gas with its extremely low sulphur content is ideal for this method. Many local authorities allow the discharge to be made at low level, above a shop doorway for instance, or into well-ventilated areas with living or office accommodation above – preferably at least 3 m above ground or pedestrian access.

Ideally, the air inlet and discharge louvres should be positioned on the same wall or face of the building. If the

Terminal

Topmost appliance should have its own separate flue or be at least 6 m below the terminal

B — B

Section through chimney cap

Main flue to lower appliances

Branch flue

Soot door at base of all main flues

Plan at BB

2.4 m min. length for branch flue to ensure adequate draught to appliance

A — A

Plan at AA

200 × 200 mm main flue (up to five appliances may be connected to one main flue)

200 × 150 mm branch flue

Main and branch flue block

Main flues to be as straight as possible with minimum number of bends in branch flues

Soot door

Branch flue gathering block

Elevation of stack

Figure 6.10 Branched flues

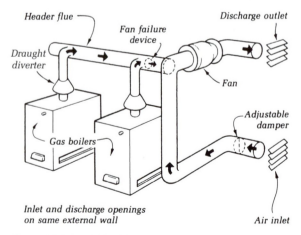

Header flue

Fan failure device

Discharge outlet

Draught diverter

Fan

Gas boilers

Adjustable damper

Inlet and discharge openings on same external wall

Air inlet

Figure 6.11 Fan diluted flues

louvres are likely to be subjected to strong wind forces, some shielding is desirable. A damper is fitted near the diluent air inlet to balance the installation and an air flow switch is fitted as a fan failure safety device on the suction side of the fan. The boilers draw in their combustion air through floor level inlets. Normal metal sheet, fibre pipe or rectangular fibre cement ducting can be used for the ducting as flue temperatures with this system are rather low, about 65 °C.

6.3 Ducts for services

Services may be distributed on or within the structural elements of a building as described for floors on pages 209 and 210 but there are, as indicated there, advantages in accommodating them in ducts formed for this purpose in the building fabric. The services are concealed and protected but they are accessible without the necessity of breaking open a floor screed, for example. Inspection and maintenance are thus made simpler and alterations and additions are facilitated. Ducts also have the beneficial effect of separating operations (see Part 1, page 14) since the installation of the services may be independent of the construction of the building fabric.

Services normally require vertical and lateral distribution within a building and in multi-storey buildings this usually results in three types of duct: main, vertical and lateral.

6.3.1 Main ducts

These link the various service lead-in points with the main controls and provide the primary horizontal distribution of services. They are usually located below the ground floor in the form of either a *subway*, a *crawlway* or a *trench*.

A *subway*, in which a person may walk, should be at least 2 m high and wide enough to provide a clear working space of 700 mm between pipes or pipe racks (figure 6.12 A). In order to minimise the overall width it is desirable to group large and small pipes on opposite sides of the subway.

A *crawlway* should be not less than 1.1 m high with the same minimum working space as a subway (figure 6.12 B). Access panels should be provided in the top at frequent intervals. Crawlways are used preferably only where the services require little maintenance.

Subways and crawlways are used where a large number of services must be distributed as in heavily serviced buildings such as laboratories and hospitals.

A *floor trench* is used when relatively few pipes are to be accommodated. Access to service pipes is from the top and the depth of the trench is generally not greater than 750 mm (figure 6.13 A).

Subways and crawlways are constructed with a concrete base and walls of either concrete or brick as shown in figure 6.12. If the site has a high water table some form of waterproofing as described for basements is essential. The floor should be laid to falls with a shallow channel on one side to convey water from possible leaks and condensation to sumps or, in suitable circumstances, to a drain through a sealed gulley.

Trenches are commonly constructed of concrete with continuous access provided by trench covers which span from side to side of the trench. The covers may be in the form of metal trays, filled to match the floor finish, set in metal frames or in the form of precast concrete slabs over which the floor finish runs as shown in figure 6.13 A. Separating strips along the line of the slabs permit the ducts to be opened up without damaging the adjacent floor areas. If manhole covers, filled as above, are used these must be placed at junctions and bends in the pipes and be long enough to permit lengths of pipe to be introduced and removed easily.

6.3.2 Vertical ducts

These lead from the main ducts to distribute the services to the various floors. Depending upon the number of pipes to be accommodated these may be casings on the face of a wall or chases cut in a wall for small pipes or cables (figure 6.13 E, F) or shallow ducts up to about 600 mm deep with access panels or doors the full width of the duct (D). A wide, shallow duct makes access to the pipes easier than a relatively deep but narrow one. Ducts large enough for a person to enter are often constructed for vertical runs of large numbers of pipes without branches crossing the risers and restricting access to them (figure 6.12 C). Access is provided either at each floor or at every other floor with internal steel ladders to working platforms of open grating floors (see page 206).

6.3.3 Lateral ducts

These lead from the main or vertical ducts to provide horizontal distribution of services at various levels. They may be in the form of wall casings, such as a hollow skirting duct for small pipes (figure 6.13 B), or a shallow trench in a ground floor (C).

In concrete upper floors it is a simple matter to form ducts in the thickness of most types of one-way spanning floors by setting a section of timber in the wet concrete, so long as they run in the direction of the span (see figure 6.14). In a solid in situ floor the necessary width of a duct is boxed out on the shuttering (A). In a hollow block floor (B) or precast rib and filler floor (C), one or more lines of blocks are omitted and in the case of a precast beam floor one or more beam sections are omitted. The holes or slots in the supporting beams through which the services pass

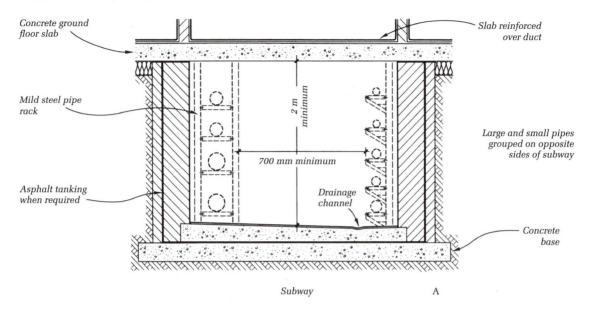

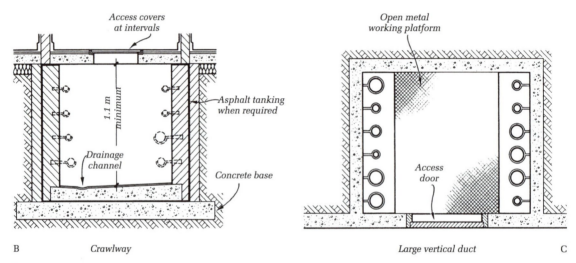

Figure 6.12 Ducts for services

should not be greater in depth than one-third of the depth of the beam and should be situated within the middle third of the span. Only small holes should be formed nearer the supports of the beam. They should be made, preferably, on the line of the neutral axis. If the latter is too low and the hole is formed at a higher level, compression reinforcement may be required in the 'bridge' over the hole.

Holes through the webs of steel beams should also be limited in size and situation as given above, although in beams acting compositely with the floor slab they may be up to two-thirds of the depth. The webs may sometimes need strengthening by means of reinforcing plates welded

round the holes. Where possible such holes, whether through concrete or steel beams, should be circular rather than rectangular in shape.

From such ducts lateral branches from the service pipes can be run at right-angles to the span in the thickness of the floor screed or rise directly to fittings when, for example, the ducts are placed relative to a run of laboratory benches.

In solid in situ slabs continuous over supporting beams it is possible to form shallow ducts at right-angles to the span on the points of contraflexure (D). These may be of a depth not exceeding about one-quarter of the thickness of

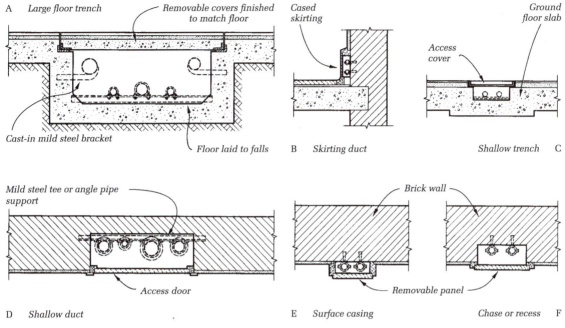

A   Large floor trench — Removable covers finished to match floor

Cased skirting

Ground floor slab

Access cover

Cast-in mild steel bracket

Floor laid to falls

B   Skirting duct

Shallow trench   C

Mild steel tee or angle pipe support

Access door

Brick wall

Removable panel

D   Shallow duct

E   Surface casing

Chase or recess   F

Figure 6.13 Ducts

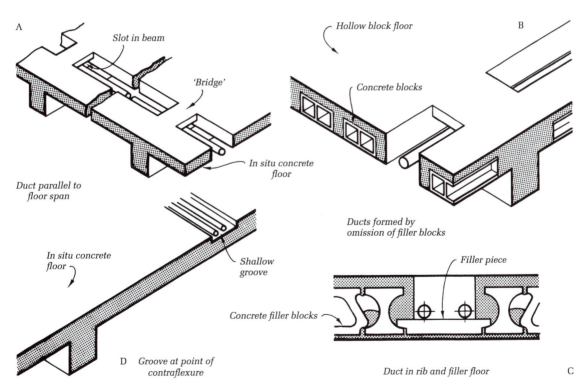

A

Slot in beam

Hollow block floor   B

'Bridge'

Concrete blocks

In situ concrete floor

Duct parallel to floor span

Ducts formed by omission of filler blocks

In situ concrete floor

Shallow groove

Filler piece

Concrete filler blocks

D   Groove at point of contraflexure

Duct in rib and filler floor   C

Figure 6.14 Ducts in concrete floors

the slab. If a deeper duct, or a complete opening through the floor, is required at any point at right-angles to the span it is necessary to form structural trimming beams on each side of the duct.

Plastic and metal ducting for telephone and electric wiring may be set within the floor screed or ducts of various diameters can be formed in floor slabs or screeds by means of expendable fibre, expanded polystyrene or cardboard tubes left in position. Alternatively, pneumatic cores in the form of long, inflatable tubes may be laid in position in an inflated state. The concrete is cast round the tube and, when set, the tubes are deflated and withdrawn. The construction of the wall of the tube is such that when it is deflated the tube twists and pulls away from the concrete. Pneumatic cores have the advantage that they can be laid in curved lines.

As mentioned on page 192, the lateral distribution of complex services in such buildings as hospitals, laboratories or some industrial processing plants often makes the provision of a service floor necessary.

References should be made to pages 331 and 333 regarding the prevention of fire spread in ducts.[11]

Notes

1 It should be noted that flues of large dimension can be accommodated inside buildings or within elements of buildings such as separating walls, the chimney being formed as a duct within the structure of the building (see page 215).

2 See Building Regulations, Approved Document J: *Combustion Appliances and Fuel Storage Systems*, sections 2.4 to 2.5. Also, BS 5854: *Code of Practice for Flues and Flue Structures in Buildings*.

3 Moler brick and moler concrete are lightweight building materials produced using diatomite aggregates. These originate from the hollow siliceous skeletons of minute marine organisms. See also page 219.

4 The minimum efflux velocity by natural draught convection movement is generally taken to be 6 m/s, and where draught is induced by motorised fan, 7.5 m/s. An inadequate efflux velocity may cause 'downwashing' on the leeward side of the chimney. 'Inversion' may also arise where cold air enters the top of the chimney and is heated by the warm flue gases to cause cooling of the chimney lining. This results in condensation and emission of acid smuts.

5 To enable flue gases to continue in the same direction after leaving the flue they should travel at about one and a half times the wind speed likely to be encountered at a given site. The speed of flue gases can be calculated by reference to inlet temperatures, area and shape of flue and the frictional resistance of the flue.

6 The Clean Air Act was introduced following the effects of post-war smoke and fog pollution in industrial and densely populated communities. In particular, the London smog that lasted five days during December 1952, taking the lives of some 4000 people. Since then, technological advances relating to the dispersal and dilution of flue gases combined with the decline of manufacturing industries in the UK have considerably reduced the possibility of anything similar recurring. Additionally, under the Act, local authorities are empowered to create smoke control areas and to consider planning applications for provision of chimneys.

7 This information and end note 5 are reproduced from 'Chimneys and flues: design and construction' by Anthony Collins, *RIBA Journal*, October 1973. For more up-to-date information see Building Regulations, AD J, BS 5854, BS EN 1443: *Chimneys. General Requirements* and CIBSE Guide B as referred to on page 215.

8 BS 4076: *Specification for Steel Chimneys*, Appendix B gives useful design information in connection with this problem. The BS covers general requirements for steel chimneys, linings and claddings and supporting structures.

9 See BN EN 62305: *Protection against Lightning*.

10 This list has been extracted from that given in the *Gas Handbook for Architects and Builders* published by the Gas Council.

11 See *MBS: Environment and Services*, chapter 19, and BS 8313: *Code of Practice for Accommodation of Building Services in Ducts* for information on the arrangement and support of pipes within ducts.

7 Stairs, ramps and ladders

Means of vertical circulation form the subject of this chapter. After a brief reference to timber stairs, reinforced concrete stairs, in situ cast and precast, are described followed by descriptions of metal stair construction. Ramps and the requirements to be met in their design and factors relevant to the design and fabrication of metal ladders are then discussed.

7.1 Stairs

The subject of stairs has been introduced in Part 1, chapter 10, where the functional requirements, basic design factors and the different types of stair are described. An introduction to concrete stairs is given and the construction of timber stairs is described in detail. These are considered further in this chapter.

Loading Timber stairs for small domestic buildings, as pointed out in Part 1, are normally constructed on the basis of accepted sizes for the various parts. Other types must be designed having regard to the imposed loadings laid down in BS 6399-1 for buildings of different occupancies in terms of a minimum uniformly distributed load, and a concentrated load, the latter to be applied in positions as for floors (see page 193). Distributed loadings range from 1.5 kN/m² for individual dwellings to 4.0 kN/m² for most other occupancies and places of public assembly and up to 5.0 kN/m² for stairs in grandstands. The superficial area on which these loads are assumed to act is measured horizontally.

The loading on balustrades or other forms of guarding, assumed to be acting at the top of the guarding, ranges from a uniformly distributed load of 0.36 kN/m run for stairs in individual dwellings and residential buildings to 3.0 kN/m run for stairs in places of public assembly where

crowds can panic and exert much greater forces. In all cases BS 6399 gives uniformly distributed loads and point loads which are required to be applied to the infill of the balustrade. Balustrades to ladders, which this Standard defines as a stair not more than 600 mm wide, must resist a force of 0.22 kN/m run.

It should be noted that a figure of 0.74 kN/m run must be taken for parapets or other guarding to roofs and external balconies.

7.1.1 Timber stairs

Newel and ladder type stairs are described in Part 1. As indicated in that volume stairs of considerable span and width may be constructed by the use of laminated timber. By the adoption of glued and laminated techniques (see page 251) using horizontal laminae, cranked strings may be formed in similar manner to reinforced concrete, and cantilever treads may be built up in tapered form to permit treads of considerable width and overhang beyond the strings.

Geometrical stairs Geometrical stairs in timber must be constructed with strings which are helical in form for circular stairs or have circular or wreathed portions linking straight sections in rectangular stairs.

The curved strings may be constructed by the traditional method using narrow vertical staves, or strips of timber with radiating joints, glued together and forming a core between two plywood laminates (see technique for circular column forms, figure 10.20) or by the use of continuous vertical board laminates in the form of normal glued laminated construction, the thickness of the boards depending upon the radius of the curve. The traditional method is used for small radii curves.

In both cases a vertical circular jig or former is required of the appropriate radius and strong enough to remain stable whilst the laminae are wrapped around it. As in concrete stairs, a helical timber stair may be constructed with a single wide laminated spine string over which the treads cantilever on each side. In this case the jig is provided with set-off steps to form a bed on which the string may be clamped to hold its helical shape as it is built up.

The steps may be constructed in normal closed- or open-riser form between pairs of strings, or open-riser construction with the treads bearing on recesses cut in the tops of the strings. With a single spine string the cantilever treads may similarly bear on recesses in the string or may be supported on some form of metal plate or strap bracket screwed to the top of the string, similar to the examples shown in figure 7.10.

Handrails for geometrical stairs may be laminated in the same manner as for the strings.

7.1.2 Reinforced concrete stairs

Concrete stairs are widely used because of their high degree of fire resistance and the relative ease with which a variety of forms may be produced. In situ cast stairs will be described first followed by a continuation of material on precast stairs given in Part 1.

In situ cast stairs may be designed with or without strings. In the latter case the stair flight itself, acting as an inclined reinforced slab, becomes the major supporting element.

String stair The strings may span between landing trimmers or be cranked to span beyond the landings to take a bearing at the perimeter of the stair (figure 7.1 A). This type of stair will be thinner than a slab type and therefore somewhat lighter in weight. With half-space landings as illustrated the inclined strings, which are reinforced as normal beams, can bear on reinforced concrete trimmers at the landings, and the flight slab will span between the strings. The landings will span between the trimmers and the enclosing staircase wall or frame. The strings may be upstand or downstand and, in the case of the latter, the effective depth will be from the soffit of the string to the internal junction of the treads and risers. The waist[1] thickness of the flight need only be about 75 mm. An upstand string is useful both from a functional and an aesthetic point of view. It prevents dropped articles and cleaning water from falling over the sides of the stair and it gives weight and smoothness of flow to a stair designed to appear as a slab flowing between floors. When an intermediate flight is incorporated, producing quarter-space landings, it is necessary to use cranked strings, and as these must run across the flight it is necessary to use downstand beams.

For free-standing stairs, a single substantial central string may be designed to carry the flight which cantilevers on each side. The flight may be cast in situ or be made up of precast elements bolted or tied into the in situ cast string, to give a smooth inclined soffit or a stepped soffit. Precast concrete treads or laminated timber are often used to produce an open-riser stair in this form.

Inclined slab stair Unless the span of the flight is very long, or strings are required for visual reasons, the stair can be designed without strings, the flight being designed to act as a slab spanning between the trimmers (figure 7.1 B). In this case the span of the flight is the horizontal distance between the centres of the trimmers. The effective depth is the waist thickness of the slab, which is designed on the same basis as a floor slab. In the case of slabs designed to span in the direction of the flight, a side of which is built at least 110 mm into a wall, BS 8110-1 permits a 150 mm wide strip next to the wall to be deducted from the loaded area (see figure 7.1).

Cranked slab stair In this stair there are no trimmers and the top and bottom landings, together with the flight, are designed as a single structural slab spanning between enclosing walls or frame (figure 7.2 A). The appearance is clean and the thickness of the slab is not unduly great if the flight is not too long. This form of stair is useful when there are no side supports available for trimmer beams, as in the case of completely glazed sides to a projecting staircase. Should supports be available at the ends of the landings so that the latter may be made to span at right-angles to the direction of the flight, the landing slabs may be considered as beams supporting the flights, in which case BS 8110-1 requires that the effective span of the flight should be taken as the going of the flight plus half the width of the landing, subject to a maximum of 900 mm, at each end.

Monolithic cantilever stair In this stair the flights and landings are cast in situ and cantilever out from a wall, either the enclosing wall to the staircase or a central spine wall as illustrated in figure 7.2 B. The soffit may be smooth or stepped, the latter resulting in a fairly thin slab, the compression zone of which is stiffened by the folded form. When the stair cantilevers from a central spine wall, it becomes completely self-supporting and, if projecting from the face of a building, may be fully glazed all round. This form of stair also provides a useful solution to problems of sound insulation when the stair, for this reason, is required to be separated from the surrounding structure, since it is possible to leave an insulating gap all the way round at all points (see plan). In these cases the half-space landing would be partially supported by the end of the wall

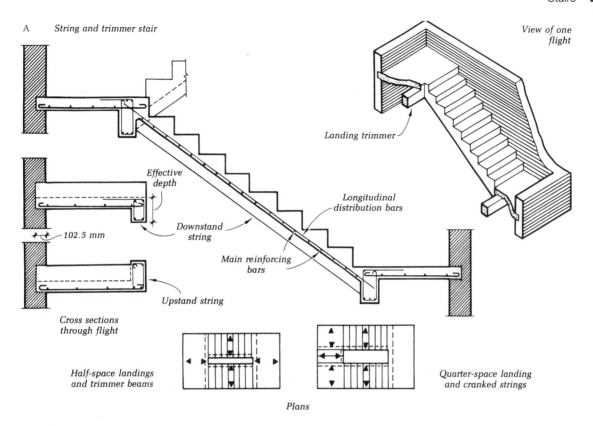

A   *String and trimmer stair*

View of one flight

Landing trimmer

Effective depth

102.5 mm

Longitudinal distribution bars

Downstand string

Main reinforcing bars

Upstand string

Cross sections through flight

Half-space landings and trimmer beams

Quarter-space landing and cranked strings

Plans

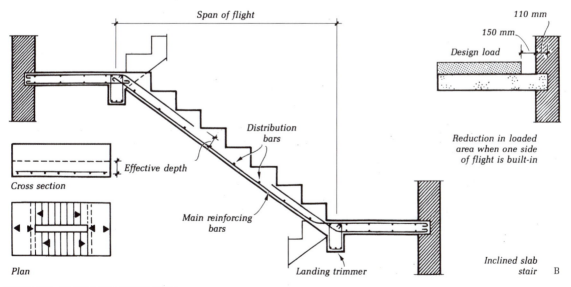

Span of flight

110 mm

150 mm

Design load

Effective depth

Cross section

Distribution bars

Reduction in loaded area when one side of flight is built-in

Main reinforcing bars

Plan

Landing trimmer

Inclined slab stair   B

Figure 7.1   In situ cast concrete stairs

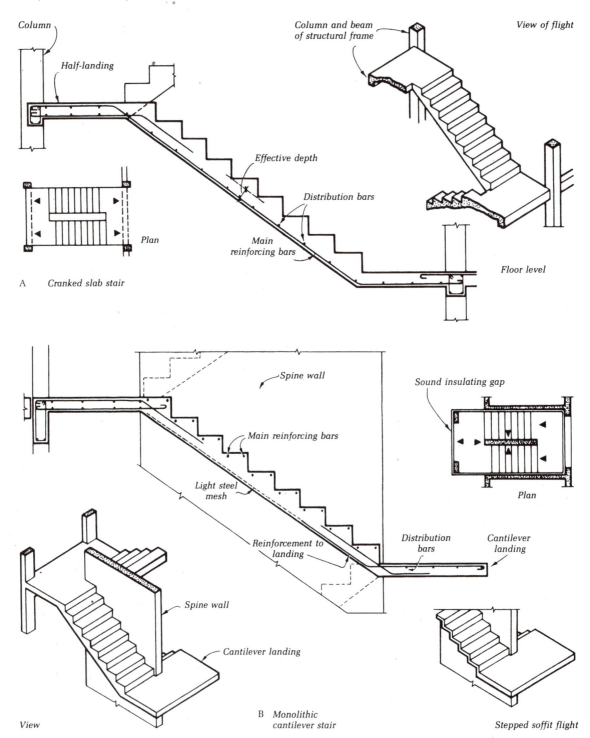

A   Cranked slab stair

B   Monolithic cantilever stair

View

Stepped soffit flight

Figure 7.2 In situ cast concrete stairs

and partly by the ends of the adjacent flights as shown, and reinforcement would be designed accordingly.

Continuous slab stair   This is a double-flight stair which receives support only at the floors above and below (figure 7.3 A). It consists structurally of a continuous slab, monolithic with the floors, which runs from one floor to the landing level, turns on itself and continues without any support to the next floor. This is not a cheap stair to construct since not only the normal stresses of bending and shear, but also those of torsion have to be resisted. The slab may be reduced in width to form a wide shallow beam carrying open-riser cantilever treads. It may be placed centrally under the treads or eccentrically, so that they cantilever entirely over one side. With materials strong in tension, such as reinforced or prestressed concrete or laminated timber, the latter arrangement is practicable even with broad flights, although torsion stresses in the slab are increased.

In the types of stairs without trimmer beams the relationship of the end risers of the flights at a half-space landing affects the positions of the intersections of the sloping soffits and the landing soffit, as well as the form of the handrail turn. If the top riser of the lower flight is set back to line approximately with the second riser of the upper flight, it is possible to make the intersections of the sloping soffits and the landing coincide on the same line, which can be the face edge of the landing between the flights, without the landing being made excessively thick (see figure 7.3 A). This gives a clean appearance on the underside, simplifies detailing of applied finishes and permits a satisfactory handrail turn. This point is also illustrated and discussed relative to balustrading in *MBS: Internal Components*, chapter 8, where this subject is fully considered.

Spiral stair   The newel type of spiral stair in its smaller form is usually constructed of precast concrete and is described in Part 1; the larger stairs are constructed on a large core as a monolithic cantilever stair (Part 1, figure 10.1). The core may be solid or hollow in the form of a duct.

The open well, or helical stair, figure 7.3 B, although visually very fine if well designed, is complicated in structural design and construction. A large proportion of steel is required to resist the bending, shear and torsion stresses, and the shuttering is expensive. The slab usually varies in thickness from top to bottom, increasing towards the bottom, and may vary in thickness across the width. There are two or three sets of reinforcement with top and bottom layers in each: continuous bars running the length of the spiral, cross or radial bars and sometimes diagonal bars laid tangential in two directions to the inner curve. The large amount of steel reinforcement and the complicated shuttering make this an expensive stair to construct.

The helical stair may also be designed with closed strings with the flight slab spanning between. The effect of a helical stair depends upon the free flow of the curve from one floor to another. In many cases the limitations on the number of steps in a flight makes it impossible to design such a stair without intermediate landings, which interrupt the flow of the stair.

Precast concrete stairs   The concrete stairs discussed above are primarily in situ cast stairs. Although precast concrete has long been used for simple solid or open-riser steps, or for small utilitarian spiral stairs, some examples of which are shown in Part 1, the precasting of large stairs has not been common. However, with the general use of cranes on building sites, many large stairs which once would have been constructed only in situ are now precast. These may be precast in the separate parts of strings and steps or they may be cast in complete flights and landings, depending upon the nature of the particular job and the size of the crane to be used. Many types are available as proprietary standardised systems.

Figure 7.4 shows a 1.37 m wide cantilever stair in which the steps are in the form of an 'L' with only the building-in end as a solid rectangle. This considerably reduces the weight of the stair. Masonry walls should be built in a strong cement mortar for at least 300 mm above and below the line of a cantilever stair such as this.

Figure 7.5 shows a cut-string and open-riser stair in which the strings are stepped to take the treads, in this case of timber, which are screwed on. Precast concrete treads could be used and bolted to the strings in a similar way, or they could be secured by projecting rods and a small amount of in situ concrete cast in mortices or grooves left for the purpose.

A closed-string type is shown in figure 7.6, in which the strings are precast and post-tensioned and on the inside of which are cast stepped bearings to take the ends of the precast treads. To position the treads and to avoid subsequent movement a stub is cast on each end which drops into an accommodating mortice in the stepped bearings on the string.

An alternative to the casting of the strings as separate elements is to cast them as a pair braced apart the required distance to form an open frame, so that the whole can be hoisted by crane and set in rebates formed on the edges of the landings. This is shown in figure 7.7 A. The steps, precast individually, are positioned and grouted in recesses formed to take them in the top faces of the strings.

When a slab stair without strings or trimmers is required this can be cast in elements consisting of a flight and parts of the top and bottom landings, the ends of the landings bearing on rebates in the staircase wall or frame. The half-space landings are completed by in situ concrete

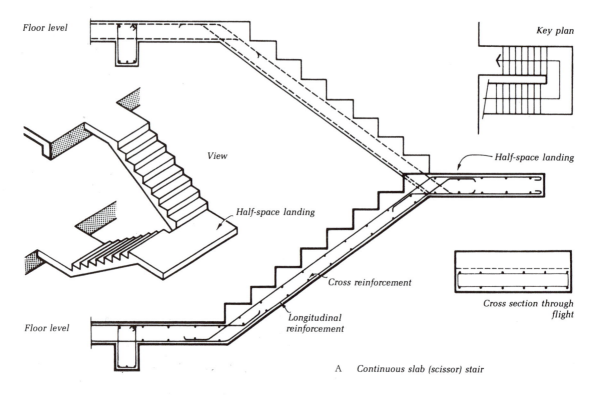

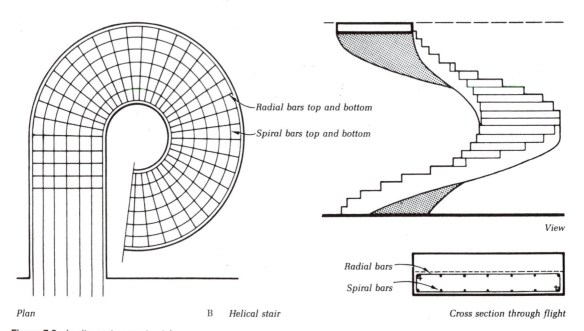

Figure 7.3 In situ cast concrete stairs

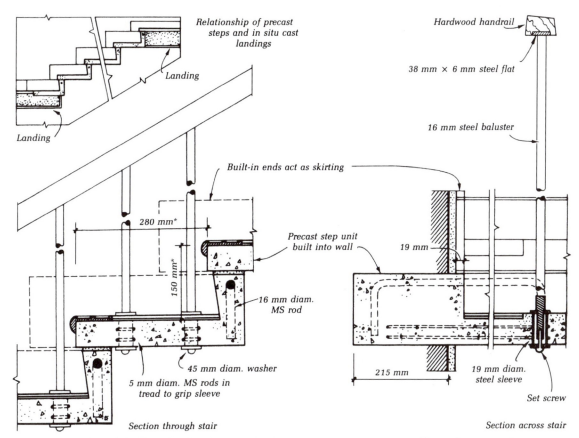

Relationship of precast steps and in situ cast landings

Landing

Landing

Hardwood handrail

38 mm × 6 mm steel flat

16 mm steel baluster

Built-in ends act as skirting

280 mm*

150 mm*

Precast step unit built into wall

19 mm

16 mm diam. MS rod

45 mm diam. washer

5 mm diam. MS rods in tread to grip sleeve

215 mm

19 mm diam. steel sleeve

Set screw

Section through stair

Section across stair

* For going and rise dimensional limitations, see Building Regulations, AD K: Stairs, Ladders and Ramps, section 1.

Figure 7.4   Precast concrete cantilever stair

filling placed around rods left projecting for this purpose at the sides of the landing sections, as shown in figure 7.7 B. Such elements are heavy and require a crane of sufficient capacity for hoisting.

7.1.3 Metal stairs

Cast-iron escape stair   The oldest type of metal stair is probably the external cast-iron and steel fire-escape stair. These are made up of standard strings, about 175 mm by 10 mm mild steel, to which are bolted perforated cast-iron or mild steel chequer-plate treads. Perforated cast-iron risers can also be fitted if required. The landings are formed of 13 mm cast-iron or mild steel chequer plates and the stair and landings are usually carried on a structure of rolled-steel beams and channels. Details are shown in figure 7.8.

Spiral stair   Standard newel-type spiral stairs of a similar nature can also be obtained. They are manufactured in standard size components to satisfy the Building Regulations and are often used especially for small internal stairs. These may be formed in cast-iron, the steps being similar to those for the concrete stair shown in Part 1. They are threaded on to a metal newel-post. Alternatively, mild steel T-, angle- or channel-brackets are welded at the necessary points to a mild steel tube newel, as shown in figure 7.9, the timber treads being bolted or screwed to the cantilever brackets. In the cast-iron type, the tread may be cast with a recess in the top face to take an infilling such as carpet or plastic tile.

Helical stairs may be formed with cut or closed strings of mild steel plate (figure 7.10 A). If the diameter is large and the rise is considerable, this type of stair is likely to require intermediate support.

String stair   This can be constructed in various ways with strings of mild-steel tube, rolled-steel beams or channels and treads of steel, timber or precast concrete. These form open-riser or closed stairs, examples of which are

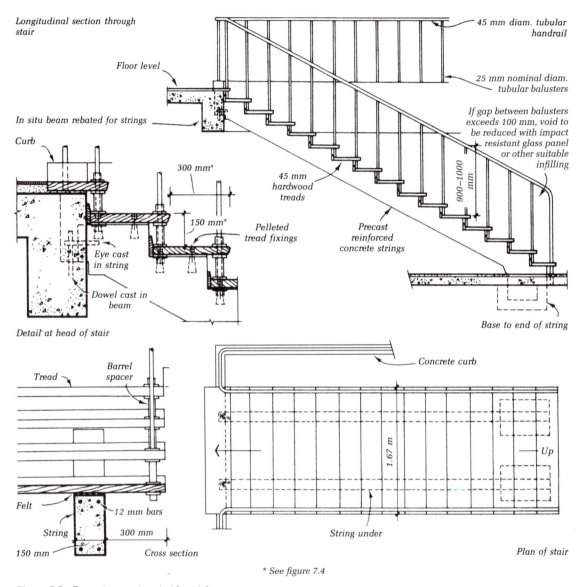

Figure 7.5 Precast concrete cut-string stair

shown in figure 7.10. The treads are bolted to mild-steel plate seating brackets which are bolted or welded to the strings (B and D), or the brackets may be formed of small diameter bar in the form of an inverted flat 'U', the legs of which are welded to the string (C). Very careful detailing of the brackets is necessary in order to obtain a satisfactory appearance in the finished stair.

Figure 7.11 A shows a string stair in which the boxed channel string is raised above the treads to the level of a guardboard. The treads are suspended from the strings by square or rectangular section steel balusters, screwed to the outside face of the strings and connected to each other

at the bottom by crossbars which support the treads. A stair of very light appearance resulting from the use of small section steel strip as a string is shown in figure 7.11 B. It should be appreciated that this construction is, of course, suitable for short flights only, such as that shown.

Where some degree of fire resistance is required in the case of stairs likely to be used internally for escape purposes, the above types of stair would not be suitable, since in many cases the risers are open and steel is exposed. Where, in such cases, it is considered desirable to use steel strings, these must be provided with some protective coating. Illustration E at the bottom of figure 7.10 shows a

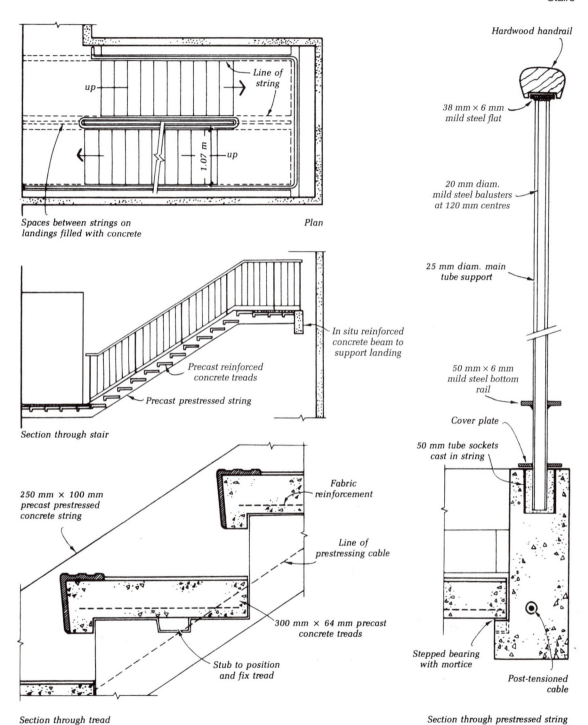

Hardwood handrail

up

Line of string

1.07 m

up

Spaces between strings on landings filled with concrete

Plan

38 mm × 6 mm mild steel flat

20 mm diam. mild steel balusters at 120 mm centres

25 mm diam. main tube support

In situ reinforced concrete beam to support landing

Precast reinforced concrete treads

Precast prestressed string

Section through stair

50 mm × 6 mm mild steel bottom rail

Cover plate

50 mm tube sockets cast in string

250 mm × 100 mm precast prestressed concrete string

Fabric reinforcement

Line of prestressing cable

300 mm × 64 mm precast concrete treads

Stub to position and fix tread

Stepped bearing with mortice

Post-tensioned cable

Section through tread

Section through prestressed string

Figure 7.6 Precast concrete closed-string stair

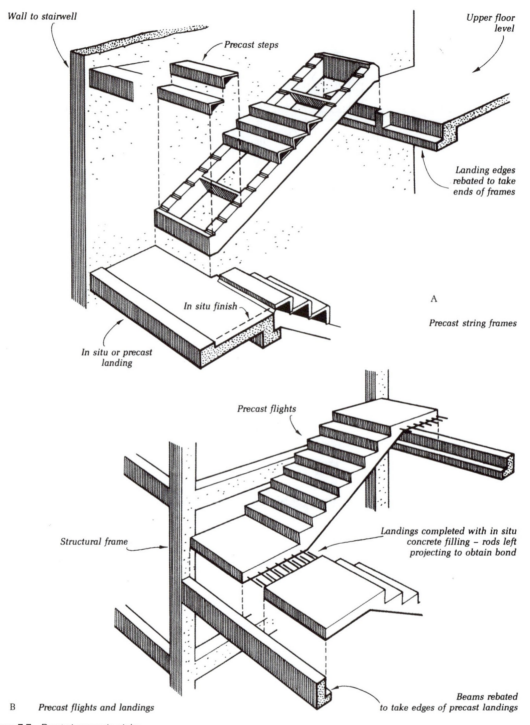

Wall to stairwell

Upper floor level

Precast steps

Landing edges rebated to take ends of frames

A

In situ finish

Precast string frames

In situ or precast landing

Precast flights

Landings completed with in situ concrete filling – rods left projecting to obtain bond

Structural frame

Beams rebated to take edges of precast landings

B   *Precast flights and landings*

Figure 7.7   Precast concrete stairs

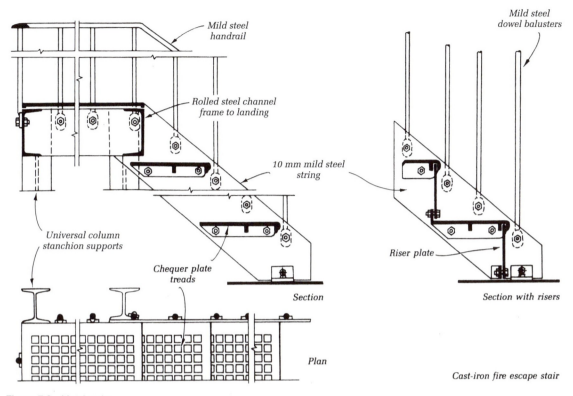

Figure 7.8 Metal stairs

staircase with mild steel channel strings enclosed with 25 mm vermiculite plaster on expanded metal and carrying precast concrete steps. The risers return at the ends in line with the outside face of the strings and the treads cantilever a considerable distance at each side.

Pressed steel stairs These are produced from light gauge sheet steel pressed components (figure 7.12) comprising the tread and riser, secured at each end to a pressed steel closed-string or a deep boxed channel member, according to the span of the flights. Going, rise and pitch angle are standardised to Building Regulations requirements. The treads are designed to take a filling of granolithic, terrazzo or similar material, to form the finished surfaces. Alternatively, the tread (and riser as well if desired) can be covered with timber or marble so that the stair becomes simply the structural element. The landings are constructed from dovetail steel sheeting, which gives a rigid structure and provides a good key for any wet filling. The soffits of flight and landing are usually fitted with steel clips to which expanded metal can be fixed for plastering or the lining methods shown may be adopted.

7.1.4 Universal stairs

Staircases can be specifically designed for particular situations to suit the variation in floor to floor height in different buildings. Designing and producing bespoke stairways for these situations is very expensive in time and materials, therefore, for estate housing and other uniform applications, the use of standard stair flights and components is far more economical. The overall rise with these stairs will determine the floor to floor height. Standard or universal steps have also been developed by means of which the rise and going of a flight may be varied within wide limits, and by the use of which staircases of any height and pitch can be constructed. This is accomplished by providing a sloping joint between tread and riser as shown in figure 7.13, so that, as the steps slide backwards or forwards for adjustment, every change in the going is accompanied by a proportional change in the rise. The proportion of rise to going is governed by the angle of the slope and is commonly governed by the rule that twice the rise plus the going should be between 585 and 610 mm.[2] The steps can be made of any suitable material and the same principle

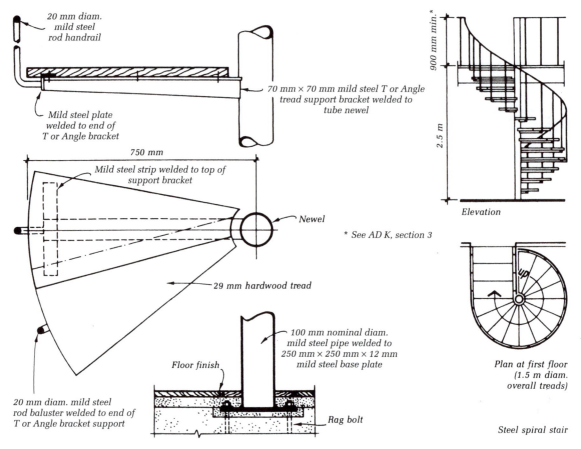

Figure 7.9   Metal stairs

can be applied to a universally adjustable mould for casting in situ concrete stairs.[3]

7.2 Ramps

These are mainly used for the passage of vehicles rather than for pedestrians, since they take up a large amount of space compared with a normal stair. They can, however, be used with good effect both functionally and architecturally where the space is available. They are essential for access and circulation by disabled persons.[4]

Car ramps   Ramps for cars generally should have a slope of about 1 in 10 although they may be as steep as 1 in 7, especially if the ramp is short. The radius to the centre line of curved ramps should be not less than 7 m, based on the turning circle of the average size car. Curved ramps should be slightly banked and the whole surface should be roughly ramped if concrete, or be treated in some other way to give a good hold for tyres. The width of the

ramp will depend upon whether it is for one- or two-way traffic. With a minimum radius of 7 m, a minimum width of 3.65 m should be allowed for the former and not less than 7.30 m for the latter, which allows for a central separating curb 300 mm wide and a width of 3.35 m on the outside of the ramp where the radius is greater.

Pedestrian ramps   Ramps, if properly designed, probably provide a safer means of pedestrian movement between different levels than the normal stair, since they do not necessitate the accurate placing of the foot. The safe slope of a ramp is limited by the risk of slipping, which is influenced by the nature of the surface and whether or not the ramp is internally or externally situated. The following considerations must be taken into account.

The Building Regulations limit the maximum gradient to 1 in 12 generally. For ramps not exceeding 10 m in length which are likely to be used by disabled persons, the gradient should not be steeper than 1 in 20 with a maximum rise of 500 mm. The maximum gradient for a 2 m

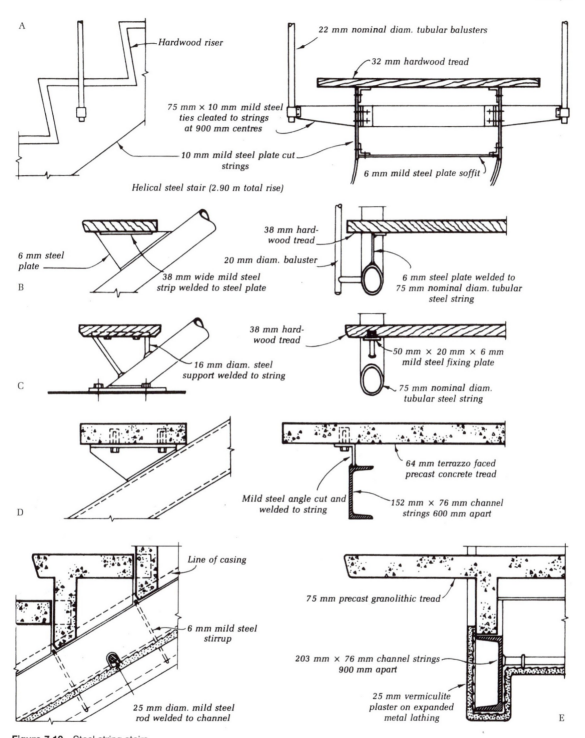

A

Hardwood riser

22 mm nominal diam. tubular balusters

32 mm hardwood tread

75 mm × 10 mm mild steel ties cleated to strings at 900 mm centres

10 mm mild steel plate cut strings

6 mm mild steel plate soffit

Helical steel stair (2.90 m total rise)

B

6 mm steel plate

38 mm wide mild steel strip welded to steel plate

38 mm hardwood tread

20 mm diam. baluster

6 mm steel plate welded to 75 mm nominal diam. tubular steel string

C

16 mm diam. steel support welded to string

38 mm hardwood tread

50 mm × 20 mm × 6 mm mild steel fixing plate

75 mm nominal diam. tubular steel string

D

Mild steel angle cut and welded to string

64 mm terrazzo faced precast concrete tread

152 mm × 76 mm channel strings 600 mm apart

Line of casing

6 mm mild steel stirrup

25 mm diam. mild steel rod welded to channel

75 mm precast granolithic tread

203 mm × 76 mm channel strings 900 mm apart

25 mm vermiculite plaster on expanded metal lathing

E

Figure 7.10 Steel string stairs

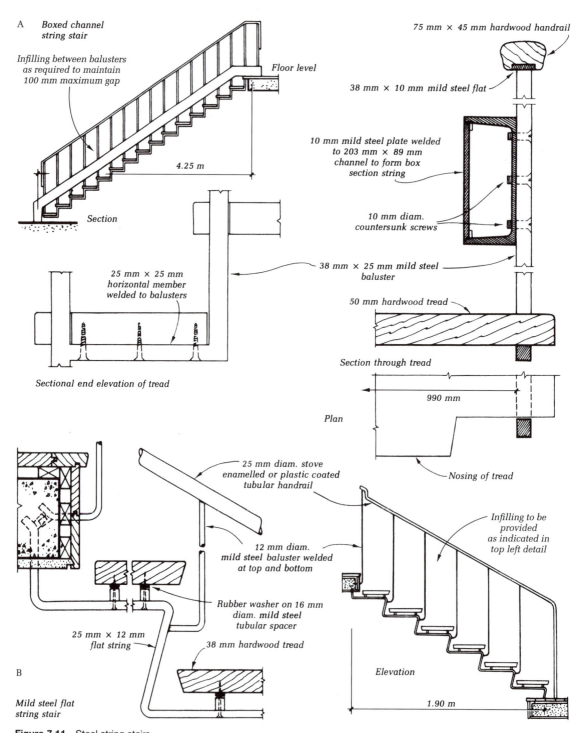

A Boxed channel string stair

Infilling between balusters as required to maintain 100 mm maximum gap

Floor level

Section

4.25 m

25 mm × 25 mm horizontal member welded to balusters

Sectional end elevation of tread

75 mm × 45 mm hardwood handrail

38 mm × 10 mm mild steel flat

10 mm mild steel plate welded to 203 mm × 89 mm channel to form box section string

10 mm diam. countersunk screws

38 mm × 25 mm mild steel baluster

50 mm hardwood tread

Section through tread

Plan

990 mm

Nosing of tread

25 mm diam. stove enamelled or plastic coated tubular handrail

12 mm diam. mild steel baluster welded at top and bottom

Rubber washer on 16 mm diam. mild steel tubular spacer

25 mm × 12 mm flat string

38 mm hardwood tread

Infilling to be provided as indicated in top left detail

Elevation

1.90 m

B Mild steel flat string stair

Figure 7.11 Steel string stairs

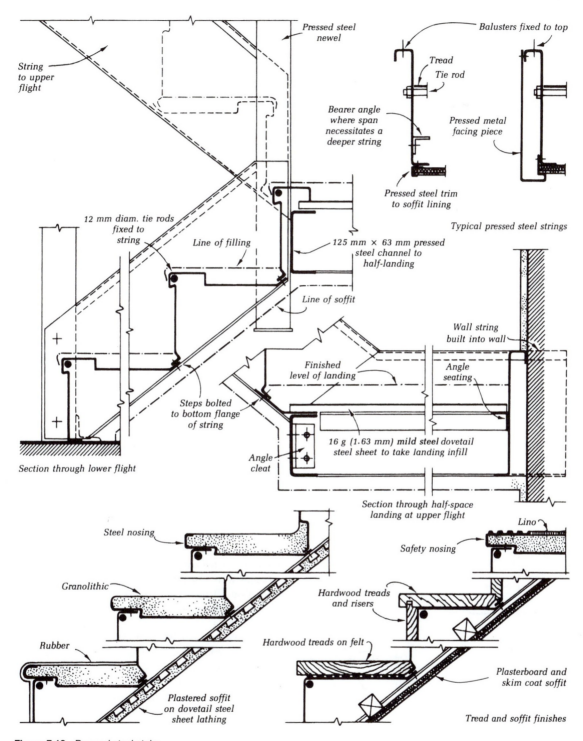

Figure 7.12 Pressed steel stairs

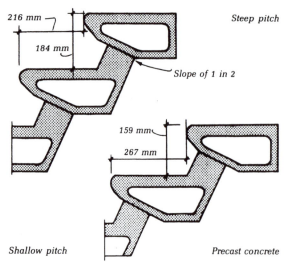

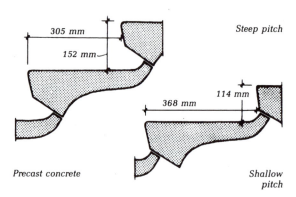

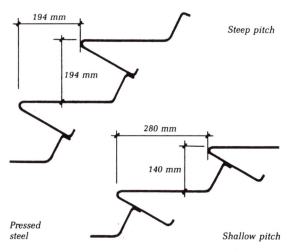

Figure 7.13 Universal stairs

Table 7.1 Guide to spacing of cleats on ramps in industrial premises

Slope of ramp	Recommended spacing	
	If load is carried (mm)	No loads carried (mm)
1:6	355	455
1:5	330	430
1:4	305	405
1:3	280	380

going or ramp length is 1 in 12 with a maximum rise of 166 mm. For a going of between 2 and 10 m, the maximum rise can be interpolated. Ramps considerably steeper than this are sometimes required in certain circumstances, as in factories or other industrial situations where, for example, a footway may be required to follow an inclined conveyor. The slope should never exceed 1 in 3. These steeply sloping ramps should always be provided with evenly spaced cleats across the ramp, spaced apart to suit natural walking. The cleat spacing will depend upon whether or not the pedestrians using them are likely to carry loads. Table 7.1[5] indicates suitable spacings for slopes up to 1 in 3. As cleats require accurate foot placement, ramps on which they are used lose their main safety features and should be avoided if possible.

The minimum width of a ramp to be used by one person should be the same as that for stairs. Greater widths will be required when two or more persons will pass. This will depend upon the nature of the building in which the ramp is situated and the amount of pedestrian traffic likely to use the ramp. Where access or escape of disabled persons is a factor the width should not be less than 1200 mm, and there should be a kerb at least 100 mm high on any open side. A landing at least equal in width and length to the width of the ramp should always be placed at a change of direction in the ramp.

Ramps to be used by disabled persons should be no longer than 10 m with horizontal landings 1.5 m long between each flight and top and bottom landings not less than 1.2 m long.

Handrails between 900 and 1000 mm in height are required on ramps rising more than 600 mm: on one side at least if less than 1 m wide and on both sides if the ramp is 1 m or more in width. Guardings 900 mm high must be provided at the sides of the flights. At the sides of landings they must be 1100 mm high. If the length of a ramp likely to be used by disabled persons is more than 2 m handrails and guardings are required on both sides.

Surface finishes   A ramp should always be constructed with a good slip-resistant surface. Cement or granolithic surfaces may be finished with a wood float or swept with a stiff broom while still green. This exposes the particles of sand and provides a rough surface which does, however, wear smooth after a time, although the granolithic surface will give a more lasting result because a greater degree of roughness may be obtained initially. Abrasive grit materials may be added to the surface mix to increase the friction and reduce the wear. Slip-resistance is further increased by the provision of transverse grooves formed in the surfaces. If a suitable aggregate is included in the top, asphalt can provide a good slip-resistant surface. Wood, in a dry unpolished state, is reasonably slip-resistant, but can become slippery when wet. Metal surfaces are not altogether satisfactory as even when formed with figured surfaces these soon lose their pattern and become slippery.

7.3 Ladders

A ladder is defined in BS 5395-3 as a stair having a pitch greater than 65 degrees.[5,6] These are usually of metal and

are used as a means of access to roofs and other high places and sometimes as a means of escape.

Ladders should be steep enough to make the user face them when descending, but vertical ladders should be avoided if possible because they are less safe and harder to climb. Whenever possible the minimum pitch should be used. Ladders with a pitch greater than 75 degrees should have rungs: these are known as *fixed ladders*. Those with a pitch from 65 to 75 degrees should have flat steps and are known as *companion way ladders*. The latter should be used on short rises of 3 m or less rather than a fixed ladder (figure 7.14).

7.3.1 Fixed ladders

The width of a fixed ladder should be from 380 to 450 mm between strings and its height should not exceed 6 m without intermediate landings. Landings should always be provided at the top of ladders and it should be remembered that when descending it is safer to step sideways rather

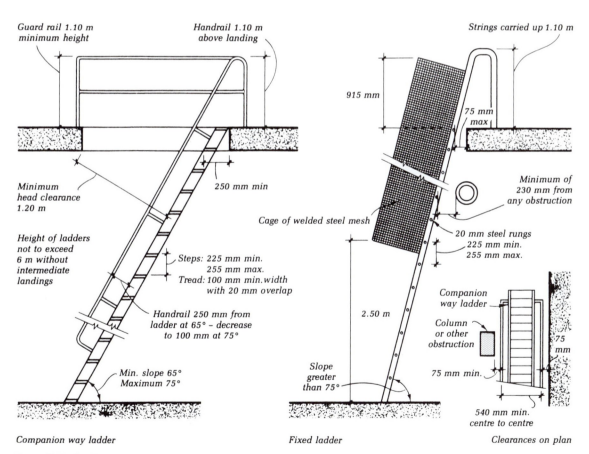

Companion way ladder   Fixed ladder   Clearances on plan

Figure 7.14 Ladders

than backwards on to a steep ladder. Ladders from which a user could fall 2 m or more should be enclosed with a safety cage. Cage clearances for different pitches are given in BS 5395-3. The strings, which serve as handrails, should rise above the top platform as shown in figure 7.14 and should widen out above the platform level to at least 600 mm and not more than 700 mm.

Fixed ladders may be used as a means of escape from fire where a normal stair is not practicable provided they are not for use by members of the public and provided they are constructed of non-combustible materials.

7.3.2 Companion way ladders

The width of this type of ladder should be from 450 to 550 mm between strings. Handrails should be provided on both sides, the distance between which should not be less than 540 mm centre to centre. This should widen out to at least 610 mm above the platform level. The top tread should be level with the platform with no gap between as shown in figure 7.14.

Details of recommended clearances, heights and sizes for both types of ladder are given in figure 7.14.

Notes

1 See Part 1, section 10.4.3.
2 See Part 1, page 208. The Building Regulations, AD K, section 1, indicates an overall tolerance of between 550 and 700 mm.
3 A step of this type was used many years ago by Alvar Aalto (centre of illustration). In Great Britain the design of universal stairs is protected by patent.
4 See Building Regulations, AD M: *Access to and Use of Buildings.*
5 This table is extracted from Industrial Data Sheet S3, 'Safe Access above Ground Level', issued by the Australian government. Closer to home and more up-to-date information is in BS 5395-3: *Stairs, Ladders and Walkways. Code of Practice for the Design of Industrial Type Stairs, Permanent Ladders and Walkways.*
6 There are other definitions of a ladder: e.g. AD K – a means of access to another level formed by a series of rungs or narrow treads on which a person normally ascends or descends facing the ladder.

8 Roof structures

The first half of the chapter reviews the characteristics of materials used and the various structural forms adopted for medium and large spans and the principles on which they are based, covering framed roof structures, shells of different types, grid structures and tension structures, both membrane and cable forms. The second half in concerned with the constructional methods used in the application of different materials to these structural types. In conclusion the control of movement due to settlement and thermal changes is briefly considered.

The functional requirements of the roof are discussed in Part 1, chapter 7 and reference is made to the economic and structural significance of the materials to be used in its construction. These will now be discussed in more detail.

Materials for roof structures

Steel, aluminium, reinforced concrete, timber and plastics are all commonly used for the construction of roofs. With all these materials constructional forms have been developed in roof structures which take advantage of the particular characteristics of the materials. A brief comparison of those properties which are particularly relevant to their use in roof structures is made below.

Steel

Strength Steel has high strength in both compression and tension and a small amount of material is able, therefore, to carry large loads. Working or design yield strengths for mild steel of 275, 355 and 450 N/mm² are permitted[1] for all normal structural members by BS 5950-1: *Structural Use of Steelwork in Building. Code of Practice for Design. Rolled and Welded Sections.*

Elasticity A structural material under stress should not stretch or contract to an excessive degree. This is particularly important in horizontal members where large deflections due to loading must generally be avoided. The ratio of stress to resultant strain, known as Young's Modulus or the modulus of elasticity,[2] indicates the extent to which the material will resist elastic deformation. If its resistance is high the material is stiff, the deformation under stress will be low and the deflection of a beam under load will therefore be small. Since the minimum depth of a beam is often dictated by deflection rather than by the strength of the materials used, a high modulus of elasticity permits either a shallower beam section for a given deflection or a greater span for a given depth of beam. This can be seen clearly in the expression for the deflection of a beam,

$$d = \frac{\text{constant } (c) \times w \times l^4}{EI}$$

See page 138. Depending on the quality and grade, steel has a modulus of elasticity of about 205 kN/mm², indicating that it is a stiff material.

Ductility Structural materials should be able to withstand large deformations without suddenly failing and cracking. In structural frames high stresses are often induced over restricted areas at some points and deformation will occur. Provided the material is sufficiently ductile it will not crack, but what is known as plastic flow will take place and the load will be transferred to the surrounding material, so that at no point is the failing stress reached. Steel is a ductile material and, with a yield point of 240 up to 450 N/mm², undergoes considerable strain after the elastic limit and before ultimate failure. This can be seen in the diagram of stress/strain curves in figure 8.1.

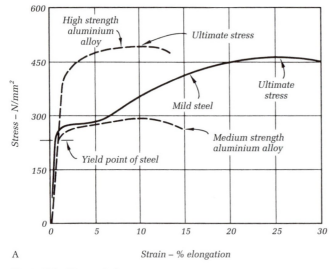

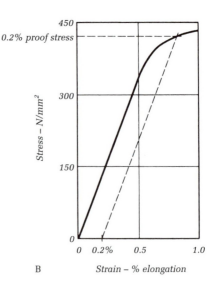

A — Strain – % elongation

B — Strain – % elongation

Figure 8.1 Stress–strain curves

General properties The properties of steel are such that the dead/live load ratio of steel members is small. That is to say they are able to carry heavy live loads at the expense of a comparatively small dead or self-weight.

Solid and hollow structural steel sections, including cellular beams, can now be bent to curves relatively inexpensively and successfully, made possible by developments in bending techniques which have taken place in recent years. These are now being increasingly used for roof construction providing design opportunities beyond the traditional triangle.

Aluminium alloys Pure aluminium is quite soft and is alloyed with other elements to make it a suitable structural material. Aluminium alloys have the advantage of being corrosion resistant and only about one-third of the weight of steel, but they have the disadvantage of being more expensive in first cost.

Strength The stress–strain curves of aluminium alloys exhibit no sharply defined yield point (figure 8.1), so that there is no clear indication of the elastic limit and no obvious point to which the working stress can be related. A 'proof' stress is, therefore, specified to aid this purpose. This is the tensile stress which produces a non-proportional extension of a defined amount of the original length.[3] This is 0.2 per cent. It will be seen from figure 8.1 that the proof stress on which the working stress is based can lie close to the ultimate stress of the alloy. One of the commonly used structural alloys has an ultimate stress of 310 N/mm², a proof stress of 270 N/mm² and a working stress in bending of 162 N/mm². While stronger alloys are

available, unfortunately the aluminium-copper alloys which have greater strength have an inferior corrosion resistance.

Elasticity The modulus of elasticity of aluminium alloys, 69 kN/mm², is about one-third that of steel. An aluminium alloy beam will, therefore, under the same load and support conditions and neglecting self-weight, deflect three times as much as a similar steel beam. The stiffness of a beam is measured by its flexural rigidity (EI), and in order to maintain a given deflection when the modulus of elasticity decreases, the moment of inertia must be increased (see expression for deflection on page 249). This may be done by increasing the depth of the section to between $1\frac{5}{8}$ and $1\frac{3}{4}$ of the depth of the steel section. In addition, or as an alternative, the cross-sectional area may be increased.

The lack of elastic stability of thin members and the danger of local buckling must be guarded against when using aluminium sections. The web of a beam may buckle sideways if it is too thin or insufficiently stiffened, lateral buckling due to torsion may occur in an unrestrained slender beam, and local buckling may take place in a thin compression flange at the points of maximum bending moment. These dangers are avoided by using sections with greater flange and web areas or with a stiff cross-sectional shape such as box sections or sections with lipped edges as shown in figure 8.2.

Ductility The proof and ultimate stresses of aluminium alloys lie close together and there is very little elongation before failure occurs. In areas of high stress there will,

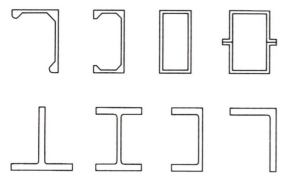

Figure 8.2 Typical aluminium structural sections

therefore, be little plastic flow taking place to permit the load to be transferred to surrounding areas. This has to be borne in mind in the design of aluminium structures.

Generally   In normal conditions aluminium is highly resistant to corrosion. It has a coefficient of expansion about twice that of steel, so that provision for greater temperature movements must be made in aluminium structures.

Aluminium alloys may be applied to structures in a similar way to steel but result in a considerably lighter structure. They are most economically used in cases where the weight of the structure itself forms a substantial part of the load and the imposed loads are comparatively light. They are, therefore, obviously suitable materials for roof structures, particularly those of long span where the dead load rises rapidly and the dead/live load ratio is likely to be high. Because of their characteristics, aluminium alloys are best used in structures which are inherently stable and resistant to torsion and where the members can be kept short and directly loaded as in grid structures.

Reinforced and prestressed concrete
Strength   The compressive strength of concrete, together with its deficiency in tensile strength, has been referred to on page 163. Methods of making good this deficiency by reinforcement are also described, and the advantages of pre-stressing are described on page 180.

Elasticity   Concrete has a low modulus of elasticity ranging from 21 to 35 kN/mm², but due to creep effects the effective modulus for long term loading may fall to only one-half of these values.

Generally   Over wide spans reinforced concrete used in beam and slab form has a high dead/live load ratio. It is most effectively used in structural forms which can take advantage of its monolithic character, particularly three-

dimensional forms such as shells, doubly curved slabs and folded slabs. The pre-stressing of concrete brings about a considerable reduction in depth of spanning members and depth/span ratios from 1:30 to 1:120 are possible. The reduction in depth together with a reduction in the overall cross-sectional area, made possible by the increased resistance of prestressed concrete to shear, results in a considerable reduction in dead weight and results in satisfactory dead/live load ratios over wide spans.

Timber   Timber was one of the earliest materials to be used for structural purposes, but our knowledge of its behaviour and capabilities is comparatively new. Developments in methods of joining timber, particularly by means of metal connectors and glues, developments in the strength-grading of timber and research into improved methods of design have resulted in the more efficient use of the material. Timber is a comparatively light material and the species used for normal structural purposes have weights approximately one-sixteenth that of steel.

Strength   Timber is an organic material and the knots and faults brought about during growth or seasoning constitute zones of weakness. The technique of strength grading is a means of establishing the loadbearing capacity of a piece of timber either visually in terms of knots and other visible faults by which, on the basis of the strength of clear wood of the same species, a reduced allowable working stress is determined,[4] or mechanically by measuring the modulus of elasticity of the piece and allocating an allowable stress. The basic or grade stresses in flexure and compression parallel to the grain range from around 3 to 11 N/mm², depending upon the species and grade of timber, with values rather more than half of these in tension.

The problem of joints is an important factor in timber design, and earlier forms of joints had an efficiency of no more than 15 to 20 per cent relative to the timber entering the joint. Two methods of increasing their efficiency are by means of connectors and synthetic glues.

Timber connectors consist of various forms of metal plates and rings through which a bolt passes (see figure 8.23 and Part 1, figure 7.30). Their effect is to increase the strength of the joint, particularly in tension and shear.

Glues made from synthetic resins produce joints as strong and even stronger than the timber joined. A considerable number of these glues is available, all of which have different characteristics and some of which are immune to attack by dampness or decay. Glues can be used for lattice construction but are more advantageously used in building up laminated timber members, which can be much stronger than the same size section in solid wood. A further advantage of laminated construction is that different qualities of timber may be used in the same section, the better quality being limited to the more highly stressed zones. By means of gluing, continuity of structure is obtained and it is a simple matter to form curved members and portal frames with the greatest depth where the stresses are highest. The cost of manufacture, however, makes the use of glued laminated members economic only when it is clear that the requirements are beyond the range of solid timber.

Elasticity   The modulus of elasticity of normal structural softwoods is around 4 kN/mm^2. Although some timbers have a somewhat higher value the modulus of elasticity of timber is low compared with that of other materials. In spite of this, however, having regard to the light weight of the material, comparative analysis with other materials shows that, in terms of flexural rigidity and strength, timber is the most efficient from the point of view of weight and cost of material.

Generally   The stiffness of timber relative to its weight and cost makes it particularly suitable for structures in which the load-carrying capacity is determined by its flexural rigidity (*EI*), such as structures which are large in relation to the load they carry. This includes roofs of all types, floors bearing moderate or light loadings and single-storey buildings, particularly those of large height and span. As a rough guide, structures which are liable to fail through elastic instability, and for which timber is the most suitable material, are likely to be those with a load–intensity ratio, $\sqrt{P/l}$, of less than about 1.50, where *P* is the compressive load in Newtons and *l* is the effective length of the member in millimetres.

In addition to its use in framed and laminated structures, timber may be used in the form of stressed skin plywood panels built up as folded or prismatic slabs or in the form of planks built up as doubly-curved shells. Very high strength to weight ratios are attained in this way, with weights per square metre of floor area covered as low as 25 kg. As a structural material timber has the advantage of ease of working and fabrication. Because of its comparatively light weight, built up members can be easily handled. When properly used it is a permanent material and it has good thermal insulating properties, which is a further advantage when it is used in stressed skin forms of roof structure. Although it burns freely in thin sections, when used in the sizes normal in lattice construction it remains structurally stable during a fire for a greater length of time than steel (see chapter 9, section 9.3.1).

Plastics   The large range of plastics now available have diverse characteristics and properties.[5] The widest application of plastics in building is for non-structural purposes, but considerable development of these materials for structural purposes has taken place in certain types of structure where advantage can be taken of their particular properties.

Plastics are light in weight and on an average weigh only about one-sixth of the weight of steel.

Strength   The tensile strength of unreinforced plastics is only about 60 N/mm^2, but when reinforced with suitable material, such as glass fibre, a tensile strength of 160 N/mm^2 is developed, using a randomly oriented fibre of fibre volume fraction 30 per cent. The compressive strength is of the same order as that of the tensile strength.

Elasticity   Thermo-setting plastics have a low modulus of elasticity, much the same as those for timber (3.50 kN/mm^2), but the modulus rises to 8.00 kN/mm^2 for a material similar to that described above.

Ductility   Plastics are not ductile materials. Little plastic flow, therefore, takes place in areas of high stress and, as in the case of other materials of low ductility, this must be borne in mind in the design of structures incorporating plastics.

Generally   Plastics have a high coefficient of expansion, about eight times that of steel, but when glass fibre reinforced they have a coefficient about the same as that of steel. The light weight of plastics gives a favourable strength/weight ratio so that they are particularly suitable for roof structures, provided these are of the type in which the low stiffness of the material is overcome by the inherently stiff form of the structure. The types of structure, therefore, to which plastics can most advantageously be applied are space structures of the stressed skin type, in which the strength is derived more from the geometry of the form than from the properties of the material. Plastics have been used in folded plate form and in the construction of many geodesic domes and in grid structures combining plastic tetrahedra, or doubly curved sheets, with grids of metal rods.

Roof loading

- The dead load consists of the self-weight of the structure itself[6] and of the roof claddings, coverings and internal linings.
- The superimposed load consists of the weight of snow, any incidental loads applied during the course of maintenance work and, in the case of flat roofs used as roof gardens, play areas or for other purposes, additional loads according to the purpose for which the roof is used.
- Wind forces. The dead load is usually based on the weights of materials specified in BS 648: *Schedule of Weights of Building Materials*.

Superimposed loads on flat roofs　For flat roofs, that is up to 10 degrees pitch, with no access provided other than that necessary for normal maintenance purposes, BS 6399-3: *Loading for Buildings. Code of Practice for Imposed Roof Loads* recommends an allowance for:

(i) a uniformly distributed load of 0.6 kN/m^2 measured on plan or
(ii) a concentrated load of 0.9 kN or
(iii) a uniformly distributed snow load or
(iv) a redistributed snow load,[7]

whichever produces the worst load effect.

When access in addition to that necessary for maintenance purposes is provided, the figures are:

(i) a uniformly distributed load of 1.5 kN/m^2 or
(ii) a concentrated load of 1.8 kN or
(iii) a uniformly distributed snow load or
(iv) a redistributed snow load,[7]

whichever produces the worst load effect.

When the roof is to be used for specific purposes the distributed and concentrated loads above are to be replaced by the appropriate floor loading recommended in BS 6399-1: *Loading for Buildings. Code of Practice for Dead and Imposed Loads* reduced, as for floors, in relation to the roof area where it is applicable (see page 193).

Superimposed loads on pitched roofs　The superimposed load allowance on roofs with a pitch greater than 10 degrees, to which no access is provided other than for maintenance purposes, should be:

● a uniformly distributed load of 0.6 kN/m^2 as for flat roofs for pitches of 30 degrees or less, reducing to no allowance for pitches of 60 degrees or more (values for pitches between 30 and 60 degrees shall be obtained by interpolation)
● whatever the pitch a concentrated load of 0.9 kN, a snow load and a redistributed snow load must also be considered as for flat roofs.[7]

Wind forces　Some indication of the variations in wind pressure and suction over roof surfaces is given in figure 3.2 in this volume and figure 7.1 in Part 1. The allowances to be made for wind pressures normal to the surface of flat and pitched roofs are assessed by the methods given in BS 6399-2: *Loading for Buildings. Code of Practice for Wind Loads*. As in the case of walls greater suctions and pressures occur at gable ends, near the eaves and near the ridge of a pitched roof (see figure 7.1 in Part 1). Fastenings for roof sheeting near these points should therefore be designed to take account of these greater forces which will be exerted.[8]

8.1 Types of roof structure

The classification of roof structures into two- and three-dimensional forms has been described in Part 1 (page 140) and as indicated there, these may be constructed of a number of different materials using different constructional techniques. At this point the various structural forms used for roofs of medium and large span will be reviewed, with particular reference to the principles on which they are based and which govern their structural behaviour, and to the practical considerations which have led to their development. Later in this chapter these are discussed in terms of constructional methods and details in different materials (see page 285).

8.1.1 Trusses and girders

Trussed roofs　The classification of roofs constructed of two-dimensional members as single, double and triple roofs according to the number of stages necessary economically to transfer the loads to the supports has been described in Part 1, section 7.2. The primary structural member in triple pitched roof construction, as explained in Part 1, may be in the form of a roof truss or a rigid frame, the former implying a double-pitched triangulated structure and the latter a structure with continuity between vertical and spanning members. Small span trussed and rigid frame roofs are described in Part 1, chapter 7.

Trussed roofs are widely used for single-storey and shed-type buildings (see figure 8.3). Considerations affecting the triangulation of the truss are discussed in Part 1, pages 48 and 166.

Light types of factory building requiring a clear internal height of about 3.65 m and a span from 9 to 12 m can be economically constructed with steel trusses spaced from 3 to 3.65 m apart bearing on the tops of stanchions as shown in figure 8.3 A. The normal fixing at the feet of the truss produces a reasonably unrestrained joint with the stanchion. The normal methods of fixing the column feet to the foundation pads[9] produces a comparatively rigid joint at this point. Side wind pressure will set up bending stresses in the columns as they react as vertical cantilevers, so that the stress distribution will be zero at the top of the columns increasing to a maximum at the foundation, which it will tend to rotate (F). Within the limits of sizes given, the bending stress at the base will be comparatively small and the rotational tendency on the foundation will be slight. Thus columns of comparatively small cross section can be used and, with foundation pads sized to take the vertical loads, the stresses on the soil due to wind pressure can usually be kept within safe limits.

When the span is greater than 12 to 15 m and for functional reasons the columns must be high, the column

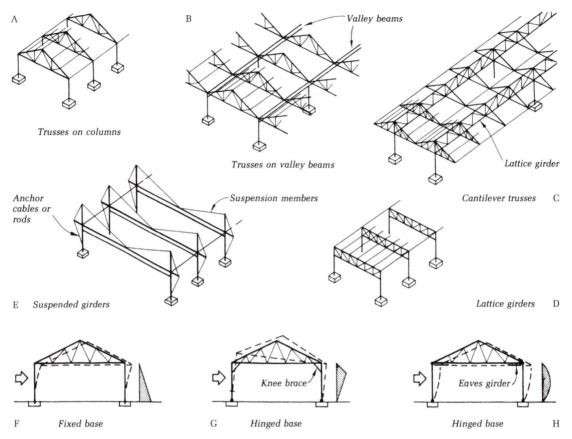

A Trusses on columns

B Trusses on valley beams — Valley beams

Lattice girder

Cantilever trusses C

Anchor cables or rods — Suspension members

E Suspended girders

Lattice girders D

F Fixed base

G Hinged base — Knee brace

H Hinged base — Eaves girder

Figure 8.3 Truss and girder roofs

section increases rapidly and large turning moments are applied to the foundation slabs. In these circumstances, to avoid uneconomic foundations a non-rigid or hinged joint between the column foot and the foundation may be introduced to relieve the latter of any rotational tendency, since no bending stresses can be transferred through such a joint (see also page 258). In order to provide the necessary rigidity against lateral wind pressure a knee brace may be introduced to provide a stiff joint between the columns and the roof truss (G). The stress in the columns will then be somewhat reduced and will be zero at the foundation and a maximum at the knee brace. Because of the rigid joints, some bending will be transferred to the feet of the truss which must be designed to withstand it, but the foundation slabs may be limited to the size required solely by the vertical loading. The introduction of rigid and hinged joints in this way results in the structure acting as a whole under side pressure of wind, with a tendency to 'uplift' on the windward side, which may be marked in the case of light structures.

An alternative method, which reduces bending in the columns and avoids bending stresses in the truss, is to introduce horizontal eaves girders instead of knee braces (H). These are lattice girders on a horizontal plane, running the length of the building and supported at the heads of the columns and the bottom ties of the trusses. Lateral rigidity is provided by these girders which pick up the wind pressure on roof and walls through the columns and transmit it to the ends of the building or to cross walls or braced frames. Intermediate foundations are thus relieved of any vertical component due to wind pressure. This is an economic method for high buildings provided the building is not too long or is divided at intervals by walls or braced cross frames which can transfer the wind forces from the girders to the soil. The columns are designed with unrestrained top and bottom joints so that the bending stresses are a maximum at the centre and zero at top and bottom. Induced bending stresses in foundations and truss are therefore avoided.

Roof trusses may be constructed in steel, aluminium alloy or timber, in spans up to more than 60 m when required. In the case of very large spans the pitch is kept low in order to avoid excessive internal volume and to reduce the area of roof to be covered and the weight of the structure.

Roof girders Trussed or lattice girders are widely used for medium and large spans when a flat or low pitch roof is required (figure 8.3 D). Universal steel beams are not economic for spans much above 10.50 m, although this can be extended by the use of castellated and cellular beams (see figure 4.1). Reinforced concrete beams have an economic limit of about 9 m. Prestressed concrete, however, is very suitable for wide span roof beams since small depth/span ratios are possible.

The economic depth of trussed girders is from one-sixth to one-tenth of the span. Over very wide spans the depth of the girders may be reduced by providing support at points along their length by means of tension members connected back to the tops of the supporting columns which are extended above the girder level (figure 8.3 E). The tops of the columns, of course, require anchoring back to the ground to provide resistance to the pull of these tension members or the columns must be made stiff enough to provide this resistance.

Trussed girders may be designed with parallel chords or with the top chord double-pitched or curved where a low pitched roof is required.[10] This is normal for large spans, but in the case of smaller spans beams as a whole are sometimes pitched in the middle to give a low double-pitched roof suitable for low-pitch roof coverings.

Girders may be used as valley beams in multi-span trussed roof structures to permit the wider spacing of internal columns, the girders supporting a number of trusses (figure 8.3 B). Where very wide column spacing is required the depth of the girders will increase excessively the height of the building. In this case what is known as cantilever truss or 'umbrella' truss construction is used. In this the girders are made the full depth of the truss and are placed in the line of the ridge so that the truss cantilevers out on each side of the beam, the feet of adjacent trusses meeting at the valleys as in C.

Due to the large depth of beam at the junction with the supporting columns, the beam to column joint can be comparatively rigid. Suitable materials for the construction of trussed girders are steel, aluminium alloys and timber.

Vierendeel girders without diagonal members, but with rigid joints between the chords and the vertical members, are only occasionally used in special circumstances in single-storey structures. These are described in chapter 4.

Beyond a span of about 12 m, depending upon the standard of lighting required, a reasonable degree of natural lighting through the walls is not likely to be achieved unless the structure is very high. In order to provide satis-

factory lighting to the interiors of extensive single-storey buildings a number of roof forms have developed.

North light roofs This type of roof may be in shell form, which is described later, or in trussed form. In lattice construction it is an asymmetrical truss, the steeper and shorter side of which is glazed and is sited to face north as shown in figure 8.4 A. As in the case of symmetrical trusses, the supporting columns may be placed at greater distances apart, say three bays, with the intermediate trusses picked up on valley beams. Where greater column spacing is required and the valley beams become excessively deep, lattice girders in the line of the ridge are used to form cantilever north light construction. To obtain wide spacings of the main lattice beams the construction may be in the form of trussed rafters spanning between the beams, with the plane of glazing running from the ridge, that is the top of the lattice beam, on to the 'back' of the adjacent trussed rafter (B). An alternative to this is the use of castellated or cellular beams, which are suitable for long spans carrying light loads, spanning between the top chord of one lattice girder and the bottom chord of the adjacent girder. The plane of glazing is the depth of the main beam (C).

Monitor roof A monitor is a mono-pitch lantern light with glazing at the sides only. The side facing north is usually large in area and that facing south is small. This roof provides very even lighting at the working plane for a comparatively small volume of roof; good lighting may be achieved with quite low ceilings, the spacing of the monitors being arranged for any given height to provide an even distribution of light. When spans are not great and the columns may be spaced about 4.50 m apart, the monitor frames may be built off Universal beams or shallow lattice beams (D) or, alternatively, the monitor frames may be formed as integral parts of a cranked beam of welded steel or in situ or precast concrete (E). Where wide column spacing is essential, deep lattice beams are used spaced 6 to 7.50 m apart, according to the spacing required for the monitors, which support shallow lateral beams on the bottom chords spaced about 4.50 m apart. These secondary beams carry the monitor frames which straddle the top of the main beams (F). Instead of separate secondary beams and monitor frames, cranked welded steel beams may be used spanning from the top chord of one main beam to the bottom chord of the adjacent beam.

8.1.2 Rigid or portal frames

Over large spans, deep lattice girders and pitched roof trusses, particularly the latter, may result in excessive volume within the roof space of the building which, because

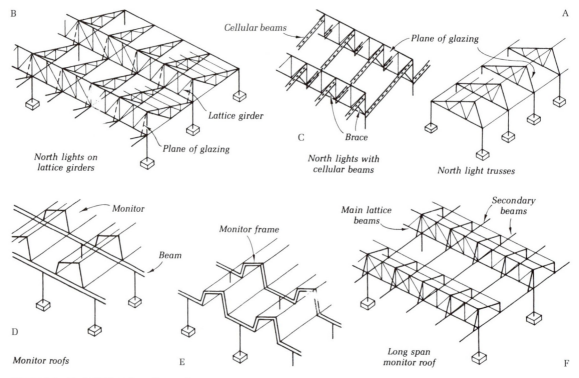

B

Cellular beams

Plane of glazing

A

Lattice girder

Plane of glazing

C

Brace

North lights on
lattice girders

North lights with
cellular beams

North light trusses

Monitor

Secondary
beams

Main lattice
beams

Monitor frame

Beam

D

Monitor roofs

E

Long span
monitor roof

F

Figure 8.4 North light and monitor roofs

of the obstruction by beams and ties, may not always be useful space. Further, with an increase in span the extra material necessary to provide adequate strength must be added to a simple beam or truss at the points where its own dead weight will increase the bending moments in the structure. The use of rigid frame construction overcomes these disadvantages to a very large extent.

The characteristic of the rigid frame is continuity of structure due to the stiff, or restrained, joints between the parts, and because of the nature of the stress distribution within such frames less material is required at the centre of the spanning elements than in a comparable simply supported beam. With increase in span, the whole of the necessary extra material is not required to be placed in the beam element, so that the maximum economic span is much greater. The smaller depth of the beam elements results in comparatively unobstructed, usable space for the full height of the building.

The difference between these two forms of construction in this respect can be seen in figure 8.5 A, which shows a rigid frame and a trussed roof construction spanning very much the same distance; the overall height of the rigid frame structure, the space within which is wholly utilised, is approximately the same as the overall height of

the truss which encloses a large volume of space lying above the volume of the building below. In some types of building, this space is, of course, valuable for housing services.

Effects of continuity   The result of the continuity arising from the introduction of stiff or rigid joints between the parts of the frame is illustrated in figure 8.6, where a portal frame rigidly fixed to its foundations is compared with a beam structure simply supported on two columns. It can be seen that the bending in the beam of the portal frame is transferred through the rigid joints to the columns. The resistance to this bending offered by the column results, however, in a reversed bending at the ends and a reduction of bending at the centre of the beam.[11] The nature of the deflections in each structure can be seen. There is little or no bending in the columns of the beam structure and only a single curve deflection in the beam, but there is considerable variation in curvature in the rigid frame. The points at which the direction of curvature changes, that is the points of contraflexure, are points at which there is no bending moment and at which the bending stresses in the members change 'signs'. This can be seen in the bending moment diagrams, which also show that the stiff junctions of beam and columns in the rigid frame are zones at which

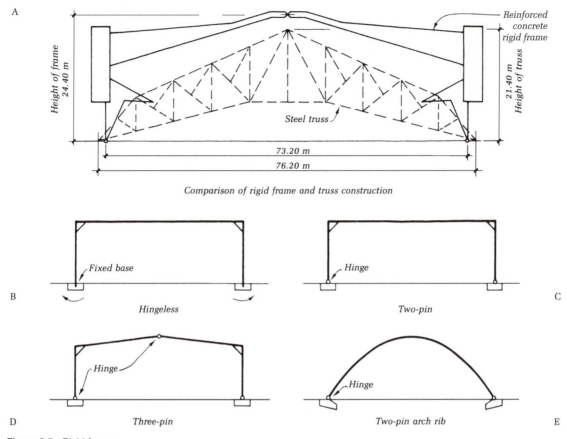

Comparison of rigid frame and truss construction

Figure 8.5 Rigid frames

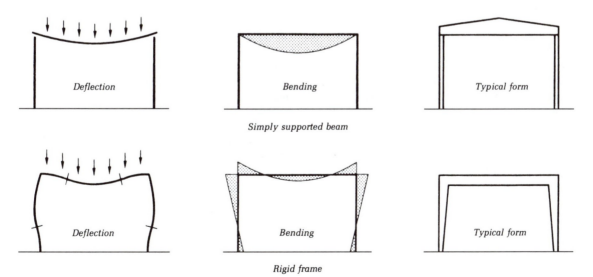

Figure 8.6 Comparison of rigid frame and beam construction

the bending moments are large. In contrast, there are no bending moments at the unrestrained junctions between columns and beam in the beam structure.

These differences in stress distribution produce differences in form. In the beam and column structure the columns may economically be of a uniform section, but the beam, with its maximum bending moment at the centre, will in large spans be most economically constructed in terms of material if greatest in depth at that point.[12] In the rigid frame, the stresses at the top of the columns will often be greater than those at the base, requiring a greater amount of material at that point. In the horizontal member there will be less disparity between the stresses at the ends and mid-span, so that a member of uniform depth will be economic. The relative proportion of end and mid-span moments depends on the relative stiffness of beams and columns. If the columns are slender compared with the beam they will provide little fixity at the ends and the mid-span moment will approach that of a simply supported beam; the end moments will be small. In some cases where the bending moments at the ends or haunches are considerably greater than those at mid-span, the depth of the beam at the centre may logically be less than that at the ends.

It can thus be seen that continuity, because of the transfer of stresses from one part to another, results in all parts of the structure providing resistance to the stresses set up by the load with a consequent reduction in bending moment at particular points. As mentioned earlier, less depth is required at the centre of the spanning member of the rigid frame, thus reducing the bending moment further by the amount of dead weight saved at this point. Although more material has been added at the ends to take up the stresses due to the stiff junctions, it lies over or close to the columns and does not greatly increase the bending moment. This is an important consideration in large span construction where the dead weight of the horizontal member is such an important factor in the design, particularly in cases where the available depth is small. Some form of rigid frame will usually provide a more economic structure than a simple combination of unrestrained columns and beams.

As with the knee-brace truss construction described above, the stiff joints between columns and beam in the rigid frame provide lateral rigidity and make it possible to introduce hinges when necessary.

Hinged joint   This is also referred to as a non-rigid, unrestrained, pivoted or pinned joint. In structural frames a hinge implies a junction between two parts that can transmit a thrust and shearing force but not a bending moment, since it permits free rotation as explained in Part 1, chapter 3. This, for the designer, simplifies the analysis of the structure by making it statically determinate. Direct stresses only exist at such joints and, since these can be resisted efficiently with a closer concentration of material than bending stresses, the shape of the structure can be varied accordingly and produces what are now typical forms in rigid frames with hinged joints (figure 8.7). For practical purposes hinged joints may facilitate the site erection of prefabricated frame components since they may be simply executed in comparison with the forming of continuous joints, particularly in concrete frames. In addition, pin jointed base hinges may be used as fulcrum points when lifting half-frames into position. They also serve, as explained already, as a means of relieving foundations of all tendency to rotate under the action of wind or other imposed loads on the frame. The actual form of hinge will depend upon its main purpose and upon the extent of freedom from restraint it is required to give. It need not necessarily be a true hinge or pivot provided the rigidity of the structure at the hinge point is low since the movements and degree of rotation are comparatively small. The 'split' hinges without metallic parts used in in situ concrete work are, therefore, feasible. Methods of forming hinged joints of various types are illustrated in figures 8.30, 8.31 and 8.32.

It has already been pointed out that rigid structures of all types are sensitive to differential settlement of the foundations and to movements due to changes in temperature. The effects of these movements must be borne in mind at the design stage, the former by considering the superstructure and foundation design together, and the latter by the provision of expansion joints or hinges where necessary.

Rigid frames can be constructed in steel in lattice or solid web form; in aluminium alloy in lattice form; in timber in lattice or solid web form, and in concrete.

Types of rigid frame   The fundamental and constant characteristic of the single-storey rigid frame in all its forms is the stiff or restrained joint between the supporting and spanning members. Apart from this the form can vary in a number of ways. The spanning member may be horizontal, pitched or arched; the junctions of the vertical members with the foundations may be restrained or hinged; a hinge may be introduced in the middle of the spanning member, and the structure itself may be solid or latticed. The members may be regular in cross section or may vary in shape according to the distribution of stresses within them.

Fixed or hingeless portal   This is a fixed-base frame with the feet rigidly secured to the foundation blocks and with all other joints rigid. Bending moments are less and more evenly distributed in this than in other types, but a moment or rotational tendency is transferred to the foundations (figure 8.5 B and figure 8.31 A).

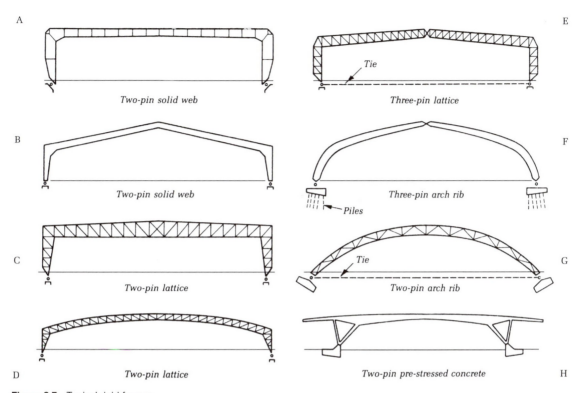

A — Two-pin solid web
B — Two-pin solid web
C — Two-pin lattice
D — Two-pin lattice
E — Three-pin lattice (Tie)
F — Three-pin arch rib (Piles)
G — Two-pin arch rib (Tie)
H — Two-pin pre-stressed concrete

Figure 8.7 Typical rigid frames

Two-pin rigid frame   In this form hinged joints at the base are introduced to relieve the foundations of any tendency to rotate (figure 8.5 C). This is necessary where base moments are high, especially where ground conditions are weak. Bending moments in the vertical members are greater than in a fixed portal because of the absence of negative bending moments at the feet. Long-span frames of this type with curved or pitched spanning members may be tied at the eaves to reduce the stresses at the haunches resulting from the tendency of the frame to splay outwards under load. Typical forms are shown in figure 8.7 A, B, C, D, H.

Three-pin rigid frame   In this form a further hinged joint is introduced at the crown or mid-point of the spanning member (figure 8.5 D). The moments in this type and the deflection at the crown are greater than in the other two because of the absence of rigidity at the centre and the resultant reduction in positive moments in the spanning members (compare F with E, figure 8.8). It will, therefore, be less economical in material, unless the centre point is considerably higher than the eaves. The presence of the three hinges, however, makes this form of frame statically determinate and simpler to design. In precast concrete or

lattice forms this type is usually easier to erect than the portal or two-pin frame.

Rigid frames impose horizontal thrusts of some magnitude on the foundations due to the tendency of the column feet to splay outwards as the spanning member deflects. This must be met by adequate resistance in the soil or, where necessary, by the use of inclined foundation slabs, piled foundations or ties between the foundations (figure 8.7 E, F, G). This is of particular importance in fixed and two-pin frames in which even a small horizontal movement will cause considerable redistribution of moments in the frame, with possible adverse effects on the structure as a result.

The horizontal thrust will vary with the stiffness of the frame and with the relative proportion of spanning member to vertical members. It will, therefore, be large with three-pin frames, and in all types the greater the span relative to the height the greater will be the horizontal thrust (figure 8.8 D, E, F).

Arch rib   This is a rigid frame but has no vertical members as such (figure 8.5 E). It may be fixed, two-pin or three-pin in form. A fixed arch rib exerts bending moments on the foundations as well as vertical and

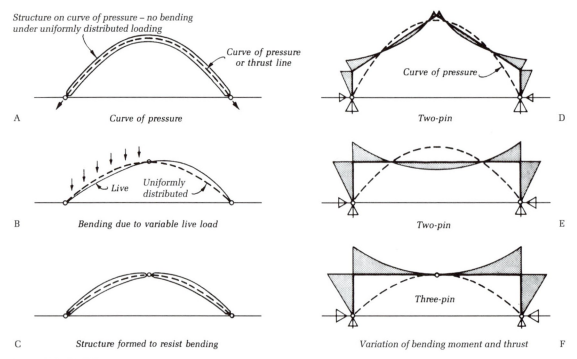

Structure on curve of pressure – no bending under uniformly distributed loading

Curve of pressure or thrust line

A   *Curve of pressure*

Live   Uniformly distributed

B   *Bending due to variable live load*

C   *Structure formed to resist bending*

Curve of pressure

Two-pin   D

Two-pin   E

Three-pin

Variation of bending moment and thrust   F

Figure 8.8   Rigid frames

horizontal thrusts and is used only where the soil can offer adequate resistance. Changes in temperature and, in the case of concrete, shrinkage, affect the stresses in the structure. The function of the base hinges in a two-pin arch rib is the same as those in normal rigid frame (figure 8.7 G). The introduction of a crown hinge with hinges at the feet produces a statically determinate structure (F). Small differential settlements of the foundations will not affect the thrusts on the foundations and, therefore, stresses in the structure will not be altered. Some form of tying in or stabilising of the foundations is often necessary with arch ribs since the horizontal thrust is often greater than in other types of rigid frame. The intensity of these horizontal thrusts will vary not only with the loading on the arch but also with its slope: as the curve becomes less steep so will the horizontal thrust increase.[13]

Curve of pressure   Forces acting on a structure will set up a 'natural line of forces' through which the loads would, in theory, be transferred most economically to the bearings. For example, a rope or chain, uniformly loaded, will take up a catenary or near-parabolic curve through which all the forces act directly in tension, with no rotation or bending in the rope. The depth of the curve will depend upon the resistance at the ends of the rope. A similar 'thrust line' or 'curve of pressure' is set up in an arch

under uniform load through which the load is transmitted to the abutments. It is, in fact, the bending moment diagram for the loading. If the arch form follows closely this parabolic curve it will be in direct compression at all points with no bending stresses in it (figure 8.8 A). Loading, however, is not always uniform and the arch may be subject to appreciable wind loads or other variable loads, so that in practice it may not be possible to eliminate bending entirely (B). But, by designing the arch to follow as closely as possible the bending moment diagram due to the particular loading, bending can be kept to a minimum (C). In many circumstances, however, it may not be economical to shape an arch in this way, nor indeed may the arch form be so suitable as the normal rigid frame.

The magnitude of the bending moments induced in an arch or rigid frame which does not follow the 'thrust line' will vary as the modified shape diverges from the curve of pressure (figure 8.8 D, E, F). Any point of an arch rib or frame lying on the curve of pressure will be free of bending moment, as at the points of contraflexure or at all hinged joints which, as they are incapable of transferring bending, must always lie on the curve of pressure. Those parts lying furthest from the curve will be under greatest stress (F). Whilst the use of a crown hinged joint will automatically produce a zero bending moment at the joint and simplify analysis of the structure, the frame will be far stiffer and a more even stress distribution will occur if continuity is maintained and hinges are restricted to the base connections (E). This will be clear from the illustrations if it is borne in mind that the moments at various points in the frame are in direct proportion to the distance from the 'curve of pressure'.

It will be seen that the arch or near-arch form is usually subject to smaller bending moments than the normal rigid frame because its shape lies closer to that of the curve of pressure (D). It is, therefore, particularly useful for very wide spans in which economy in dead weight is a critical factor in the design. When a curved shape running right down to the ground level is not suited to the function of the building, the arch form is often used as the spanning member of a normal rigid frame so that the curve springs from a point at a reasonable distance above the floor level as in figure 8.7 D.

8.1.3 Spacing of main bearing members

The spacing of main bearing members, unless determined by other requirements, is fixed by the most economical combination of frames and purlins. For the common types of lattice truss and beam construction in all materials, the spacing lies between 3.70 and 7.50 m, although for very wide spans it may be economical to increase the spacing to as much as 15 m. For rigid frames of various types the spacing lies between 4.50 and 12 m.

Beam and frame With beam and frame structures primarily subject to bending stresses as distinct from direct stresses, the wide spacing of wide span members usually has economic advantages because the cost of the smaller number of more heavily loaded members with deeper purlins is not so great as that of a greater number of more lightly loaded members which would result from a closer spacing. The reasons for this are connected with the different rates at which bending moments and deflection vary with variations in the span and in the unit loading.

Increase in the height of the columns carrying the roof structure increases the overall cost of the structure and when columns are tall, economies can similarly be effected by increasing the spacing, since, within limits which will vary for each case, the cost of fewer more heavily loaded columns, as in the case of the spanning members, will be less than that of a greater number of more lightly loaded columns. The reason for this is that in tall columns buckling is usually the critical factor in design and not the loadbearing capacity. These aspects are discussed more fully in chapter 4 on page 138.

Steel truss and purlin In simple steel truss and purlin structures, the closer the trusses are spaced the lower will be the overall steel content of the structure. This is because variations in the spacing of the trusses has a much greater effect on the weight of the purlins than on the weight of the trusses. An increase in spacing necessitating the use of the next largest angle section for the purlin results in the increased cost of the purlins more than offsetting the saving arising from the smaller number of trusses and columns required. There is, however, a practical limit on the spacing of the trusses. This is reached when the trusses are so lightly loaded that the smallest standard steel sections are too large to permit the members to be reduced to the areas required for design purposes. This limit occurs at spacings just under 3 m. Within limits the weight of the purlins per m^2 of floor area is largely independent of the truss span and is nearly constant for any particular spacing of trusses. For spans ranging from 12 to 21 m a spacing of 3.80 m is often satisfactory in practice, the optimum span being about 15 m.

For maximum overall efficiency of the structure, however, particularly in the wider spans, the spacing of the main members will not be the same for all spans. When spans are large, fabrication costs of the main members increase, and for spans greater than about 21 m it is economically desirable to space the trusses at wider centres, in some cases to distances at which lattice purlins can economically be used. In some circumstances where a wide column spacing is required, rather than space the trusses further apart with larger purlins, it may prove cheaper to carry a number of trusses on valley beams supported on widely spaced columns, even though the direct transfer of load from truss to column is normally the most economical method.

In some cases the use of steel decking permits purlins to be omitted if the frames are sufficiently closely spaced for this purpose. The omission of one stage in the supporting structure in this way may reduce the overall cost of the structure. This can sometimes be done by using bearing members which are particularly suited to light loads over long span. These include castellated and cellular beams and cold formed lattice beams.

8.1.4 Rigidity and stability of framed roofs

Rigidity against wind pressure must be provided in all forms of framed structure. In roof structures lateral rigidity may be provided by stiff joints either between the vertical members and the spanning member or between the frame and the foundations. The significance of these alternatives and the use of eaves beams have already been discussed in section 8.1.1. Wind pressure in a longitudinal direction would cause 'racking' or tilting over of the frames. This, in structures with only light purlins spanning between the trusses or frames, is resisted by bracing which

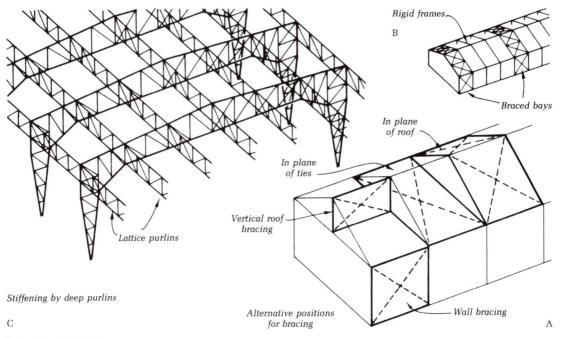

C   Stiffening by deep purlins   Lattice purlins

In plane of ties   Vertical roof bracing   Alternative positions for bracing   Wall bracing

Rigid frames   B   Braced bays   In plane of roof   A

Figure 8.9 Wind bracing

may be placed either in the plane of the roof or, in the case of roof trusses, in the plane of the ties (see figure 8.9 A). This wind bracing should be placed at each end of the building and, in long structures, at intervals of 30 to 60 m (B). With large span trusses it is advisable to brace continuously at tie level along the length of the structure. Vertical bracing in the plane of the 'king rod' between the gable end and one or two end trusses gives increased rigidity; alternatively, vertical bracing can be provided in the wall panels. In large structures with greater spacing of trusses or frames, the purlins will be deep and may provide adequate longitudinal rigidity without separate wind bracing (C).

The stiff joints at the haunches of rigid frames or between roof trusses and columns make the structure act as a whole in a lateral direction, and hinged joints at the feet result in a tendency for the whole structure to rotate about the leeward hinge and to lift up at the windward hinge (see figure 8.3 G). In large, comparatively light structures, this uplift may be so great that the dead weight of the structure is insufficient to anchor it down, and very large concrete foundation blocks may be required for this purpose, or tension piles or ground anchors may be used. The base must then be designed with long holding down bolts at the joint and heavy reinforcement in the foundation blocks.

The structures discussed so far are all two-dimensional structures. There are numerous forms of three dimen-

sional, or space structures, and these may be constructed of slabs or plates of solid material or of lattice framework, either form of which can be plane or curved in shape. They are discussed in the following pages.

8.1.5 Shell roofs

The term 'shell' is usually applied to three-dimensional structures constructed with a curved solid slab or membrane acting as a stressed skin, the stiffness of which is used to transfer loading to the points of support. They may be considered as single curvature shells, based on the cylindrical or parabolic form, and double curvature shells, based on the spherical and other more complicated forms. The term 'doubly curved shell' is commonly applied only to forms other than the spherical shell.

The main characteristic of a shell construction is the very thin curved membrane. This thin membrane is made structurally possible by providing restraint at the edges such that bending stresses in it are so small as to be negligible or are completely eliminated. The membrane is then subject only to direct stresses within its thickness (figure 8.10 A). Many examples of the shell form are to be seen in nature. For example, the bamboo rod, the crab shell and the bird's egg, all of which are exceedingly light but exceedingly strong. A blown egg has been known to support more than 45 kg distributed load. The efficiency of many modern shell constructions can be judged from the

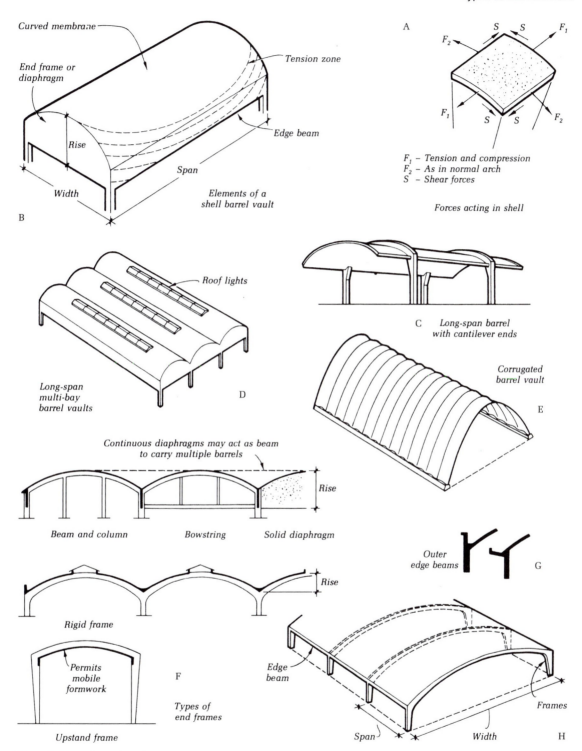

A

F_1 – Tension and compression
F_2 – As in normal arch
S – Shear forces

Forces acting in shell

Curved membrane

End frame or
diaphragm

Tension zone

Rise

Edge beam

Span

Width

Elements of a
shell barrel vault

B

Roof lights

Long-span
multi-bay
barrel vaults

D

C   Long-span barrel
with cantilever ends

Corrugated
barrel vault

E

Continuous diaphragms may act as beam
to carry multiple barrels

Rise

Beam and column   Bowstring   Solid diaphragm

Rise

Rigid frame

Permits
mobile
formwork

Upstand frame

F

Types of
end frames

Outer
edge beams

G

Edge
beam

Span   Width   H

Frames

Figure 8.10 Shell barrel vaults

fact that they are relatively thinner than the shell of an egg. Long span concrete barrel vaults may be as thin as 57 to 63 mm thick for spans up to over 30 m. Short span barrels will be somewhat thicker than this and domes of 45 to 50 m span may be as thin as 90 mm. Thicknesses as small as 38 mm are possible over spans of 30 m with some double curvature forms such as hyperbolic paraboloids.

In practice, for application to building work at least, the basic form of sphere or cylinder, for example, must be cut. The cut or free edges represent zones of structural weakness because direct stress can then only be transmitted in a direction parallel to the cut edge. In order therefore to make full use of the structural properties of the remaining parts of the curved membrane, it is necessary to strengthen the edges by means of ribs or edge members which are called edge beams, or in the case of a dome, a ring beam. The latter will be subject to direct tensile stresses and, if the dome is supported on columns, to bending stresses as well. The edge beams of a cylindrical shell vault will be subject to bending stresses. In addition to edge beams, stiffening members are required at the open ends of cylindrical shells (figure 8.10 B). These end frames, as they are called, stiffen the cut end of the shell against buckling, the maintenance of the shape of the shell being essential in order to develop the membrane stresses within it.

Single curvature shells These are barrel vaults of which there are two forms: long- and short-span barrels, typical examples of which are shown in figure 8.10 D, H respectively.

Long-span barrel vaults These act primarily as a 'beam', the span of which is the length of the vault. The shell constitutes the compression member and the edge beams the tension members or flanges. Although the direct stresses in a shell are mostly compressive, shear forces are set up near the supports which give rise to diagonal tension (figure 8.10 B). In the case of reinforced concrete shells, this must be resisted by reinforcement placed at 45 degrees across the corners of the shell as shown in figure 8.35 A. The width of the barrel should be one-half to one-fifth of the span and the rise from the underside of the edge beams to the crown of the vault about one-tenth of the span for single span, and one-fifteenth for continuous span vaults. The depth of the edge beams is usually about half of the total rise, but in the smaller spans this may sometimes be reduced. In multi-bay buildings, the edge beam may be eliminated altogether, provided the necessary rise is obtained by increased curvature of the shell membrane. In these circumstances the fold in the shell constitutes a beam as shown in the rigid end frame example in figure 8.10 F. In the case of a concrete shell it is a disadvantage to have an excessive rise to the shell as this prevents easy placing

of the concrete. Various ways of forming edge beams and end frames are shown at F and G.

The width of a long-span barrel is usually not more than 12 m with a maximum practicable width of 15 m. The maximum economic span is about 30 to 45 m. When the end frames are as far apart as this, equilibrium without bending moment in the shell is generally not possible, but where such a span is necessary, a satisfactory solution can be obtained by a suitable choice of rise to span and depth of edge beams. In order to avoid excessive rise of the structure, the radius of curvature of the shell increases as the span and, therefore, the width increases. For small spans a radius of about 6 to 7.5 m is used, for spans from 15 to 30 m a radius of 9 m, and for spans over 30 m a radius of 12 m. Lighting openings may be formed in the crown of a shell vault provided they are kept clear of the ends and are not more in width than about one-fifth the width of the barrel in reinforced concrete shells and about one-third in timber shells. Circular lighting openings in reinforced concrete shells can be formed in various parts of the shell, but should not exceed about 1.20 m in diameter and must be kept clear of the bottom edges and the ends, particularly the corners, of the shell. The edges of all openings must be strengthened with edge ribs and in the case of a long opening in the crown of the vault, cross ribs at intervals along the opening are required.

Since a shell barrel vault acts as a beam along its length, it is possible to cantilever the structure beyond the end frames. The external edge of the curved shell will usually require stiffening with a rib unless the projection is quite small (figure 8.10 C).

By pre-stressing the tension zones of a long-span barrel vault (figure 8.35 D) the rise may be reduced to one-twentieth of the span.

Short-span barrel vaults These are used when the clear span is beyond the practicable and economic limits of a long-span barrel indicated above, or where the interruption of roof space by the valleys of a succession of narrow long-span shells would be a disadvantage (compare D, H in figure 8.10).

End frames in the form of arch ribs, rigid frames or bowstring frames are usually spaced 9 to 12 m apart, sometimes up to 18 m. The depth of the edge beams should be about one-fifteenth of the span. The total rise should not be less than one-tenth of the span or the chord width, whichever is the greater. Because the spans of the shells are generally less the edge beams are shallower than in long-span barrels. The width of short-span barrels is often great and the radius of curvature is therefore large. With shells of over 12 m radius it is usually necessary to prevent buckling of the curved membrane by introducing stiffening tabs running across the width of the vault at 3 to

6 m centres. These can be placed either above or below the curved shell and need not necessarily continue down to the edge beams, since buckling will occur in the upper compression zone of the shell. Due to the interaction of the shell and the stiffeners, bending stresses often occur, and because of the thinness of the shell they may be very high. Provision for these must be made either by a local increase in the thickness of the shell or by special reinforcement.

North light and cantilever barrel vaults Many variations of the simple barrel vault are possible, especially in reinforced concrete, and examples of these are shown in figure 8.11. These include asymmetrical forms such as north lights and cantilever shells (A, B) and double-cantilever shells (C, D). The total vertical rise of these should be

not less than one-eighth of the span and the rise of the arc should be at least one-twenty-fifth of the span. Edge beams will normally need to be not less than one-eighteenth of the span in depth, but this may vary according to conditions. When two shells meet at a considerable angle, as at the middle of a double-cantilever shell, the edge beam may be omitted as indicated above, since the fold in the shell constitutes a beam. The upper edge beam of a north light shell is usually very small as it is normally supported by struts at 1.80 to 2.75 m centres in the plane of the north light glazing. These struts bear on the side of the valley gutter which is therefore designed as a deep 'L' edge beam to the bottom of the adjacent shell (A and figure 8.35 C). In cantilevered shells the end frames or diaphragms carry the whole of the roof load and their

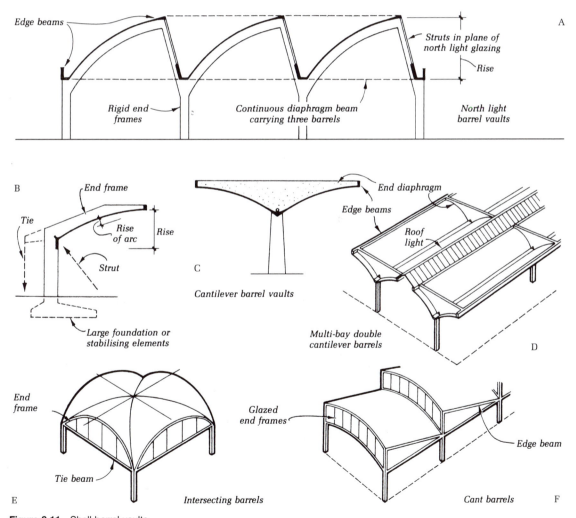

Figure 8.11 Shell barrel vaults

depth must therefore be substantial. In the case of a single-cantilever shell the foundations must be designed to prevent overturning, or struts or ties must be provided to ensure stability (B). End frames in all forms of shell may be placed either above or below the curved slab. Cylindrical vaults may intersect to form cross vaults or be 'canted' as shown in E, F. This permits glazed lights to be formed between the stiffening frames of adjacent barrels.

Barrel vaults may also be constructed as parabolic or elliptical shaped shells springing direct from the foundations and stiffened with ribs at intervals. Alternatively, in concrete a corrugated shell can be used in which the stiffening effect of the ribs is provided by corrugations in the surface of the shell which, when repeated, forms a continuous corrugated surface as shown in figure 8.10 E (see also page 295).

Double curvature shells Double curvature adds to the stiffness of a stressed membrane and is not limited to the normal rotational dome. Geometrical surfaces with double curvature are divided into two main groups:

1 those in which the curvature is in the same direction in sections cut at right-angles, that is to say, either both concave or both convex, as in a sphere
2 those in which the curvature is opposite in sections cut at right-angles, as in hyperbolic paraboloids and hyperboloids of revolution. In these the surface appearance resembles a saddle.

Shell domes are included in the first group and the second group are commonly called doubly curved slabs or shells (see figure 8.12 A, B).

Shell domes The simplest form is the spherical dome which has been constructed in concrete over spans of 45 m or more. The shape may vary according to the plan shape to which the edges are cut. All cut edges must be stiffened with edge beams. To avoid horizontal thrust in circular shell domes it is necessary to construct the shell of approximately elliptical cross section with a rise of about one-sixth of the span, otherwise a ring beam or ties must be provided at the base of the dome to take up the thrust (figure 8.13 A, B). In the case of domes square or triangular shaped on plan (C, D, E), the edge beams must be designed to take up this thrust and transfer it to the bearings.

The ring and edge beams serve also to resist stresses set up by temperature changes in the shell, which may sometimes need thickening near the beams and also round any openings formed in the shell.

Early concrete shell domes were formed by the intersection of a number of cylindrical shells resulting in a polygonal form. The finest example of this is probably the Market Hall at Leipzig, covered by three octagonal domes each 75.50 m span, with a shell thickness of 90 mm (F). Each dome is formed by the intersection of four cylindrical shells, the ridges at the intersections replacing the rigid frames in a normal barrel vault. G shows intersecting

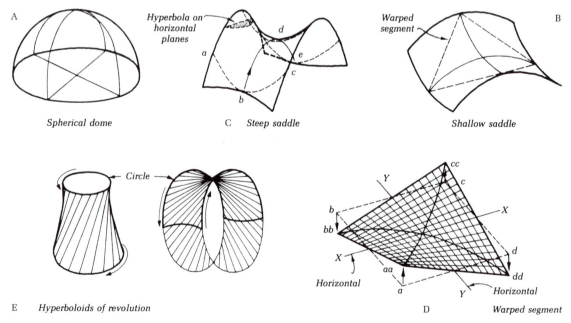

Figure 8.12 Double curvature shells

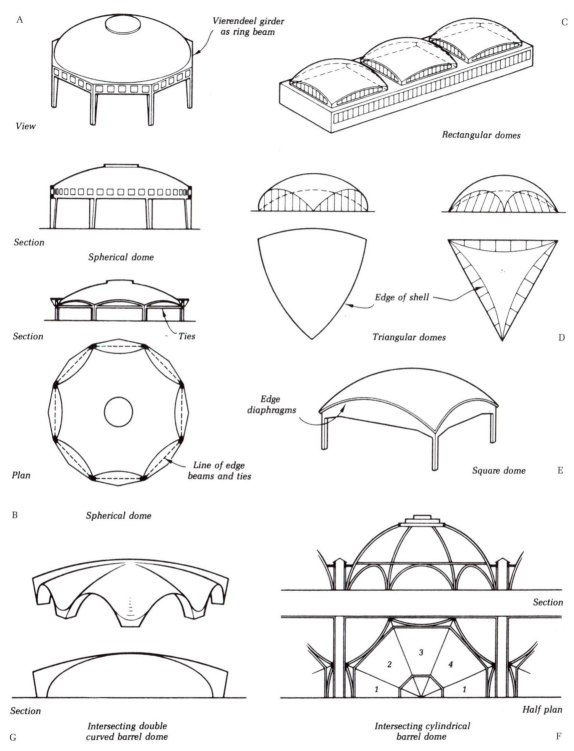

A

Vierendeel girder
as ring beam

C

View

Rectangular domes

Section

Spherical dome

Section   Ties

Edge of shell   Edge of shell

Triangular domes   D

Edge
diaphragms

Plan

Line of edge
beams and ties

Square dome   E

B   Spherical dome

Section

Section

Intersecting double
curved barrel dome

Half plan

Intersecting cylindrical
barrel dome   F

G

Figure 8.13 Shell domes

double-curvature barrels which result in a 'dome' form. Double-curvature barrels may also be used either as a means of providing increased stiffness to the shells in short-span vaults or in order to obtain curved edge beams in long-span vaults when pre-stressing is to be applied (see figure 8.35 D).

Doubly curved shells   Of this group of three-dimensional geometrical surfaces that most used is the hyperbolic paraboloid, or saddle shaped form, so named because, when cut, some sections reveal hyperbolas and others parabolas as shown in figure 8.12 C. It is formed by moving the vertical parabola *abc* through the parabolic trajectory *bde*. This type of surface has greater resistance to buckling than dome forms because of its shape. In practice part of a hyperbolic paraboloid is most frequently used, in the form of a warped segment or parallelogram, although the saddle-shape and conoidal forms are appropriate in some circumstances. The relationship of a warped segment to the saddle-shape is indicated in B.

The basic construction of a warped segment or parallelogram is illustrated in D. Points *a* and *c* of the horizontal plane *a b c d* are raised to new higher position *aa*, *cc* and points *b* and *d* are depressed to *bb*, *dd*. The sides are divided into an equal number of sections and the corresponding opposite points are joined. The resulting net of straight lines defines part of a hyperbolic paraboloid surface. The characteristic feature of this surface is that, although all cross sections cut parallel to the edges are straight lines, cross sections parallel to the diagonals are parabolas. That on the diagonal running through the raised corners, and all sections parallel to it, being concave upwards, and that on the diagonal running through the lower corners being convex upwards. That is to say, the upper surface of the segment curves in concave form between the higher points and in convex form between the lower points.

Stresses in doubly curved shells   In spite of its complicated shape the stresses in a paraboloid shell can be more easily analysed than in most other surfaces. The shell may be considered as a series of arches and suspension cables intersecting each other at right-angles; the shell is in direct compression in directions parallel to the convex or 'arched' section and in direct tension in directions parallel to the concave or 'cabled' section (figure 8.14 C). Because of the great stiffness given by the double curvature, and as all the stresses are direct, the shell may be exceedingly thin. For spans in the region of 30 to 40 m a reinforced concrete shell will be from 38 to 50 mm thick. The forces exerted on the edges by the 'arches' and 'cables' are the same, and since they act also at equal angles to the edge but in opposite directions, they resolve into shear forces along the edge with no component perpendicular to the edge to cause bending stresses. Since the principal stresses are equal, the shell itself is in a state of uniform shear and, as the stress is the same over the whole surface, the practical design in terms of reinforcement in a concrete shell or the boarding in a timber shell is simplified.

Edge beams are required to carry the edge shear forces. The depth of these should be disposed equally above and below the shell membrane. They may be placed wholly above or below the shell but the cross section of the beam will need to be larger. The overall rise of the shell, that is the difference in height between the low and high corners, is important. If this is small the shell will be shallow and if too shallow it will tend to buckle. The ratio of the rise to the diagonal span is, therefore, a significant factor and this should be as large as possible and never less than one-fifteenth.

Combination of panels   Doubly curved panels may be used singly or in combination and the number of supports required will depend upon the way in which the panels are related to each other. In the case of a single rectangular panel only two supports are required. If the roof is supported at the two lower corners the edge beams will be in compression and a horizontal outward thrust will be exerted on the supports. This thrust can be resisted either by a heavy buttressed support or by a tie rod between the lower corners of the shell (figure 8.14 A, C). If the panel is supported at the two high points the edge beams will be in tension and an inward pull will be exerted on the supports. This must be resisted by a strut running between the high points (B). The tie is the cheapest of these three methods but the reduction in clear headroom it causes and the possibility that it may spoil the interior appearance are disadvantages. The buttressed supports may be costly if the columns are high because of the high bending stresses which will be set up by the outward thrust of the shell. Some provision must be made for tying down the unsupported corners against wind forces, particularly if the sides of the building are not enclosed (C). When the building is enclosed at the sides the edge beams are fastened to the heads of the enclosing walls.

Some combinations of warped segments are illustrated in figure 8.14. It will be seen in some cases that the edges of some panels lie alongside those of adjacent panels, so that the forces in each balance each other. In certain combinations of panels the inclined edge beams of adjacent panels can be integrated with the corner supports to form a rigid frame in order to obviate the use of a tie. The bounding edges of warped segments need not necessarily be straight, but may be curved in parabolic form. An example of this is shown. The development of the conoidal form, using both straight and curved edges, is shown in E and the use of the full saddle shape in D.

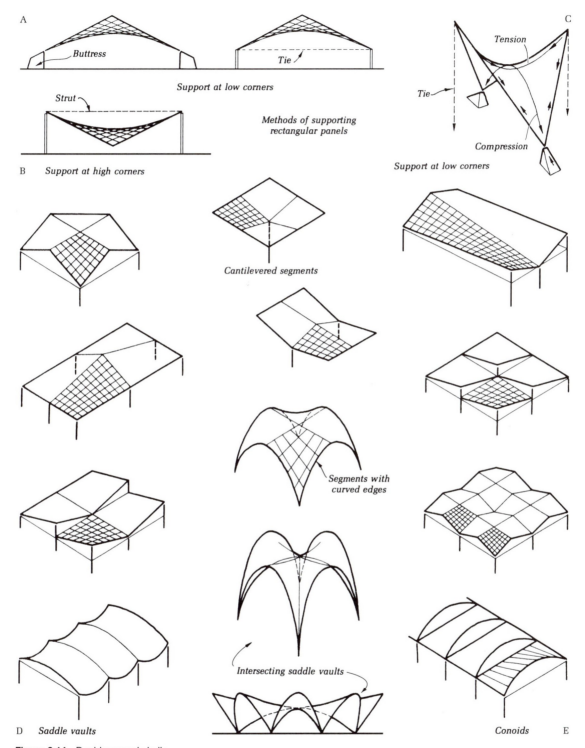

A

Buttress

Tie

Support at low corners

*Methods of supporting
rectangular panels*

Strut

B   *Support at high corners*

C

Tension

Tie

Compression

Support at low corners

Cantilevered segments

*Segments with
curved edges*

Intersecting saddle vaults

D   *Saddle vaults*

Conoids   E

Figure 8.14   Doubly curved shells

The hyperboloid of revolution referred to earlier is based on the circle and its development is illustrated in figure 8.12 E.

Shells can be constructed in reinforced concrete, timber and occasionally in steel plate.

8.1.6 Folded slab roofs

This form of construction is also called folded plate and, when there are a large number of facets, prismatic structure. It is another form of stressed skin or membrane structure in which the stiffness of the skin is used to distribute the loading to the points of support.

If a flat slab is folded or bent it can behave as a beam spanning in the direction of the fold with a depth equal to the rise of the folded slab (figure 8.15 A). When loaded, compression and tension stresses will be set up at the top and bottom of the section respectively and shear stresses in the slabs on each side of the fold. Each slab spans between the folds and must be thick enough to span this distance and to have sufficient stiffness to distribute the loads longitudinally. End frames or diaphragms must be provided to collect the forces in the slabs and transfer them to the supports (A, C). The shape of the roof may vary from a simple pitched roof of two slabs to a multi-fold or prismatic form involving several plates, some examples of which are given in figure 8.15.

The span and width of each bay govern the overall depth. As an approximate guide this should not be less than between one-tenth and one-fifteenth of the span, or one-tenth of the width, whichever is greater. The width of each slab is limited only by the requirements of adequate

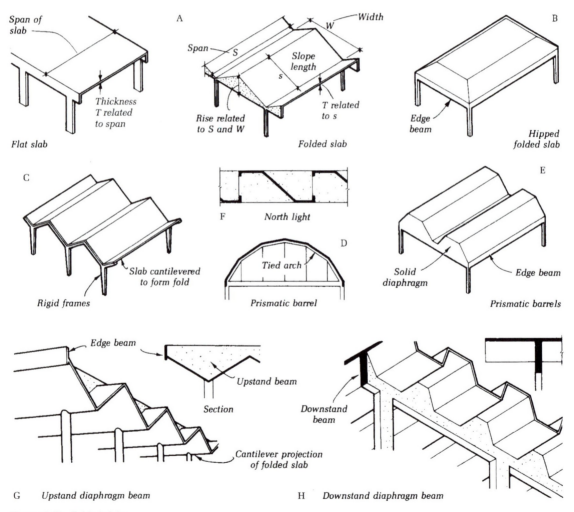

Figure 8.15 Folded slabs

lateral stiffness which dictate the thickness. In practice, it is often cheaper to use a large number of narrow slabs rather than a few wider slabs. This is because, although a greater number of folds must be formed, thinner slabs can be used and the amount of material and the dead weight of the structure is less. When 'barrels' are formed of a large number of 'facets' or slabs (D), the influence of the rigidity of the folds will be proportionately greater than when fewer and, therefore, wider slabs are employed. One-twentieth to one-twenty-fifth has been given as a reasonable value for the thickness/slope length ratio for the slab element. The slab thickness will generally be thicker than that in comparable cylindrical shells for spans in excess of about 4.5 m. Folded slab construction is competitive with these shells, however, because the flat shuttering is comparatively inexpensive. It is possible to use composite construction using repetitive precast concrete slab elements as permanent shuttering to an in situ structural topping (see figure 8.38).

Diaphragm beams   There are various methods of retaining the folds in position at the points of support where the loads are taken down to the foundations. A solid diaphragm beam, a lattice truss or a rigid frame may be used, to which the ends of the slabs are rigidly secured. When vertical supports down to the ground can be permitted, columns may be placed under each fold carrying a

diaphragm beam which follows the s [...] slabs (figure 8.15 A, G). Diaphragms may [...] or below the slabs as required (G, H). The d[...] not necessarily be vertical in order to fulfil its [...] may be sloped to form a hipped end, the angle [...] serving to transfer the loads to the supports (B). Wi[...] diaphragms may carry a number of folded elements w[...] wide column spacing is necessary (E, F, H).

The free edges of folded slabs should be stiffened. They must either be supported along their length, be provided with edge beams or be cantilevered a short distance to form a fold (B, G, C).

The slabs may be perforated where light is required through the roof and, provided sufficient slab is left on each side of the perforations and adjacent to the folds, the slab can be reduced to a series of struts. It becomes then, in effect, a Vierendeel girder, the top and bottom flanges of which are the folds in the slabs. Folded slabs can be used over continuous spans, with depth/span ratios sometimes as low as one-fortieth, and as cantilevers (G, H).

Use of counterfolds   This form of construction can be used to extend down to the ground in arch or rigid frame form as well as for simple span roofs supported on columns or walls (figure 8.16 A). A great variety of complicated and interesting forms can be obtained by the introduction of reverse or counterfolds at various points (B, C). Dome and

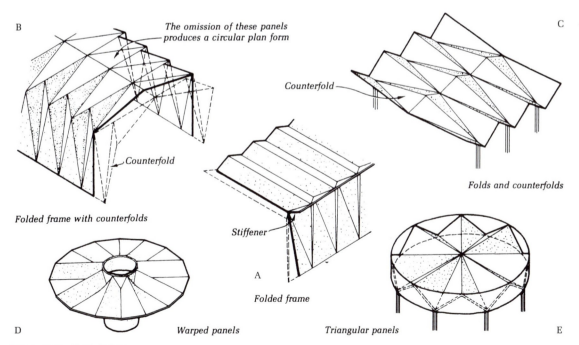

Figure 8.16   Folded slabs

vault forms can also be constructed. Circular shapes can be covered with folded slabs developed to form a horizontal roof structure as distinct from the dome form. This necessitates variation in the slab widths, the introduction of counterfolds, the use of triangular shaped slabs or the warping or twisting of the slabs, which then become hyperbolic paraboloids. Either the centre or perimeter supports may be omitted and the structure then becomes cantilever in form with tension and compression rings replacing the stiffening diaphragms. Two examples are shown in D and E.

Folded slab structures derive most of their strength from their shape. The folded form gives great rigidity and makes possible the efficient use of material of high elasticity such as plastics and aluminium alloys, the use of which in other forms of construction often results in excessive deflections. Concrete, timber and plastics may be used in this form of construction.

8.1.7 Sheet steel arches

Curved trapezoidal profiled sheeting, already referred to in relation to claddings and floors, can be used structurally in roof construction as arches with a high strength to weight ratio. Using profiles with a depth of 60 mm a single skin arch can span up to 13 m and a double skin with insulation between up to 22 m, with a minimum rise from springing of one-tenth of the span. Tie rods at springing level are necessary unless the supporting structure is capable of resisting the thrust of the arch. Current bending technology permits reverse bending to be made on a sheet to produce an 'S' shape with which undulating roof forms can be constructed.

8.1.8 Grid structures

Grid structures, apart from single layer flat grids, are three-dimensional or 'space structures'. Unlike shells or folded slabs, however, they are constructed not with solid membranes but with lattice or grid frameworks. In some systems, however, the grid is partially formed by the folds and edge junctions of bent or folded sheet panels of suitable material in which the skin strength of the sheet element forms a very large proportion of the total strength of the structure.

These structures usually provide a simple and economic method of covering very large areas without internal intermediate support. As they permit the prefabrication and standardisation of the component parts, and as the dead weight is small compared with many other forms of structure, there are considerable savings in construction costs, the savings increasing with the increase in span. The eco-

nomy of these structures as far as prefabrication and standardisation are concerned is related to the simplicity, or otherwise, of the jointing technique and the multiplicity, or otherwise, of members of similar length and section.

Grids are very stiff so that they are very useful for situations where deflection and not load is the criterion. That is, where the span is great and the load is small, as is often the case in roof construction. The structural depth can be relatively small because of the stiffness of the structure. Apart from single layer flat grids the stresses in all the members of a grid structure are direct, except for some slight transverse bending moments in diagonal members. Weights as low as 50 to 150 kg/m^2 of floor area covered have been achieved generally and in some cases far less.

Grid structures may be constructed in metal, reinforced concrete, timber or plastics. The majority at the present time are constructed in metal which lends itself to the prefabrication and standardisation of the component parts, and for which comparatively simple methods of joining the parts have been developed. In situ cast reinforced concrete is suitable for single layer grids, but is less suitable for double layer grids where the stiffness is provided primarily by triangulation and in which prefabrication and standardisation of the parts are logical. Although this can be done with concrete in the form of precast elements post-tensioned together, it is not an ideal medium for this purpose. Plastics may be incorporated in this type of structure in the form of three-dimensional folded or curved elements.

Grids can be applied in many arrangements to flat, curved or folded roofs. Their application to floors in the form of rectangular grid and diagonal beam construction has already been mentioned in chapter 5. The application to roofs can be broadly classified as follows:

- Space frames
- Flat grids
- Folded grids
- Folded lattice plates
- Braced barrel vaults
- Braced domes.

Space frames Although all the structures under consideration, apart from single-layer flat grids, are in fact space frameworks, the term 'space frame' is usually applied to a hollow section or three-dimensional lattice beam (figure 8.17 A and figures 8.39 and 8.40). The hollow shape confers great lateral rigidity on the beam whilst retaining the economic advantages resulting from optimum depth/span ratios. A convenient and common cross-sectional shape for such a beam is a triangle, since this is an inherently stable shape not requiring additional bracing. The longitudinal

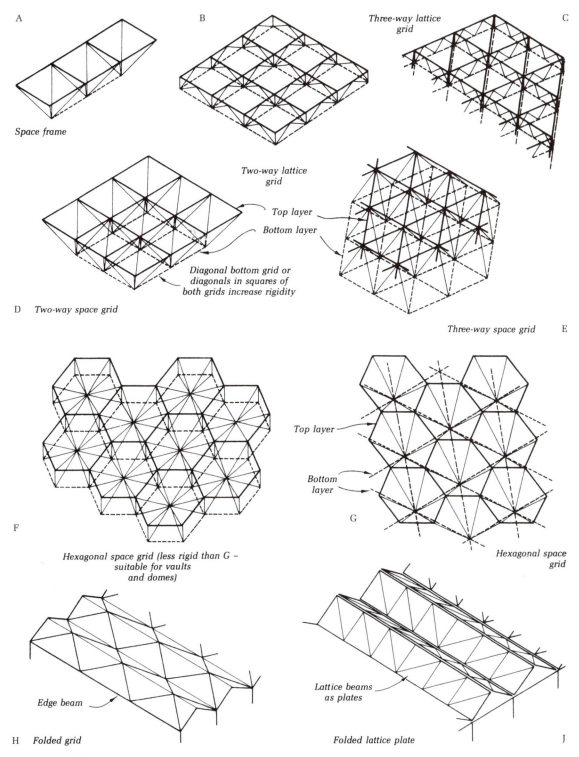

A

Space frame

B

Three-way lattice grid   C

Two-way lattice grid

Top layer

Bottom layer

Diagonal bottom grid or diagonals in squares of both grids increase rigidity

D   Two-way space grid

Three-way space grid   E

Top layer

Bottom layer

F

Hexagonal space grid (less rigid than G – suitable for vaults and domes)

G

Hexagonal space grid

Edge beam

H   Folded grid

Lattice beams as plates

Folded lattice plate   J

Figure 8.17   Grid structures

members will be in tension or compression and the shear stresses will be taken by the diagonal members in the sides of the triangle.

Flat grids   These may be single-layer or double-layer grids, each of which will be considered in turn.

A single-layer flat grid   This is, in fact, a two-dimensional structure, but is considered here rather than earlier because of its grid nature and certain characteristics which it has in common with double-layer grids, particularly the ability to disperse heavy concentrated loads throughout all the members of the grid. It has two or more sets of parallel beams intersecting at right- or oblique-angles and the beams are rigidly connected at all intersections, which produces bending and torsion of all the members (figures 5.3 and 5.4).

These grids are ideally suited to structures which are to carry heavy concentrated loads. Because of the interconnection of the parts, a concentrated load is distributed between all the members of the grid, decreasing the high stresses in the directly loaded area. The stress distribution in grid frameworks under heavy concentrated loads is therefore comparatively even.

The interconnected beams may be arranged in various layouts and the boundaries of the grid may be rectangular or circular. The rectangular grid, although widely used, is not the most efficient in terms of stress distribution since there are no members in the corners and these are in fact the most highly stressed zones. In theory, the beams could be arranged in such a way as to follow the trajectories of the principal stresses in the slab. Thus they would be where they are most needed. As the beams would have to be curved, such a layout is not likely to be economic, particularly as far as reinforced concrete is concerned, although this could be done using glass-fibre reinforced cement or plastic pans as shuttering. A close approximation to this arrangement can, however, be obtained by a diagonal grid layout. Because the beams tend to follow the lines of principal stresses, the stress distribution in this type of layout is much more even than in rectangular grids. Triangular or three-way grids are extremely strong and lead to very uniform stress distribution in the structure. These layouts are used for the larger spans within the economic range of 15 to 24 m. Depth/span ratios as low as one-thirtieth for rectangular grids and only one-fortieth for the others are often possible.

The square and diagonal grids are described in chapter 5 under 'Rectangular grid' and 'In situ diagonal beam' floors.

Double-layer flat grids   These may be lattice grids, that is, they are formed by intersecting lattice beams, or they may

be space grids. In lattice grids, each set of bottom horizontal members lies immediately under the top set in the same vertical plane, as in normal lattice beams, but in double-layer flat space grids they do not lie in the same vertical plane. Thus, in its simplest rectangular grid form the space grid consists essentially of two sets of interconnected triangular space frames (figures 8.17 D and 8.42 A). Lattice grids may be two- or three-way grids (figure 8.17 B, C), and the space grids, of which there are a great number of variations, in addition can be hexagonal (E, F, G). Lattice grids are not so stiff as space grids.

These structures are usually fabricated from circular or rectangular section metal tubes which may be welded together or joined by connectors at the junctions, although reinforced concrete precast compressive members can be used together with tensile members in steel. Precast concrete members post-tensioned together can also be used, but compared with tubular metal structures concrete results in large dead weights. Double-layer grids, as well as other types of space frameworks, lend themselves to prefabrication and there are many commercial prefabricated systems on the market. In the ideal system, all the framing members would be of the same length, joined together with identical connectors, so designed as to require the minimum of fastenings at the end of each member entering the connector.

Prefabricated parts   In addition to systems consisting of individual bars and connectors joined together on the site, some systems consist of individual prefabricated units, such as pyramids, made up of bars, of which the edge members of the flat bases are joined together on the site and the apices tied together by tie bars in each direction. These are fixed to bosses on the apex of each pyramid (see figure 8.42 B, C). Instead of open pyramids formed of bars or tubes, pyramids formed of thin sheets of aluminium, plastics or plywood can be used, fixed in the same way along the edges by means of bolting, riveting, welding or gluing. The apices of all units are connected together in the same way by bar or tube members to form a grid. Stressed skin space grids such as this are very light and economic and have a high load-carrying capacity (figure 8.43).

Double-layer flat grids are generally not competitive with other systems below about 21 m span. They have been used for spans of over 90 m but greater spans than this are economically possible. The depth should be one-twentieth to one-thirtieth of the span so that the depth of a double-layer grid for spans around 30 m would be about 1.0 to 1.50 m. This allows ample working space and space for services. Spans in the region of 90 m require a depth of about 3 m. These wide spans and the very large cantilever projections which can be used are possible because of the light weight, strength and great rigidity of this type of structure. It does not collapse if one part fails. Relatively large areas can be removed when required or be omitted in the original design without destroying the stability of the remainder of the structure.

Folded grids These are space grids in the form of bent or folded diagonal plane grids, the folds usually corresponding to the valleys and ridges of the roof, as shown in figure 8.17 H. The folded grid is therefore a series of continuous intersecting beams cranked at the folds. Folded grid roofs are usually of the multiple ridge and valley type or north light type and cover very large areas without internal support. Structures of this type can span distances up to 90 m in the direction parallel to the ridges and almost unlimited lengths in the other direction. Folded grids can also be applied to the hipped roof form. Longitudinal members are required at the folds to give rigidity to the structure, particularly on wide spans. Although over small spans they might be omitted, rigidity would not be great and, in any case, some members would be required to sup-

port the roof covering and its substructure. Edge beams are required at all boundaries. Folded grids are usually constructed in steel using ordinary rolled steel sections for the diagonal members and channels for the edge beams.

Folded lattice plates In principle, this is the same as folded slab construction in reinforced concrete or timber, but the planes or plates are constructed as lattice beams, usually of steel, although occasionally of timber. The adjacent edges are interconnected at the folds, usually at intervals at the node points (figure 8.17 J). The roof profiles to which it can be applied and all other considerations are the same as for folded slabs. Folded lattice plate structures in steel can span greater distances than folded slabs in reinforced concrete on account of their lighter weight.

Braced barrel vaults Although similar in form to reinforced concrete shells, braced barrel vaults, being an assembly of bars, are non-homogeneous (figure 8.18 A, B). The members can be arranged to follow directly the

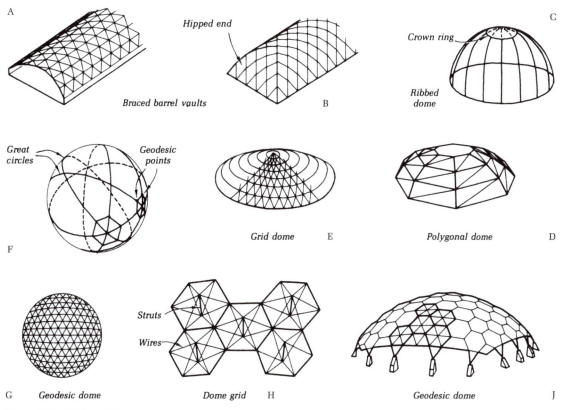

Figure 8.18 Grid structures

lines of maximum stresses. In practice, several types of bracing are used each of which results in a different type of structural behaviour. The shape of the barrel is of great importance in relation to the stress distribution and although the cylindrical form is not the best shape, it is very useful from a practical point of view since prefabrication is facilitated by the fact that the slope of all the members is the same, the length of the bars can be the same and identical connectors can be used at all joints. In some cases the barrel is formed of curved members but more frequently the 'curves' are made up of short straight members.

The rise of the barrel should be between one-eighth and one-twelfth of the span. These figures relate to multi-barrel structures and the lower figure is only suitable for continuous spans or barrels which cantilever considerably at the ends. Greater rises should be used for single-width barrels, whether of single span or continuous span form. Where these rises cannot be obtained, edge beams may be used, except in the case of single-width barrels, and their depth included in the rise as in the case of reinforced concrete shells. Small rise shells are likely to produce large horizontal thrusts and, while these are mutually resisted in the valleys of multi-barrel structures, resistance is not easily provided in a single-width barrel without the use of cross-ties. The radius of the barrel should not exceed 11 m since buckling tends to occur in the shell with larger radii. The greatest practicable span for single layer vaults appears to be about 36 m. End frames of some form must be provided at the ends of the vaults. In place of end frames, hipped ends can be used and these increase the strength of the barrel considerably (B).

One particular form of curved grid system, the Lamella system, utilises a large number of identical members called *lamellas* which are joined without the use of a connector component. The grid is arranged in diamond or rhombus pattern and one member or lamella is continuous through each joint, each lamella being twice the length of the side of the diamond. These are joined by a single bolt (see figure 8.45).

By means of double-layer grids, vaults may be constructed with spans up to 90 m or widths up to 30 m.

Braced domes   These are constructed either with curved members lying on a surface of revolution, or of straight members with their connecting points lying on such a surface. Some forms consist of ribs running from the base to the crown or of inclined bars, also running from base to crown, but connected by a number of horizontal polygonal rings. Other types are constructed as space grids similar to those used for barrel vaults.

The first group includes ribbed domes, consisting of meridional ribs connected only at the crown or at a crown compression ring when there is a top opening (figure 8.18 C and figure 8.46). The other types in this group are polygonal in form and have horizontal rings and diagonal stiffening members and differ in the number of diagonals, the relationship of the polygonal rings to each other in terms of rotational position and in the number of sides making up the dome (D). In addition to these there is the stiffly jointed 'framed' dome which consists of continuous meridional ribs and polygonal horizontal rings rigidly connected at all intersections. This type is constructed in welded steel but is not often used in practice because it is not easily amenable to modern prefabrication techniques. Three-pin ribbed domes are frequently used because the ribs, being identical, can be easily prefabricated and erection is simple, only a small central tower being required to provide temporary support for the crown ends of the ribs until they are all interconnected.

Space grid domes constitute the second group and may be constructed as lamella or as two- or three-way or hexagonal grids. Lamella domes are constructed of lamella ribs, producing the typical diamond or lozenge shaped pattern as described under braced barrel vaults, and domes of over 90 m span have been built in this way. The other types of grid dome systems vary in the grid pattern, the material used, the connecting system, which may be by welding, bolting or specially designed node connectors or clamps, and in the degree of prefabrication adopted (E). As with lamella domes, they are extremely economical and spans of well over 90 m have been covered. For these very large spans, double-layer three-way or hexagonal grids are used to give greater stiffness against buckling.

Geodesic domes   Another widely used grid form is the geodesic dome. Points on a sphere lying on a 'great circle' are called geodesic points, a great circle being any circle running round the surface of the sphere having the same radius as that of the sphere itself. Lines joining such points are called geodesic lines (see figure 8.18 F). Since a sphere encloses the maximum volume with minimum surface area, a true geodesic dome would enclose the maximum volume with the minimum amount of structural material. This necessitates the use of curved members lying on great circles (G), but surface coverage is usually achieved by repetitive straight-edged patterns, using triangles and hexagons, in which the nodes only are geodesic, these being connected by straight members which are not geodesic (H, J). As with any other form of braced structure geodesic domes may be single-layer or double-layer, the latter being extremely rigid and suitable for very large spans. They may also be of stressed skin construction in which the covering acts as an integral part of the structural system, or of what may be termed formed surface construction, in which flat sheets of suitable material are bent and interconnected along their edges to form the main structural grid of the dome (figure 8.47).

A pin-jointed triangular frame is rigid and this shape is therefore used, or hexagonal grids are broken down into triangles. It is impossible to cover a complete sphere entirely with hexagons, a certain number of pentagons being essential. Apart from exceptionally simple layouts, the members will be of different lengths and, further, a level alignment of the members at the base of such domes is not possible. This needs careful architectural consideration.

Double layer geodesic domes have been built over spans of more than 116 m.

8.1.9 Tension roof structures

Tension construction uses either a continuous membrane as structure and roof covering or a network of cables or pin-connected links to support a separate cladding or covering material. These, by stretching or suspending, are put into a state of tension.

Structures which have to resist bending are basically inefficient since the stresses involved are complex. Even when the structure is designed so that the individual components are directly stressed, as in lattice girders or space grids, there are certain members in compression which need to be stiffened against buckling. Shell or arch structures represent attempts to use materials in direct stress but it is impossible to eliminate all bending in any but the smallest forms of these structures. The advantage of using a tension system for a roof is that the only direct stresses which occur are tensile. This avoids the buckling effect associated with other structures.

As indicated in Part 1 (pages 5–7) these structures may be formed as a network of cables suspended from supports, as a membrane stretched over supports on the principle of a tent, or as a membrane tensioned into shape by compressed air, known as an air stabilised or pneumatic structure.

Cable network structure   This consists basically of a set of catenary cables suspended from supporting members (figure 8.19 A) and secured by guys as shown or anchor elements (F) or from compression rings or arches (C, D, E). One of the problems associated with the application of this technique is that of 'flutter'. This occurs because the roof is comparatively light in weight and frequently assumes shapes which induce greater suction or 'lift' at certain wind speeds (A). To reduce this effect it is desirable to prestress the principal suspended cable system by another system of cables curved in the opposite direction. In principle there are three methods: (i) a doubly curved system of cables in which the main cables are prestressed by another set at right-angles (B, E), (ii) bracing ties which

tension the two cables in concave form (F), (iii) spreaders which tension the two cables in convex form (G). The effect of suction will be to reduce the tension in the principal cables following the catenary paths and to increase the tension in the restraining cables following 'arched' paths. Conversely, the application of loads will reverse this stress distribution. Methods (ii) and (iii) can be applied to both 'beam' (F) or grid (H) arrangement. A net or linked membrane as shown in H will assist in distributing the stresses throughout the system. These can be fabricated of continuous cables or of similar length elements or links.

A single layer net may be put into tension by stretching between the tops of column supports and ground anchors as shown in J.

This type of roof provides a very light structure for spanning large areas because the absence of buckling due to compression permits the minimum amount of material to be used, and because the use of high-strength, high-tensile steel permits wires or linked rods of small diameter to be used. It has the further advantage that it can be erected without scaffolding. Against this must be placed the following disadvantages: the expensive supports required, whether in the form of compression ring or props and anchors and, secondly, the complications in cladding.

Membrane structure

Tent structure   In this type of membrane structure the membrane can be put into tension by (i) stretching over compression supports and anchoring to the ground or to the heads of shorter braced columns in the manner of the traditional circus tent on its poles (figure 8.20 A) or (ii) by suspending in various ways by tension supports, usually cable hangers, and anchoring to the ground in the same way (B). By the use of either system complex forms can be developed. In large tent structures where the primary stresses are high, cables are either incorporated within the panel and edge seams or are positioned externally to the membrane to which they are connected by link components at close intervals along the seams and edges.

A thin tension member carrying a uniform non-axial load, such as its own weight, takes up a curved or catenary shape (see also page 260). When a thin suspended membrane has to resist both positive and negative forces caused by snow and by wind suction or uplift it must, in order to function satisfactorily, be doubly curved in form as in the doubly curved systems of cables shown in figure 8.19 B, C, D and E. The membrane is prestressed in both directions by the initial tension set up by stretching it over or from its supports and, as in the case of a cable network (see above), the application of a load produces an increase in tension in the direction of one curve and a reduction in tension in the direction of the other curve.

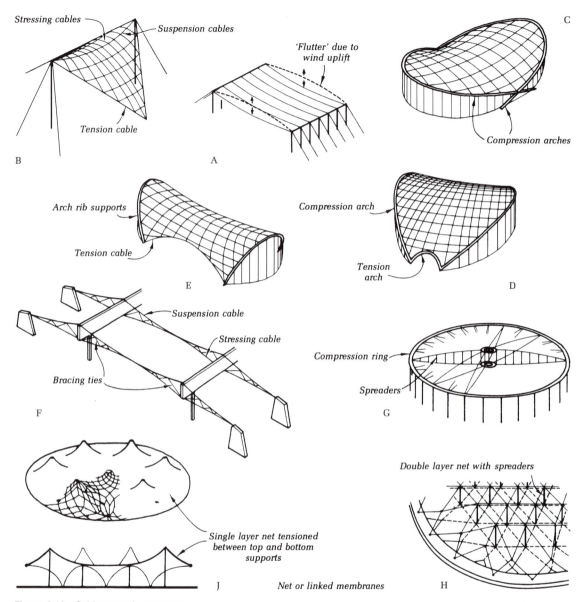

Figure 8.19 Cable network structures

The membrane is attached at the edges to boundaries as in cable networks which may be walls, beams, arches or, more generally, catenary cables by means of which the membrane loads are transferred to rigid structural supports. The overall form of the tent structure is produced by these boundaries and by the difference in the prestressing forces in the two main directions. The simplest form is the hyperbolic paraboloid (figures 8.19 B and 8.14 C) but more complex forms may be developed by the use of ridge

and valley cables, internal supports (figure 8.20 A) and by the introduction of holes with cable boundaries.

This type of structure is not capable of spanning the distances which can be covered by cable networks or air stabilised structures.

Air stabilised or pneumatic structures These consist essentially of a thin flexible membrane stabilised by air under pressure which induces tensile stresses in the

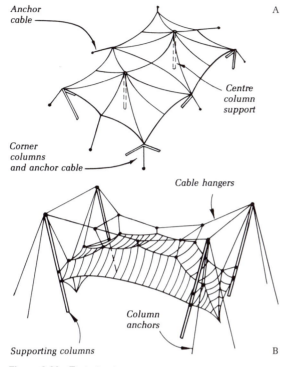

Anchor cable

A

Centre column support

Corner columns and anchor cable

Cable hangers

Column anchors

Supporting columns

B

Figure 8.20 Tent structures

membrane enabling it to support loads. The pressure must be such that under all conditions of loading the development of compressive stresses in the membrane is prevented.

There are two types of such structures: air supported and air inflated.

Air supported structures consist of a single space enclosing membrane supported by air at a pressure slightly above that of the atmosphere, in practice between 15 and 25 mm of water pressure (figure 8.21 A). The membrane causes uplift forces at ground level which must be counteracted by adequate anchorage. A constant air supply must be provided to make up for leakage through access points and to some extent through joints in the membrane. Air supply is maintained by low pressure fans and air loss through large access openings is usually minimised by the use of air locks or air curtains.

The strength of the air supported structure depends upon the internal air pressure, the volume of air contained within the membrane, the characteristics of the material and the form of the structure. The greater the air pressure and the greater the volume of air the greater will be the rigidity of the structure. Air supported structures enclose relatively large volumes of air and, therefore, only small air pressures are necessary to achieve stability. For this reason this type has a greater spanning capacity than the air inflated structures described later.

The ideal form is the sphere or, for practical reasons, the three-quarter sphere, but these are not so easily fabricated as the cylindrical form. The latter may have square or spherical ends, but circumferential stress differences in the cylinder and sphere result in folding at the junction of the two forms which can be avoided by a square end (figure 8.21 B).

The tensile stress (T) developed in the membrane of this type of structure by the internal air pressure (P) is directly proportional to the radius of curvature (R) of the membrane: $T = PR/2$ (figure 8.21 C). Thus for equal areas covered and for equal air pressures a structure having a greater radius of curvature will develop proportionately greater membrane stresses and will require a stronger membrane and heavier anchorages than one with a smaller radius.

Cables or ropes can, however, be employed to divide the membrane into smaller elements each with a small radius of curvature, thus reducing the membrane stresses and permitting a thin membrane to be used. The cables form grooves in the membrane and take up the main forces. By the use of different cable and cable net arrangements a wide variety of forms can be achieved (8.21 D).

Internal membrane walls, partitioning the space within the structure, can similarly influence the form of the membrane by anchoring it to the ground, to form indents resulting in smaller curvatures in the membrane (E). Vertical cable ties are an alternative which do not sub-divide the interior. The use of cables and membrane walls results in concentrated anchorage forces which, in large structures, can be considerable.

Air inflated structures are forms of air stabilised structures in which air is used to inflate tubes and double membrane elements to form stiff structural members such as arches, beams, columns and walls capable of transmitting loads to their points of support (figure 8.21 F). The strength of these structures depends upon the same four factors referred to in relation to air supported structures. Larger spans are achieved with higher pressures and larger volumes. With tubes of 0.5 m diameter spans of up to 20 m can be achieved. Air replenishment at intervals is necessary to make up air losses due to the slight porosity of most membrane materials.

There are two types of this structure: (i) the inflated rib structure, which consists essentially of a system of pressurised tubes which support an enclosing membrane in tension, the latter providing stability to the frame of tubes (figure 8.21 F, H), and (ii) the inflated dual membrane structure in which air is contained between two membranes which are held together by diaphragms or 'drop threads' (figure 8.21 G, and J). These can be compartmentalised to reduce the risk of total collapse.

As with air supported structures adequate ground anchorage is required and similar forms of anchors may be used.

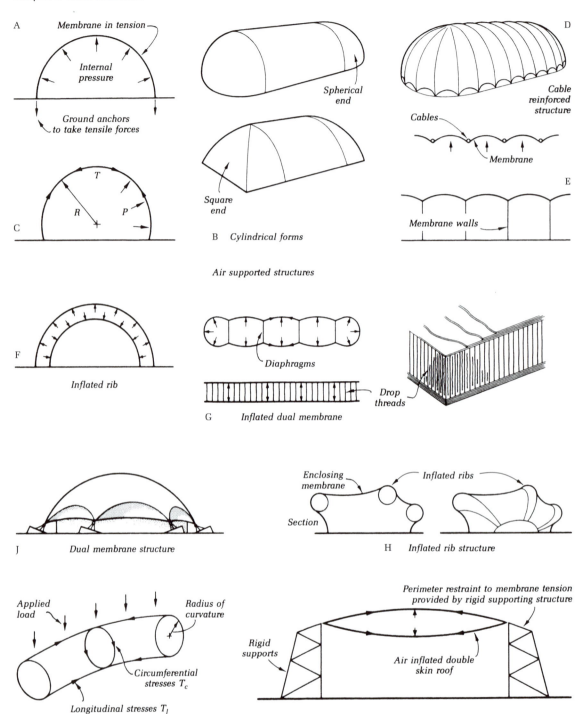

A

Membrane in tension

Internal pressure

Ground anchors to take tensile forces

C

T

R

P

B   Cylindrical forms

Spherical end

Square end

Air supported structures

D

Cable reinforced structure

Cables

Membrane

E

Membrane walls

F

Inflated rib

Diaphragms

G   Inflated dual membrane

Drop threads

J   Dual membrane structure

Enclosing membrane

Inflated ribs

Section

H   Inflated rib structure

Applied load

Radius of curvature

Circumferential stresses T_c

Longitudinal stresses T_l

K   Air inflated structures

Perimeter restraint to membrane tension provided by rigid supporting structure

Rigid supports

Air inflated double skin roof

Hybrid structure   L

Figure 8.21   Air stabilised structures

The tensile stresses in the membrane of a tubular beam or rib are longitudinal and circumferential and for any internal pressure are directly proportional to its radius of curvature:

$$T_1 = \frac{PR}{2} \text{ and } T_c = PR$$

(figure 8.21 K).

The bending moment due to an applied load is calculated in the normal way and the stresses set up by this in the member are established, using the section modulus value for the particular beam section. To avoid the folding of the membrane material the compressive stresses due to bending must not exceed the pre-tensioning stresses in the upper fibres of the tube, nor must the total tensile stress in the lower part exceed the safe permissible stress for the material used. Shear stresses can also cause folding at the sides of the beam, especially in the region of point loads.

The loadbearing capacity of an air inflated structure increases in proportion to the internal air pressure, increases with its cross-sectional area (in the case of dual membrane components with the distance between the membranes) and decreases, as with conventional components, with an increase in span. The arch and dome forms are stiffer and carry greater loads than straight beam and flat slab forms.

There are many forms of combined or hybrid construction using air stabilised structures, some of which integrate the two forms described, taking advantage of the superior insulating properties of the inflated dual wall construction and of the greater spanning capability of air supported construction. Others combine air stabilised construction with conventional forms of structure as in the example L shown in figure 8.21.[14]

8.2 Choice of roof structure

The most appropriate form of structure will depend upon the type of building, foundation conditions, spans to be covered, nature and magnitude of loads, lighting requirements and accommodation for services, the possibility of future alteration and speed of erection, as well as aesthetic considerations. For some types of buildings a few solutions only are possible, as in the case of an auditorium with rigid requirements of space, acoustics and volume and ventilation. In others, industrial buildings for example, particularly those of small and medium span, a wide range of structure is possible.

As indicated in Part 1, page 141, short span construction will usually be cheapest but the span of the structure is usually fixed by the type of building as this will dictate the minimum areas of unobstructed floor space required. The minimum spans must therefore be compatible with requirements of clear floor area.

The clear internal height of the building is another factor governed largely by the use of the building. In industrial buildings sufficient clearance must be provided for the installation and maintenance of plant, and the widespread use of fork-lift trucks requires minimum headrooms for efficient working. Much ancillary equipment is hung from the roof structure and the provision of sufficient headroom below this will govern the minimum internal height.

Light and heat considerations Lighting requirements may have a profound effect upon the form of structure used, particularly in the case of very wide buildings, in which the central areas cannot adequately be lit from the side walls. Roof lighting must then be provided and this will affect the shape of the roof. Generally speaking, with a height of 5.50 m reasonable natural light cannot be obtained from wall windows if the distance between them exceeds about 12 m. Various types of roof structure designed to provide adequate lighting for the interior have already been described, and the suitability of a particular form will depend not only upon lighting requirements but on other functional requirements.

Apart from considerations of lighting and the other considerations mentioned above, the roof structure must be thought of in relation to the heating requirements of the building. From this point of view the structure which, while fulfilling other functional requirements, restricts the internal volume of the building to a minimum and at the same time offers the minimum exposed roof surface area, will be the most efficient.[15] The reduction in volume reduces the amount of space to be heated and the reduction in roof surface area keeps heat losses through the roof to a minimum. From this point of view alone, therefore, the flat roof is the most acceptable form.

Although it may not always be acceptable for other reasons, it may be generally accepted that as far as lighting and heating requirements are concerned the larger the span the lower should be the pitch of the roof. To achieve this, where functional requirements demand large span roofs and wide column spacing, the use of forms of construction which are more expensive than others may be economically justified. When rigid frames are used to decrease the volume of enclosed space it is often economically advantageous to pitch the spanning members sufficient to permit the use of forms of sheet roof covering which require a fall of a few degrees.

Equipment support and adaptability The need to suspend equipment from the roof must be considered, particularly if the loads to be carried are heavy and are point loads. Shell roofs, which are essentially designed to carry comparatively light uniformly distributed loads, may not for this reason be suitable. Point loads of any magnitude should, wherever possible, as in other structures, be restricted to the main stiffening beams and end frames of shell structures. If, however, it is essential to suspend from the shell, all loads greater than about 50 kg should be spread by plates or cross reinforcement. In addition to suspended equipment, provision may have to be made for ventilating or extract ducts and, particularly if these are large, they will have to be related to the roof structure at an early stage.

The choice of structure will sometimes be weighted by considerations of adaptability and maintenance. Steel construction is generally more adaptable than reinforced concrete if alterations are necessary, and it is comparatively easy to strengthen the structure, although fixing to matured concrete can be made by metal fixing studs applied by cartridge hammer or to threaded inserts cast in when the concrete is placed.

Concrete generally requires less maintenance than steel, which must be painted periodically. Steel lattice construction of all forms offers a large surface area for corrosion and is difficult to paint. Rigid frames built up of welded steel plates, on the other hand, expose less surface to the air and are easier and cheaper to paint. In highly corrosive atmospheres it may be more economical to use resistant alloys or zinc coated steel (galvanised) instead of mild steel in order to reduce maintenance costs, even though the cost of the structure itself may be higher than in mild steel. Timber structures do not require protective coatings internally, unless for fire insulation purposes, and maintenance costs are therefore considerably less than for metal structures.

Very often speed and cost of erection will have a bearing upon the nature of the structure to be used. This is discussed in more detail later.

8.2.1 Span and type of structure

Although short-span construction will usually result in the least expensive structure, functional requirements often dictate medium or large spans. For these the most economic form must then be selected.

Of the wide variety of roof structures available, some are economic only over large spans, some only when there is considerable repetition of the structure and its parts on one particular job. As far as the materials of the structure are concerned, some, such as aluminium alloys, are only really economic when used in large-span construction. Care must be taken in design, therefore, not to use materials which can be exploited most favourably only over wider spans when, in fact, large spans are not justified by other requirements, nor to select forms of structure the economic use of which depends on repetition, when the job is too small to provide adequate repetition.

As with all types of structure the characteristics of the soil on which the building rests must be considered. In circumstances of weak soil requiring expensive foundations, it may be cheaper to use widely spaced, wide-span construction in order to limit the number of foundation points.

Research on the comparative costs of different forms of roof structure in different materials has shown that the variation in cost of structural frames alone of the same span and loading, and fulfilling the same requirements, is generally small. Bearing in mind that the cost of the structural frame may be as low as one-sixth of the total cost of the building, it will be appreciated that quite large variations in this element will have only a relatively small effect on the total cost. Nevertheless, in large buildings in particular, one-sixth of the total cost may be a substantial sum, in which case consideration must be given to the problem of keeping this figure to a minimum.

Truss and column frame   For simple shed-type buildings where lighting is not important, the truss and column frame is probably the most cost effective structure for spans up to about 30 m. The space above eaves level, however, is obstructed by the trusses and over the wider spans this type of structure encloses a considerable volume of space. For spans above 30 m 'umbrella' construction or north light roofs with lattice girders in the plane of the ridge can be used up to about 45 m span, at spacings not exceeding 18 to 21 m. The unit weight and cost of a steel folded lattice plate and an 'umbrella' roof of the same span and spacing of supports are about the same, but the folded plate construction has the advantage of unrestricted space under the roof.

In trusses or girders and in structural systems relying upon 'beam effect' such as space frames, economy in the use of any material will result when increased moments of inertia are produced by increased depth of structure. Structures which are more closely related to the space which they enclose, such as rigid frames or arch forms, will frequently be more expensive span for span than those incorporating deep beams, girders or trusses, but may compensate for this by reducing the volume of the building and the area of cladding and finishes.

Rigid frames   Rigid frames in welded steel and reinforced concrete either 'square' or 'arched', although not always economic over smaller spans, are frequently used because of the advantages which have been mentioned above. Reinforced concrete rigid frames can be economic over 18 to 30 m spans. For spans of over 30 m, and certainly over 45 m, rigid steel frames with lattice structure are likely to be cheaper than those with solid web designed on the elastic theory, but not necessarily those designed on the plastic theory. For spans of 60 m or more, arch rib construction is likely to be the most economic.

Prestressed concrete is usually competitive in cost with uncased steel for solid web frames, so that where lattice frames are not suitable there is likely to be an economic advantage in using prestressed concrete.

Over spans of more than 24 m the saving in weight over steel in an aluminium structure may be as much as four-fifths. In spite of this, the cost of an aluminium structure may not be lower than that of steel, and aluminium only begins to be competitive with steel over very large spans of 60 m or more.

In buildings where the aesthetic value of exposed timber is desired, advantage can be taken of the material's high strength to weight ratio.[16] Bowstring trusses with laminated timber top chords can span up to 70 m or more, and lattice rigid frames have been constructed over spans of 30 m or more. Laminated timber 'bents' forming rigid frames and laminated timber arches are used over spans of more than 60 m.

Medium-span structures in reinforced concrete can be competitive with steel, if constructed of standardised pre-cast members. When construction involves non-standard members and profiles cast on the site, the flexibility of design and adaptability of concrete are very advantageous.

Shell structures   Shell concrete construction can be economic in medium- and wide-span roofs, using long-span barrels up to spans of 30 to 45 m, above which short-span barrels would be used. There will not be much difference in the relative cost of steel and shell structures provided that the shell is repeated more than four to five times. Although shell construction may not always be competitive in cost with steel, nor usually in speed of construction, it is in other respects equal to steel and superior to it as far as maintenance is concerned.

Shell vaults can be prestressed, with a consequent reduction in the rise to span ratio, but the structural advantage of pre-stressing must be weighed against the increased cost and complication of placing ducts, anchorages and cables and it is unlikely that this will be justified for spans much below 33 m. Above this span the advantage of pre-stressing increases.

Shell construction in laminated timber produces a very light structure. A timber hyperbolic paraboloid shell 18 m square will be about 50 mm thick and weigh approximately 25 kg/m^2; a comparable concrete shell would be about 63 mm thick weighing about 150 kg/m^2. Laminated timber domes and vaults show similar weight advantages. Elliptical shell domes have been constructed with diameters over 45 m and barrel vaults over spans of up to 30 m.

Sheet metal arch construction gives a roof structure with a high strength to weight ratio thus keeping the loading on the supporting structure to a minimum. Spans up to 22 m are possible.

Grid structures   Grid structures in flat or curved form are particularly useful over very wide spans where the advantage of rigidity and light weight are important. Single-layer flat grids are economical up to 15 to 24 m span. Braced barrel vaults have the advantage over reinforced concrete shell vaults of considerable reduction in weight and of rapid erection on the site. They require no shuttering and may be partially or completely prefabricated. For spans above about 18 m, they are about 10 per cent cheaper than reinforced concrete shells or even 20 per cent in some cases. By the use of folded grids and double-layer flat grids, spans of over 90 m can be covered economically. Where the circular or polygonal plan form is acceptable, braced domes of various types can be used over spans of up to 90 to 120 m. As explained on page 251, aluminium alloys are advantageously employed in grid structures.

The relative advantages of steel, aluminium alloys, concrete and timber are discussed on pages 249–51. The metals are produced not only as 'I', channel and other common sections in hot-rolled steel or extruded aluminium alloy, but also as tubes, hollow sections and, in the case of steel, as cold-formed sections produced from thin strip steel.[17]

Cold-formed steel structures   The manufacture of cold-formed steel sections and the design considerations involved in their use are described on page 158. Although they are used for two- or three-storey structural frames as described in chapter 4, they are used to greatest advantage in roof structures where a low dead/live load ratio is important. For spans up to about 15 m the weight of cold-formed steel construction per m^2 of floor area is about 40 to 60 per cent less than that of similar hot-rolled steel construction, the greater savings being over the shorter spans. This is largely due to the fact that when hot-rolled sections are used for short-span structures, sections for the various members cannot always be obtained small enough to match the minimum area required for stress resistance, as the choice is limited by the smallest sections rolled. In addition, the wide variety of shapes which can be cold-formed makes it possible to avoid eccentric loading of the members and the use of gusset plates, and this further decreases the weight of the structure.

As far as cost is concerned, beyond very small spans the increased cost of fabrication, and sometimes of the material, is likely to make the price equal to or more than that of ordinary hot-rolled construction, but against this must be placed the advantages of the comparative ease with which sections of different shapes can be produced, the simple fixing of roof claddings and ease and speed of erection. Light lifting equipment only is required for erection purposes because the fabricated units are much lighter than those in hot-rolled construction. Savings in erection costs are given as around 15 to 25 per cent for trusses and 10 per cent for purlins and bracings. Provided that detailing is simple, construction in cold-formed sections can be very economical when used on standard units or with a high degree of repetition. Over medium spans with light loading, cold-formed lattice beams show savings over hot-rolled lattice beams although these savings are not so great as in the case of trusses. Because of the efficient distribution of metal in cold-formed sections and the advantages shown in their repetitive use, considerable economies result when they are used as purlins, particularly as zed and sigma sections, in conjunction with main roof members of hot-rolled steel.

Welded structures Roof structures constructed from welded steel tubes will be considerably lighter than similar structures constructed in ordinary steel sections. The weight of a tubular framework per m^2 of floor area may be from 25 to 40 per cent less than that of a hot-rolled structure for spans up to about 15 m. As in the case of cold-formed sections the greater savings are made in the shorter spans. Actual savings will depend upon the simplicity of detailing. Although there will be a saving in weight by the use of tubular construction, a reduction in cost compared with that of normal steel construction is unlikely unless simple detailing is used and gussets are reduced or eliminated. However, if these points are watched and there is considerable repetition, tube frameworks will generally be cheaper in overall cost, or at least competitive with hot-rolled construction.

Efficient large-span construction can be obtained by using both tubes and hot-rolled sections together: the latter as tension members and the former as compression members. This is because tubes provide a more efficient section in compression, and are designed to a lower basic tension stress than, for example, hot-rolled angles. Erection takes rather longer than with ordinary steel frames in spite of the fact that tubular frames are lighter and stiffer in lifting. This is largely because additional access facilities will be required as tubular components are less easy for carrying out fixing operations on.

If welding instead of bolting of normal steel sections is employed there will be an overall saving in the weight of the structure, but not necessarily a saving in cost unless there is considerable repetition of the work and simple detailing. This is because the saving in weight is offset by the increased cost of fabrication. In order to reduce the fabricating cost to a minimum, welding should be restricted to the assembly workshop or to the floor on the site. Where possible, any connections required to be made when the structure is erected in position should be bolted.

8.2.2 Speed and economy in assembly

As already indicated, the cost of assembly in terms of the speed and facility with which it may be carried out is an important factor in the overall cost of a roof construction. Metal and timber structures are generally quicker to erect than those in in situ concrete. Reference has already been made to the ease with which structures fabricated from cold-formed sections are erected. The use of precast concrete, which can be assembled fairly quickly, reduces the time disadvantage of concrete. Construction time with wide-span concrete structures can be reduced by the use of prestressing with precast concrete elements.

The pre-casting of concrete roof frames from 6 to 40 m or so in span, the large ones being cast in three to four units, bolted together on the site, is common practice. By this means they can be easily and quickly erected by the use of simple derricks or mobile cranes. The erection of large rigid frames cast in three or four sections is carried out with the use of a timber trestle, either on a track or moved by crane into each frame position, which supports the separate units until the joints between them are bolted up. Three-pin frames in lattice construction, particularly in aluminium, may be erected in two halves which are swung over into position using the base connections as hinges. In the case of large span concrete structures, erection can be speeded up and a great amount of expensive shuttering avoided by the application of post-tensioning to small precast units. These can be stressed into a single element and then hoisted into position by cranes or hydraulic jacks.

Although rigid concrete frames of various types are those most commonly carried out in precast construction, beyond certain heights it may sometimes be cheaper to precast small shell vaults and domes on the ground and then hoist them into position. Shell concrete domes 7.50 m square weighing 13.70 tonnes each have been precast on site and lifted into position by a pair of cranes. Lightweight concrete shell barrel vaults over 18 m long and weighing 20 tonnes each have similarly been cast and hoisted into position. In each case a number of such units was cast for the project. In the case of the domes twelve, and in the case of the barrel vaults six units were required.

In each of these examples the weight of the units was within the lifting capacity of a crane or pair of cranes, but in certain circumstances it may be cheaper than other forms of construction to cast very much larger units and lift by hydraulic jacks. Groups of prestressed concrete barrel vaults linked together by end frames into units measuring 56 m × 33 m overall and weighing about 1500 tonnes have been cast on the floor of a building, with all internal roof finishings and external roof coverings applied at ground level, and then raised nearly 15 m by means of hydraulic jacks. Two pairs of jacks were positioned at each of the four supporting column positions, the columns being built up on the jacks with specially designed interlocking concrete blocks placed in position at the completion of each lift by the jack.

Grid structures of all types are fabricated on the site as a whole, or in large sections from prefabricated units, and are then lifted into position usually by crane, or by hydraulic jack in the cast of very large structures raised as a whole. Steel lattice grids more than 90 m square and 3 m deep, weighing over 1000 tonnes have been lifted in a similar manner to lift slab construction in reinforced concrete.

8.3 Constructional methods

The basic principles and economic applications of roof structures have been described in the preceding pages. It now remains to described the constructional methods adopted in the application of different materials to the various types of structure.

8.3.1 Beams

These are constructed in timber, steel and concrete.

Timber

Glued and laminated beams These are described in Part 1 (page 130) and may be in the form of 'glulam' or laminated veneer sections. As stated there the laminations of glued and laminated sections may be arranged vertically or horizontally (figure 8.22 A). Horizontal lamination allows a more economical use of timber, particularly for larger section beams, since the depth of the beam may be formed of multiple laminates and is not limited by the width of any given plank. Furthermore, horizontal lamination facilitates shaping (B). By selecting quality laminates of timber, the inherent defects of solid sections such as knots and shakes can be rejected. The thickness of each laminate is usually about 50 mm for straight members, reducing to 16 mm according to the radius of curve for shaped

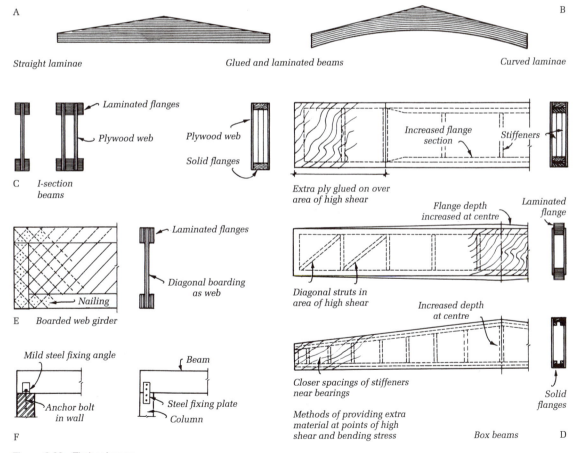

A Straight laminae   Glued and laminated beams   Curved laminae   B

C   I-section beams — Laminated flanges — Plywood web

Plywood web — Solid flanges

Increased flange section — Stiffeners

Extra ply glued on over area of high shear

E   Boarded web girder — Nailing — Laminated flanges — Diagonal boarding as web

Diagonal struts in area of high shear

Flange depth increased at centre — Laminated flange

F   Mild steel fixing angle — Anchor bolt in wall   Beam — Steel fixing plate — Column

Closer spacings of stiffeners near bearings

Increased depth at centre

Methods of providing extra material at points of high shear and bending stress   Box beams   D

Solid flanges

Figure 8.22 Timber beams

members. The shapes of beams may be varied to give sloping upper surfaces, slight cambers and sections of varying depth. Beams much over 18 m in length are not usually formed in glulam construction. Beams may be secured to masonry or be supported on solid or laminated timber columns and secured by metal angles, plates and bolts as shown in F.

Web beams   For spans of up to 12 or 15 m a more economical distribution of material is obtained by the combined use of solid timber flanges and webs of plywood (C). Such beams may be stiffened by using a number of webs to form a hollow box section and additional cross-sectional area may be added to the flanges between the webs to accommodate larger bending moments (D). Web stiffeners are required and at points of high shear stress these may be more closely spaced or alternatives may be adopted as shown. For spans over 18 m and up to 30 m the use of laminated glued and nailed beams of I-section will produce a more economical structure than solid rectangular laminated beams. The webs and flanges are formed of boards approximately 25 mm thick fixed together with glue and carefully calculated nailing (E). This system enables short pieces of timber to be used economically, each flange being composed of a number of overlapping lengths and the web formed from diagonal boarding braced where necessary by the addition of vertical stiffeners glued and nailed to it.

See also Part 1, section 8.4.3, where steel nail plates separate timber top and bottom flanges to create composite beams. This type of beam has become very popular for joists spanning up to about 8 m, purlins and rafters.[18]

Steel   British Standard Universal beams may be used where head room is critical and minimum beam depth is desirable. Reference should be made to chapter 4 where the merits of mild and high-tensile steel sections and castellated and cellular beams are discussed (pages 142 and 143; see also page 255).

Concrete   For spans much over 6 to 9 m reinforced concrete is not likely to be economical for beams, but wide spans can be covered economically if the beams are prestressed (see page 251).

8.3.2 Trusses and lattice girders

Timber   Girders and trusses may be fabricated with connectors and bolts. The two commonly used connectors are the toothed plate and split ring, illustrated in Part 1, figure 7.30. Toothed connectors are useful for joining sawn timber. Split-ring connectors carry greater loads than toothed connectors but require accurately machined grooves cut in the timber and the timbers are usually faced on four sides and prepared in the fabrication workshop.

As indicated in Part 1, gussets give greater fixing areas with all methods of jointing and permit members to butt against each other rather than overlap. When steel plate is used for this purpose (figure 8.23 A), or for heel straps (B), with bolted joints a shear plate connector is used as shown in A and C, the flat side being in contact with the steel.

In order to permit the development of the full allowable load at each connector, minimum spacings and distances from the bolt centre to the edges and ends of the members being joined must be ensured. These vary with the type and size of connector. Typical distances are given in table 8.1. For smaller distances the maximum allowable load must be reduced.

Table 8.2 indicates the relative efficiency of the types of connection used.

Girders up to about 45 m and trusses to about 60 to 75 m may be constructed in timber. The upper chords of

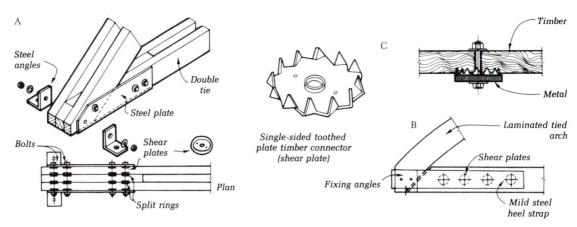

Figure 8.23   Timber trusses and girders

Table 8.1 Bolted joints – spacing and edge distances

Diameter and type	Edge distance mm		End distance mm		Spacing mm	
	A	B	A	B	Angle of connector axis to grain	
					Parallel to grain	Perpendicular to grain
64 mm Toothed plate	38	38	44	95	Load parallel to grain 95 Load normal to grain 76	Load parallel to grain 76 Load normal to grain 95
64 mm Split ring	44	70	102	140	Load parallel to grain 171 Load normal to grain 89	Load parallel to grain 89 Load normal to grain 108

A = unloaded edge or end   B = loaded edge or end (force from connector acts towards it)

Table 8.2 Relative efficiency of connections

Relative efficiency of connections	% Efficiency	
	Single lap	Double lap
Screws in prebored holes	2.82	5.65
Bolts – 1 at 25 mm diam.	21.30	21.30
Toothed connector – 64 mm with 12 mm bolt	21.10	19.10
Split ring connector – 64 mm diam	38.50	–

$$Efficiency = \frac{Max.\ working\ load\ in\ joint}{Max.\ working\ load\ in\ member}$$

girders may be curved, parallel or inclined, and are usually laminated in the larger spans. Some examples are shown in figure 8.24.

Steel   Girders fabricated from hot-rolled sections are described and illustrated on pages 145–48. Conventional steel trusses are described in Part 1, page 168. These may be bolted or welded complete in the fabricating shop or, if too large for transport, are made in two halves with the necessary connecting plates and provision for attaching connecting members as shown in the welded example in figure 8.25. Assembly, by means of bolts, is carried out at ground level prior to hoisting frames to position. Trusses and girders fabricated from welded tubes possess the advantage of lightness and stiffness. This form of construction is discussed on page 284, where the need for simplicity of detailing and, as far as possible, the avoidance of gussets is emphasised. Typical joints are shown in figure 8.26.

The jointing of members by direct welding is shown in A (see also figure 8.27 C). This is facilitated by the use of rectangular hollow sections and these are often used for the major members of a truss or girder for this reason. The use of diaphragms at the junction of tubes of equal or near equal diameter facilitates the jointing (figure 8.26 B). When gusseted joints are adopted the tubes may be cut and welded to them as shown in C. In large trusses and girders some site joints are always necessary and examples of purlin, bracing and tie joints are shown in D.

The method of supporting roof girders along their length by tension members as a means of reducing their depth to a minimum is described on page 284. The junction of the tension, or suspension, members with the girders and columns is usually by means of pin joints, a typical example of which is shown in figure 8.28.

Light lattice beams formed from angles and flats are used for lightly loaded roof structures up to spans of 12 m. Such beams frequently incorporate high-tensile steel bottom chords (figure 8.27 A). The use of cold-rolled steel sections welded or riveted together in lattice beams is discussed on page 283. Examples of these beams are illustrated in figures 5.11 and 8.27 B and in Part 1, figure 6.10.

Aluminium   Aluminium and steel trusses and girders are generally similar in form. As explained on page 250, care must be taken to prevent local buckling of members in compression, and while typical aluminium sections will be used some members may need to be doubled up to produce adequate stiffness.

Concrete   Modern light concrete trusses have resulted from the introduction of prestressing which enables small members to be drawn together by the stressing cables, so that the tension members of the trusses are pre-compressed (see page 180). Some typical examples are shown in

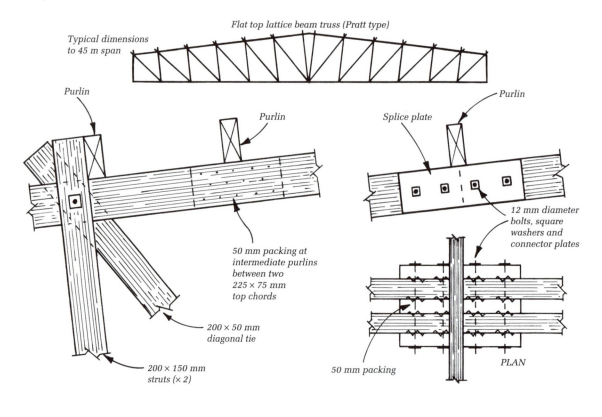

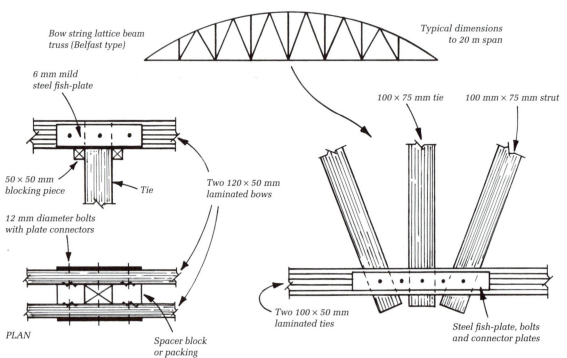

Figure 8.24 Further timber trusses and girders

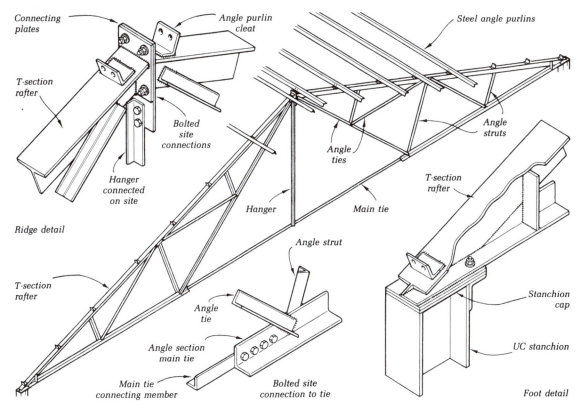

Figure 8.25 Steel trusses

figure 8.29. A shows a built-up truss with the bottom boom tensioned by cables. B shows a large concrete bowstring girder and C a lattice beam formed in a similar way from precast elements stressed together. D shows an example of a concrete beam with exposed stressing cables, forming a trussed beam.

8.3.3 Rigid frames

Timber The techniques used in fabrication of timber beams, trusses and girders are also applicable to the construction of rigid frames. Lattice frames are built up with bolted or glued joints on the same lines as trusses and girders. Figure 8.30 shows examples of rigid frames constructed in hollow box, solid laminated, and built up 'I' form. Where glued laminations are employed in frames with curved angles or in arch ribs (B, C, E) the laminates must be thin enough to bend easily. The ratio of the radius of curvature to the thickness of the laminates must not be less than 100, and 150 is normal practice. The cost of fabrication rises as the laminates get thinner, therefore sharp curves should be avoided where possible. Metal fittings

are employed to connect the members, and details of connections will vary according to the degree of rigidity required at to joints. Base connections are formed by cleats (A, D) or by locating the feet of the frames in metal shoes bolted to concrete foundations (C, E). The degree of rigidity provided by a connection will vary with the size of the frame and the nature of the connection. That shown in G and the similar example in E provide considerable flexibility. Those shown in A, C and F and the box connection in E will be stiff or rigid connections. In larger frames hinged metal bearings may be bolted to the feet of the frames (H). Crown joints are either bolted through the apex (C, D) or secured by splice plates (B, K) to provide a rigid connection, or are provided with a hinged bearing (E, J). Boxed plywood frames are used for spans up to about 18 m, glulam frames up to about 24 m and lattice and built-up I section frames up to about 45 m. Arch ribs are constructed in glulam up to spans of more than 60 m.

Steel Rigid frames constructed in steelwork may be of welded solid web construction or of lattice construction. Some typical forms are shown in figure 8.31. The lattice

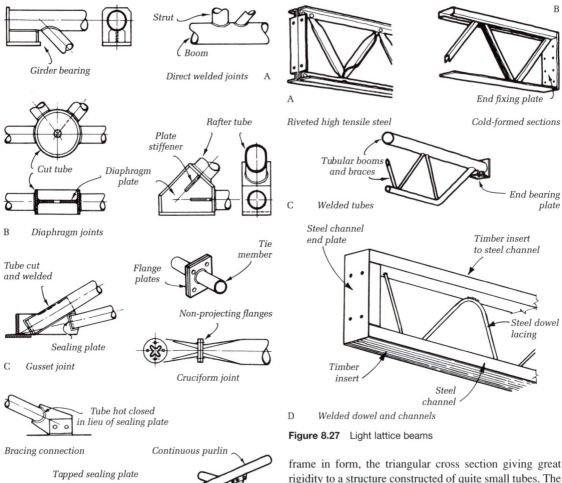

Girder bearing

Strut

Boom

Direct welded joints   A

Cut tube

Diaphragm plate

Plate stiffener

Rafter tube

B   Diaphragm joints

Tube cut and welded

Flange plates

Sealing plate

C   Gusset joint

Tie member

Non-projecting flanges

Cruciform joint

Tube hot closed in lieu of sealing plate

Bracing connection

Tapped sealing plate

Rafter

Threaded stud

Single bay purlin

Continuous purlin

Sealing plate

Site joints   D

Figure 8.26   Welded tube connections

A   Riveted high tensile steel

B   End fixing plate

Cold-formed sections

Tubular booms and braces

C   Welded tubes

End bearing plate

Steel channel end plate

Timber insert to steel channel

Steel dowel lacing

Timber insert

Steel channel

D   Welded dowel and channels

Figure 8.27   Light lattice beams

form is commonly employed in the construction of large-span frames. A number of lightweight pre-fabricated frames have been developed for short to medium spans which are of welded angle, tube or cold-rolled steel construction. Figure 8.31 A shows an example of a welded steel solid web fixed portal. Where, as in this case, the foundations contribute to the stability of the structure, the feet of the frames must be rigidly fixed to the concrete bases. The illustration shows one method involving embedding the foot in the concrete. Another method uses a heavy gusseted base bolted to the concrete slab. The welded tubular steel three-pin frame shown in B is a space

frame in form, the triangular cross section giving great rigidity to a structure constructed of quite small tubes. The nature of the junction between frame and closely spaced purlins (C) gives a rigid joint and provides lateral rigidity to the whole structure. It also results in considerable membrane action so that the whole of the roof structure will act to some extent as a lattice grid. The crown and base connections for this frame are shown in detail.

Crown and base joints vary in complexity and rigidity. In all of the hinge connections shown it will be seen that the frame is free to rotate (one permits rotation in all directions) but is held in position. Both crown hinges (D) and base hinges (E, F) can be formed as rockers or true pin joints. It should be noted that rocker base hinges must incorporate long bolts as shown in order to resist vertical tension and horizontal forces. At a crown rocker a bolt is required to resist shear forces.

Aluminium   Aluminium is normally used in lattice form, and rigid frames in this material are basically similar to steel lattice frames, although they will differ in detail because of the structural characteristics of the metal.

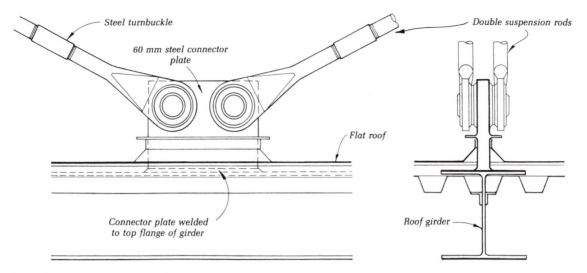

Steel turnbuckle

60 mm steel connector plate

Double suspension rods

Flat roof

Connector plate welded to top flange of girder

Roof girder

Figure 8.28 Suspended steel girders

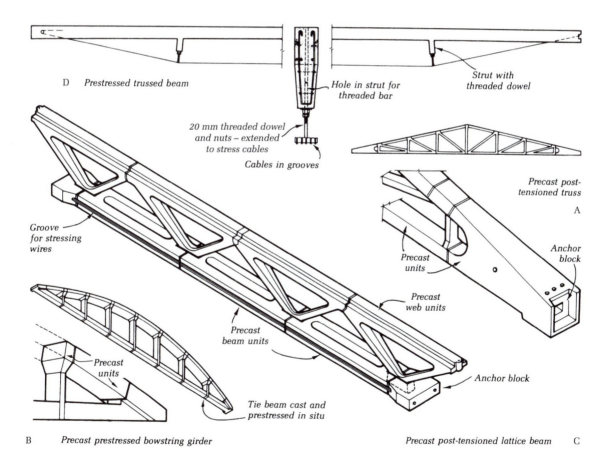

D   Prestressed trussed beam

Hole in strut for threaded bar

Strut with threaded dowel

20 mm threaded dowel and nuts – extended to stress cables

Cables in grooves

Precast post-tensioned truss

A

Groove for stressing wires

Precast units

Precast web units

Anchor block

Precast beam units

Precast units

Precast units

Anchor block

Tie beam cast and prestressed in situ

B   Precast prestressed bowstring girder

Precast post-tensioned lattice beam   C

Figure 8.29 Prestressed concrete trusses and girders

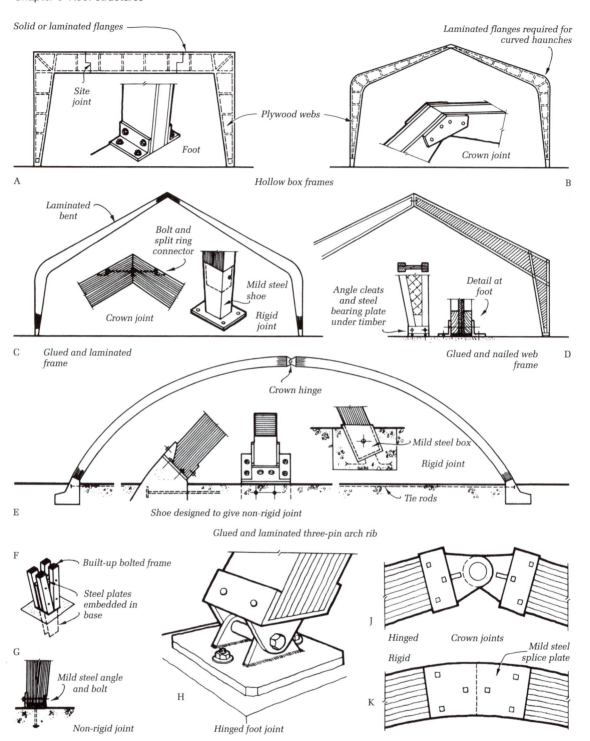

Solid or laminated flanges

Site joint

Foot

A

Hollow box frames

Laminated flanges required for curved haunches

Plywood webs

Crown joint

B

Laminated bent

Bolt and split ring connector

Crown joint

Mild steel shoe

Rigid joint

C   Glued and laminated frame

Angle cleats and steel bearing plate under timber

Detail at foot

Glued and nailed web frame   D

Crown hinge

Mild steel box

Rigid joint

Tie rods

E   Shoe designed to give non-rigid joint

Glued and laminated three-pin arch rib

F   Built-up bolted frame

Steel plates embedded in base

G   Mild steel angle and bolt

Non-rigid joint

H   Hinged foot joint

J   Hinged   Crown joints

Rigid   Mild steel splice plate

K

Figure 8.30 Timber rigid frames

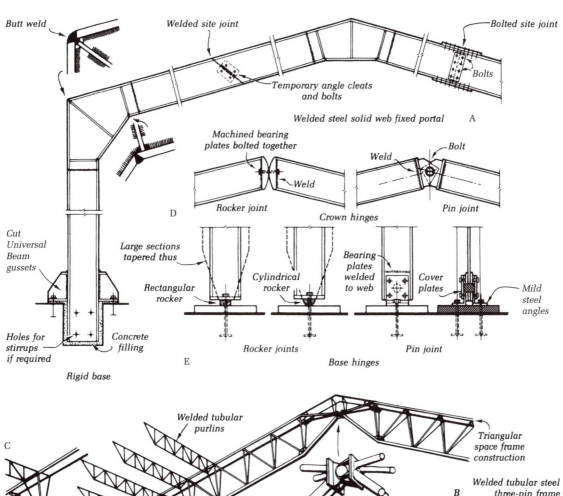

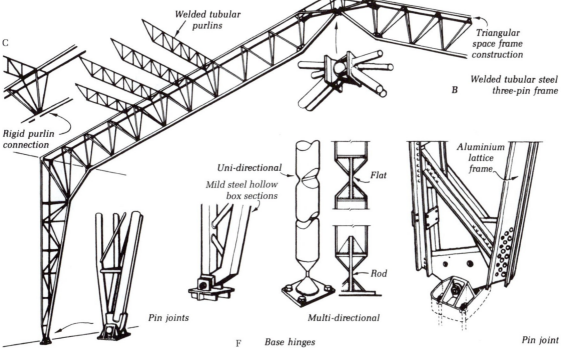

Figure 8.31   Metal rigid frames

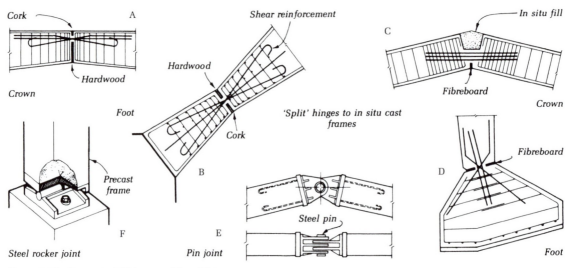

Figure 8.32 Concrete rigid frames – hinge joints

Concrete Concrete rigid frames are generally cast in situ when the span is large. Medium-span and small-span frames are most frequently precast and joined on site. 'Split' joints in in situ concrete form a hinge by reducing the member to a very small section through which passes 'bundled' reinforcement which holds the members is position and resists any shear stresses. Crown and base joints may be formed in this manner. Some typical examples are shown in figure 8.32 A, B, C, D. The principle of reducing the section of a member, and thus its stiffness, to form a hinge is also illustrated in the two joints shown in the tubular metal member in F, figure 8.31. Figure 8.32 also shows a metal pin joint (E) and a rocker joint (F) formed by welding steel flange plates to the reinforcement.

As explained in Part 1, precast concrete frames fabricated in large sections are joined on site at points of low stress or at the haunches, and bolted connections at these points are illustrated in that volume in figure 7.31. In multi-bay construction adjacent frames may be bolted together as shown in figure 8.33 A. A haunch junction effected by concreting-in connecting reinforcement left projecting from each section is shown in B. Precast frames may also be formed of prestressed concrete sections, stressed by cables or rods connecting the 'column' and 'beam' members as shown in the same figure.

8.3.4 Shells

Timber Some details typical of timber shell construction are shown in figure 8.34. Site constructed barrel vaults use layers of boarding nailed and glued together. A shows details of a 30 m span barrel vault the board membrane of

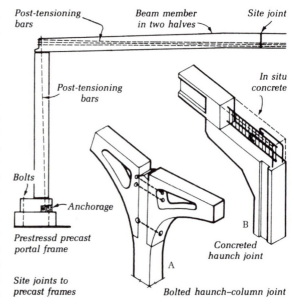

Figure 8.33 Concrete rigid frames – site joints

which is stiffened by ribs at 12 m centres. Small barrels lend themselves to prefabrication and light forms of construction. B shows a shell prefabricated in 1.20 m units, constructed as a sandwich panel of ply and paper honeycomb core. Edge beams to barrels may be of built-up box form or laminated form as shown.

Timber shell domes are normally built up with boards glued and nailed or screwed together, with laminated edge beams on the lines shown in C. In hyperbolic paraboloids,

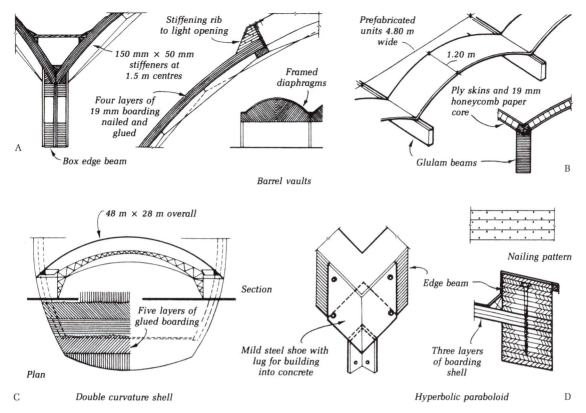

Figure 8.34 Timber shell construction

constructed in the same way, the majority of layers of boarding were, in early examples, laid parallel to the edge beams and at right-angles to each other with some laid diagonally but it has been found that greater rigidity results from laying the majority diagonally. A detail of the edge beam and shoe bearing of a hyperbolic paraboloid is shown at D.

Concrete A typical reinforcement layout for a long barrel vault is shown in figure 8.35 A, and it will be seen that the bars follow the lines of principal shear and tension stresses in the corners and towards the edge beams. Typical edge beam details for long barrels and north light shells are also shown (B, C). It can be seen that the upstand to the gutter at the base of a north light shell forms its edge beam and also supports struts in the plane of the glazing which, by reducing its span, permits the top edge beam to be quite small.

Reference has been made earlier to the possibility of prestressing long barrel shells and alternative ways of doing this are shown in D. When the edge beams are very shallow the post-tensioning cables are placed in the lower part of the shell. They are, however, placed more easily

within edge beams. A double curvature barrel gives the advantage of straight cables. The reasons for curving cables or curving the beams are given on page 185.

Corrugated barrels, such as that shown in figure 8.10 E, may be constructed on a framework of ribs of tubes, angles or timber shaped to the curve of the shell, braced apart and covered with hessian or other suitably strong fabric. Concrete is applied to the fabric in two or more coats, the fabric sagging between the supports under the weight of the first coat. In the cast of large shells the concrete is applied by means of a cement gun. For small shells up to about 12 m span no reinforcement is used and the thickness of the shell varies from 19 to 38 mm. The width of the corrugations would be from 900 mm to 1.80 m and the depth about one-fifth of the width. For large span vaults the structure is cast on formwork with angular corrugations in the form of folds. Alternatively it may be precast in sections and formed into a monolithic structure on the site as shown in figure 8.35 E.

Details of a cantilever hyperbolic paraboloid are shown in figure 8.36. The fact that a hyperbolic paraboloid surface can be developed by two groups of mutually perpendicular straight lines has the practical advantage of

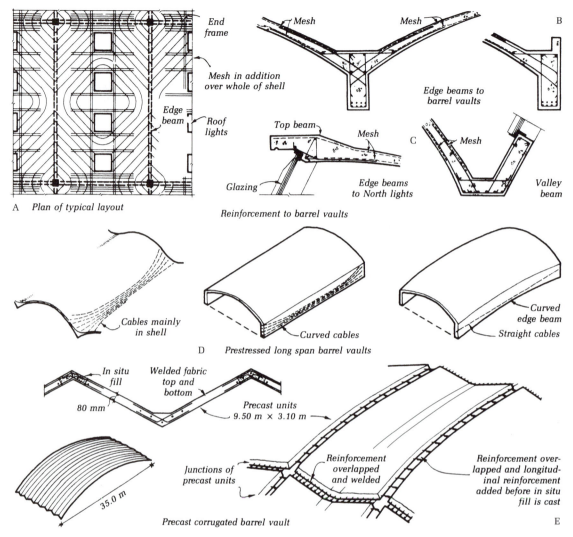

A   *Plan of typical layout*

Reinforcement to barrel vaults

Edge beams to barrel vaults

Edge beams to North lights

Valley beam

D   *Prestressed long span barrel vaults*

Cables mainly in shell

Curved cables

Curved edge beam

Straight cables

Precast corrugated barrel vault

E

Figure 8.35   Concrete shell construction

simplifying the shuttering for this type of concrete shell, since it consists wholly of straight timbers. The straight supporting bearers are placed at the appropriate varying angles parallel to the edge beams and are covered with straight planks at right-angles to them to produce the final doubly curved surface on which the concrete is cast (see figure 10.30). As all the stresses in these shells are direct, they may be as thin as 38 to 50 mm for spans of 30 to 40 m. In practice the minimum thickness of a concrete shell depends largely on the method employed to place the concrete, and slabs as thin as this make necessary the 'guniting' or spraying-on of the concrete.

Due to the thinness of reinforced concrete shells it is generally necessary to provide an insulating lining and this can often be conveniently achieved by using insulating building boards as permanent shuttering. Alternatively, the interior of the shell may be sprayed with vermiculite, this method having the advantage of allowing inspection of the underside of the concrete. Insulation may be provided externally by such materials as woodwool, cork or vermiculite screed and in this position has the advantage of protecting the concrete from solar heat and consequent thermal expansion.

Plastics   Shells, as noted on page 252, are appropriate structural forms in which to use plastics but, although much research and development work has been carried out in the structural use of plastics, their application in this

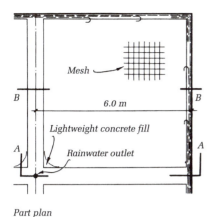

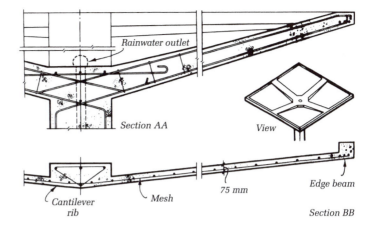

Figure 8.36 Hyperbolic paraboloid shell

field has not developed to the extent of that in the field of claddings.[19]

8.3.5 Folded slabs

Timber Folded slabs may be constructed in timber as (i) framed panels, (ii) hollow stressed skin panels, or (iii) laminated board panels. These are illustrated in figure 8.37. It will be seen that in each case the folds are stiffened either by additional members or by joining the edge members of adjacent slabs. The framed panels shown (A) have a non-structural covering of woodwool slabs and are joined at the hips. Figure 8.37 B is an example of a folded slab using 16 mm plywood stiffened on the underside by ribs. The top members are splay cut and are bolted together, and the ends of adjacent panels are bolted together over the supporting beams which are at 8 m centres. In the valleys, support and fixing to beams is given by 450 mm long brackets formed from steel angles welded toe to toe. The stressed skin panels of thin ply (C) are coach screwed at the folds, thus uniting adjacent edge members. When the slabs are built up with layers of boarding as in D they must be stiffened at the folds with laminated beams.

Concrete Concrete folded slabs may be cast in situ or be precast. Precasting may be carried out in two ways:

1 by casting full size panels and uniting a pair at the valley joint before hoisting, the ridge joint then being formed with in situ concrete uniting projecting reinforcing bars as in the vault in figure 8.35 E
2 by using smaller precast trough units spanning from fold to fold as shown in figure 8.38, steel being placed at the longitudinal folds and mesh and shear reinforce-

ment on the top before a structural topping is cast to form a composite construction.

Plastics Because of their geometrical form, folded slabs, like shells, are suitable forms for the structural application of plastics, but the remarks made under shells apply here.

8.3.6 Space frames

Timber The main problem in applying timber to space frame[20] construction is one of connection between timber members in several planes. One solution is to use three lattice girders and connect them by metal lugs at the nodes, as shown in figure 8.39 A. An alternative is shown in B using steel lugs bolted to the ends of each member by means of which they are fixed to pressed or cast metal 'multi-directional' connectors.

Steel Steel space frames may be constructed of cold-rolled sections, angles or tubes riveted, bolted or welded together or to suitably shaped gusset plates or connectors. Tubes are very suitable since they may be more easily joined at any angle, and due to their better performance in compression will produce lighter structures, particularly over large spans. Figure 8.40 A, B, C shows three examples of steel space frame construction in angles and tubes together with details of joints. The types of connectors shown in figure 8.42 may also be used in the fabrication of space frames instead of welding.

Aluminium As stated earlier, the direct stressing of members and the inherent resistance to torsion in space frames makes them suited to construction in aluminium. Rods or round and rectangular tubes are commonly used,

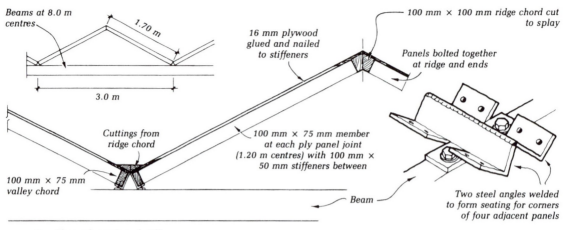

Beams at 8.0 m centres

1.70 m

3.0 m

16 mm plywood glued and nailed to stiffeners

100 mm × 100 mm ridge chord cut to splay

Panels bolted together at ridge and ends

Cuttings from ridge chord

100 mm × 75 mm member at each ply panel joint (1.20 m centres) with 100 mm × 50 mm stiffeners between

100 mm × 75 mm valley chord

Beam

Two steel angles welded to form seating for corners of four adjacent panels

B   Plywood panels and stiffeners

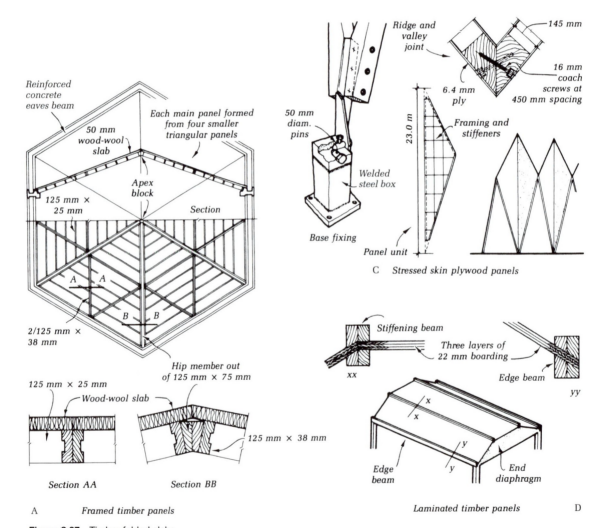

Reinforced concrete eaves beam

50 mm wood-wool slab

Each main panel formed from four smaller triangular panels

Apex block

125 mm × 25 mm

Section

A   A

B   B

2/125 mm × 38 mm

Hip member out of 125 mm × 75 mm

125 mm × 25 mm

Wood-wool slab

125 mm × 38 mm

Section AA   Section BB

A   Framed timber panels

Ridge and valley joint

145 mm

16 mm coach screws at 450 mm spacing

6.4 mm ply

Framing and stiffeners

50 mm diam. pins

Welded steel box

Base fixing

23.0 m

Panel unit

C   Stressed skin plywood panels

Stiffening beam

Three layers of 22 mm boarding

xx

Edge beam

yy

x

x

y

y

Edge beam

End diaphragm

Laminated timber panels   D

Figure 8.37   Timber folded slabs

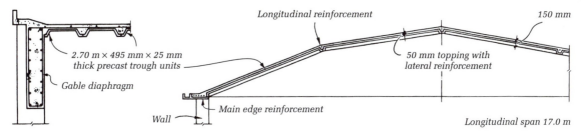

Figure 8.38   Precast concrete folded slab

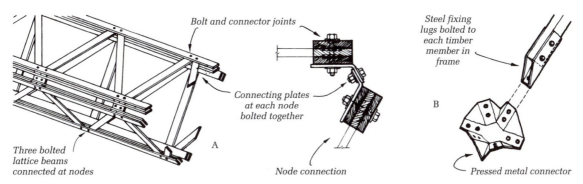

Figure 8.39   Timber space frame

often together with specially extruded sections, as shown in figure 8.40 E. Connectors of various types are also used into which the tubes fit, one of which is illustrated in D. Other suitable connectors are shown in figure 8.42.

Concrete   The development of pre-casting and prestressing techniques has made possible the use of concrete members of comparatively light weight and small cross-sectional area which may be used in the construction of space frames. The members may be connected together with site bolting and grouting in of reinforcement in situ. Some details of a concrete roof constructed of space frames of triangular section closely spaced and carrying light decking are shown in figure 8.40 F, and indicate the general arrangement of the members and joints.

8.3.7 Single-layer grids

Steel   Steel flat and folded grids are constructed of I-beams welded together at the intersections. One element may be continuous, to which the intersecting elements are welded to produce structural continuity in both directions. Alternatively, the beams are 'halved' to each other before welding to facilitate positioning of the beams (figure 8.41 A).

Aluminium   The structural limitations of aluminium mentioned earlier preclude its economic use in single layer flat grids.

Concrete   Concrete grids have commonly been cast in situ and this remains the likely method for heavily loaded grids, the accurate casting of the ribs being facilitated by the use of glass-fibre reinforced cement pans as permanent shuttering (figure 8.41 B). Steel or polypropylene pans can also be used, but as temporary shuttering with numerous re-uses. Solid expanded polystyrene blocks may also be used. Additional information can be found on pages 198 and 362.

For lightly loaded roofs of long span a lighter structure can be obtained with post-tensioned precast concrete units as shown in C.

8.3.8 Double-layer flat grids

Timber   The problem of jointing is identical with that in space frame construction and the methods used are the same. Two-way space grids may be constructed with lattice beams in a similar way to the space frame illustrated in figure 8.39, but with the horizontal lattice replaced by top lateral members connected at the node points which,

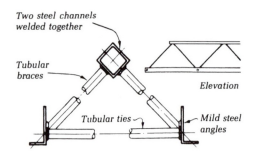

Two steel channels welded together

Tubular braces

Tubular ties

Mild steel angles

Elevation

A   Space frame of welded steel

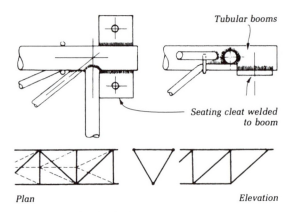

Tubular booms

Seating cleat welded to boom

Plan   Elevation

B   Tubular steel space frame

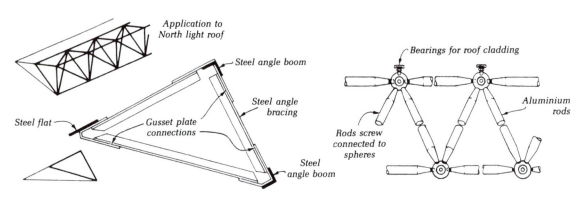

Application to North light roof

Steel angle boom

Steel angle bracing

Steel flat

Gusset plate connections

Steel angle boom

C   Steel space frame with gusset plates

Bearings for roof cladding

Aluminium rods

Rods screw connected to spheres

D   Aluminium space frame with node connectors

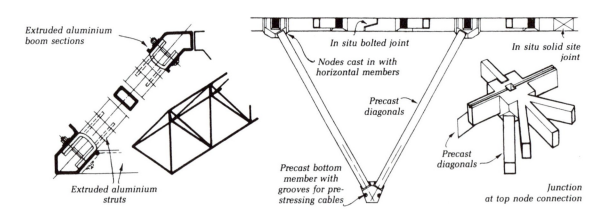

Extruded aluminium boom sections

Extruded aluminium struts

In situ bolted joint

Nodes cast in with horizontal members

Precast diagonals

Precast bottom member with grooves for pre-stressing cables

In situ solid site joint

Precast diagonals

Junction at top node connection

E   Aluminium space frame

Precast concrete space frame   F

Figure 8.40   Space frames

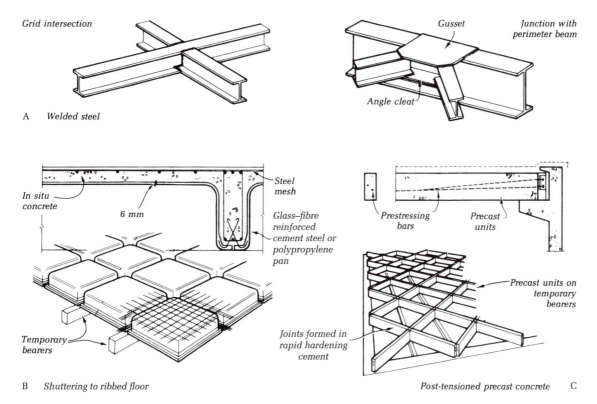

Grid intersection

A   Welded steel

Gusset   Junction with
               perimeter beam

Angle cleat

In situ concrete

6 mm

Steel mesh

Glass–fibre reinforced cement steel or polypropylene pan

Temporary bearers

Joints formed in rapid hardening cement

Prestressing bars

Precast units

Precast units on temporary bearers

B   Shuttering to ribbed floor

Post-tensioned precast concrete   C

Figure 8.41   Diagonal single layer flat grids

together with bottom lateral members form an interconnected system of triangular space frames as shown in
figure 8.42 A.

Steel and aluminium   Double-layer grids are constructed
in steel and aluminium in basically the same way as
described under space frames but invariably making use of
connectors at the nodes.

   Various types of connectors are used for joining separate members of steel and aluminium tube and cold-rolled
channels, some of which are shown in figure 8.42. Steel
tubes are welded to that shown at D, most of the welding
being carried out at ground level before the whole, or large
sections of the grid are hoisted into final position. E and F
show bolted forms of node joints, the latter requiring specially formed tube ends with captive bolts. The cast aluminium connector in G requires specially crimped ends to
the aluminium tubes but necessitates only 'hammer fixing'
in fabrication.

   B and C illustrate typical arrangements of members
in grids based on prefabricated inverted pyramids built
of tubes and angles which are joined together on site as
shown. Instead of 'open' pyramids, sheet aluminium or

plastic can be used to form them as shown in figure 8.43,
and they are joined together in a similar way but with node
connectors to join them to a top grid of tubes or extruded
sections.

8.3.9 Folded lattice plates

Timber   Two examples are shown in figure 8.44. In A,
each of the lattice beams is constructed from prefabricated
rectangular frames, each braced with diagonal steel bars or
stressed wires. They are bolted together and the whole
roof is prestressed by post-tensioned bars at the eaves. The
other example in B shows a roof with wide plates formed
of rafters, trussed to give stiffness to the plates, carrying
purlins covered with diagonal boarding. Lattice girders are
formed in the plane of the slab at the eaves, from which
tension cables run to the ends of the ridge which acts as a
strut.

Steel and aluminium   Lattice plates can be constructed
of angle sections joined by gussets or in tubular welded
construction. Each lattice is prefabricated and the adjacent
chord members joined by site welding or by bolting

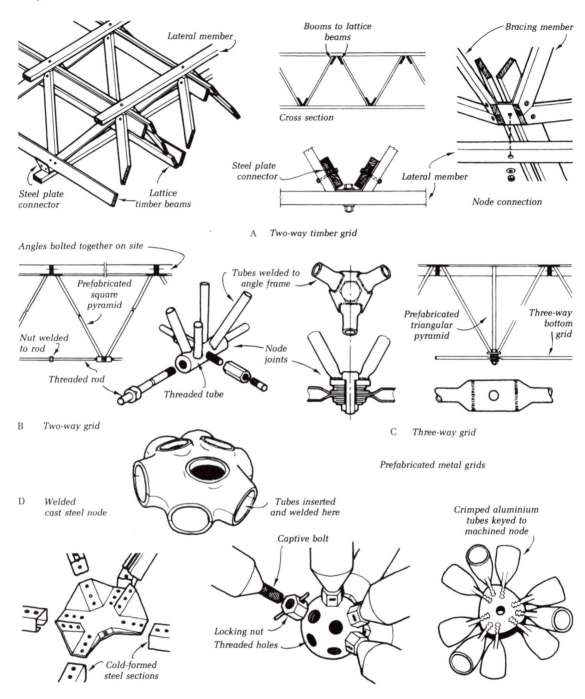

Lateral member

Booms to lattice beams

Bracing member

Cross section

Steel plate connector

Steel plate connector

Lattice timber beams

Lateral member

Node connection

A   Two-way timber grid

Angles bolted together on site

Prefabricated square pyramid

Tubes welded to angle frame

Prefabricated triangular pyramid

Three-way bottom grid

Nut welded to rod

Node joints

Threaded rod

Threaded tube

B   Two-way grid

C   Three-way grid

Prefabricated metal grids

D   Welded cast steel node

Tubes inserted and welded here

Crimped aluminium tubes keyed to machined node

Captive bolt

Locking nut
Threaded holes

Cold-formed steel sections

E   Pressed steel node

F   Screw connected node

Key connected node   G

Figure 8.42   Double-layer space grids

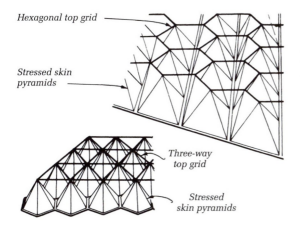

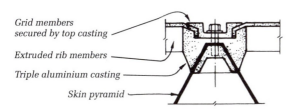

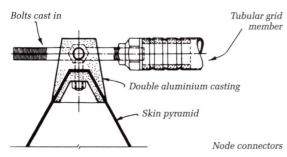

Figure 8.43 Stressed skin grid construction

together suitable flanges welded to them as shown in figure 8.44 C. Metal glazing bars may be screwed or welded to the framework where roof lighting is required, and solid areas may be covered with prefabricated panels clipped to the members. Alternatively, the frame may be covered with concrete on expanded metal with the internal surfaces sprayed with vermiculite plaster.

8.3.10 Braced barrel vaults

Timber 'Lamella' construction, referred to on page 276, can be used for braced barrel vaults. This is illustrated in figure 8.45, where the lamellas are shown joined with bolted connections. With the simplest form of single bolted connection, the ends of the two opposite lamellas butting on to the continuous lamella at the joint are slightly

staggered to permit the fixing bolt to pass through the central hole and pick up the ends of the lamellas as shown. Decking or purlins are used to triangulate the diamond to make the structure stable in the plane of the surface. A modified form, also shown, has a more complicated joint involving steel plates, but which preserves the visual continuity of the ribs.

Steel and aluminium Braced barrel vaults are constructed in these metals on the lines described for other grid systems. The methods of covering have been described above under lattice plates. Steel and aluminium lamella construction can also be used, incorporating cold-formed or extruded sections for the lamellas.

8.3.11 Ribbed domes

Two examples are illustrated in figure 8.46 showing in each case the methods of forming the junction at the crown, by connector (A) or by compression ring (B), and details of the bracing. The timber example shown is for a span of over 90 m and is covered with woodwool or similar insulative slabs. Smaller, lighter ribbed domes may be constructed with laminated ribs covered with plywood or layers of tongued and grooved boarding.

8.3.12 Grid domes

Timber 'Lamella' construction or triangular stiffened ply or framed panels similar to those in figure 8.37 may be used for grid domes.

Steel Tubular steel grid domes may be constructed with node connectors into which the tubes are inserted and welded, similar to those used for double layer grids, but using six-way instead of eight-way connectors to form a three-way grid, as shown in figure 8.47 A.

Aluminium Aluminium grid domes may be constructed with members joined by types of connectors already described. Another form is shown in figure 8.47 B, in which channel section members are bolted to a cast hexagonal connector. An alternative for two-way grids using rectangular aluminium tubes cross-halved to each other is also shown (C). The tubes are extruded or built up from extruded sections according to the size of the dome.

The use of sheet material by means of which the grid members may be formed, and are strengthened against buckling by the membrane action of the sheets, has been described. One system (D) uses prefabricated diamond-shaped sheet aluminium panels each with a tube cross strut, three of which form a basic hexagon when joined together. When totally fabricated the struts form an

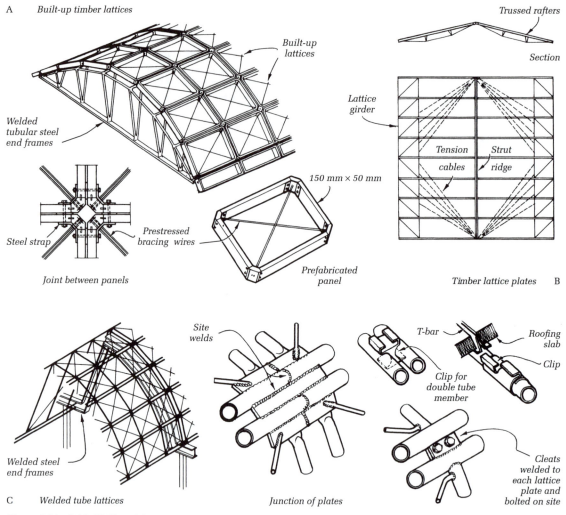

Figure 8.44 Folded lattice plates

external hexagonal grid. A similar but double-layer system is shown in E.

Concrete Grid domes may be formed in concrete in situ either by normal means or by the method shown in figure 8.41 B, for plane grids.

Plastics Aluminium is not the only material which can be used in sheet form for structures of this type. Sheet steel would be satisfactory. But as explained earlier in this chapter, materials which lack great stiffness, such as aluminium or plastics, can most effectively be used in this way. Complete pyramidal hexagonal units with flanged base edges may be formed of plastic, bolted together at the flanges and connected at their nodes to a top grid of tubes

(see figure 8.43). Alternatively the units may be formed of flanged triangular panels bolted together and to node connectors as in figure 8.47 F, to form a geodesic dome.

8.3.13 Tension structures

As indicated earlier in this chapter these fall into two categories, cable network and membrane structures.

Cable network structure Details of this form of construction vary widely. Various means are adopted to connect the cables to supports and common methods are illustrated in figure 8.48 A. The two sets of cables are secured at their crossings either by U-bolts or by connectors designed to fulfil both this function and that of

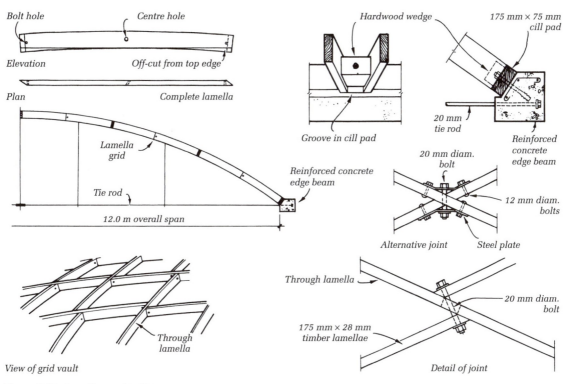

Figure 8.45 Lamella construction

supporting the cladding elements. Both are illustrated in B.

Cable structures can be clad with light metal sheets such as corrugated aluminium, tongued and grooved timber boards, plywood panels, lightweight slabs or panels, 'gunite' concrete sprayed on to expanded metal lathing or with sprayed-on plastic of a suitable type. Some of these methods are shown in figure 8.48 B.

Temperature changes affect the length of the cables and some provision for movement must be made in the cladding, similar to that shown at the joint between the plywood panels.

Membrane structures

Membranes Coated fabrics, plastic films, woven metallic fabrics or metallic foils may be used, the first being most commonly employed. These may consist of a terylene or nylon fabric coated on one or both sides with a plasticised elastomer such as neoprene or vinyl, or of woven glass fibre coated with *Teflon* or silicone which, because of its superior characteristics, is now widely used. Some forms are highly translucent. Single-skin tent membranes are suitable only over areas which do not require heating to habitable standards. It is possible to produce a two-skin membrane with wool or mineral fibre insulation between but its use would result in a loss of translucency. Foil 'cushions' have been used to provide insulation which consist of two foils of plastic held apart by air pressure.

The tubes in air inflated tube structures are normally in the form of a circular woven fabric sleeve with an inner airtight elastomer lining and an outer weather resistant coating of similar material. Small tubes requiring very high pressures necessitate high performance standards for the fabric, joints and sealing, since failure of the tube results in a dangerous explosion.

Jointing methods depend on the membrane material used and consist of sewing through a double folded seam, cementing or heat welding. The first method, although necessitating sealing of the stitches, is the cheapest but produces joints only about 75 per cent as strong as those made by welding or cementing.

Anchorages Direct or positive anchorage to the ground is used for structures which are to be in position for any length of time. The most reliable form for air supported structures is that in which the membrane is secured throughout its perimeter to a continuous concrete foundation.

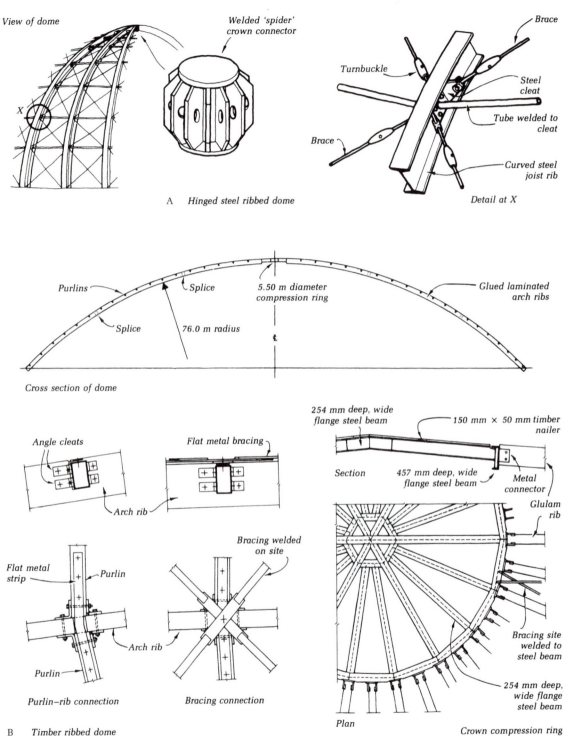

View of dome

Welded 'spider' crown connector

X

A   Hinged steel ribbed dome

Brace

Turnbuckle

Steel cleat

Tube welded to cleat

Brace

Curved steel joist rib

Detail at X

Purlins

Splice

5.50 m diameter compression ring

Glued laminated arch ribs

Splice

76.0 m radius

Cross section of dome

Angle cleats

Flat metal bracing

Arch rib

254 mm deep, wide flange steel beam

150 mm × 50 mm timber nailer

Section   457 mm deep, wide flange steel beam

Metal connector

Glulam rib

Flat metal strip

Purlin

Bracing welded on site

Arch rib

Bracing site welded to steel beam

Purlin

Purlin–rib connection

Bracing connection

254 mm deep, wide flange steel beam

Plan

Crown compression ring

B   Timber ribbed dome

Figure 8.46   Ribbed domes

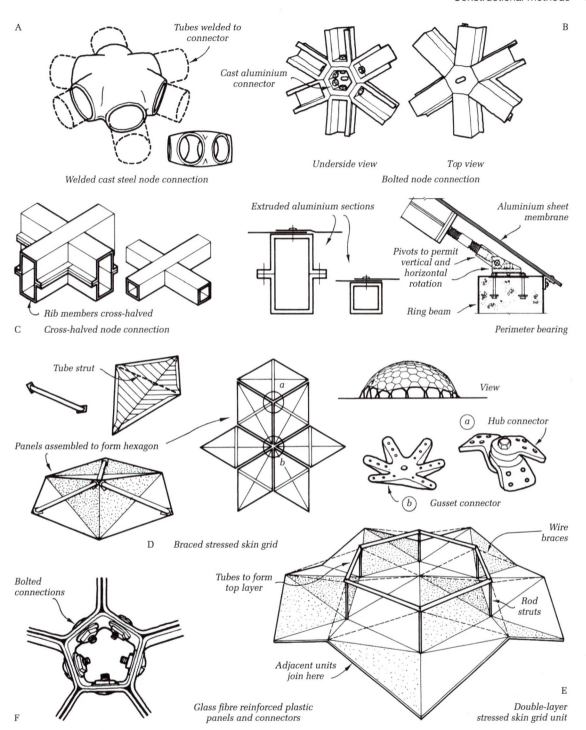

A

Tubes welded to
connector

Cast aluminium
connector

B

Underside view

Top view

Welded cast steel node connection

Bolted node connection

Extruded aluminium sections

Aluminium sheet
membrane

Pivots to permit
vertical and
horizontal
rotation

Rib members cross-halved

Ring beam

C   Cross-halved node connection

Perimeter bearing

Tube strut

View

Panels assembled to form hexagon

a

b

(a) Hub connector

(b) Gusset connector

D   Braced stressed skin grid

Bolted
connections

Tubes to form
top layer

Wire
braces

Rod
struts

Adjacent units
join here

F

Glass fibre reinforced plastic
panels and connectors

E

Double-layer
stressed skin grid unit

Figure 8.47   Grid domes

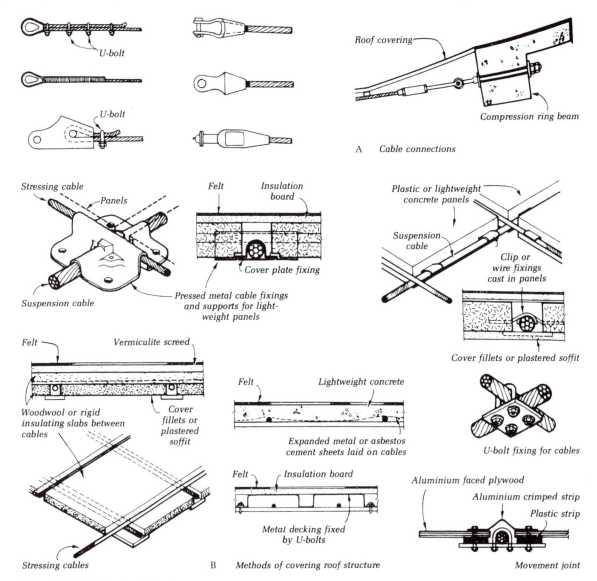

A Cable connections

B Methods of covering roof structure

Figure 8.48 Cable network structures

The bottom edge of the membrane is finished with a rope welt and is clamped to a continuous timber cill piece by coach screws or bolts passing through steel angles or timber battens as shown in figure 8.49 A. The cill piece is rag-bolted at intervals to the concrete strip foundation.

Alternative methods of positive anchorage use pipes or cables, accommodated in fabric sleeves or hems attached to the membrane, which are secured to anchors at intervals round the perimeter. Figure 8.49 B shows the first type in which sections of pipe are inserted in a hem round the base of the membrane and are attached to ground anchors at cut-

outs in the hem at approximately 1 m intervals. The second type is shown in C. In this a rope or cable passes through catenary shaped fabric sleeves sewn to the base of the membrane and is secured at intervals to the ground anchors.

These two forms of anchorage do not require a continuous concrete ground anchor but can be used in conjunction with steel spiral screw anchors (D) or expanding head anchors installed in previously bored holes or bored piles, placed at the appropriate distances apart and of a depth sufficient to provide the necessary resistance to the membrane uplift. These forms of anchors are also suitable for tent

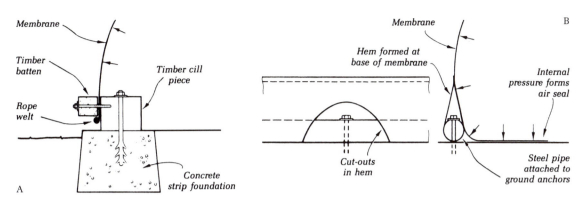

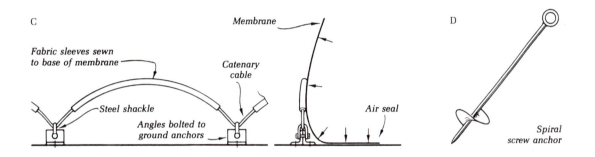

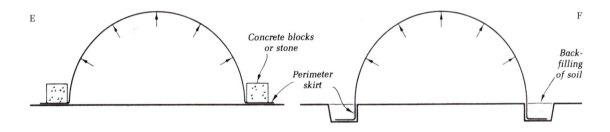

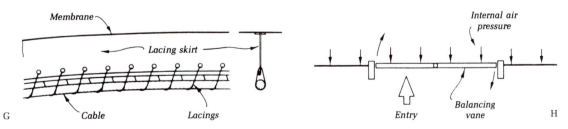

Figure 8.49 Air supported structures

structures. In this latter type of membrane structure jacks on the supporting columns and turntables on the tying-down cables are used to stress the membrane into its final form.

For air supported structures of a temporary nature ballast anchorages may be used. These may be formed in a number of ways: (i) with stones or concrete slabs placed on a perimeter skirt laid flat on the ground (figure 8.49 E), (ii) with soil backfilling on to a skirt placed in a perimeter trench (F), or (iii) with a split ballast tube at the base of the membrane laced up after filling with gravel, sand or soil. Water may also be used as ballast but has the disadvantage that at points of high local lift-up forces it may flow away leaving no anchorage at the point where it is most required.

Cables and ropes   Steel cables or nylon or polyester ropes used to divide air supported membranes into smaller elements (see page 279) can cause chafing and wear on the membrane as the two move relative to each other. This problem can be overcome by placing the cables internally, the membrane being attached to them by dropped skirts from the membrane and lacings as shown in figure 8.49 G. Similarly, the chafing which can be caused by ridge and valley cables in tent structures can be avoided, in the case of valley cables in the manner described here and in the case of ridge cables by placing the cable externally along the line of the ridge and linking the membrane to it by metal connectors. The methods used to terminate cables at

connections to the supporting structure are similar in principle to those used in cable networks (figure 8.48 A).

Access openings   As stated on page 279, large openings in air supported structures require air locks or air curtains. These are not essential for small openings but, in order to overcome the problem of the force required on a normal door to overcome the pressure differential, a door counterbalanced by a vane should be used as shown in figure 8.49 H. The vane, reacting to the same internal pressure as the door, provides a counterbalance and makes opening an easy operation. Where traffic is heavy such a door would result in high air loss because it would be open for too great a length of time. In such circumstances a revolving door should be used. These are not suitable for escape purposes and emergency exits of outward-opening single- or double-leaf doors set in an independent rigid framework should be positioned adjacent to all revolving access doors.

8.4 Movement control

8.4.1 Settlement movement

Where settlement joints are provided in the structure, as described in chapter 4, the joint in the roof slab must be designed to permit relative movement. Methods of forming and weatherproofing the joint will in principle be similar to those adopted for expansion joints (see figure 8.50 D).

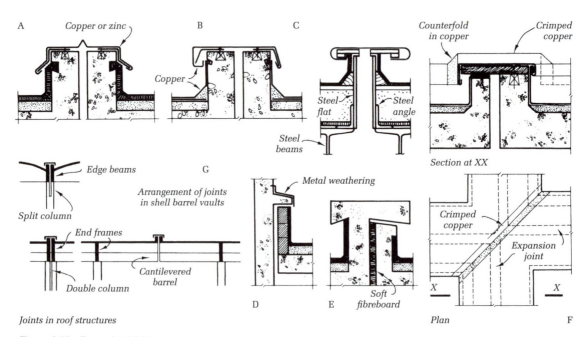

Joints in roof structures

Figure 8.50 Expansion joints

8.4.2 Thermal movement

Thermal insulation placed on top of the roof slab may be sufficient to keep movement to a negligible amount in small buildings, but in large buildings, although movement will be reduced, expansion joints will usually be required. Long parapets should always be provided with expansion joints. Typical details of expansion joints in roofs are shown in figure 8.50 A to E. Extensive roofs will require joints in both directions and a method of forming a four-way junction at an intersection is shown in F.

In multi-bay barrel vault roofs expansion joints are made between edge beams and, in continuous barrels, between end frames using double or split columns as shown in figure 8.50 G. Alternatively, the joint in the length of the barrel can be made between cantilevered sections of the barrel as shown, to avoid the use of double or split columns. Provision for movement in long beams to framed roof structures is made by means of hinged, roller and rocker bearings.

When a long roof slab is supported on loadbearing masonry walls, damage to the walls may be caused by the outward thrust at the ends of the roof unless (i) continuous expansion joints are formed in the roof and walls, or (ii) expansion joints are formed in the roof and a sliding bearing on the wall is incorporated. Cracking in top floor partitions caused by roof movement (figure 4.36 C) can be avoided by isolating the top of the partition from the roof slab. Plaster should not in this case be carried over the junction of walls or partition with the roof slab.[21]

Reference is made in chapter 4, page 187 to the magnitude of thermal movements and to constructional details of expansion joints generally.

Notes

1 See end note 7, chapter 4.
2 See Part 1, chapter 3.
3 See also *MBS: Materials*, chapter 9.
4 See BS 4978: *Specification for Visual Strength Grading of Softwood*.
5 See *MBS: Materials*, chapter 13.
6 The significance of the self- or dead-weight of the structure, especially with increasing span, is referred to in Part 1, section 6.2.
7 This term has reference to the moving by wind of snow from one location to another on a roof, producing high local snow loads due to the accumulation of drifted snow in valleys and against parapets or other obstructions such as chimneys. Guidance is given in BS 6399–3 in determining this loading relative to the shape of the roof and the relationship of its parts and to adjacent buildings and other obstructions. For general purposes, the snow load on roofs of 30° pitch and over is taken as 0.75 kN/m². For roof pitches less than 30°, 1.5 kN/m².
8 See Part 1, page 139, on pressure and suction distribution on roofs and on the effect of suction on lightly clad structures.
9 Described in Part 1, chapter 6.
10 See Part 1, figures 3.41 and 3.44.
11 See Part 1, *Beam action*, page 43.
12 Although in overall terms costs of fabrication may make it cheaper not to vary the depth.
13 See Part 1, page 36.
14 For a more detailed consideration of air stabilised structures see *Air Supported Structures* published by the American Society of Civil Engineers, 1997.
15 See comparison of truss and rigid frame structures on page 255.
16 See also page 251 on the use of timber in this context.
17 See BS EN 10162: *Cold Rolled Steel Sections. Technical Delivery Conditions. Dimensional and Cross Sectional Tolerances*.
18 Further examples of composite beam applications are found in chapter 4.3 of *Construction Technology* by R Chudley and R Greeno, Pearson.
19 See BRE Digest 442, *Architectural Use of Polymer Composites*, and Information Papers IP 10/03, IP 11/03 and IP 5/04 on fibre-reinforced polymers.
20 For definition of this term as used here see page 272
21 See BRE Digest 361, *Why do Buildings Crack?*

9 Fire safety

This chapter first introduces the concept of Fire Safety Engineering and then examines the manner in which fire develops and spreads through a building. The grading of elements of construction and of buildings relative to their reaction to fire is explained and the reaction of various materials is considered together with the classification and use of combustible materials. Factors affecting the design of buildings generally in relation to fire are then outlined followed by an examination of design and construction associated with means of escape and with the limitation of fire spread. The chapter concludes with a consideration of the fire resistance and fire protection of steelwork.

Fire safety is concerned with the protection of the occupants, contents and structure of a building from the risks associated with fire. The subject is of vital importance and is so linked with the design and construction of buildings that it is essential to have an understanding of the factors which influence the nature and form of fire protection and of the principles which are the basis of the various regulations.[1] It is for this reason that it is discussed in this chapter in relatively broad terms, involving as it does considerations of planning as well as of construction, rather than in terms merely of the constructional aspects of protection which, in fact, are conditioned by, and cannot be separated from, planning aspects.

Fires are almost always the result of negligence. The rate of growth, ultimate severity and nature of the risks involved in the event of fire depend largely on the use to which the building is being put. For example, a theatre or concert hall, because of the large number of people accommodated, involves a high life risk even though the combustible content may be low. But a large warehouse storing much combustible material involves a consider-

able risk of extensive damage to structure and contents but a low risk to occupants because their number is likely to be small. The degree of damage finally sustained will be influenced by the structure, its effectiveness in confining the fire and its ability to remain stable both at the seat of the fire and remote from it.

It is the purpose of fire safety, therefore, to protect life, goods and activities within a building. These aims are achieved by inhibiting the combustion of the materials from which the building is constructed and by preventing the spread of fire within the building and between buildings. In addition, by ensuring that the elements of construction fulfil their functions for a sufficient length of time during a fire, the occupants are able to escape and the fire brigade is given time to deal effectively with the outbreak and thus limit the total damage. Protective measures involve suitable forms of construction, suitable planning of the building internally and in relation to adjacent buildings, and satisfactory planning and construction of the means of escape.

Fire safety engineering The requirements of Building Regulations in respect of fire safety and protection have, up to the present time, generally been met by following the guidance given in the Approved Documents relating to the Regulations, especially in Approved Document B, volumes 1 and 2: *Fire Safety*. Volume 1 refers solely to *Dwellinghouses* and volume 2 *Buildings other than dwellinghouses*.[2] More recently, however, another approach has been emerging, now called fire safety engineering, in which each aspect of fire safety and protection in a particular building, rather than being considered in isolation, is considered in relation to the others so that changes in one can have a beneficial effect on others. Indeed, this is pointed to in AD B itself in that it permits, for example, a reduction in fire resistance periods if sprinklers are installed. This document also makes reference to fire safety engineering as an alternative approach and suggests that 'it may be the only practical way to achieve a satisfactory standard of fire safety in some large and complex buildings' and goes on to outline some of the factors which should be considered and some of the measures which could be taken in adopting a fire safety engineering approach to satisfy the Building Regulations.

This approach takes into account such matters as the anticipated risk of a fire occurring and its likely severity and rate of growth, means of detection and warning of fire, the degree to which the spread of fire and smoke will be controlled and limited, the likely danger to people, means of escape and speed of evacuation, the fire resistance of the structural materials, the installation of active measures for fire protection such as sprinklers and smoke detectors and the provision of good maintenance and staff training during the life of the building.

Computerised quantitative methods are being used in relation to many of these considerations, but as yet fire safety engineering is not an accurate science and until recently there has been little guidance on calculations in this field. However, the British Standards Institution has produced the published document PD 7974: *Application of Fire Safety Engineering Principles to the Design of Buildings*, in seven parts. The Building Research Establishment has also published numerous papers on methodology, calculation methods and data which enable studies in this area to be carried out with greater accuracy. Furthermore, the Loss Prevention Certification Board, in conjunction with BRE Certification Ltd, have published their own Loss Prevention Standards for construction. These tend to be favoured by building insurers, as many aspects of fire protection and safety exceed those of the Building Regulations.

In the field of steelwork a Code of Practice for the fire resistant design of steel construction, BS 5950-8, permits a fire safety engineering approach (see also page 334).

9.1 Growth of fire in buildings

The origin of a fire is usually the result of negligence, ranging from direct acts such as lighted cigarettes left burning to more indirect causes such as poor installation and maintenance of electrical wiring and the ineffective control of vermin.

The growth of a fire depends on the amount and disposition of combustible material within the building, either in the form of unfixed materials or parts of the fabric such as wall and ceiling linings, which will contribute to the fire.[3] At the beginning of a fire materials near the source of ignition receive heat and their temperatures rise until, at a certain point, inflammable gases are given off. Ignition of the material then occurs and it begins to produce heat instead of merely receiving it. This 'primary' fire then preheats the remaining combustible material and raises its temperature to ignition point. After this a spark or, for example, a flame burning along a floor board, will in a moment start an intense fire involving the whole of the contents. The time taken for this instantaneous spread of the fire, or 'flash-over' as it is termed, to occur depends on the proximity of combustible materials to the source of the fire and on the presence or otherwise of an adequate supply of oxygen. In full-scale tests on domestic living rooms this 'flash-over' occurred after about 15 minutes. At an early stage the structure at the seat of the fire becomes involved and the strength of some non-combustible materials such as steel, for example, is adversely affected.[4]

The further spread of fire occurs by the usual methods of heat propagation, that is, by conduction, convection and radiation, together with the process of flame spread across the surfaces of combustible materials.

Depending on the construction and design of the building, heat can be conducted through walls causing temperatures high enough to ignite spontaneously any combustible materials stacked on the side remote from the fire.

Any unprotected flue-like apertures such as stair-wells, lift shafts, light-wells or wall cavities may permit the spread of fire by the passage of convection currents and flying brands.

Heat can be radiated to ignite combustible materials at some distance from the source. In this way fire is spread to other areas as well as to nearby buildings.

The spread of fire is more rapid where combustible linings are used, and particularly where air spaces exist behind them as this provides two surfaces over which flame can spread.[5]

The later stages in the progress of a fire are mostly affected by the behaviour of the particular structure involved. The deformation of supporting columns and beams in a framed building structure may cause apertures in enclosing walls and floors through which direct flame spread can occur. Ultimately, the complete collapse of a building may happen as a result of the weakening of the structure and the gutting of all but the non-combustible parts such as brick walls.

9.1.1 Hazards associated with fire

These may be considered in order of importance as:

Personal: the hazard to the occupants of the building
Damage: the hazard to the structure and contents
Exposure: the hazard due to the spread of fire to other buildings.[6]

The hazards to building occupants are due to the following factors:

Reduction of oxygen This is due to the consumption of oxygen by the fire and is accompanied by toxic or asphyxiating gases evolved by the fire, particularly carbon monoxide. An associated additional hazard is smoke which results from incomplete combustion. Staircases providing means of escape and corridors giving access to them, where exposed to any particular risk, should be protected by fire-resisting partitions and self-closing fire-resisting doors. Whenever possible staircases should be ventilated to the open air in all storeys. See page 326, *Smoke control*.

Increase in temperature Breathing is difficult above a temperature of 149 °C, and since this temperature will be reached well in advance of the path of the fire, it is essential that automatic alarms be provided, designed to operate at given temperatures (49 °C to 70 °C).[7]

Spread of flame The risk here is of burning by physical contact with flame and should be minimised by enclosing escape routes with non-combustible materials, or materials of low flame spread.

The degree of hazard to occupants is influenced by the distance between points of escape, size and number of exits and stairs, and the existence or otherwise of a sprinkler system.

The practical requirements in dealing with these hazards are discussed in more detail in later sections of this chapter.

9.2 Fire-grading

The term 'fire-grading' has a two-fold application: (i) it is applied to the classification or grading of the elements of structure of buildings in terms of their degree of resistance to fire; and (ii) with a broader meaning, it is applied to the classification of buildings according to the purpose for which they are used, that is, according to occupancy. This grading of buildings is considered from two points of view, firstly in terms of damage and exposure hazard, for which the protection is mainly provided by structural precautions, and secondly in terms of personal hazard, for which protection is provided primarily by easy means of escape. The first is considered here and the second later under the sections on fire escape.

The severity of a fire depends largely upon the amount, nature and distribution of combustible material in a building. Thus in determining the requisite degree of fire protection it is necessary to take into account the use and size of a given building, and by this means to assess the probable amount and type of combustible material which would contribute to a fire.

9.2.1 Fire load grading

The assessment of the severity of a fire due to the combustible materials in a building is made by reference to what is known as the 'fire load', which is the amount of heat, expressed in kJ, which would be generated per square metre of floor area of a compartment of the building by the complete combustion of its contents and any combustible parts of the building. The fire load is determined by multiplying the weight of all the combustible materials by their calorific values and dividing by the area of the floor; it is based on the assumption that the materials are uniformly distributed over the whole area of the floor.

The calorific value is the property of a material which indicates the amount of heat which will be generated by a particular quantity of that material and it governs the ultimate severity of a fire. Thus the maximum heat is evolved from materials having highest calorific values. The process for establishing calorific values is specified in BS EN ISO 1716: *Reaction to Fire Tests for Building Products. Determination of the Heat of Combustion*. Some examples include:

bitumen	35 355 kJ/kg	petrol	46 520 kJ/kg
cork	16 747 kJ/kg	rubber	39 542 kJ/kg
paper	16 262 kJ/kg	wood	18 508 kJ/kg

An office, therefore, with 2.5 kg/m^2 of combustible (wood) furniture and papers would have a fire load of 406.55 to 462.70 MJ/m^2.

The fire load is used as a means of grading occupancies, as shown in table 9.1. The numerical grading of 1, 2

Table 9.1 Fire load grading

Grade of occupancy	Low fire load MJ/m^2	Moderate fire load MJ/m^2	High fire load MJ/m^2
Fireload MJ/m^2	Not exceeding 1150	1150 to 2300	2300 to 4600
Building types	Flats, offices, restaurants, hotels, hospitals, schools, museums, public libraries	Retail shops, e.g. footwear, clothing, furniture, groceries Factories and workshops generally	Warehouses, etc., used for bulk storage of materials of non-hazardous nature
Fire resistance	1	2	4

and 4 corresponds with the minimum fire resistance in hours required for protection of structural elements.[8]

Materials of the same calorific value can give rise to differences in fire risk according to ease of ignition, rate of burning and whether or not, for example, they are explosive or emit dangerous fumes. These would constitute exceptional risks. In addition, certain processes such as paint spraying with inflammable materials, or the application of heat to combustible materials also constitute exceptional risks and will contribute significantly to the fire growth rate.[9] Any building in which these risks are likely to arise or in which the fire load is greater than the maximum for High Fire Load grading must be considered separately.

Work on the special problems associated with large retail premises has led to the concept of a 'largest likely fire' where there is a sprinkler system. This fire is of 5 megawatt heat output and occupies an area of 3 m × 3 m in size, otherwise expressed as a heat release rate of 550 kW/m². Figures for other premises include offices 290 kW/m², hotel room 250 kW/m² and industrial premises 260 kW/m². This criterion can also be used as the basis for calculation when designing smoke control systems in shopping malls.[10] (See also *Smoke control*, page 326.)

9.2.2 Fire resistance grading

It has been stated earlier in this chapter that protective measures against fire include means for limiting the spread of fire together with the provision of structural elements capable of fulfilling their functions during a fire without the risk of collapse. The first, to some extent,[11] and the second entirely, depend upon the use of elements of structure[12] of an appropriate degree of fire resistance.

Fire resistance   The term 'fire resistance' used in connection with fire protection may be defined as 'the ability of a component or construction of a building to satisfy for a stated period of time some or all of the criteria specified in BS 476: Parts 20, 21 and 22, specifically integrity and insulation'.[13] It is applied to elements of structure, not to a material. Fire resistance depends on the way in which materials are used in an element and not solely on whether they are combustible or not.

The *fire resistance grading* of elements of structure is determined by tests carried out in accordance with this Standard by the Loss Prevention Certification Board and Building Research Establishment (LPCB/BRE Certification Ltd). The tests are applied to elements of structure whether composed of one or more materials and the elements are graded according to the length of time during which, while exposed to the test fire, they satisfy the conditions laid down in the Standard.

It must be understood that the grading is applied to the structural element as a whole and to a precise specification of that element. What may appear to be minor changes in the details of construction of an element may result in a great change in its fire resistance.

Measurements have been taken of the severity of fires caused by different fire loads. This information has enabled the degrees of severity of fires due to known fire loads to be expressed in terms of periods of exposure to the standard test fire. For design purposes the equivalent severities shown in table 9.1 were adopted by the Joint Committee on Fire Grading of Buildings.[14] This means that if a building is to contain a fire load of say 1400 MJ/m² then walls, floors and other elements of construction having a fire resistance of 2 hours would resist the effects of fire without collapse or penetration of the fire even if all the material within it burned.

The determination of the necessary fire resistance of every structure on the assumption that it must withstand the complete burn-out of the contents would pay no regard to the effect of other means of protection, such as fire fighting and automatic sprinklers, for example, and would render impracticable many sound forms of construction and make the cost of the structure high. As indicated in the section on fire safety engineering on page 312, lesser degrees of resistance relative to fire load become practicable, by assuming a rational combination of all methods of fire protection and by having regard to the fact that in smaller buildings escape and fire fighting are easier, and fires can more easily be brought under control with less chance of the structure collapsing.

In practice, therefore, the necessary fire resistance of a building (or any part of it if subdivided by suitable walls and/or floors into compartments) in any particular case is determined with reference to types of users, fire load, the areas of floors and the height and cubic capacity of the building. Building Regulations express the varying characteristics of differing building types by designating 'purpose groups' to which fire-resisting requirements are related.[15]

The notional periods of fire resistance ascribed to forms of construction of many kinds are given in BRE report 128, *Guidelines for the Construction of Fire Resisting Structural Elements*. Some guidance is given in table 9.2. The special case of achieving fire resistance in structural steelwork is dealt with in section 9.5.4.

9.3 Materials in relation to fire

9.3.1 Effect of fire on materials and structures

Non-combustible materials   These are materials which if decomposed by heat will do so endo-thermically, that is

Table 9.2 Guidance on periods of fire resistance for some common forms of construction

Floors: *timber joist*

Half hour	1		At least 21 mm tongued and grooved (t&g) boarding or sheets of plywood or wood chipboard, floor joists at least 38 mm wide at 600 mm maximum spacing
			Ceiling 12.5 mm plasterboard* with at least 2 mm neat gypsum plaster finish
1 hour	2		At least 21 mm t&g plywood or wood chipboard, floor joists at least 50 mm wide
			Ceiling Not less than 30 mm plasterboard* in two layers, joints staggered and taped with 2 mm gypsum plaster finish

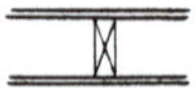

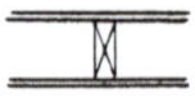

Floors: *concrete*

1 hour	3	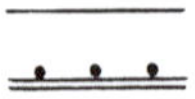	Reinforced concrete floor not less than 95 mm thick, with not less than 20 mm cover on the lowest reinforcement

Walls: *internal*

half hour loadbearing	4		Timber framing members at least 44 mm wide† and spaced at not more than 600 mm apart, with lining (both sides) of 12.5 mm plasterboard* with all joints taped and filled
	5		75 mm reinforced concrete wall** with minimum cover to reinforcement of 25 mm
1 hour loadbearing	6	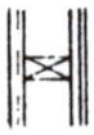	Framing members at least 44 mm wide† and spaced at not more than 600 mm apart, with lining (both sides) at least 25 mm plasterboard* in 2 layers with joints staggered and exposed joints taped and filled
	7		Solid masonry wall (with or without plaster finish) at least 90 mm thick (75 mm if non-loadbearing)
			Note: For masonry cavity walls, the fire resistance may be taken as that for a single wall of the same construction, whichever leaf is exposed to fire
	8		120 mm reinforced concrete wall** with at least 25 mm cover to the reinfocement

Walls: *external*

Half-hour loadbearing wall less than 1 m from the relevant boundary	9		100 mm brickwork or blockwork external face (with or without a plywood backing); timber framing members at least 38 mm wide and spaced not more than 600 mm apart with 10 mm plywood sheathing
			Internal lining 12.5 mm plasterboard* with at least 10 mm lightweight gypsum plaster finish
1 hour loadbearing wall less than 1 m from the relevant boundary	10		Solid masonry wall (with or without plaster finish) at least 90 mm thick (75 mm if non-loadbearing)
			Note: For masonry cavity walls, the fire resistance may be taken as that for a single wall of the same construction, whichever leaf is exposed to fire, e.g. 2 hours = 100 mm, 4 hours = 200 mm

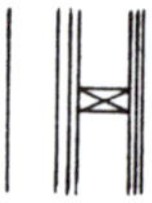

Reinforced Concrete: *columns*

Free standing	11	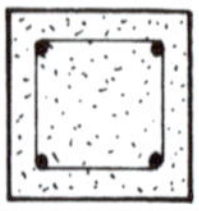	Minimum lateral dimension: 300 mm = 2 hour fire resistance 450 mm = 4 hour fire resistance Reinforcement cover = 35 mm minimum

Table 9.2 (Cont'd)

Reinforced Concrete: *columns (cont'd)*

| Attached to compartment wall | 12 | 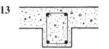 | Maximum 50% of column sides exposed
Minimum exposed lateral dimension:
    200 mm = 2 hour fire resistance‡
    350 mm = 4 hour fire resistance
Reinforcement cover = 25 mm minimum‡
otherwise = 25 mm minimum |

Reinforced concrete: *beams*

| Integrated with 'fire break'
    floor slab (2 hour fire
    resistance minimum) | 13 | | Beam width:
    150 mm minimum = 2 hour fire resistance
    240 mm minimum = 4 hour fire resistance
Concrete cover to reinforcement:
    50 mm = 2 hour fire resistance
    70 mm = 4 hour fire resistance |

Notes
* Whatever the lining material, it is important to use a method of fixing that the manufacturer says would be needed to achieve the particular performance. For example, if the lining is plasterboard the fixings should be at 150 mm centres as follows (where two layers are being used, each should be fixed separately):
9.5 mm thickness   −30 mm galvanised nails
12.5 mm thickness   −40 mm galvanised nails
19 mm–25 mm thickness   −60 mm galvanised nails
† Thinner framing members such as 38 mm or light steel framing may be suitable depending on the loading conditions.
** Depending on the concrete quality, may provide a greater fire resistance than indicated.
‡ Not 'fire break'. A 'fire break' wall is provided between compartments and has a fire resistance of at least 4 hours. A 'fire break' floor has a fire resistance of at least 2 hours. The term 'fire break' is used by insurers to describe elements of separation or compartments.

The forms of construction shown in this table are only a selection of representative examples. Many other forms, using materials other than those indicated in the table, are capable of providing a required fire resistance. Specialist manufacturers' catalogues should be consulted.

with the absorption of heat, or, if they oxidise, do so with negligible evolution of heat. Also included are those materials which require a temperature beyond the range of most fires before they react in any way.

Non-combustible materials do not contribute to the growth of a fire but are damaged when the temperature is reached where decomposition, fusion or significant loss of strength occurs. When incorporated in the structure the loss of strength during a fire may be such that they no longer maintain the integrity of the structure. Examples of non-combustible materials are metal, stone, glass, concrete, clay products, gypsum products and some fibre cement products.[16]

Apart from marble and gypsum, which liberate free lime under severe heat, the majority of these materials do not decompose chemically under the action of fire. However, certain natural stones, concrete, gypsum and asbestos products (prohibited in new buildings, but exist in many older structures) decompose by losing their water of crystallisation, and in so doing acquire pronounced fire endurance. Although metal and glass suffer negligible

decomposition these materials lose considerable strength at high temperatures, and glass and some metals such as aluminium and lead fuse or soften under heat.

Steel It should be noted that steel loses strength and rigidity above a temperature of 299 °C and at 427 to 482 °C, a temperature well within the normal range of building fires, there is a loss of strength of up to 80 per cent. This potential weakness, together with the expansion which takes place under heat, means that in the early stages of a fire unprotected steelwork will bend, buckle and expand. Deformation of a supporting steel member or frame causes walls and floors to fall away and leaves the fire free to spread into other areas which might not have been affected had the structure been maintained. This is an example which underlines the fact that non-combustible materials are not necessarily fire-resisting and this may be further emphasised by considering timber, which although combustible will, if of adequate section, fulfil its structural function longer than mild steel.

Aluminium   The poor performance of aluminium structures in this respect should also be noted. Aluminium has a much lower critical temperature than steel and for elements under load this is about 200 °C as against 500 °C for steel. The potential weakness and disadvantages of steel relative to the normal temperature range of building fires referred to above are, therefore, more pronounced in the case of aluminium.

Concrete   The behaviour of concrete under action of fire depends largely upon the type of aggregate used. Flint gravel expands greatly and causes spalling of the concrete. Other stones, apart from limestone, behave similarly to a lesser degree, but crushed clay brick and slag do not cause spalling. Provided spalling does not occur, disintegration is slow. Thus additional protection can be given to reinforcement or steel members by increasing the concrete cover so long as non-spalling concrete is used. If a spalling type is used the cover tends to break and fall away.

In terms of fire resistance, aggregates are often classified as follows, those in *Class 1*, which have been exposed to high temperatures in their manufacture and/or are of lower density, being non-spalling types:

Class 1   Foamed slag, pumice, blast furnace slag, crushed brick, burnt clay products, well burned clinker, crushed limestone, expanded slag, expanded clay, sintered pulverised fuel ash and pelleted fly ash, etc.

Class 2   Flint, gravel, granite and all crushed natural stones other than limestone.[17]

At temperatures higher than those required to produce spalling if free lime is present in the cement it is converted into quicklime after which, if the concrete is exposed to water, or even moist air, the lime slakes and in expanding causes complete disintegration of the concrete.

Combustible materials   These are materials which, within the temperature range associated with fires, will combine exothermically with oxygen. That is to say, in their reaction with oxygen considerable heat is evolved and they flame or glow.

Such materials, whether forming part of the structure or the contents of the building, are responsible for the growth of a fire and its ultimate severity. Examples of such materials are wood or wood products, vegetable products, animal products and manufactured products such as fibreboard and strawboard. Within this classification are flammable materials which ignite readily and react vigorously, producing rapid flame spread. Examples of these materials are volatile liquids (petroleum distillates), certain plastics and certain paints based on nitrocellulose products.

Timber   Although timber is a combustible material it will, as pointed out earlier, function as a structural member for a relatively longer period than one of metal provided it is of adequate section. If of sufficient size timber is extremely difficult to burn. Some species such as teak, iroko, jarrah and others are highly resistant to fire.

Wood boarding up to 9.5 mm thick ignites relatively easily and continues to burn. But timbers about 150 mm thick will char in depth and this inhibits rapid combustion of the wood beneath. Tests have indicated that for any given cross-sectional area and load, a beam having a square cross section will have the longest fire endurance, that is, the time elapsing before collapse occurs.

The known rates of charring of different timbers, established by BS 476 furnace tests, permit the residual sections of unburnt timber remaining after a fire to be predicted for structural members. As there is little significant loss of strength in the unburnt timber the initial size of structural members can be calculated so that they will continue to support their design loads after exposure to fire for different periods of time. BS 5268-4: *Structural use of Timber. Fire Resistance of Timber Structures* gives guidance on the method of calculation. Reference should also be made to *MBS: Materials*, chapter 1 where this subject is also covered and tables of charring rates are given.[18]

The degree of combustibility of timber can be reduced by brush or spray treatment with proprietary fire-retardant paints. Pressure impregnation gives better results than brush applications. Timber should be worked to finished sizes before impregnation as treated timber is difficult to work.

Plastics   The fire properties of plastics vary widely. Although they usually require higher temperatures than most natural materials before ignition occurs they have higher calorific values and, therefore, a much greater heat output weight for weight than, for example, wood.

Plastics generally generate large quantities of smoke and toxic gases. Thermoplastics soften at temperatures ranging from 75 °C to 150 °C and will melt during burning, giving rise to possible spread of fire by flaming droplets. Most thermosetting plastics char like timber at temperatures above 400 °C and they burn at 700 °C to 900 °C. Further information on plastics generally and on their behaviour in fire is to be found in *MBS: Materials*, chapter 13.

The widely varying nature and properties of plastics precludes the possibility of classifying them in terms of total fire hazard and this presents the designer with serious problems in specifying them for safe use in buildings. Much research has, however, been carried out on their behaviour in fire and many papers and reviews on the subject are available and reference should be made to these.[19]

The following characteristics of combustible materials will influence the precautions necessary in providing adequate protection.

Ignitability   A measure of the ease of ignition expressed as the minimum temperature at which the material ignites under given atmospheric conditions, e.g. wood, wood products and cellulose materials 221 to 298 °C, plastics 260 to 482 °C, synthetics of nitro-cellulose origin upwards of 138 °C, bitumen upwards of 65 °C, petrol distillates 204 to 482 °C. The method of test for ignitability is specified in BS EN ISO 11925-2: *Reaction to Fire Tests. Ignitability of Building Products Subjected to Direct Impingement of Flame.*

Flammability   The property of a combustible material which determines the severity of flame and flame spread. It is related to volatility and the vigour with which volatile gases react with oxygen. In addition to actual volatile liquids which may be stored within a building, many organic natural and synthetic materials exhibit this property, particularly when well dispersed as in fabrics.

Calorific content   The rate at which heat is generated in the reaction between all combustible materials and oxygen is related directly to the flammability of a material, and the ease with which the reaction occurs is related to the ignitability of the material.

It will be obvious from what has been said regarding the effect of fire upon materials and structures that in the design of buildings it is essential to have some means of assessing the likely behaviour in fire of various materials and forms of construction, particularly at a time when new materials and combinations of new materials to form constructional elements are constantly being introduced. BS 476, as already indicated, provides means for grading or classifying the structural elements of a building for this purpose. It also provides means for doing the same for materials. The different parts of this Standard specify the nature and details of the various tests to be used in classifying materials under different characteristics and reference should be made to the many appropriate parts to establish the coverage of each.

9.3.2 Classification of combustible materials

Some materials are obviously combustible. Others are not obviously so because they burn so slowly. Yet in a fire they could contribute to its severity. Further, wall and ceiling linings, which present large surfaces, provide an easy means for the spread of fire when constructed of combustible materials. Particularly this is so where air spaces or cavities exist behind such linings, so permitting the

rapid and undetected spread of fire. For practical design purposes some method is required to determine whether or not a material is combustible, and also the ease with which flame is likely to spread over its surface. BS 476 specifies two tests in this connection: (i) combustibility; and (ii) surface spread of flame. The first, carried out in accordance with Part 4, *Non-combustibility Test for Materials*, indicates whether or not a material will burn or contribute to a fire, although no degree of combustibility is defined; the second, carried out in accordance with Part 7, *Method of Test to Determine the Classification of the Surface Spread of Flame of Products*, compares the rate and extent of flame spread along the surfaces of different materials and classifies the results as follows:

Class 1   Surfaces of very low flame spread
Class 2   Surfaces of low flame spread
Class 3   Surfaces of medium flame spread
Class 4   Surfaces of rapid flame spread.
  It should be noted that materials in this class are unacceptable as wall and ceiling linings under the provisions of AD B as indicated in table 9.3, which is reproduced from this document.

There is also a *Class 0*. In performance terms this supersedes *Class 1* and, although the BS tests apply to wall

Table 9.3   Spread of flame classifications for internal surfaces

Location	European class*	BS 476-6 class
Small rooms of area not more than 4 m² in a residential building and 30 m² in a non-residential building Domestic garages of area not more than 40 m²	D-s3, d2	3
Other rooms (including garages)	C-s3, d2	1
Circulation spaces within dwellings		
Other circulation spaces, including the common areas of blocks of flats	B-s3, d2	AD B class 0

It is permitted for part of the wall surfaces in rooms to be of a lower class than that given in this table (but not lower than Class 3) provided the total lower-classed area in any one room does not exceed one half of the floor area of the room subject to a maximum of 20 m² in a residential building and 60 m² in a non-residential building.
* BS EN 13501-1: *Fire Classification of Construction Products and Building Elements. Classification Using Test Data from Reaction to Fire Tests.*

The European classifications given are not necessarily equivalent to the British national standards unless separately tested to satisfy these.

and ceiling linings, *Class 0* is not a BS classification. It is defined as a separate classification in AD B, appendix A:

Class 0 (a) is composed of non-combustible material or materials of limited combustibility,[20] or
  (b) has a surface material of *Class 1* which has an index of performance (I) of not more than 12 and a sub-index (i_1) of not more than 6 in accordance with BS 476-6. The face of any thermoplastic material should only be regarded as *Class 0* surface if either:
  (i) the material is bonded throughout to a non-thermoplastic substrate and the material and substrate together comply with (b) immediately above; or
  (ii) the material complies with (b) above and is used as a lining to a non-thermoplastic *Class 0* surface.[21]

The results of these tests and the results of fire resistance tests under Parts 21 and 22 should not be confused. A material which satisfies the combustibility test might fail under the conditions for fire resistance test; a building-board having Class 1 flame spread classification may or may not make a significant contribution to the fire resistance of a composite structure in which it is used. Table 9.4 shows the surface spread of flame classifications of a number of common lining materials. Test results for proprietary materials may be obtained from manufacturers' promotional literature, trade associations, The Building Research Establishment and The Fire Protection Association.

It will be noticed that the classification of a material may be altered by the application of intumescent paint or fire-retardant solution, but too great a dependence should not be placed on these treatments. Some materials achieving a particular surface classification in this way may,

Table 9.4 Surface spread of flame classification

Class 0	*Class 2*
Mineral fibre tiles with cement or resin binding	Synthetic resin bonded paper and fabric sheets
PVC/steel laminates	Standard hardboard with certain decorative treatments
600 680 720 density flame retardant polyurethane finished flooring grade wood particle board	Compressed straw slabs with painted distemper finish
PVC faced fibre cement board	
Any material of limited combustibility	
Brickwork, blockwork, concrete and ceramic tiles	
Plasterboard	
Woodwool cement slabs	

Class 1	*Class 3*
Metal faced (including all edges) plywood	Timber and plywood density more than 400 kg/m³
Flame retardant hardboard and medium hardboard	Wood particle board
Flame retardant insulation board	Compressed straw slabs with manilla or impregnated cardboard covering
Flame retardant hardboard with wood grain veneering	Glass fibre reinforced sheet polyester
Standard hardboard with intumescent paint	Standard hardboard
Compressed straw slabs with surface treatment of vermiculite and wood chippings	Medium hardboard
Compressed straw slabs plastered	Fibre insulating board with certain decorative treatments
Hardwood with surface impregnation treatments	
Glass fibre reinforced plastic sheets to BS 4154-1 with flame retardant additives	
Resin bonded chipboard with proprietary finishes	
600 680 720 flame retardant wood particle board	
Melamine impregnated wood veneer laminate (can be class 1 on order)	
Melamine faced plastic laminate – flame retardant grade	*Class 4*
Melamine faced hardboard	
PVC faced fibre cement board	Plywood and timber density less than 400 kg/m³
Flameproofed decorative veneers on plywood backings	Fibre insulating boards
Composite boards of urethane foam/faced both sides with plasterboard	Compressed straw slabs with one or both faces embossed with PVC co-polymer foil 0.25 mm thickness
PVC/steel laminates	Acrylic sheets (polymethyl methacrylate)

in fact, under certain conditions aid the development of a fire to a greater extent than those achieving the same classification without treatment. This has been shown by the results of fire tests in rooms fully lined with various materials.

These tests also indicate that in terms of aiding the development of fire there is not a great difference between this type of *Class 1* lining and *Class 3* linings. In view of this, therefore, when *Class 1* linings must be used it is probably advisable to limit the choice to those achieving this class without surface treatment, as is required by the former Government *Department of Education and Science, Building Bulletin* no. 7, 'Fire and the Design of Educational Buildings' (The Stationery Office). This is now superseded and published by the Department for Education and Skills as Building Bulletin 100: *Designing and Managing Against the Risk of Fire in Schools*. Further, the behaviour of these paints and solutions during the use of a building is not fully known, and it is possible that their effectiveness may be reduced by heat and condensation, for example. For this reason insurance companies prefer inherently low surface flame spread materials if non-combustible alternatives are out of the question for reasons of economy, and this is reflected in the premiums charged.

The fact that materials of similar surface spread of flame classification can, under the same conditions, assist the growth of fire to a different extent has resulted in the development of the Fire Propagation Test as specified in BS 476-6.

Furthermore, certain types of insulating materials such as foamed plastics which melt at temperatures lower than those used in the surface spread of flame test are difficult to classify by that test and the fire hazard of such materials may better be determined by the Fire Propagation Test which is more discriminating amongst materials of lower fire hazard than the surface spread of flame test.

It has already been pointed out that different materials of the same calorific value can give rise to differences in fire risk according to their other characteristics, one of which is ease of ignition. This is tested in accordance with BS EN ISO 11925-2.

9.3.3 The use of combustible materials

As has previously been mentioned, combustible linings are potentially more dangerous when there are air spaces behind them. This condition arises when these materials are used as a wall lining on battens, as a cladding to timber or metal studs in partitions, as false ceilings to mask beams and services or in the construction of raised flooring. It often occurs in light factory-type buildings where the inner insulating lining is separated from the outer cladding by the structural framing supporting both (figure

9.1 A). The hazard is particularly great in the case of roofs, since the lateral spread of flame causes the main fire to spread by radiation and by the fall of burning pieces of the lining itself.

Linings Linings should therefore be of at least *Class 1* spread of flame and non-combustible or underdrawn with a non-combustible sheeting if combustible, as at B. To avoid continuous air spaces, the lining should be placed directly below the roof covering, and cavity barriers or non-combustible infillings provided where corrugated or other profiled sheeting is used, as at C. Further cavity barriers can be formed in roof spaces by enclosing roof trusses with fire-resisting building boards or by infillings at suitable intervals of fire-resistant textiles (see figure 9.1 E and G). Where a decking is used, combustible insulating linings should be placed above the deck and below the roof covering to avoid the creation of air spaces, as at D. However, advantage should be taken of non-combustible linings by placing them below the deck in order to protect it from fire below. To avoid continuous air spaces above false ceilings formed with combustible linings, the lining should be placed directly below any downstand beams tight against the soffits and adequate cavity barriers should be installed between the ceiling and floor above in the plane of all walls and partitions below (see figure 9.2 J).

An example of good all-round performance in this situation is that of woodwool slab decking, which is self-spanning, of good insulation value and *Class 1* spread of flame classification. See AD J, section 8 for other acceptable materials. Subject to specific requirements, the minimum provision is for cavity barriers to have a 30 minute fire resistance, spaced at a maximum of 20 m in any direction.

Roof coverings In addition to the fire hazard of insulating lining materials, roof coverings may contribute to a fire. Although regulations require non-combustible coverings generally, they permit the use of combustible coverings on a non-combustible base or in circumstances where the building in question is well away from any adjacent property. Combustible coverings used are mastic asphalt, bitumen felts of both mineral and vegetable base and bitumen protected metal. Mastic asphalt can be considered as of low fire risk due to the large proportion of inert material in it, but bitumen burns easily and should only be used directly on a non-combustible and fire-resistant decking or layer, since collapse of the deck would spread fire into the building from outside and fire to the roof from the inside. In the form of protected corrugated metal sheeting, bitumen coverings are particularly hazardous when combined with combustible linings. One of the chief factors contributing to the spread of fire with this type of construction

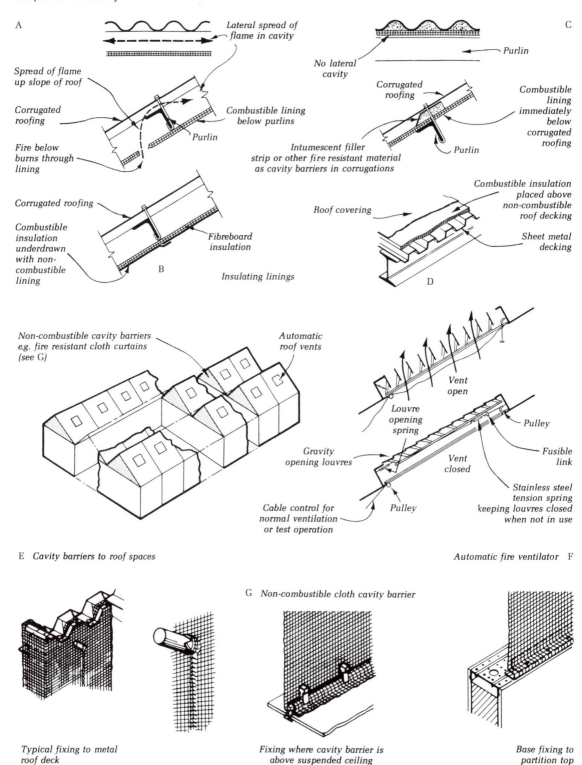

A

Lateral spread of flame in cavity

C

Spread of flame up slope of roof

Corrugated roofing

Purlin

No lateral cavity

Corrugated roofing

Combustible lining immediately below corrugated roofing

Fire below burns through lining

Combustible lining below purlins

Purlin

Intumescent filler strip or other fire resistant material as cavity barriers in corrugations

Corrugated roofing

Combustible insulation placed above non-combustible roof decking

Roof covering

Combustible insulation underdrawn with non-combustible lining

Fibreboard insulation

Sheet metal decking

B

Insulating linings

D

Non-combustible cavity barriers e.g. fire resistant cloth curtains (see G)

Automatic roof vents

Vent open

Louvre opening spring

Gravity opening louvres

Pulley

Vent closed

Fusible link

Stainless steel tension spring keeping louvres closed when not in use

Cable control for normal ventilation or test operation

Pulley

E   *Cavity barriers to roof spaces*

Automatic fire ventilator   F

G   *Non-combustible cloth cavity barrier*

Typical fixing to metal roof deck

Fixing where cavity barrier is above suspended ceiling

Base fixing to partition top

Figure 9.1   Fire protection to single-storey buildings

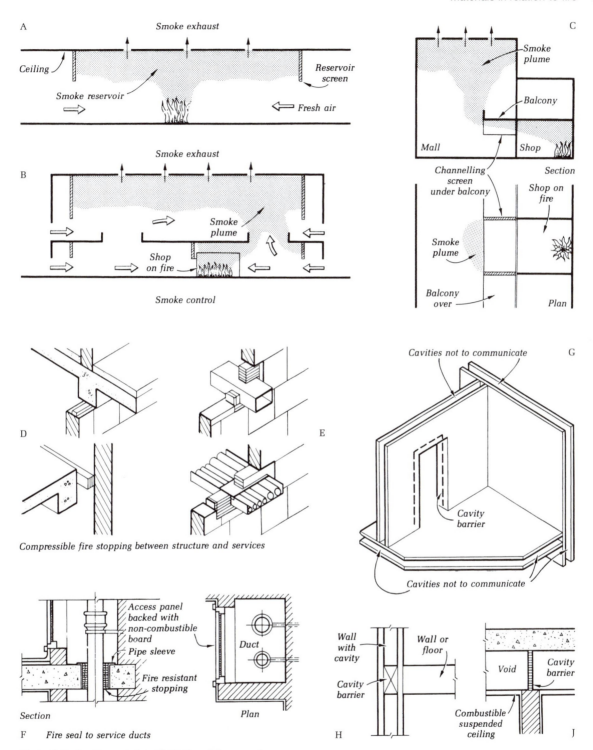

Figure 9.2 Smoke control and limitation of fire spread

Table 9.5 Notional designations of roof coverings

Pitched roofs covered with slates or tiles

Covering material		Supporting structure		BS 476-3 designation	BS EN 13501-5 classification
1	Natural slates	**1**	Timber rafters with or without underfelt, sarking, boarding, woodwool slabs, compressed straw slabs, plywood, wood chipboard, or fibre insulating board	AA	$B_{ROOF}(t_4)$
2	Fibre-cement slates				
3	Clay tiles				
4	Concrete tiles				
5	Bitumen felt strip slates *Type 2E*, with underlayer of bitumen felt *Type 2B*	**2**	Timber rafters and boarding, plywood, woodwool slabs, wood chipboard, or fibre insulating board	BB	$C_{ROOF}(t_4)$
6	Strip slates of bitumen felt *Class 1* or *2*	**3**	Timber rafters and boarding, plywood, woodwool slabs, compressed straw slabs, wood chipboard, or fibre insulating board	CC	$D_{ROOF}(t_4)$

Any reference in this table to bitumen felt of a specified type is a reference to bitumen felt as so designated in BS 747: *Reinforced Bitumen Sheets for Roofing. Specification*. See also BS EN 13707: *Flexible Sheets for Waterproofing. Reinforced Bitumen Sheets for Roof Waterproofing. Definitions and Characteristics*; BS 476-3: *Fire Tests of Building Materials and Structures. Classification and Method of Test for External Fire Exposure to Roofs*; and BS EN 13501-5: *Fire Classification of Construction Products and Building Elements. Classification Using Data from External Fire Exposure to Roof Tests*.

is the fall of flaming bitumen and lining material into the building well in advance of the main fire.

In addition to wood wool already mentioned, the following insulating materials may be used without fire hazard: plasterboard, vermiculite, glass fibre and mineral wool (the last two without a bitumen binder or paper covering).

BS 476-3 specifies a test to assess the capacity of a roof construction to prevent the penetration of fire from outside and the degree to which the covering will spread fire. The type of construction permitted for a roof depends on the purpose group and size of the building and its distance from the boundary. Types of construction are specified by the two letter designations from BS 476-3. The first letter refers to flame penetration:

A Not penetrated within 1 hour
B Penetrated in not less than $^1/_2$ hour
C Penetrated in less than $^1/_2$ hour
D Penetrated in preliminary flame test

The second letter refers to the surface spread of flame test:

A No spread of flame
B Not more than 533 mm spread
C More than 533 mm spread
D Continuing to burn for 5 minutes after withdrawal of the test flame, or with a spread of more than 381 mm across the region of burning in the preliminary test.

Table A5 of AD B designates pitched and flat roof coverings according to the materials used and their supporting structure (see table 9.5 which gives part of that table) and

uses these designations in setting the limits of proximity of a building to the site boundaries.[22]

Roof lights   These may be constructed of rigid thermoplastic materials which may be used over any space other than a protected shaft/stairway (see page 333). Any internal linings or surfaces exposed between the light transmitting part and the ceiling below must comply with the flame spread designation of the ceiling. Limitations on the area of such roof lights, their distance apart and their distance from relevant boundaries, having regard to the classification of the upper and lower surfaces of the plastic and the use of the spaces below, are laid down in the Building Regulations.[23]

Suspended ceilings   Suspended ceilings in circulation spaces or rooms are permitted to contain one or more panels of plastic material and tables 2 (volume 1) and 11 (volume 2) in AD B set limits on the size and disposition of these, which are the same as for plastic roof lights. This concession does not apply where the ceiling provides fire protection to the structure, unless the plastic elements have been satisfactorily tested as part of a fire protecting ceiling assembly.

9.4 Factors affecting design

9.4.1 General design and planning

The broad approach to planning for fire safety is to design the elements of construction to withstand the action of fire for a given period dependent on the size and use of the building, to compartmentalise the building so as to isolate the fire within a given section or area, to separate specific risks within the building and generally to prevent the uncontrolled spread of fire from its source to other parts of the building. Further, a building must be planned to allow the occupants to escape by their own unaided efforts. Suitable separation must be provided to prevent fire, hot gases and smoke from spreading rapidly by means of common spaces such as corridors, staircases and lift shafts, thus trapping the occupants and causing panic.

Compartmentation and separation   The structural elements used to prevent horizontal and vertical spread of fire are the walls and floors, together with any structure necessary to support them, all of which must be of adequate fire resistance.[24]

Any openings within these elements of separation must be protected in a manner which does not nullify the effect of that element during a fire, whilst affording access during normal use. Such protection may be given by self-closing fire-resisting doors or steel doors or roller shutters held open by fusible links and arranged to close automatically in the event of fire.[25] Glazing must be fire resistant where light is required to penetrate the separating elements.

Compartment walls and floors   As already indicated in chapter 3, section 3.8, buildings are sub-divided into compartments of restricted floor area or cubic capacity by fire-resisting walls and floors in order to prevent rapid and extensive spread of fire within the building. Table 12 in AD B (volume 2) shows the maximum permitted dimensions for buildings or compartments according to the use and height of the building.

In addition to restricting the spread of fire, compartmentation has advantages relative to fire fighting. It has been found in practice that where fires occur in large undivided spaces within a building, the greater intensity of heat and volume of smoke generated prevent fire fighters from attacking the fire at its source. They are driven out of the building, where their hoses become less effective. It is generally considered that 7000 m³ is the maximum volume of any one compartment which could reasonably be tolerated from a fire-fighting standpoint. This figure will vary in practice according to the use to which a building is put and according to such circumstances as the availability of fire-fighting services and the provision of automatic sprinklers.

Compartment walls and floors must have the appropriate fire resistance applicable to the purpose group and height of the building under consideration. In certain circumstances compartment walls and floors in hospitals are required to be constructed of materials of limited combustibility[26] if they have a fire resistance of 1 hour or more.

Junctions of compartment walls and floors with external walls or other compartment walls should be bonded or fire-stopped to maintain the fire resistance of the compartment. Junctions with the underside of roof coverings should be fire-stopped to prevent the spread of fire over the wall in the event of fire penetrating the roof near the wall head. AD B, sections 5 (volume 1) and 8 (volume 2), lays down specific requirements for the roof covering over a distance of 1.5 m on each side of the wall. Alternative to these requirements the wall may be taken up at least 375 mm above the surface of the roof.

The only openings permitted in compartment walls or floors are one or more of the following: an opening for a fire door fitted in accordance with Appendix B, AD B or for a protected shaft or a refuse chute of non-combustible construction and openings for pipes, ventilation ducts, chimneys, appliance ventilation ducts or ducts encasing one or more flue pipes complying with the provisions of AD B, sections 7 (volume 1) and 10 (volume 2), which deals with the protection of such openings.

Suspended ceilings, if of appropriate construction and surface lining, can contribute to the fire resistance of a compartment floor. The requirements of the Building Regulations in this respect are given in AD B, sections 3 (volume 1) and 6 (volume 2). If such a ceiling is required to have a fire resistance of 1 hour or more it, and any insulation which may be above it, should be of a material of limited combustibility.

Separating walls   A form of compartmentation that has been defined in chapter 3, section 3.8. They must restrict the spread of fire from one building to another including spread via their roof spaces. All such walls must be constructed as compartment walls except that the only openings permitted are those for an escape door having the same fire resistance as the wall but not less than 1 hour, fitted in accordance with Appendix B, AD B, and those for the passage of a pipe.

External walls   These are required to restrict the spread of fire across site boundaries to adjacent buildings. Since fire spread between buildings occurs mostly by radiation through openings and *unprotected areas*[27] the risk of fire spread is related to:

● the severity of the fire – itself a function of the combustibility of the contents
● the fire resistance of the external walls
● the number and position of any openings and other unprotected areas
● the distance between the external walls and the site boundaries.

The minimum permitted distance between the side of a building and its relevant boundary depends upon the purpose group of the building and the total extent of unprotected areas. Certain small openings or other unprotected areas are considered to present a negligible risk of fire spread and may be disregarded in assessing the separation distance from the boundary. The extent and relationship of these are indicated in AD B, sections 9 (volume 1) and 13 (volume 2) and, provided they are the only unprotected areas in a wall which is fire resistant from both sides, the wall may be within 1 m of the relevant boundary or actually coincident with the boundary. If the separation distance is sufficiently great, and this varies with the type of building, the unprotected area may be 100 per cent of the wall area, in which case any walling need have no fire resistance. Methods for determining the acceptable unprotected area in walls 1 m or more from a site boundary are given in AD B, sections 9 (volume 1) and 13 (volume 2). Alternatively, other methods described in BRE Report 187, *External Fire Spread: Building Separation and Boundary Distances*, 1991, may be used for this purpose.

External walls, as well as requiring sufficient fire resistance to restrict fire spread across a site boundary, must have the combustibility of the outer surfaces restricted in order to reduce the danger of ignition from an external source and consequent fire spread up the external face of the building. The Building Regulations (AD B, sections 8 (volume 1) and 12 (volume 2)) set limits on the amount of combustible material permitted on the external wall surface relative to the height of the building, its proximity to site boundaries, the type of building and the spread of flame classification of the particular material. Similar limits apply also to the inner surface of the cladding in rainscreen construction.

In buildings with a storey more than 18 m above ground level any insulation material used in the construction of an external wall should be of limited combustibility. An exception is where the wall contains two leaves of masonry, the cavity around openings is closed, the wall cavity and roof interface is closed or the cavity is fully filled with insulation.

Shafts and ducts Flue-like apertures such as shafts, ducts and deep lightwells should be avoided in the general design of buildings, but where they are necessary, as in the case of lift shafts and staircase enclosures, they should be vented at the top to allow smoke and hot gases to disperse to the atmosphere (see page 329).

Smoke control The large single-storey shed-type building housing continuous factory processes, which cannot easily be subdivided, presents a particular problem. In such buildings the unconfined spread of smoke and carbon monoxide fumes resulting from incomplete combustion after the initial supply of oxygen has been consumed constitutes a hazard to fire fighters and prevents them from reaching the seat of the fire. Also, when trapped in a building, the heat generated by the fire causes high temperatures and renders materials more flammable by preheating them well in advance of an approaching conflagration.

Automatic ventilation In view of the above it is desirable to make provision for the removal of heat, smoke and fumes as quickly as possible by a simple automatic self-operating means of ventilation. Although this supplies more air and possibly intensifies the fire it will, nevertheless, confine it and produce less smoke thus assisting fire fighters to see and approach nearer the seat of the fire. It has been shown by experience that to be effective, such automatic ventilation must be above the fire because cross-ventilation may only serve to drive heat and smoke in a particular direction, possibly towards the fire fighters. By providing automatic fire vents in the roof, therefore, smoke, heat and fumes are enabled to rise quickly out of the building and draughts are created which draw air towards the fire, thus helping to contain it.[28] The effectiveness of roof vents is enhanced by subdividing the roof space above the ties of lattice trusses into compartments by fire-resisting curtains or non-combustible board cavity barriers fixed within the roof space (figure 9.1 E, G). They should extend down to at least the level of the tie members and automatic fire vents should be placed within the bays thus created. Automatic vents may be in the form of opening roof lights, the sashes of which open automatically in the event of fire, or in the form of special roof vents such as shown at *F*, which also permit the requirements of normal ventilation to be fulfilled.

The area of vents required in any particular situation depends on such factors as the fire risk and the height and area of the building in question.[29]

Smoke control systems In the case of buildings containing large undivided volumes such as warehouses, shopping malls and, possibly, atria buildings, contemporary design centres around the provision of smoke ventilation control systems in which the principles underlying the use of vents and fire curtains referred to above are developed. The intention is to keep the smoke in the upper reaches of the building leaving clean air near the floor to allow people to move freely. The smoke, mixed with air entrained with the hot gases, rises to be contained within a smoke reservoir near the ceiling formed by the walls, screens or other features (figure 9.2 A), from which removal to atmosphere is achieved either by natural buoyancy or by mechanical means. Sufficient air must be allowed to enter the building below the smoke layer to replace the extracted gases for the system to work. One of the factors to be considered in the design of smoke ventilation systems is that of inducing the smoke to flow in such patterns that an acceptable smoke plume is achieved. Figure 9.2 B, C show differing patterns of smoke movement and ways in which the smoke plume may be channelled towards the appointed places of extraction.[30]

Smoke and fumes from basement fires tend to rise up stairways making access difficult for the fire fighters. For this reason the provision of mechanical smoke extraction or an appropriate area of natural smoke vents with any related outlet ducts or shafts enclosed by non-combustible fire-resisting construction is required by the Building Regulations (see AD B, section 17 (volume 2)).

Smoke removal from fire escape stairs and lift shafts is also necessary and reference to this is made on this page and page 329.

In addition to the specific sections in AD B devoted to fire-resisting construction it should be noted that sections in other Approved Documents contain requirements which affect the construction of a building in relation to protection against fire. These are those sections concerned with the stability of walls (AD A), chimneys, fireplaces and heat producing appliances (AD J).

Access for fire fighting When buildings are of considerable height and area they must be carefully sited to provide access for the heaviest fire-fighting appliances to approach close to the building. The extent and siting of the access depends on the size and height of the building and whether or not 'dry' or 'wet' rising mains are provided.[31]

Access routes designed to take 12.5 tonnes in relation to a normal pump appliance are considered to be capable of carrying without damage the much greater weight of high reach appliances because of the infrequent use for this purpose and of the fact that the greater weight is distributed over a number of axles. Structures such as bridges should have 17 tonnes carrying capacity. Minimum required widths of access routes and gateways and diameters of turning circles are given in AD B, sections 11 (volume 1) and 16 (volume 2).

In the case of tall buildings external access by ladders can only serve a limited purpose, since floors above 30 m are beyond the reach of most fire service ladders. Further, when a tall building is designed on a 'podium' of much larger area two or three storeys high, no external access for fire appliances is possible for the upper floors, so that internal access must be designed as an integral part of the building. This is provided in the form of at least one lobby approach staircase, regarded for fire-fighting purposes as an extension of the street. The staircase must be separated from the accommodation on each floor by an enclosed lobby in which fire mains and all the necessary fire-

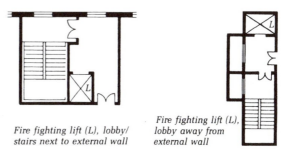

Fire fighting lift (L), lobby/ stairs next to external wall

Fire fighting lift (L), lobby away from external wall

Figure 9.3   Staircases, lobbies and lifts for fire fighting

fighting equipment is installed (see figure 9.3). Both staircase and lobby should be sited next to an external wall and be provided with adequate ventilation to ensure freedom from smoke logging. If this is not possible they should be properly ventilated, by mechanical means if necessary (see this page and page 328). In AD B, section 17 (volume 2) these protected shafts are called 'firefighting shafts' and their number and location depend on the floor area of the largest storey and its height above ground level and whether or not a sprinkler system is installed.

In addition to such a staircase a firefighting lift must be provided off the lobby, as in figure 9.3, in buildings with a floor 18 m or more above or with a basement 10 m or more below fire vehicle access level. The enclosing walls to staircase, lift and lobby should have the fire resistance appropriate to the compartment and external walls of the building according to its purpose group and height. Openings giving access to the building must be provided with self-closing fire-resisting doors.[32]

9.4.2 Means of escape

The object in providing means of escape is to permit unobstructed egress from within a building by way of definite escape routes (exit ways, corridors and stairs) to a street or an open space or to an adjoining building or roof from which access to the street may be obtained.

The primary danger to occupants is that staircases and corridors leading to them from rooms or compartments may be filled with hot gases and smoke, trapping the occupants and causing panic. It is, therefore, important that such escape routes should be enclosed with adequate fire-resisting enclosures. In addition they should be separated by fire-resisting doors planned in strategic positions to prevent the spread of smoke and fire from storey to storey via lift shafts and staircases, and to keep the latter free from smoke when used as an escape route.

Smoke removal from staircases is normally by permanent ventilation or an automatic mechanical system of smoke control at each storey or landing level. Where this is not practicable a system of pressurising the staircase against smoke entry may be used.

Air pressurisation   A properly designed system to pressurise a staircase continuously or as activated by smoke detectors, can keep it free of smoke so long as the doors are kept shut. Pressurisation can overcome adverse pressures developed in a fire and those due to adverse weather conditions and it increases the fire resistance time for doors. Short period opening of doors does not worsen the smoke conditions in the staircase. Escape routes leading to pressurised staircases should be well sealed to minimise loss of pressure.

The air for pressurisation is supplied preferably by a combination of conventional ventilation system and pressurisation system or by individual fans provided to force air into each escape circulation section which, if not located next to an exterior wall, would have short lengths of duct to connect each fan directly with the atmosphere.[33] Systems using long lengths of ducting are not recommended as there is danger of circulating smoke if a main duct is damaged, allowing smoke to penetrate and affect the whole installation. Furthermore the ducts would have to have equal fire resistance to that of the escape routes they serve.

Air flow control   Another system which has been shown to work effectively is the *Automatic Air Flow Control System*.[34] This system involves the use of smoke detectors which are connected to the retaining mechanisms of top heavy centre pivot hung windows which open inwards. When the windows are freed in response to the smoke detectors they swing in at the top to form large louvres which allow external air to enter at the bottoms of the openings and circulate from floor to ceiling and to leave at the tops of the openings, thus clearing smoke rapidly and assisting fire fighters.

The requirements of the Building Regulations in respect of means of escape vary with the size, construction, use and height of the building in question. These requirements, relating to escape routes and stairs and their enclosure and to the number and proximity to each other of escape exits based on maximum permitted travel distances, are dealt with in AD B. The requirements for dwelling houses and flats are given specifically and the requirements common to buildings other than dwellings are given generally, together with references to various parts of BS 5588: *Fire Precautions in the Design, Construction and Use of Buildings* as alternative approaches to the provision of means of escape in assembly buildings, shops, shopping complexes and offices. Means of escape in school buildings should be in accordance with the Department for Education and Skills, Building Bulletin

100: *Designing and Managing Against the Risk of Fire in Schools* and in hospitals and other residential care buildings in accordance with the Department of Health 'Firecode' documents, e.g. Health Technical Memorandum (HTM) 05-02, *Guidance in Support of Functional Provisions for Healthcare Premises*. Guidance on means of escape for disabled persons is given is BS 5588-8: *Code of Practice for Means of Escape for Disabled People*.[35]

9.5 Fire-resisting construction

9.5.1 Design and construction associated with means of escape

Staircases and ramps   Staircases and ramps which form part of the structure of a building are required to provide safe passage for users. The provisions described in AD K: *Protection from Falling, Collision and Impact* apply to all stairs apart from those of limited rise (see Part 1, page 207). Stairs forming part of a means of escape may need to meet requirements additional to those given in this particular Approved Document. For example, minimum widths are required based on the number of people likely to use the stair, the type of building which the stair serves and the number of escape stairs to be provided. When the width has to be more than 1800 mm the stair should be provided with a central handrail, the width on each side then being considered separately in assessing the stair capacity. If the stair rises more than 30 m it is considered to be less hazardous in use if the width does not exceed 1400 mm unless it is made at least 1800 mm with a central handrail.

An escape stair and its landings should be constructed of materials of limited combustibility if it serves a basement or any storey more than 18 m above ground level; if it is external (unless it only connects a ground floor with a flat roof or a floor not more than 6 m above or below ground level); if it is the only stair serving the building (but excluding a two- or three-storey flat or maisonette block); if, as already indicated, it is a fire-fighting stair.

In buildings with more than one escape route available from a storey one of the routes may be via an external stair within certain limits of height of stair[36] and user, provided it is adequately protected from fire in the building by adjacent fire-resisting external walls and self-closing fire-resisting access doors (see AD B, sections 2 (volume 1) and 5 (volume 2)).

Reference should be made to Part 1 for appropriate dimensions for staircases, handrails and guardings and to chapter 7 of this volume for constructional details.

Enclosures   The enclosures to protected stairs, landings, corridors, passages, lobbies and doorway recesses within the staircase enclosures and exits from staircases should have a standard of fire resistance of not less than one half-hour except where the requirements for loadbearing elements of structure, compartment walls and other parts of the building referred to in AD B may necessitate a higher standard. This approved document permits the unrestricted use of glazed elements in fire-resisting enclosures provided they satisfy the criteria of both integrity and insulation (see note 12, page 341). Limitations are imposed on the use of glazing which does not satisfy that of insulation.

Doors Doors on escape routes should generally open in the direction of escape clear of steps (other than a single step on the line of the doorway) and must do so if the number of persons likely to use them is more than 60.[37] Doors opening toward an escape route should be recessed so that when open they do not encroach on landings, corridors or the public way. In assembly buildings of any type exit doors and gates should be of two leaves of equal width.

Doors on escape routes which are hung to swing both ways or where they sub-divide corridors must have a panel of clear glazing at sight level (see above concerning the type of glazing). Where buildings are frequented by disabled persons these glazed panels must give a zone of visibility from a height of 500 mm up to 1500 mm above floor level.

Revolving doors, automatic doors and turnstiles should not be provided on escape routes unless side-hinged doors of adequate width and opening in the direction of escape are provided immediately adjacent to the revolving doors with an indication that they are the means of fire exit.

Sliding doors may be permitted on escape routes not normally used by more than 20 persons (such as in factories and warehouses) and must be marked to indicate the direction of opening. Wicket doors should be provided in large sliding doors[38] with an indication that they are the means of fire exit.

Doors on escape routes should never be locked during occupation of the premises and any fastenings should be such that the doors are readily openable. The fastenings should, therefore, be simple and quickly operated manually without the use of a key. In places of assembly doors should be kept free of fastenings other than panic bolts or other approved fastenings.[39]

AD B, Appendix B, volume 2 (*Buildings other than Dwellinghouses*) requires all fire doors to be fitted with an automatic self-closing device.[40] Volume 1 (*Dwellinghouses*) only requires a self-closing device on a fire door between a dwelling house and its integral garage. The minimum fire resistance required for fire doors in various locations in a building is given in AD B, Appendix B. Provision for a protected stairway with fire doors generally applies in dwellinghouses and flats that have a floor over

4.5 m above ground level. Section 2 of AD B should be consulted for guidance on specific situations.

Notwithstanding the necessity to prevent fire spread by closure of fire doors, the need for access by the rescue services and the application of reasonable force to open a door by escapers including those limited by disabilities must be incorporated in the provision.

Windows For purposes of escape and rescue from dwellings and flats a reasonable number of windows on floors up to first floor (and up to second floor if this is a loft conversion to a dwelling), facing a street or open space to which fire appliances have access, should be made to open at cill level. Such windows should have an unobstructed opening at least 0.33 m^2 in area and not less than 450 mm high and 450 mm wide, the bottom of which should be not more than 1100 mm above the floor. These limits apply also to dormer and roof windows except that openings in the roof slope may go down to 600 mm above the floor. Alternatively, direct access should be provided to a fire protected stairway.

Lifts Lifts, whilst not considered a fire hazard if properly installed, are not considered safe as part of a means of escape and must never be used for the evacuation of persons under fire conditions because of the danger of being trapped should the electricity supply fail.

Lift wells, if located so as to prejudice a means of escape, should be enclosed with fire-resisting construction or be contained within a protected stairway/shaft, that is, enclosed with fire-resisting walls.[41]

Lift motor rooms should, preferably, be located at the top of lift shafts and should be enclosed by and separated from the shaft by fire-resisting construction. Where the lift is located within a protected stairway which is the only stair serving the building, and the motor room cannot be at the top, it must be located outside the stairway rather than at the bottom of the shaft in order to avoid smoke rising up the stairway from a fire in the motor room.

A smoke outlet not less than 0.1 m^2 in area opening direct to the open air should be provided near the top of a lift shaft, enclosed with solid construction and protected with an open metal grill or widely spaced louvres.[42]

9.5.2 Design and construction associated with the limitation of the spread of fire

Walls The function of walls in providing fire protection and the regulations relating to them have been discussed already. Traditionally, fire-resisting walls were constructed of non-combustible materials. However, when distances from boundaries are sufficiently great and when, as in the case of the Department for Education and Skills,

Building Bulletin 100: *Designing and Managing Against the Risk of Fire in Schools*, the requirements for escape are stringent, the use of combustible materials and light claddings is permitted. In some cases, however, the use of light external claddings and, in particular, curtain walling, which are functionally appropriate to framed buildings, present a number of practical difficulties.

Curtain walling   The constructional aspects of curtain walling have been considered in chapter 3. In considering the system from the point of view of fire protection it is necessary to have regard to the requirements of building regulations as far as external walls are concerned. These have been discussed on pages 325–6 and it will be clear that any system of frame and panel walling such as the curtain wall must act as a whole in fulfilling these requirements, but at the same time be light and economical of space if it is to be successful.

As indicated on page 326, if the separation distance between an external wall and its relevant boundary is sufficiently great and other dimensional criteria and building function are satisfied (AD B, sections 8 (volume 1) and 12 (volume 2)), the unprotected area of the wall may be 100 per cent of the wall area, in which case any curtain walling used need have no fire resistance. Where the separation distance requires a reduction in the percentage of unprotected area this might be achieved by non-loadbearing spandrel panels of fire-resisting construction attached to the structure. This can be done, firstly, by building up off the edge of the structural floor an independent non-combustible back-up wall of appropriate fire resistance extending up to cill level or, secondly, by incorporating an appropriate fire-resisting standard panel within the curtain wall thickness but secured directly to the structure by, for example, mild steel angles connecting it to the floor edge. In the event of failure of the curtain walling this panel would continue to function. These are illustrated in figure 9.4 A and B respectively.

Alternatively, a fire-resisting curtain wall system may be employed. Since the framing is non-loadbearing the system need only satisfy the criteria of integrity and insulation in order to prevent the penetration of fire and the transfer of excessive heat. Using steel framing, laminated glass (see this page) and fireproof silicone seals at junctions of panels and framing, fire ratings of up to two hours can be attained.

The following points should be borne in mind when selecting or designing a curtain wall system of this type: (i) the panel and the frame must act together as one fire-resisting element and be tested as a whole under BS 476; (ii) fixings must be protected from fire and should not conduct heat to a vulnerable material. They should be designed to allow for exceptional expansion in a fire.

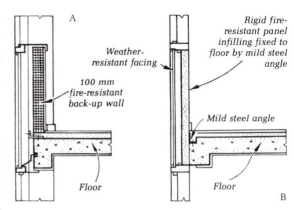

Figure 9.4   Curtain walling

Openings in walls   Doors set in fire-resisting walls must be fire-resisting in order to maintain the integrity of the walls. Such doors, together with doors used to resist the passage of smoke and fumes, are called *fire doors*.

A fire door and its frame form one unit or element and fire tests are made on this unit. Timber doors require deep rebates or the use of intumescent strips to prevent flame penetration between the door and frame. A steel frame is necessary if a steel or composite door is to develop its full fire resistance.[43]

Fire doors are described as *smoke control, fire check* or *fire-resisting*. The first serves to resist the passage of smoke and fumes, but not an actual fire. The last two are classified according to their *stability* period, that is the time elapsing until the specimen collapses, and their *integrity* period, that is the time elapsing until such cracks or other openings develop which permit the passage of flames or smoke, or until the unexposed face begins to flame. For any given fire resistance period both provide the same degree of stability, the last a higher degree of integrity. These doors are described in greater detail in *MBS: Internal Components*, chapter 6.

Glazed openings may be formed in fire doors with the same limits as for enclosures (see page 329). A construction such as that shown in figure 9.5 A can achieve a half-hour fire resistance and that shown in B one hour if the metal frame has a melting point not lower than 900 °C. A wired 'safety' glass incorporating heavier gauge wire can achieve two hours fire resistance.

The use of glazing based on laminations of clear sheet glass and an intumescent material permits longer periods to be achieved: under tests in accordance with BS 476: Parts 20 and 23 some systems of laminated glass set in suitable frames have attained up to 2 hours fire rating (see figure 9.5 C and D). Fire-resisting glazing must, in most cases, be fixed shut.

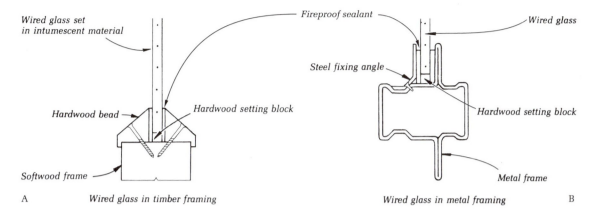

A   Wired glass in timber framing

Wired glass set in intumescent material

Fireproof sealant

Wired glass

Steel fixing angle

Hardwood bead

Hardwood setting block

Hardwood setting block

Softwood frame

Metal frame

Wired glass in metal framing   B

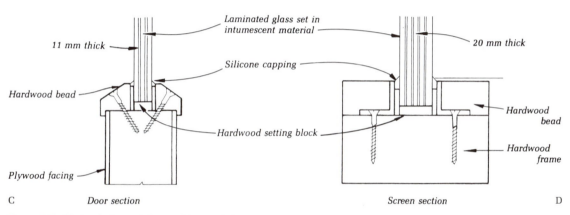

11 mm thick

Laminated glass set in intumescent material

20 mm thick

Silicone capping

Hardwood bead

Hardwood setting block

Hardwood bead

Hardwood frame

Plywood facing

C   Door section

Screen section   D

Figure 9.5   Glazing in fire-resisting partitions and doors

Glass blocks built into a wall if properly set in recesses and reinforced with light wire mesh every third horizontal joint, attain half-hour fire resistance with panels limited to 2.4 m maximum width or height.

Where a fire resistance of more than 1 hour is required doors are usually of steel or steel and fire-resisting fibre board or, on occasion, steel-encased timber. Uninsulated iron and steel doors and shutters are effective in preventing fire spread for up to 2 hours although they transmit a lot of radiant heat. Doors and shutters fitted on each side of an opening are accredited with double the fire resistance of one door only. Any doors provided in addition to steel shutters must be arranged so that they do not interfere with the normal operation of the shutters whether open or closed. Doors should not be placed between double shutters.

Typical details of steel doors and shutters are shown in figure 9.6. These embody the requirements of the Fire Offices Committee[44] regarding construction, together with the main requirements regarding the openings in which they are set.

Automatic fire barriers   Automatic fire and smoke barriers are required where compartment walls are penetrated by conveyors or ducts. These are generally of the 'garrotting' type comprising a fire door which is released by the action of heat on a fusible link or of smoke on a smoke detector to fall across the track of the conveyor or across the duct. These doors, or dampers, may have to be double in some cases. Mechanical dampers with moving shutters or diaphragms must be positioned to afford easy access for maintenance of both dampers and their associated detectors.

Intumescent honeycomb dampers set within ducts will reduce maintenance problems but will allow the passage of low temperature gases in the early stages of the development of a fire and may not be appropriate in all circumstances.

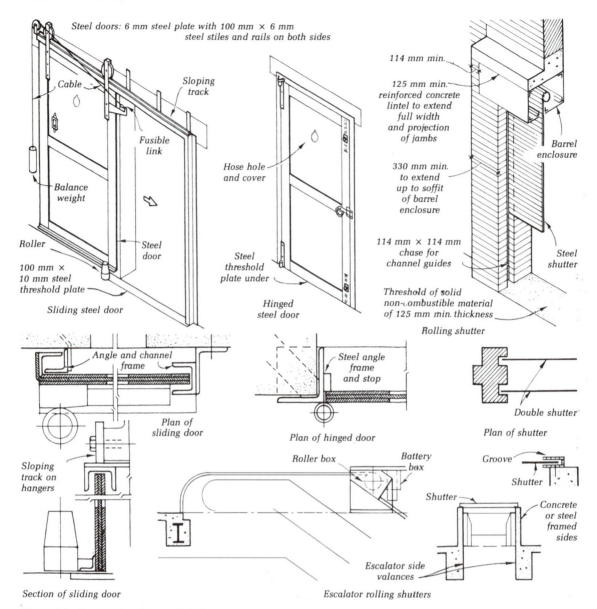

Figure 9.6 Fire-resisting doors and shutters

As an alternative to these methods protection may be provided by a tunnel incorporating a water drencher system automatically operated in the event of fire.

Where a pipe, ventilation duct or flue passes through an opening in a compartment wall or floor or a protected shaft (except an opening wholly enclosed within a protected shaft) or through a cavity barrier the penetration must be sealed. AD B, sections 7 (volume 1) and 10 (volume 2), specifies maximum permitted diameters of pipes passing through such openings and the means of fire-stopping or sealing[45] to be used. Figure 9.2 D, E show examples of intumescent seals which may be used for these purposes.

Upgrading timber floors The work of rehabilitating and altering existing buildings which have timber floors frequently requires improvement in the fire resistance of the latter. This may be achieved by the addition of protection to the underside of the ceiling, over the existing floor boarding or between the joists of the floor, and often by quite simple means. For example, a floor which achieves only a half-hour rating because the ceiling is of only 12.5 mm plasterboard may, by the simple addition of 19 mm standard plasterboard nailed to the underside, be brought up to a full hour rating. The actual materials and combinations of materials to be used for this purpose depend upon the nature of the floor and ceiling construction and the period of fire resistance required. Various types of plaster and insulating board may be used for ceiling protection; high density hardboard and plywood for applying to existing floor boards; fibre insulating material or lightweight plaster supported on mesh or expanded metal for between joist protection. The last method is especially useful where the ceiling is of historic architectural value and must not be disturbed or covered over. Where it is desired to leave the joists exposed the added protection must be either on top of the floor boarding or between the joists in the form of plaster or insulating board. The joists in this case must, of course, be of such a size that their structural stability is ensured even after charring has occurred in a fire (see page 318).[46]

Openings in floors   Apart from lifts and staircases which are enclosed throughout their height, openings in floors may occur where vertical services communicate with successive storeys.

By their nature, building services communicate with different parts of the building and, therefore, provide potential routes of fire spread as well as contributing to the propagation of fire if made of low-melting point thermoplastics such as drainage fittings, or incorporating combustible materials as in externally insulated ductwork.[47] Where such services are contained in ducts the latter should be sealed off at each floor level with a non-combustible filling to give a fire resistance equal to that of the floor. This prevents them acting as flues to spread fire vertically and may be accomplished by arranging for the floors to close the vertical shaft, openings of the correct dimension to accommodate services being left when the floor is formed. Pipes passing through the slabs can be encircled by a ferrule set in the floor and must be sealed with suitable firestopping material (figure 9.2 F). The method depicted can be expected to maintain both integrity and insulation for 6 hours in gaps of 15 mm, 4 hours in gaps of 100 mm and 2 hours in gaps of up to 860 mm. The duct enclosures are built off the floor slabs.

As an alternative to this method the opening in the floor slab may be the full area of the duct. Subsequent to the installation of the service pipes, this may be filled in tightly round the pipes with a suitable non-combustible material which can be fairly easily removed and replaced when repairs to the services are necessary. Such fillings are slag wool with retarded hemi-hydrate plaster, foamed slag concrete or vermiculite concrete, all of which can be laid on non-combustible board or expanded metal. Open ducts, or chutes, passing through fire resisting floors should be provided with double steel dampers with fusible links and counterweights.

Fire stops   These should be provided where any pipes, ducts, conduits or cables pass through openings in a compartment floor or cavity barrier. Openings should be as small as possible and the fire stopping must not restrict the thermal movement of pipes or ducts.

Suitable materials for fire stopping include: cement mortar, gypsum based plaster, cement/gypsum based vermiculite mixes, crushed rock, blast furnace slag or ceramic based products (with or without resin binders), intumescent mastics or any proprietary system capable of maintaining the fire resistance of the element concerned.

Where a lift well connects compartments in a building it should be enclosed by fire-resisting walls to form a fire-resisting 'protected shaft' as defined in AD B, in which case no oil pipes or ventilation ducts are permitted within the shaft.

Escalators penetrating a compartment floor require protection unless the whole escalator system is contained within a protected shaft. Figure 9.6 shows a typical escalator shutter installation, the construction of which must provide the same fire resistance as the floor. The shutter is normally open but closes after a warning bell is sounded in the event of a fire. Automatic operation is controlled by a heat detector set in the soffit of the coil casing.

Roofs   The problems of roof construction have already been discussed in the earlier sections of this chapter (see pages 321 and 324).

Although roofs are generally not required to possess fire resistance they may on occasion be required to do so where the roof forms part of an escape route or is adjacent to an escape way or staircase, or where it is adjacent to a taller part of a building, for example, a roof over a podium below a tower block, when collapse of the roof could cause fire to spread to the higher part by radiation or flying brands.

Cavities   The need to avoid continuous cavities in wall and roof construction when formed by the application of combustible linings has been referred to on page 321. Cavities may also be formed in framed structures behind claddings and in curtain walling and rainscreen cladding. In all cases the possibility of the cavity acting as a flue for the passage of flames and smoke in the same way as a duct must always be considered. The two-storey balloon-framed timber wall in which the vertical studs run through the two floors is an example of a cavity arising from the structure used. It requires cavity barriers at first floor level, as described in Part 1, page 102.

Cavity barriers Current building control requires the provision of cavity barriers to prevent flames and smoke spreading through hidden voids in the fabric of the building. To prevent smoke and flames passing from a cavity within an element of construction into another cavity or room, the perimeter of the element and that of any opening in it must be closed with cavity barriers. The necessary barrier may, in fact, often be provided by the construction of the element without the provision of a separate component (figure 9.2 G). A cavity crossing the end of a fire-resisting element, such as a floor or wall, must have a cavity barrier in the same plane as the element in order to prevent smoke and flames bypassing that element (figure 9.2 H, J). Extensive cavities must be sub-divided by cavity barriers to prevent smoke and flames travelling long distances.

AD B, sections 6 (volume 1) and 9 (volume 2), dealing with concealed spaces, indicates where cavity barriers are to be provided and specifies the maximum permitted dimensions of cavities. Examples of suitable materials, used at thicknesses to give a half-hour fire resistance, are fibre free board, plasterboard, steel, timber, wire reinforced mineral wool, cement, plaster, polythene sleeved mineral wool or mineral wool slabs under compression. The actual material to be used in any particular situation will depend upon the cross-sectional size of the cavity and the position of the cavity barrier. Cavity barriers must not be affected by building movement, failure of fixings or collapse in a fire of any services which penetrate them.

There are, with certain provisos, exceptions to these requirements, including brick and block cavity walls and cavities below suspended ground floors (see Part 1, note 14, page 189).

Fireplaces and flues The construction of fireplaces and flues must obviously be designed to contain the fire and prevent the spread of heat by conduction or radiation to combustible parts of the structure, and to ensure that where there is any possibility of the accidental fall of hot embers a non-combustible surface is provided to receive them. Building Regulations lay down certain thicknesses and dispositions of non-combustible materials, limit the presence of combustible materials near a fireplace or flue and require that all joints in flue linings shall be properly tight against the passage of smoke or flame. Reference should be made to Part 1, chapters 8 and 9, where these matters are discussed.

9.5.3 Fire resistance of structural steelwork

With the development of fire safety engineering and its new ways of considering the whole problem of safety in fire, new approaches to this problem relative to structural

steelwork have been suggested and some of these are touched on here.

Steel structures have until now been given adequate resistance against failure in fire by means of some form of fire protection, the steelwork being designed on the assumption that during a fire it remains cold enough to retain its design strength. If, however, the steelwork were designed itself to withstand fire without applied protection by using as a basis for the structural calculations the strength of steel at some higher temperature developed in a fire, together with an appropriate factor of safety, the stability of the structure would be assured until that temperature was reached.

Another consideration which may be taken into account is the actual behaviour of steel structures *as a whole* during a fire, particularly redundant, or continuous structures, as distinct from the behaviour of their individual parts. If, in this type of structure, some members under increasing temperature lose their loadbearing ability the parts outside the immediate fire zone can provide support for the load and the structure as a whole will not fail. This, in some circumstances, gives rise to the possibility of reducing or eliminating the need for applied fire protection.

A number of expedients have been developed to eliminate or reduce applied fire protection, among which are: the fixing of lightweight concrete blocks between the flanges of a steel column, which, by keeping the web cool, will provide a 30 minute fire resistance or, if dense concrete is poured between the flanges, a 60 minute resistance; the use of the steel and concrete plate floors shown in figures 5.6 and 5.13 D (now generally known as 'slimfloors') the beams in which achieve a fire resistance of 60 minutes without protection to the exposed flange; the application of flame shielding to steel members as described on page 339.

Guidance on the fire-resistant design of steel construction, BS 5950-8: *Structural Use of Steelwork in Building. Code of Practice for Fire Resistant Design*, provides data and calculation methods to form the basis of designing for fire having regard to considerations such as those referred to here.

9.5.4 Fire protection of structural steelwork

Current building regulations in the United Kingdom require all elements of structure of a given building, with certain exceptions, to be of the same fire resistance grading irrespective of their position in relation to the building enclosure, the appropriate fire resistance of steelwork being achieved by some form of applied protection such as those described in this section. As a fire safety engineering approach becomes more widespread there will, no doubt, be relaxations of the requirements, but in some circumstances some members may still need protection.

Internal steelwork Conventional fire protection by solid concrete encasement, although still used as indicated on page 141, has tended to be replaced by lighter forms of protection which result in reduction in weight and the rapid erection of multi-storey framed buildings in particular. Furthermore the use of steel decking as permanent shuttering and reinforcement to slabs cast over them in concrete has led to protection by false ceilings and/or lightweight casings to beams and columns.

Protection by heavy encasement (i) Solid concrete may be cast and vibrated in specially designed formwork around wire fabric reinforcement to hold the cladding in position and prevent spalling under heat (figure 9.7 A). (ii) Lighter weight versions of concrete encasement include the use of lightweight block infilling between the flanges with wire reinforcement and cement rendering (A1) and the use of precast concrete hollow block claddings set around beams and columns in interlocking sets which can reduce the weight of steel by up to 10 per cent due to the lessened superimposed load of the casing compared with solid concrete (B). Brick, block or precast concrete claddings or walls may, of course, constitute fire cover (F).

Systems of lightweight fire protection
Sprayed systems (i) Mineral fibres mixed with inorganic binders and sprayed on to the steelwork with special equipment. They dry out to form a permanent homogeneous insulation which is also sound absorbent, provides a degree of corrosion protection and assists in controlling condensation. The finishes are vulnerable to impact and abrasion but can be surface finished with hardeners and cementitious overcoatings to give smoother, less vulnerable surfaces. Further information is given in BS 8202-1: *Coatings for Fire Protection of Building Elements. Code of Practice for the Selection and Installation of Sprayed Mineral Coatings.* Some manufacturers claim weather resistant finishes suitable for external use. (ii) Vermiculite/gypsum/cement premixed with additives and binders mixed with water on site. The finish is usually textured but can be trowelled or rolled smooth and can be overcoated with a decorative finish. These are very suitable finishes to steelwork above suspended ceilings (figure 9.7 G).

Boarded casing (figure 9.7 C, F, H). There are three main types: (i) fibre-cement, e.g. mineral fibres bound together with calcium silicate, cement or other bonding agent and produced as boards or batts (C); (ii) vermiculite/gypsum or non-combustible binders formed into standard sized boards and fixed to noggings wedged and glued into the webs of beams and columns with non-combustible adhesives and pins; (iii) plasterboard in one or more thicknesses fixed by lightweight galvanised strapping around the steelwork to which the boards are screwed. This is a totally dry system which can be enhanced by a plaster skim coat (C1).

Sheet steel casing This is a system of rigid casing employing steel units lined with fire-resisting material and fixed by interlocking lateral joints or by screwing to strappings. There are no intermediate noggings and services can be run in the voids. Optional finishes include: stainless steel, coated steel, coloured stelvetite or galvanised. This type of protection can also be produced for lattice beams and similar construction as shown in figure 9.7 D. At points for which it would be difficult to produce the steel casing, as at nodes, sprayed intumescent coating can be used to link the adjacent steel casings.

Preformed systems These are rigid forms of protection using ready shaped encasements of vermiculite/gypsum, reinforced with wire or scrim according to the fire rating required and providing an excellent finish ready for decoration. Fixing is by screws to galvanised steel strapping. As with the steel encasements described above, this form of protection can also be produced for lattice forms of construction (figure 9.7 E).

Intumescent coatings These embrace a number of materials, mastics and paints which expand when subjected to heat and form a protective layer of char. Fire ratings are generally limited to one and a half hours for mastic applications and one hour for paint systems. The formulation and performance of intumescents vary considerably. Care should be taken when specifying them for use externally or in aggressive environments. Checks should be made on primer compatibility and any overcoating that may be required. This work, which can now be carried out off-site in a factory, must be carried out by specialist firms and fire inspectors will regularly check and certify the depth and quality of finish and its maintenance.[48]

Figure 9.7 F shows a typical example of multistorey construction using the varying forms of fire protection referred to above, all of which contribute to a lighter more speedily erected building. The steel frame supports prefabricated fire-resistant cladding units and steel deck concrete floor construction. The perimeter floor beams are protected with dry casing protection (H) and the internal beams are protected with sprayed protection (G) within the floor space above suspended ceilings which need not necessarily be fire resisting if the floor is suitably rated.

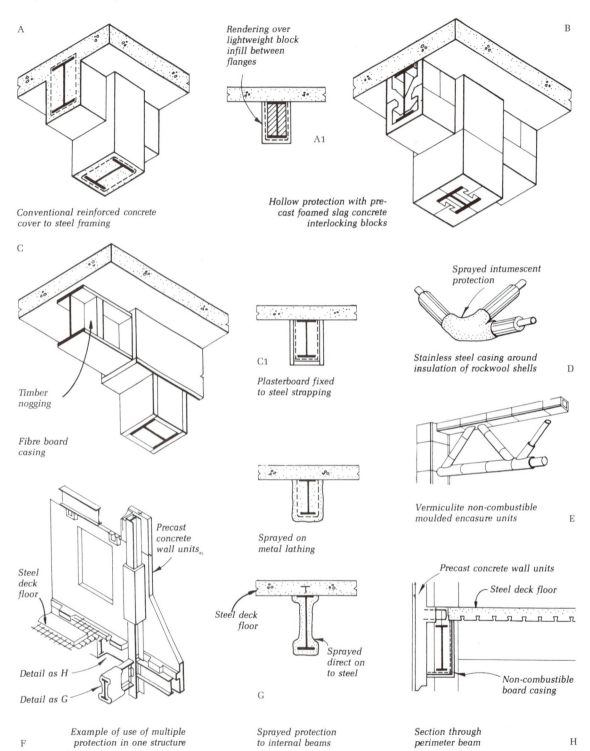

A

Conventional reinforced concrete cover to steel framing

A1 Rendering over lightweight block infill between flanges

B Hollow protection with pre-cast foamed slag concrete interlocking blocks

C Timber nogging / Fibre board casing

C1 Plasterboard fixed to steel strapping

D Sprayed intumescent protection / Stainless steel casing around insulation of rockwool shells

E Vermiculite non-combustible moulded encasure units

Sprayed on metal lathing

F Steel deck floor / Precast concrete wall units / Detail as H / Detail as G / Example of use of multiple protection in one structure

G Steel deck floor / Sprayed direct on to steel / Sprayed protection to internal beams

H Precast concrete wall units / Steel deck floor / Non-combustible board casing / Section through perimeter beam

Figure 9.7 Fire protection to internal steelwork

External steelwork The growing tendency for designers to maximise floor space, free of structure, coupled with design philosophies which engender direct expression of structural form has led to a number of developments in the protection of steelwork which may be situated outside the building façade. In this situation it is argued that structural columns and beams are to a large extent protected by the main walling and are clearly not subject to all round attack by fire as are internal columns and, therefore, a reduced fire resistance period should logically be required of them. Columns with exposed outer flanges are in little danger of overheating provided they are accommodated within the walling, their inside flanges are protected by internal fire protection and they are situated sufficiently far away from window openings (figure 9.8 A).

The approach to the fire resistance of steelwork in which the strength of the steel at temperatures developed in a fire could be used in structural calculations is referred to in section 9.5.3. The criterion for safety is the temperature of the steel structure with critical temperatures of 538 °C for columns and 593 °C for beams determined by averaging multiple temperature readings of each structural member.

The methods used or investigated to date in order to keep structural steelwork cool (i.e. below the critical temperature) have been: protective coverings (lightweight concrete, intumescent coatings), solidity of section (massive steel sections, concrete filling of hollow steel sections), water cooling (water spraying, water filling hollow sections) and flame shielding. These methods are briefly described below.

Protective coverings

Vermiculite concrete Casing is formed of exfoliated vermiculite with a Portland cement matrix reinforced with wire mesh. This has good properties of dimensional stability under conditions of shrinkage, creep, changes of temperature or structural deformation and is frost resistant and weather resistant. An excellent finish can be achieved by plastering techniques.

Thicknesses for hollow protection are 13 mm for up to 1 hour, 25 mm for 2 hours and 57 mm for 4 hours fire resistance: for profiled protection, 13 mm for half-hour, 19 mm for 1 hour and 32 mm for 2 hours fire resistance.

Partial exposure of parts of a steel frame in combination with the full encasement of other structural elements may be desirable for architectural reasons as in figure 9.8 B where the column is fully protected with insulation within a steel casing whilst the exposed beam is only protected on the vulnerable inner faces leaving the elevation exposed. The external column shown in C is totally encased and has sheet metal cladding on the outside which is an architectural feature of the façade. This degree of protection lets the adjacent windows abut the column directly.

Intumescent coating systems Some coatings can produce up to one and a half hours fire resistance and are available in a wide range of colours.[49] They are especially useful for the external protection of metal structures where preservation of the surface configuration is essential. The main disadvantages appear to be high cost in the case of one and a half hour fire resistance and the need for vigilance and maintenance of the coatings throughout the building's life and during successive ownerships.

In addition to its use as a surface coating intumescent paint may be used as a form of hollow protection to provide fire resistance periods of half and one hour, in which layers of expanded aluminium mesh coated with the paint are fixed to a welded steel wire mesh secured to the steelwork or are incorporated as linings in preformed sheet steel firecasings. The system is basically for internal use unless suitably clad for external use.

Solidity of section

Massive steel sections These have inherently great thermal capacity particularly when of low specific surface. Strength in high temperatures is related to their perimeter to section area ratio and a fire resistance period of about 50 minutes can be achieved with a ratio of about 40 at a temperature of 550 °C (see page 339).

Solid steel billets can be more economical in first costs than rolled steel sections or built up sections of similar strength.

Concrete filled sections These are hollow steel sections filled with high strength concrete which results in increased load carrying capacity and increased fire resistance. Tests on a nominal 300 mm concrete filled square hollow steel section with wall thicknesses of 10 to 16 mm showed increases from 15 minutes fire resistance to 45 minutes infilled with ordinary concrete and 50 minutes infilled with high strength concrete.

Water cooling

Water spraying Cooling of steelwork by the transfer of heat from the steel to water may be achieved by spraying directly upon the hot metal or by arranging water to run within a hollow member which has suitable holes in the walls to allow steam to escape. These holes would also act as overflows to enable a given depth of water to be

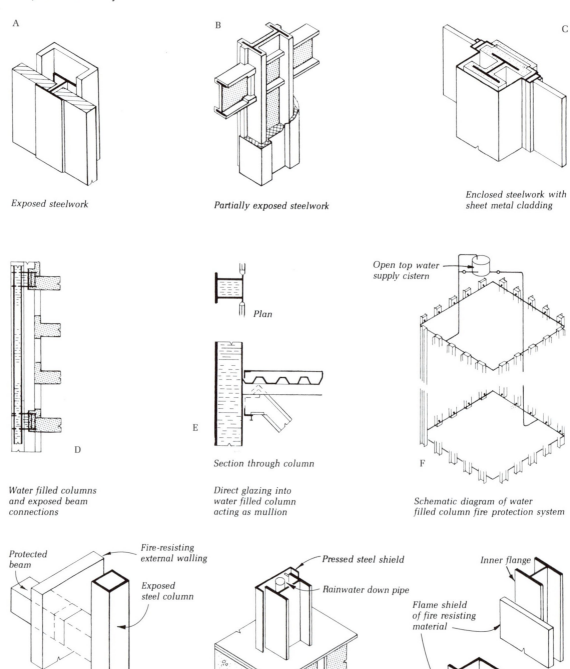

A

Exposed steelwork

B

Partially exposed steelwork

C

Enclosed steelwork with sheet metal cladding

D

Water filled columns and exposed beam connections

E

Plan

Section through column

Direct glazing into water filled column acting as mullion

F

Open top water supply cistern

Schematic diagram of water filled column fire protection system

G

Protected beam

Fire-resisting external walling

Exposed steel column

Water filled shutters automatically operated by fire detection system

H

Pressed steel shield

Rainwater down pipe

J

Inner flange

Flame shield of fire resisting material

Figure 9.8 Fire protection to external steelwork

maintained within the member. The heat transfer is maximum within the temperature range 100–150 °C and foaming agents can be released to keep the wetted areas wet for longer periods. It is theoretically estimated that the steel would remain below 550 °C so long as at least 10 per cent of its surface were to remain wet.

Water filling Systems filled with water are of two basically different forms: (i) non-replenishable and (ii) replenishable.

The non-replenishable systems are simply hollow members filled with water as in figure 9.8 D which shows a structure comprising water filled columns and exposed brackets supporting perimeter floor beams and external walling. The vulnerability of the columns and brackets can be reduced by careful positioning of windows and flame barriers in the enclosure. The example at E shows direct glazing to a water filled column acting as a mullion. The structural integrity of the building in a fire would in this case be achieved by other strategically placed fully protected structural members which gave support to the floors and mullions. The natural circulation of the heated water may be aided by the incorporation of an inner tube which promotes the circulation flow up the sides and down the centre of the member. A pressure release valve is provided to each member. The system improves fire resistance from about 15 minutes up to 30 to 45 minutes.

The replenishable systems can achieve high fire resistance so long as the source of water supply is maintained. The systems depend on natural circulation and are usually replenished from a header cistern or from the mains. A schematic arrangement of such a system of water filled external columns linked horizontally to allow cooling by thermo-syphonic action in case of a fire is shown at F. Differential circulation as between horizontal members, in which convective flow is minimal, and vertical members, in which there is maximum flow, can be reduced by arranging the structure on a diagonal elevational pattern or lattice.

Potassium carbonate may be used to prevent bursting stresses due to freezing and potassium nitrate inhibits corrosion in water cooled systems.

Flame shielding In this method structural members can be planned to occur opposite sections of fire-resisting external walling which mask them from flame attack or, alternatively, specially designed shields may be employed which intercept flames. Columns sufficiently far away from the building face or enclosing surfaces or sufficiently far away from adjacent windows to avoid flame contact do not technically require fire protection. The protective construction may be solid walling or automatically operated fire-resistant elements such as water filled shutters or louvres as shown in figure 9.8 G.

Metal shields may be fixed to a background structure or to the actual structural members they protect, using minimal attachments to minimise conduction of heat. H shows sheet metal shields interposed between the building enclosure and freestanding columns where the columns are not protected by the building enclosure. In J columns close to the building enclosure are protected by flame barriers of suitable fireproof material fixed directly behind their inner flanges and extending sufficiently on each side to prevent overheating of the steelwork by flames shooting through adjacent openings. These systems depend upon an accurate prediction of flame pattern in a wide range of climatic conditions and winds of variable force and direction.

Passive fire protection – section factors The section factor is a numerical assessment related to the time taken for a structural steel section to reach failure temperature in a fire. Data has been established by BRE Certification Ltd/Loss Prevention Certification Board in the form of Loss Prevention Standards, the frequently favoured measures for fire risk appraisal by building insurers. The time taken for failure temperature to be attained is affected by the section profile or shape. A short section with thick web and flanges will heat up less rapidly than a slender tall section. Section factors are collated into three categories: up to 90, 90 to 140, and over 140.

The required thickness of protective material encasement or enclosure to structural steel sections to satisfy various periods of fire resistance depends on the section factor. This factor is sometimes referred to dimensionally as the H_p/A factor where H_p is the perimeter of steel section exposed to fire, and A the cross-sectional area of the section. Reference to structural steel design tables may contain section factors; otherwise they can be calculated from the dimensional data found in BS 4-1: *Structural Steel Sections. Specification for Hot-rolled Sections.* Figure 9.9 shows the method for calculating the section factors for a universal beam of 533 mm × 210 mm × 101 kg/m. The first example shows solid concrete encasement with four sides exposed, and the second hollow protection with three sides exposed.

9.5.5 Fire-fighting equipment

This heading covers various forms of hand extinguishers and types of fixed installations which may be used or come into operation to contain the fire until the arrival of the fire service, together with special installations provided for the fire service in large and high buildings. For all these, reference should be made to chapter 18, *Fire-fighting equipment*, in *MBS: Environment and Services* and chapter 9, *Security and Fire Protection*, in *Building Services, Technology and Design* by R Greeno, Pearson.

Universal beam (UB) serial size 533 mm × 210 mm × 101 kg/m

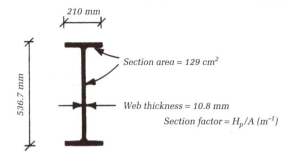

Section area = 129 cm²

Web thickness = 10.8 mm

Section factor = H_p/A (m⁻¹)

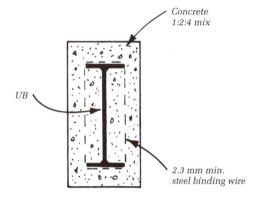

Concrete 1:2:4 mix

UB

2.3 mm min. steel binding wire

Solid encasement 4 sides exposed
H_p = $(2 \times 210) + (2 \times 536.7) + (2 \times 210 - 10.8)$
= 1891.8 mm or 1.8918 m
A = 129 cm² or 0.0129 m²
H_p/A = 1.8918 ÷ 0.0129 = 147

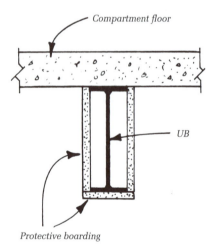

Compartment floor

UB

Protective boarding

Hollow protection 3 sides exposed
H_p = $210 + (2 \times 536.7) + (2 \times 210 - 10.8)$
= 1681.8 mm or 1.6818 m
A = 129 cm² or 0.0129 m²
H_p/A = 1.6818 ÷ 0.0129 = 130

Figure 9.9 Section factors

Notes

1 References are made to Regulations, Codes of Practice or other published documents for the purpose of giving point to the principles.

2 *Dwellinghouse*: a unit of residential accommodation; does not include a flat or building containing a flat. *Flat*: separate/self-contained residence within a building and divided horizontally from some other part of the building. A flat can have more than one storey; this is called a maisonette.

3 See 'Combustible materials', page 318.

4 'Studies of the Growth of Fire', *Fire Protection Association Journal*, Reprint no. 1.

5 See pages 319–21 and 333–4.

6 See Building Regulations, AD B4, sections 8 to 10 (vol. 1) and 12 to 14 (vol. 2).

7 See BS 5839: *Fire Detection and Alarm Systems for Buildings*.

8 See also CIBSE Guide E: *Fire Engineering*.

9 Fire growth rate categories range between slow and ultra fast as listed in Table 1 of *Fire design of steel structures*, Corus Group Plc.

10 See *Fire Surveyor*, 1981, 10 (5) pp. 23–8: 'Fire sizes and sprinkler effectiveness in shopping complexes and retail premises' by H P Morgan and S E Chandler.

11 Other factors in limiting the spread of fire are the use of non-combustible linings and planning considerations.

12 These are: a structural beam or column or any member of structural frame; a loadbearing wall; a floor, excluding the lowest floor of a building and a raised platform floor; a gallery; external and compartment walls. See AD B, Appendix E, *Definitions*.

13 BS 476: *Fire Tests on Building Materials and Structures*. Criteria relate to the ability of an element under consideration to:

resist fire penetration and capacity to bear load in a fire (*Integrity*)

resist excessive heat penetration so that fire is not spread by radiation or conduction (*Insulation*).

Table A1 in appendix A to AD B shows the method of exposure under test for various elements together with the requirements each must satisfy in terms of *Integrity* and *Insulation*.

14 The Joint Committee's role has been absorbed into the Building Research Establishment, but some of the Committee's publications although over 50 years old still retain relevance to modern thinking on fire safety. The BRE has reproduced several papers as facsimile copies, e.g. *Fire Grading of Buildings: Postwar Building Studies No. 20, Pt I* as BRE Report 236, and *Postwar Building Studies No. 29, Pts I, II and III* as BRE Report 237.

15 AD B, table D1 classifies seven purpose groups and table A2 specifies the minimum periods of fire resistance for elements of structure according to the purpose group of the building of which the element forms part.

It should be borne in mind that the Building Regulations are concerned with matters of safety and health of the occupants of a building not with property protection – Insurers' rules for fire protection embodied in the Loss Prevention Standards, published by the Loss Prevention Certification Board (see page 313), are more stringent in many respects than the requirements of the Building Regulations.

16 In relation to the growth of a fire these inorganic materials should contain not more than 1 per cent by weight or volume of organic material. See AD B, table A6 for the classification and use of this category of materials.

17 Other classifications based on density may well supersede these classes. Reference: *Guides to Good Practice* (FIP/CEB recommendations for the design of reinforced and prestressed concrete structural members for fire resistance), Fédération Internationale de la Précontrainte, Fib (UK) Group, www.concrete.org.uk.

AD Regulation 7, section 1.1.8, prohibits the use of high alumina cement from all structural work. Elsewhere it may be used only as a heat-resisting material.

18 See also BRE Digest 487/4, *Materials Behaviour: Timber*.

19 For example, BS 476: *Fire Tests on Building Materials and Structures* and BS EN ISO 10093: *Plastics. Fire Tests. Standard Ignition Sources*.

20 *Materials of limited combustibility*: a term used for materials the properties of which are assessed by reference to the test method specified in BS 476-11. See AD B, table A7 for the use of these materials. Where the Building Regulations require the use of materials of limited combustibility any appropriate non-combustible material may be used. See page 315 for this latter category.

21 Certain materials consisting of a non-combustible core at least 8 mm thick and combustible facings not more than 0.5 mm thick on one or both sides may be regarded as *Class 0* but may need to comply with any specified surface spread of flame ratings.

22 See tables 6 (vol. 1) and 17 (vol. 2) of AD B.

23 See AD B, sections 3 and 10 (vol. 1) and 6 and 14 (vol. 2) and appendix A.

24 The minimum periods of fire resistance required for the elements of the structure of a building are given in AD B, tables A1 and A2.

25 Performance requirements for fire doors and shutters are defined in appendix B to AD B.

26 See note 19.

27 *Unprotected area*, in relation to an external wall, means: a window, door or other opening; any part of the wall with less than the fire resistance required for that wall by AD B; any part of the wall with combustible material more than 1 mm thick attached or applied to its external face, whether for cladding or any other purpose.

28 See *Fire Protection of Buildings. Fire Service Manual, vol. 3: Fire Safety, section 3 – Smoke Control and Fire Venting Systems*, HM Fire Service Inspectorate, The Stationery Office.

29 See chapter 9 of *Building Services, Technology and Design* by R Greeno, Pearson.

30 See *Guidance for the Design of Smoke Ventilation Systems for Single Storey Industrial Buildings, Including those with Mezzanine Floors, and High Racked Storage Warehouses*, issue no. 3, Smoke Control Association. See also BRE Report 186, *Design Principles for Smoke Ventilation in Enclosed*

Shopping Centres; BRE Report 258, *Design Approaches for Smoke Control in Atrium Buildings*; and BS 5588-7: *Fire Precautions in the Design, Construction and Use of Buildings. Code of Practice for the Incorporation of Atria in Buildings*.

31 Requirements for the provision of vehicle access are given in AD B, sections 11 (vol. 1) and 16 (vol. 2). For dry and wet rising mains see note 28.

32 For the construction of fire-fighting shafts and lifts, see BS 5588-5: *Fire Precautions in the Design, Construction and Use of Buildings. Access and Facilities for Fire-fighting*, sections 2 and 3, respectively. See also chapter 10 of *Building Services, Technology and Design* by R Greeno, Pearson.

33 See note 28.

34 'Automatic airflow control for escape route smoke movement in multi-storey flats' by J Wilkinson, *Fire* 62 no. 772 1969. 'The Worthing "AAC" System – Automatic Airflow Control system for escape routes in new multi-storey blocks of flats': Symposium no. 4, *Movement of Smoke on Escape Routes in Buildings*, HMSO.

35 Other legislation requiring provision for means of escape are *The Fire Precautions Act, 1971; The Housing Act, 2004; The Building Act, 1984; The Regulatory Reform (Fire Safety) Order*, 2006.

36 The height of stair is not limited in buildings other than dwellings if the stair is protected from snow and ice.

37 Where special hazards exist (e.g. cellulose spraying booths) exit doors may be required to open in the direction of escape irrespective of number of occupants. Any door allowed to be opened inwards to be arranged to be locked back in such a manner that a key is required to release it. The door must not form an obstruction or reduce the required exit width.

38 See figure 9.6 for illustrations of sliding doors. A wicket door is a small door to provide pedestrian access through a larger door.

39 Requirements for fastenings may vary according to the statutory requirements of legislation, e.g. *Factories Act 1961, Offices, Shops and Railway Premises Act 1963, Fire Precautions Act 1971*.

40 Provision, however, is made in AD B, vol. 2, appendix B for the use of methods for holding open those doors where a self-closing device would hinder the normal use of a building. For example: fusible link, automatic release device actuated fire detection/alarm system or a door close delay mechanism.

41 Attention is drawn to section 48(4) of the Factories Act 1961, which requires all lifts and hoistways to be enclosed in fire-resisting material.

42 For lift installations generally, see *MBS: Environment and Services*, chapter 17, and chapter 10 of *Building Services, Technology and Design* by R Greeno, Pearson.

43 See BRE publication no. 320, *Fire Doors*, 1988.

44 The FOC was established by UK fire insurers with a sub-division that became known as the Loss Prevention Council (LPC). The LPC produced technical standards and specifications for fire prevention and control equipment in buildings. The origins of these organisations date back to the latter part of the 19th century, long before building regulations came into being. A century on, in the late 1980s the organisation was renamed Loss Prevention Certification Board (LPCB) in response to changes to the structure of the insurance industry. The LPCB incorporated the FOC and LPC technical standards and rules, now known as Loss Prevention Standards. The LPCB publish these as part of BRE Certification Ltd.

45 Such seals round pipes or elsewhere are known as *fire stops*, the function of which is to close small gaps or imperfections of fit between elements or components or to crush and close up combustible pipework in order to restrict the penetration of smoke or flame through such gaps.

46 Reference should be made to BRE Digest 208, *Increasing the Fire-resistance of Existing Timber Floors*, which reviews this subject and gives tables and diagrams of various combinations of materials related to existing construction and required periods of fire resistance.

47 See BS 8313: *Code of Practice for Accommodation of Building Services in Ducts* and BS 5588-9: *Fire Precautions in the Design, Construction and Use of Buildings. Code of Practice for Ventilation and Air Conditioning Ductwork*.

48 Refer to BS 8202-2: *Coatings for Fire Protection of Building Elements. Code of Practice for the Assessment and Use of Intumescent Coating Systems for Providing Fire Resistance*.

49 Marketed by Rentokil Initial UK Ltd as 'Albi-Steel 90'.

10   Temporary works

The chapter commences with descriptions of the different types and forms of metal scaffolding and gantries followed by descriptions of different forms of shoring in timber and steel. There follows discussion of the methods which may be used for supporting the sides of excavations for deep trenches and large excavations, then of the formwork in timber and metal required for the various parts of a concrete structure – structural frame, walls, stairs and shell vaults – together with a brief consideration of ways of treating the forms to obtain various finishes on the concrete.

The subject of temporary constructions which are often a necessary part of the total building process has been introduced in Part 1 and some aspects of timbering for excavations and of formwork have been covered. In this chapter these will be considered further and the subjects of shoring and scaffolding will be taken up.

10.1 Scaffolding

Temporary structures, constructed to support a number of platforms at different heights to enable operatives to reach their work and to permit the raising of materials, are termed scaffolds (figure 10.1).[1]

Tubular metal scaffolding is in almost universal use, although timber scaffolding may still be used occasionally in less developed parts of the world. In the UK, round poles of fir lashed together with hempen or wire rope were still in use well into the 1950s. Illustrated examples are found in old textbooks and photographs from other archive sources.

10.1.1 Tubular steel and aluminium alloy scaffolding

This has considerable advantages over timber. The small diameter and the standard lengths simplify storage and transport; if overloaded it does not suddenly break like timber, but gives ample warning by bending. Its adaptability to any purpose required on a project, such as storage racks for timber or any other material, or the framing for temporary buildings and sheds, constitutes a valuable asset.

The tubing employed is 48.3 mm outside diameter weldless galvanised steel[2] tubes, 4.37 kg/m run, or light alloy[3] tubing weighing only 1.67 kg/m run. The standard length of the unit is 6.30 m, but shorter lengths can be obtained. The shorter lengths in most common use are 1.80 m, 3.60 m and 4.30 m. The weight of a standard length of steel tube is 27.5 kg which is rather more than the CDM Regulations permit one person to handle.

Some forms of the standard couplings used to frame up the tubes are shown in figure 10.2 A. Some scaffolding systems incorporate fixed couplers the parts of which are integral with the different members as shown in B. Square or circular base plates are provided with a central pin that fits into the base of the standards. These are sufficient to take the weight of the scaffold on ordinary firm ground; for soft ground or over cellars or pavement lights stout planking is used to distribute the pressure. The base plates can be spiked to the planking, holes usually being provided in them for this purpose. Adjustable bases are available with a range of height of a few centimetres and the tubing itself can be extended to any length, by means of end to end couplers.

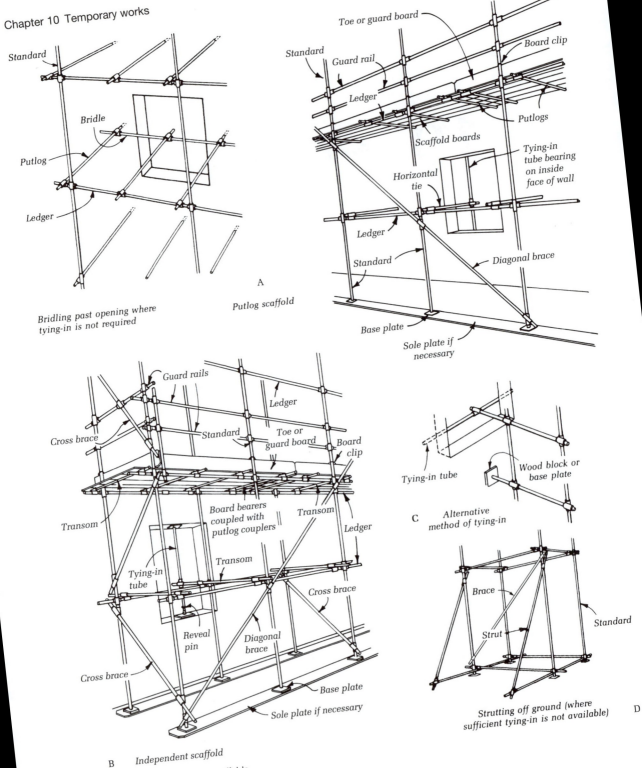

A Putlog scaffold

Standard
Bridle
Putlog
Ledger

Bridling past opening where tying-in is not required

Toe or guard board
Standard
Guard rail
Board clip
Ledger
Putlogs
Scaffold boards
Tying-in tube bearing on inside face of wall
Horizontal tie
Ledger
Standard
Diagonal brace
Base plate
Sole plate if necessary

Guard rails
Ledger
Cross brace
Standard
Toe or guard board
Board clip
Transom
Board bearers coupled with putlog couplers
Transom
Transom
Ledger
Tying-in tube
Transom
Cross brace
Reveal pin
Diagonal brace
Cross brace
Base plate
Sole plate if necessary

B Independent scaffold

Tying-in tube
Wood block or base plate

C Alternative method of tying-in

Brace
Standard
Strut

Strutting off ground (where sufficient tying-in is not available) **D**

Figure 10.1 Tubular metal scaffolds

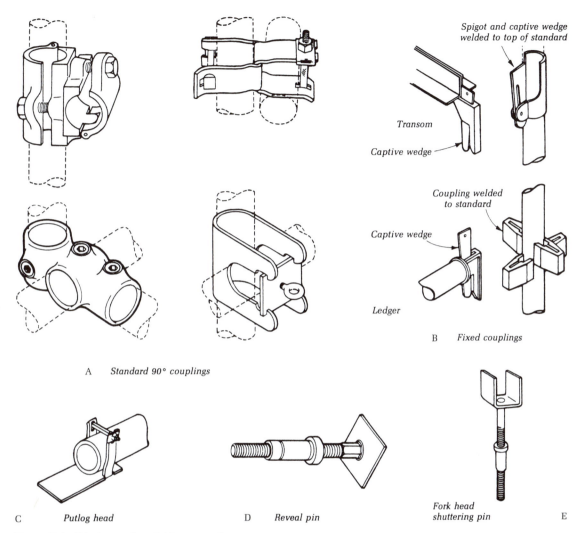

Spigot and captive wedge
welded to top of standard

Transom

Captive wedge

Coupling welded
to standard

Captive wedge

Ledger

B   Fixed couplings

A   Standard 90° couplings

C   Putlog head   D   Reveal pin   Fork head
shuttering pin   E

Figure 10.2 Tubular metal scaffolding – couplings

Two types of scaffold are used: (i) the putlog or brick-layer's scaffold and (ii) the independent scaffold.

In both types a frame of vertical tubes, called standards, and horizontal tubes, called ledgers, is erected at up to 1.40 m from the line of the building, the standards being 2 m apart and the ledgers at vertical distances of up to 1.50 m. Diagonal braces are applied to the outside of the frame to provide rigidity.

Putlog scaffold   In the putlog scaffold (figure 10.1 A) the cross members or putlogs, which carry the scaffold boards forming the working platforms, bear one end on the scaffold frame and the other on the wall which is being built, the putlog tube being flattened at one end and driven

into a joint. Alternatively, a putlog head may be used (figure 10.2 C). The scaffold is tied back to the wall at suitable openings as shown in figure 10.1 A, or as in B where the tie is coupled to a vertical or horizontal tying-in tube secured tightly between the reveals of an opening by means of a reveal pin (figure 10.2 D). This consists of a bolt with ferrule which fits the bore of a tube. The bolt may be rotated in the end bearing plate by a spanner to act as a screw jack to force the plate against the reveal.

Independent scaffold   As the name implies, an independent scaffold is self-supporting and has an inner frame erected about 150 mm from the building face (figure 10.1 B). The cross members, here called transoms, bear on and are coupled to the two frames. For tying back to the wall a tying-in tube with a reveal pin or the alternative method shown in C must be adopted. Where insufficient tying-in is available the scaffold must be strutted at ground level (D). In this form of scaffold lateral rigidity must also be ensured by the use of cross braces running from frame to frame.

Scaffold boards[4] are fixed to the putlogs or transoms by board clips and toe boards to the standards in the same way (A, B). Guard rails are at least 950 mm above the working platform and should be fixed to the outer frame and at the ends. An intermediate guard rail is also necessary to limit the gap to 470 mm between intermediate rail and toe board, and intermediate rail and guard rail.[5]

On high buildings double or treble tube standards, linked by couplers, are required for the lower part when heavy loads must be supported.

Where scaffolding is required for a short period only at some work position a *scissor lift platform* may be used. This consists of a working platform raised and lowered by a hydraulically controlled extending framework (figure 10.3). Its use can prove more economical than erecting and dis-

mantling normal scaffolding over a short interval of time. Access towers as described in the next section are also useful for maintenance and other short term work.

10.1.2 Metal scaffolding frames

In order to reduce the multiplicity of couplings and coupling operations in the framing of independent scaffolds, ledgers and transoms in some systems are prefabricated into horizontal frames of circular or rectangular hollow steel sections (figure 10.4 A). These may rapidly be secured to separate standards by lugs welded to the corners of the frame which fit into special separate couplers or into lugs bolted or welded to the standards at the appropriate intervals. Diagonal bracing is applied as in normal scaffolds.

Other systems prefabricate frames in the vertical plane as shown in figure 10.4 B, C. In B the basic frame consists of two standards spaced the normal scaffold width apart by cross tubes, the height of the frames varying from 900 mm to 1.80 m. These are built up ladder style using coupling pins or by slotting directly one into another. Bracing is bolted to the standards. In C the frames are of a single, smaller standard dimension and are built up in both directions with coupling pins to form 'towers'. Diagonal bracing is required only in the horizontal plane as shown and the 'towers' may be tied together by longitudinal tie bars. These last two systems are most useful in forming short lengths of scaffold or individual access towers where operations take place at the top level.[6]

10.1.3 Gantries

These are structures erected primarily to facilitate the loading and unloading of material, and for its storage during building operations. They consist of an elevated staging erected in front of buildings in the course of construction, designed to act as unloading platforms. They extend usually from the face of the intended structure to a short distance from the edge of the kerb, covering the footway. A gangway is provided under the staging for the convenience of the public. Some form of hoisting tackle is provided for raising material from lorries on to the platform. The use of the tower crane reduces the value of the gantry for this particular purpose, but it is still useful on sites where it is essential to have room outside the site for the storage of materials and for temporary accommodation.

Gantries are constructed of tubular scaffolding or of steel, as shown in figure 10.5. The uprights here consist of light steel beam sections and these are connected by channels at their base, which act as sleepers. Their upper ends are connected to a light lattice beam bolted on the face of the columns and resting on cleats. The two frames are connected with cross frames bolted to them. The various

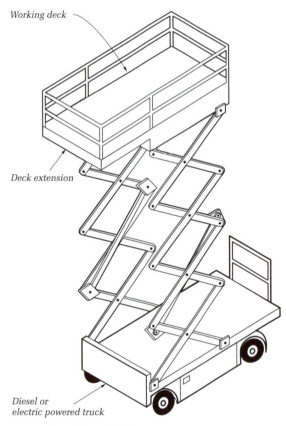

Working deck

Deck extension

Diesel or electric powered truck

Figure 10.3 Scissor lift platform

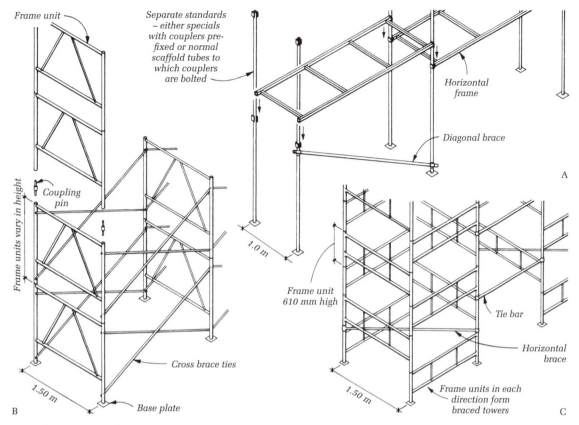

Figure 10.4 Metal scaffolding frames

parts are standardised, which ensures simplicity and rapidity in erection, and enables them to be used many times. The inner frames of both timber and steel gantries frequently have to be taken down to a basement level for a bearing. In this case either longer standards are employed, or a subsidiary frame is erected to support the upper frame, being cleated to the upper lengths of the uprights by fish-plates through the web.

If the gantry is over the public way, it must be close-boarded to prevent dust, rubbish or water falling upon pedestrians, or be under-decked with supplementary sheeting.

10.2 Shoring

Shoring is the means of providing temporary support to structures that are in an unsafe condition until such time as they have been made stable, or to structures which might become unstable by reason of work being carried out on or near them, such as the underpinning of the structure's foundations.

Timber has always been used in the past and is still used for shoring, although effective flying and raking shores are now constructed with tubular scaffolding. Steel stanchions and, particularly, steel needles are now used for dead shoring.

Classification There are three general systems of shoring: (i) raking shores, (ii) horizontal or flying shores, (iii) dead or vertical shores.

10.2.1 Raking shores

In timber these consist of inclined timbers called rakers placed with one end resting against the face of a defective wall, the other upon the ground (see figure 10.6 A). The most convenient and best angle for practical purposes is 60 degrees, but this may vary up to 75 degrees. The angle is often determined in urban areas by the width of the footway. On tall buildings these shores are fixed in systems of two or more timbers placed in the same vertical plane,

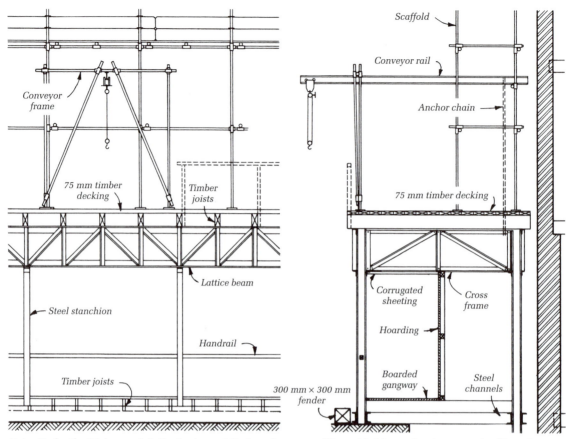

Note: Under the Highways Act, the fender and the board gangway (if provided for pedestrian access) will need illumination during darkness.

Figure 10.5 Steel gantry

inclined at different angles to support the building at varying levels, as shown in figure 10.6 A.

The purpose of a raking shore is to prevent the overturning of a wall – not, in the case of a tilting or bulging wall, to force it back.

Timber shores A wallpiece, consisting of a 50 or 75 mm softwood, is fixed to the wall by wallhooks driven in the joints of the brickwork. This receives the heads of the rakers and distributes their thrusts over a larger area of wall.

To form an abutment for the head of the raker, a needle, consisting of a piece of 100 mm × 100 mm timber about 330 mm long, and cut as shown in B, is passed through a mortice in the wallpiece, and projects into the wall at least 115 mm, a half-brick being taken out to receive it. The function of the needle is to resist the thrust

of the raker and prevent it slipping on the wallpiece and to transmit the thrust through the wallpiece to the wall.

The centre lines of rakers should pass through the centre of the bed of any wallplates in the wall (A). If the joists should be parallel to the wall, the produced centre lines of the floor, wall and raker should meet at a point.

The feet of the rakers rest upon an inclined sole plate usually embedded in the ground, the angle between the sole plate and the rakers being less than 90 degrees to permit the latter to be tightened up gradually by means of a crowbar. The shore should be forced tight, but not enough to disturb the wall. On soft ground the sole plate is bedded on a platform of timber to distribute the pressure over a greater area (A).

The horizontal distance between the systems on unperforated walls is usually not more that 2.40 m; but on

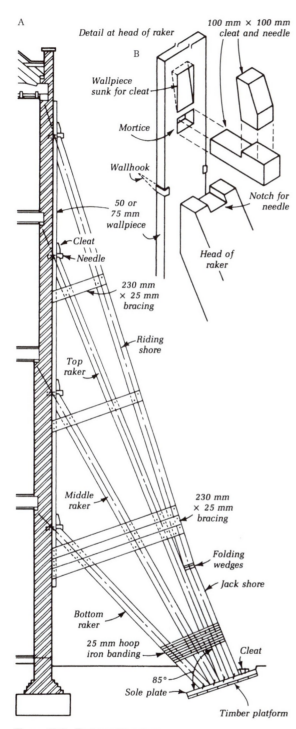

A

Detail at head of raker

B

100 mm × 100 mm
cleat and needle

*Wallpiece
sunk for cleat*

Mortice

Wallhook

*Notch for
needle*

*50 or
75 mm
wallpiece*

Cleat

Needle

*Head of
raker*

230 mm
× 25 mm
bracing

*Riding
shore*

*Top
raker*

*Middle
raker*

230 mm
× 25 mm
bracing

*Folding
wedges*

Jack shore

*Bottom
raker*

25 mm hoop
iron banding

Cleat

85°

Sole plate

Timber platform

Figure 10.6 Timber raking shore

walls pierced with windows they are placed on the intervening piers.

Steel shores When support from the ground to the face of a wall is given by steel rather than timber construction, the shore is in the form of a vertical framework of tubular steel stabilised by raking braces as in figure 10.7 A and figure 10.8 and is called a *restraint frame*. This is horizontally and diagonally braced to adjacent frames positioned between window openings in a perforated wall or at 1–2 m intervals along an imperforate wall. If only one frame is required this would be doubled and the pair braced together to give lateral stability (figure 10.7 B).

Thrust from the wall is taken by horizontal tie members at floor levels acting through plated screwjacks bearing on the wallpiece. To overcome any tendency to uplift the lowermost horizontal tie members are fixed securely to the base of the wall or, alternatively, adequate kentledge is placed on the bottom of the frame and at intervals along the ties linking to the adjacent frames. If fixing to the wall is not possible the feet of the vertical members can be secured to the base on which they bear, sometimes necessitating the provision of concrete anchor blocks at these points.

10.2.2 Horizontal or flying shores

These are used to provide temporary support to two parallel walls, where one or both show signs of failure, or where previous support, in the form of floors, is to be removed. They are used mostly in urban areas, usually where one of a number of terrace buildings is to be removed, to provide temporary support to the buildings on either side. They are erected as the old structure is being removed, and are taken down when the new building is of a sufficient height and strength to provide support.

Timber shores A single flying shore consists of a horizontal timber set between the walls to be supported, the ends resting against wallpieces fixed on the walls. It is stiffened by inclined struts above and below it at each end. These struts also provide two more points of support to each wall.

The maximum length between walls for single flying shores in timber is usually considered to be 9 m (figure 10.9). For larger spans, from 9 m to 12 m, a compound or double flying shore is necessary, framed up as shown in figure 10.10.

Steel shores Support by horizontal or flying shore in steel would be provided in the form of a lattice girder of tubular steel, placed to line with the floors if possible. Where permitted this would be bolted direct to the walls

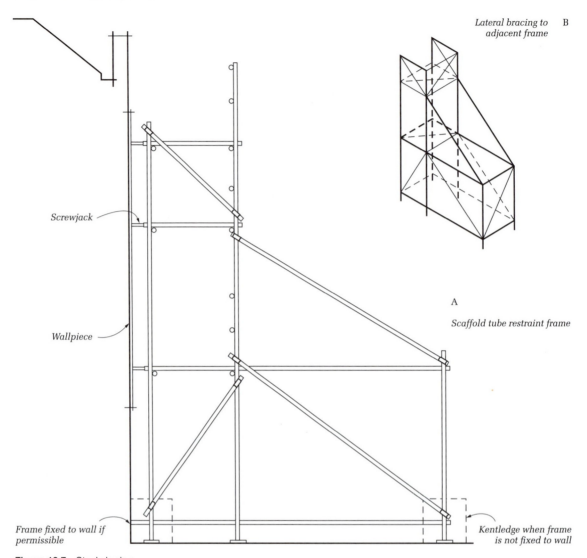

Lateral bracing to adjacent frame   B

A

Scaffold tube restraint frame

Screwjack

Wallpiece

Frame fixed to wall if permissible

Kentledge when frame is not fixed to wall

Figure 10.7 Steel shoring

on each side, otherwise wallpieces would be used to which it is fixed. Adjacent shores are laterally braced together. A span of 10 m is considered to be the economic maximum.

Where one building is higher than another it may be necessary to erect a raking shore upon a flying shore or sloping shores may be suitable.

Horizontal flying shores are usually erected at 3 to 4 m intervals on plan.

10.2.3 Vertical or dead shores

Shores placed vertically are termed dead shores and are now invariably constructed in steel. They are used for tem-

porarily supporting the upper parts of walls, the lower parts of which are required to be removed, either in the process of underpinning or reinstatement during repair, or for the purpose of making large openings in the lower parts. Where a dead shore immediately under the wall is not convenient a system of dead shores is used, comprising a pair of shores supporting a horizontal beam. The wall is then carried by the beam.

Procedure using dead shores If, for example, the lower part of a building is to be removed in order to form a large opening, the procedure would be on the lines illustrated in figure 10.11 A.

Tall restraint frame

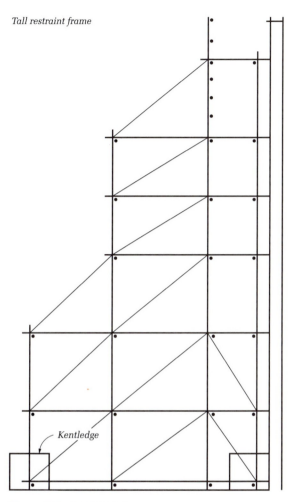

Kentledge

Figure 10.8   Steel shoring

Floor strutting   The whole of the floors, the roof, and any other loadbearing on the wall are supported by a system of strutting to relieve the wall of all weight normally taken by it. This system of strutting should be firmly supported by a sole piece on the solid ground below the lowest floor. The sole piece should be bedded continuously in mortar and be sufficiently stiff to distribute the weight over its whole length.

Insertion of needles   Perforations are next made in the wall a short distance above the line of the top of the beam that will ultimately support the wall. Through the holes horizontal beams called needles are inserted, consisting of Universal steel sections. These should not be placed a greater distance apart than 1.80 m in unperforated walls, but when there are windows the needles must be placed under the piers.

The needles are carried at each end by dead shores of proprietary steel supports in the form either of a braced tubular frame (figure 10.12) or H-section adjustable soldiers as in figure 10.11. Before the system is tightened up, a bed of cement mortar should be placed on the top of the needle at the point where it passes through the wall to ensure a proper and solid bearing of the wall on the needle.

Strutting of openings   All window openings must be strutted to prevent deformation taking place. In ordinary small windows this consists of an upright against each reveal, with two or three struts between, cut long and driven up tightly as shown in figure 10.11. In large openings a stronger framing is necessary and any arches will require support by a turning piece, or centre, made to fit, with the reveals strutted as before.

If the building is old or at all defective, raking shores or restraint frames are imperative, but it is wise in most circumstances to use them to steady the building during the progress of the works. These are fixed against the piers between the windows and close beside the dead shores.

When all the shores are fixed in position, the two end piers are built, or if the supports are to be stanchions these are erected, the minimum amount of existing wall being taken away to allow for this work, after which the remainder of the wall is removed. The new beam is then raised and fixed, and the brickwork above filled in to the underside of the old work. The new brickwork should be built in cement mortar to avoid settlement in the work.

A week at least should be allowed for the new work to set before any of the shoring is struck. The needles should be eased and removed first, then the strutting from the windows, the floor strutting inside, and, lastly, any raking shores. About two days should be allowed between each of these operations in order that the work may take its bearing gradually on the new supports.

Great care is required in carrying out these operations on a corner building. The needling would be constructed to suit the special requirements of the job and the angle of the building should always be supported by raking shores on each face (figure 10.11 B).

When support directly under the wall is required proprietary steel H-section adjustable soldiers would be used as shores, normally in pairs tied to each other and linked by horizontal and diagonal braces to a small braced tube frame at right-angles to give stability in both directions.

The lengthy procedure described above can often be avoided by the use of 'stools' or post-tensioned sectional concrete or steel beams (figure 10.11 C, D, E), inserted in the wall in sections, the lower part of the wall not being removed until the beams are completed. These, and the method of insertion, are described under 'Underpinning' on pages 70–71.

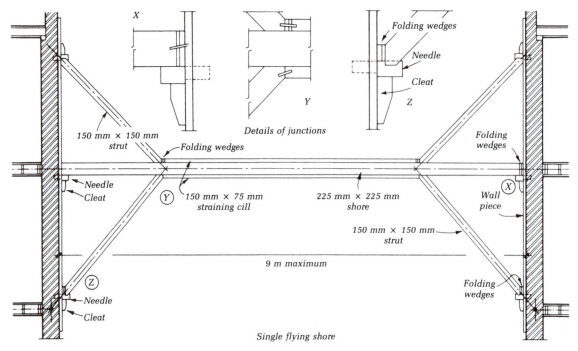

Figure 10.9   Timber shoring

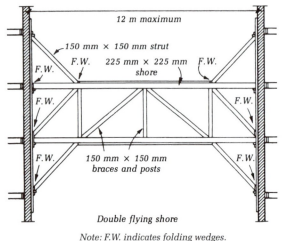

Double flying shore

Note: F.W. indicates folding wedges.

Figure 10.10   Timber shoring

10.3 Support for excavations

This subject is introduced in Part 1, Chapter 11, where the support required for shallow trenches, that is those not exceeding 1.20 m in depth, is described.

Although, as indicated in Part 1, timber is still used for excavations there are now available many alternative means of support, either proprietary systems or non-proprietary methods not employing timber. These tend to be used for excavations beyond the size and depth of the trenches described in Part 1.

10.3.1 Trenches

Considerable care must be taken in the cutting of trenches and in the selection of the methods of providing support, particularly if the trenches must remain open for some length of time, since, as pointed out in Part 1, the pressures on them can be high.

When trenches over 1.20 m in depth are required in soft soils a *plate lining* system may be used (figure 10.13) in which pairs of vertical steel soldiers are driven at intervals and strutted across the trench. As excavation proceeds steel plate units are pushed down against the excavated face between the soldiers, in which are grooves or rebates to accommodate the edges of the units. More units are pushed down on top of those already in place as excavation continues to the bottom of the trench. If a heavy hydraulically operated excavator is being used this will be employed to push down the plates.

Drag box   Where the soil is firm enough to permit its use a *drag box* may be employed. This consists of two long panels the full depth of the proposed trench rigidly strutted

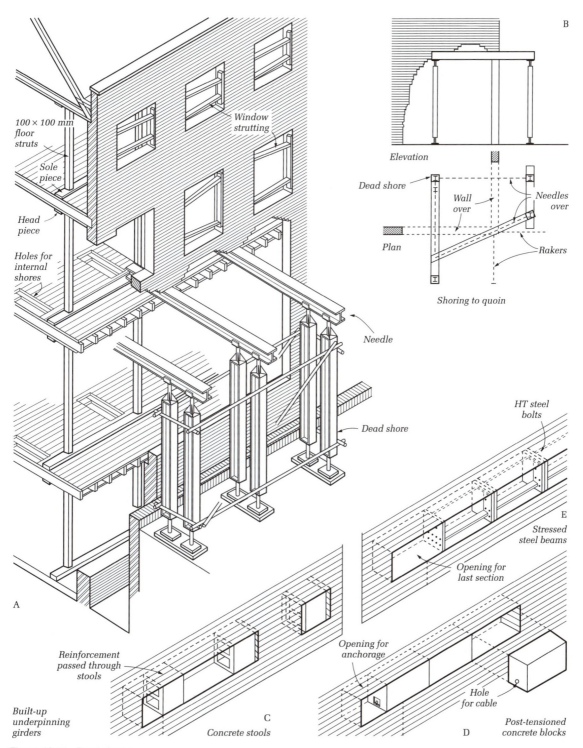

100 × 100 mm floor struts

Sole piece

Head piece

Holes for internal shores

Window strutting

B

Elevation

Dead shore

Wall over

Plan

Needles over

Rakers

Shoring to quoin

Needle

Dead shore

HT steel bolts

E

Stressed steel beams

Opening for last section

A

Reinforcement passed through stools

Opening for anchorage

Built-up underpinning girders

C

Concrete stools

Hole for cable

D

Post-tensioned concrete blocks

Figure 10.11   Dead shores

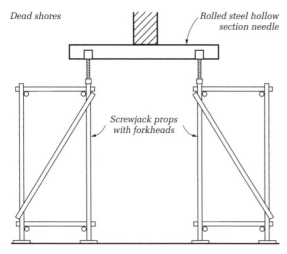

Figure 10.12 Steel shoring

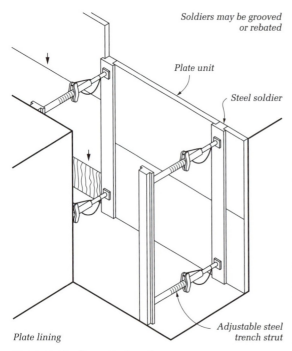

Figure 10.13 Support to trenches

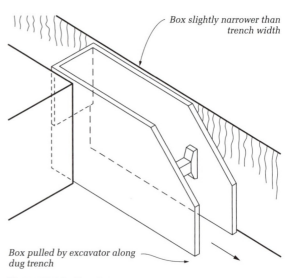

Figure 10.14 Drag box

is possible in drainage work because of the modern practice of testing in short lengths made possible by quick-action, push-fit jointing systems now in use.

10.3.2 Wide excavations

Wide deep excavations and large, deep trenches may be supported by what is known as H-piling if the soil is free of ground water or it has been lowered by well points.

Well points These are devices for forming small wells that can easily be sunk into the ground and withdrawn after use. They are driven with the aid of water pumped down the centre of the supporting tube (figure 10.15 A) and the ground water is drained through a perforated outer tube (B). For depths over 5 m they are set in stages (C).[7]

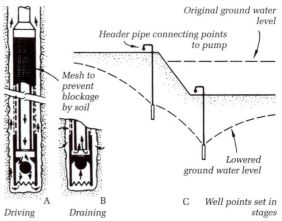

Figure 10.15 Well points

apart, the end edges of the panels usually being shaped to a knife-edge. The width of the unit used will be somewhat less than that of the trench being excavated (figure 10.14). It is lowered into the trench so that it can be pulled along by the excavator to follow the trench digging to give protection to the operatives from falling, relatively loose upper soil where trenches can be quickly backfilled. This

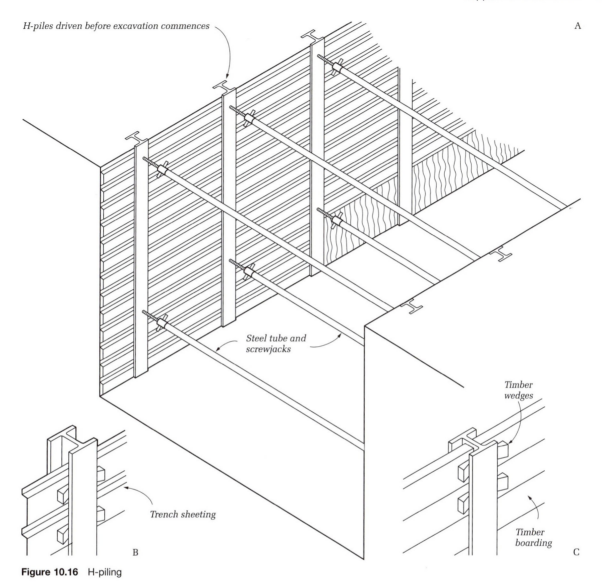

H-piles driven before excavation commences

A

Steel tube and screwjacks

Timber wedges

Trench sheeting

Timber boarding

B

C

Figure 10.16   H-piling

H-piling   Prior to excavation, steel sections, similar to Universal column sections but with thicker webs, are driven into the soil at 1.5 to 1.75 m centres to act as soldiers to the supporting timbers or horizontal trench sheeting (figure 10.16 A). As excavation proceeds the trench sheeting or horizontal timbers are placed in position one under the other against the soil face, their ends being wedged behind the front flanges of the soldiers (B, C). Where the soil is firm, not requiring continuous support, open spaced boarding would be used. If the H-piling is supporting the side of a wide excavation it may require raker or ground anchor support.

Sheet steel piling   If the site is waterlogged, sheet piling is used to support the excavation. This consists of sections of heavy corrugated steel sheet with edges formed to produce interlocking joints (figure 10.17) which in wet soil can prevent the entry of water where the ground water pressure is sufficiently low. When it is high the joints may require caulking. The piling is driven by drop hammer, diesel hammer or double-acting steam or compressed air hammer.

Alternatively, a system which vibrates the pile into the ground can be used which drives at a very high rate with much less noise, or one in which the piles are driven by

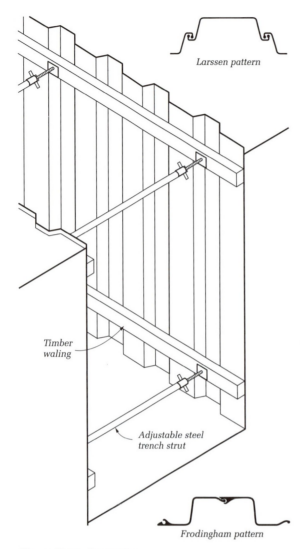

Larssen pattern

Timber
waling

Adjustable steel
trench strut

Frodingham pattern

Figure 10.17 Sheet piling

a multi-ram hydraulic driver, the reaction for which is obtained from other partly driven piles in the group being driven by the machine. This method is vibrationless and is also quiet in operation.

The piles are withdrawn by large extracting grips used in conjunction with a crane, a double acting hammer or hydraulic jacks. Normally only piling used in temporary works is extracted, and then generally only when the piles are less than about 12 m in length. Piles over this length require a very large extracting force and it is sometimes cheaper to leave the pile in position than to withdraw it.[8]

Retaining wall construction Figure 10.18 shows a method of support employed when the ground to be excavated for a basement is removed prior to the construction of the retaining wall to support the soil outside the building. For the first stage sufficient soil is removed to permit a top row of trench sheeting to be placed and strutted (A). The soil below is then removed and, if necessary, further stages of support are fixed, after which the retaining wall is constructed, the remainder of the basement excavated and the floor laid.

An alternative method is to excavate the basement for its full depth, leaving sufficient 'dumpling' to support the surrounding soil (B), and to lay the floor slab up to this. Concrete blocks can be formed along the edge to form bearings for the struts. The soil can then be removed in stages and the floor and retaining wall completed.

Construction in a trench When a deep excavation is near buildings or streets the basement area is often not at first excavated, the wall being built in a strutted trench in order to avoid the risk of movement when, as on many contracts, the trench runs right round the site. The central 'dumpling' of unexcavated soil is not removed until the retaining wall is completed and functioning.

A method of working is shown in figure 10.18 C. The trench, which will generally be the full width of the toe, is covered with 75 to 100 mm of concrete, the lower boards or trench sheeting are removed and the brick protective skin to the asphalt tanking is built up as far as the soil will permit. (In very loose soils it may be necessary to use precast concrete walings or precast concrete sheet piling which will be left permanently in position.) After the asphalt is laid on the concrete blinding and against the lower part of the brick skin, the toe to the wall is cast with mortices formed in the top face to take vertical posts or soldiers. These are later erected with their feet in the mortices and are spiked and cleated to the upper horizontal struts. To permit the pouring of the first lift of wall, the lower struts are removed and replaced by shorter struts bearing on the soldiers. The shuttering to the face of the wall is built up in panels and these are fixed to the soldiers. This process of building up the brick skin, asphalting and pouring subsequent lifts of concrete, is then repeated to the top of the wall. As the wall rises and the upper cross-struts are removed, the lower part of the soldiers can be blocked off the completed lower parts of the wall.

When no external tanking is to be applied and the wall is cast directly against the soil, the position of the vertical reinforcing bars will be close to the poling boards or sheeting and the walings will obstruct them. In these circumstances the walings are blocked off the poling boards by short blocking pieces with spaces between them at the bar

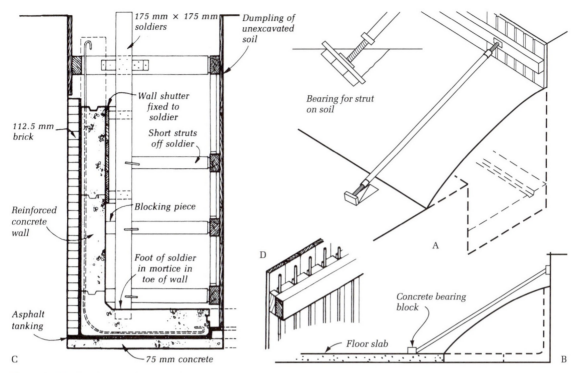

Figure 10.18 Retaining wall construction

Labels in figure:
175 mm × 175 mm soldiers
Dumpling of unexcavated soil
112.5 mm brick
Wall shutter fixed to soldier
Short struts off soldier
Reinforced concrete wall
Blocking piece
Foot of soldier in mortice in toe of wall
Asphalt tanking
75 mm concrete
C
Bearing for strut on soil
A
D
Concrete bearing block
Floor slab
B

centres to permit the bars to pass through them between the walings and the poling boards, as shown at D.

10.3.3 Shafts

It is often necessary to sink shafts for foundations. These are made from 1.20 m square and upwards, the former being the smallest size a person can work in without difficulty.

For deep shafts sheet piling would be used and to avoid the obstruction of strutting hydraulically operated metal walings, which can be up to about 5 m long, are employed (figure 10.19).

Where the excavation is wide ground anchors or raker support may be necessary, the latter of timber or steel props; if the excavation is also deep, with a high ground water pressure, a system of walings, struts and rakers of heavy Universal steel sections may need to be fabricated.

10.4 Formwork

The subject of formwork has been introduced briefly in Part 1, Chapter 11, where that for a simple slab is described. As indicated there, concrete must be given form

by casting it in a mould. These moulds are known as formwork or shuttering which will support it until it has developed sufficient strength to support itself.[9]

The structure which carries the actual formwork is called *falsework*, as exemplified in the support to floor shutters (figures 10.23 and 10.24), in which the main supporting members are vertical, together with bracing members to provide stability.

Reference has already been made in chapter 4 to the fact that the cost of the formwork may be as much as one-third or more of the total cost of the concrete work as a whole, and to the effect that the nature of the formwork can have on its cost. The formwork for any project must be considered at the design stage. Economy is more likely to be achieved if it is designed and worked out in detail before work commences on the site, taking into consideration the nature of the elements to be cast and the methods of handling likely to be used on the site. For example, handling by crane makes possible the use of much larger sizes of wall and floor shutters than if manually handled.

Design of formwork   The general requirements governing the design and construction of formwork are as follows:

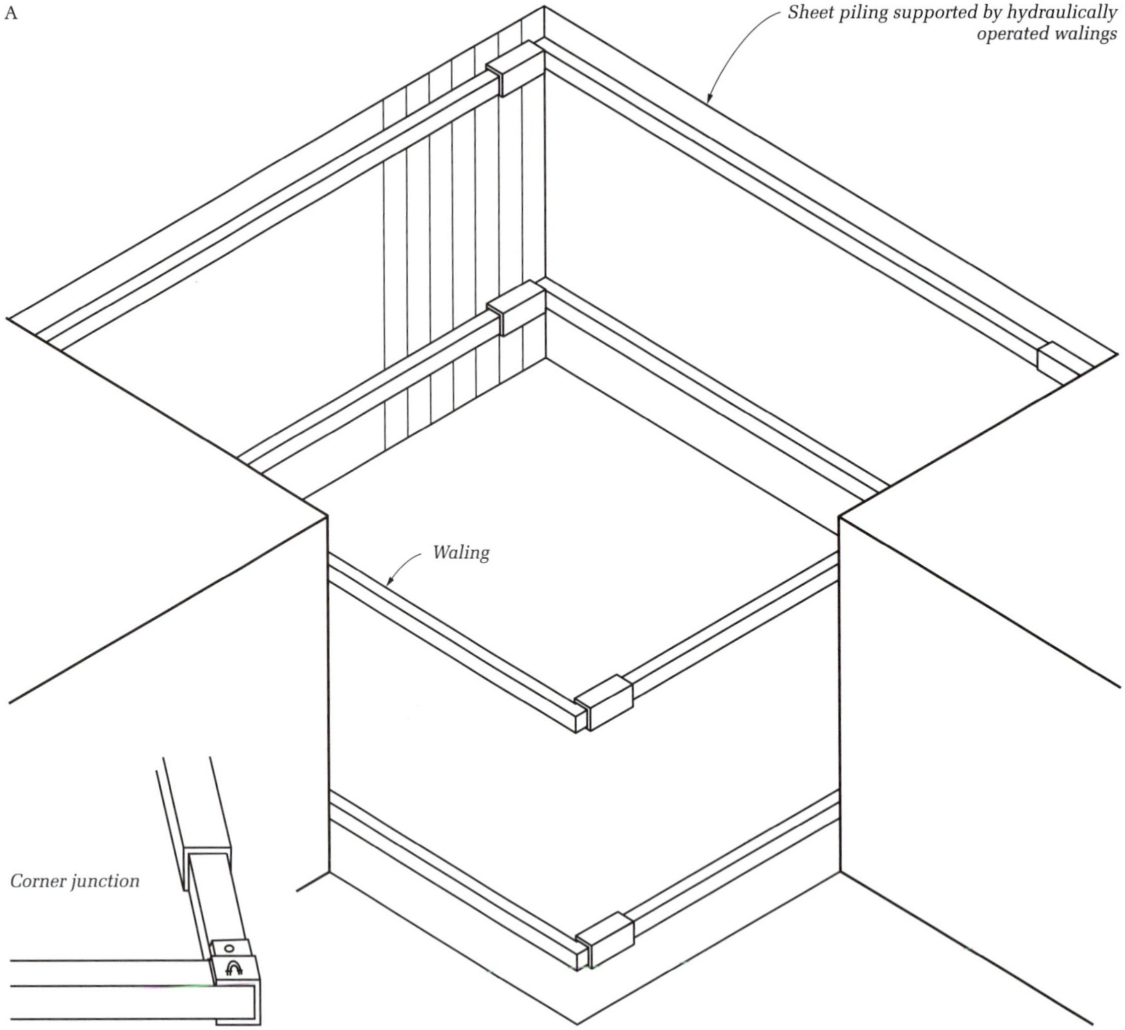

A

Sheet piling supported by hydraulically operated walings

Waling

Corner junction

Figure 10.19 Support to large shafts

- It should be strong enough to bear the weight of the wet concrete and all incidental working loads and it should be rigid enough to prevent excessive deflection during the placing of the concrete.
- The joints should be tight enough to prevent the loss of fine material from the concrete.
- It should be so designed and constructed that erection and stripping is orderly and simple and all units are of such a size that they can be easily handled. It should be possible for the side forms to be removed before the soffit shuttering is struck.
- If the concrete is to be fair-face the formwork in actual contact with the concrete should be so arranged and jointed that the resulting concrete has a good appearance.

In horizontal work the formwork must support its own weight, the weight of the wet concrete and reinforcement placed upon it, and the weight of operatives and any transporting equipment which is being used for the work. The formwork for vertical work must resist the pressure of the wet concrete pushing it outwards, and wind pressure. The outward pressure of the concrete depends upon its stiffness, the depth of concrete placed at one time and the way in which it is consolidated and will increase with increased wetness of mix and an increased height of

Table 10.1 Minimum times for striking formwork

Formwork	Surface temperature of concrete	
	16 °C	7 °C
Vertical formwork to columns, walls and large beams	12 hours	18 hours
Soffit formwork to slabs	4 days	6 days
Props to slabs	10 days	15 days
Soffit formwork to beams	10 days	15 days
Props to beams	14 days	21 days

Note: For temperatures between 0 and 16 °C BS 8110-1: *Structural Use of Concrete. Code of Practice for Design and Construction* gives formulae for calculating the appropriate times, on which those above 7 °C are based.

concrete placed. When tamping by vibrating equipment is used, joints must be tight and the whole must be sufficiently braced to prevent any movement.

The formwork must be designed and constructed so that it may easily be removed or 'struck' without damage to the formwork itself or to the hardened concrete. To facilitate this, nailing in timber forms should be kept to a minimum. Erection should be such that the formwork can be struck in the following order: one side of columns, sides of beams, bottoms of slabs and beams, the remaining sides of columns (see table 10.1).

Formwork may be constructed of any suitable material. Timber was once always used for this purpose and, by its nature, still has the advantage of the flexibility of forms which may be produced by its use. Proprietary steel and reinforced plastic formwork is available, generally in standard units of suitable sizes. It can be quickly erected and dismantled and with care can be used a greater number of times than timber forms. It is widely used where the dimensions of the elements to be formed are suited to the sizes of the standard units.

This type of formwork is designed to be quickly erected and dismantled. The forms are made up of standard panel units which can be used for column, beam, floor and wall construction as shown in figure 10.20.

They can be re-used a great many times, but if roughly handled need considerable maintenance in straightening, welding and patching up broken and cracked edges. This must be put against the savings arising from the greater number of re-uses.

10.4.1 Formwork construction

Timber should be sound and well seasoned and may be dressed on all four sides, on one side and one edge, or on one side and two edges. Timber dressed on all four sides is uniform in size and therefore more easily adapted for different purposes. The advantages arising from this often make it more economical to use than timber dressed in the other ways.

Plywood decking should be resin-bonded external grade, or if particle board is used this should be a moisture resistant type. The thickness will depend on loads to be carried and on the available supply. The latter is generally the governing factor as any ordinary size can be used by adjusting the spacing of supports.

As mentioned earlier, nailing should be kept to a minimum and, where used, the nail heads should not be quite driven home, as this makes it easier to draw them with a claw hammer or nail bar. Bolts and wedges are preferable to nailing, but are more costly.

Column forms In the case of rectangular columns, timber forms consist of four shutters or panels made up of ply or particle boards on studs to form the column casing, as shown in figure 10.20 A.

Adjustable steel clamps hold the panels in position and resist the outward pressure of the wet concrete; their spacing will be dictated by the size of the panel studs and will reduce towards the bottom where the outward pressure of the concrete is greatest. Horizontal battens are framed between the panel studs at each clamp position.

Circular column forms were traditionally constructed in timber as shown in figure 10.20 B, consisting of two semi-circular halves of thin ply backed by closely spaced vertical battens or staves supported by timber or thick ply yokes.

Proprietary formwork A proprietary method of building up a rectangular form is shown at C. The panel edges are perforated so that they can be locked to adjacent panels by the same type of connectors used for steel wall forms (figure 10.26) which permit rapid erection and striking. Circular column forms may consist of spirally wound steel, plastic or 10 mm cardboard tubes which require support to maintain the shape and verticality of the form. Alternatively, they may be fabricated in two halves preformed from steel or glass-reinforced plastic with stiffening ribs incorporated up the height.

Beam forms The beam box consists of three panels made up, as in column forms, of ply or particle board on studs (figure 10.20 D). This is carried by cross-bearers supported either directly or by an adjustable steel prop at either end or by a pair of ledgers running parallel with the beam box which are in turn supported by adjustable props (figure 10.20 D). Height and levelling of the beam form are achieved by the adjustable props which also allow for striking the form on completion.

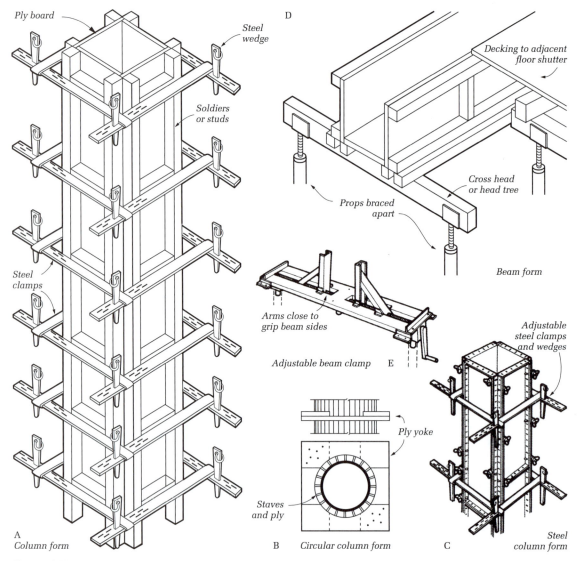

Figure 10.20 Formwork for columns and beams

The free side of an outer beam is braced at the top against horizontal concrete pressure by struts off the ends of the bearers. The tops of deep beams are similarly braced (see Part 1, figure 11.4 B) or, alternatively, steel beam clamps may be used which provide both bearer and strut support and adjust to varying widths of beam (figure 10.20 E).

When secondary beams bear on main beams openings for them are cut in the sides of the main beam box around which the box studs are framed and, if necessary, strengthened at the bearing of the secondary box. Junctions of beams and columns are formed in the same way.

When using proprietary steel components the box may be built up with three tray panels as in a column form and is supported on adjustable beam clamps and props. In some systems the box is formed from sheet steel folded to form the three sides.

Floor forms The construction of a simple floor shutter for a slab bearing on walls is illustrated and described in Part 1, Chapter 11, and as indicated there it is made up of 'decking' on which the concrete is placed, supported by joists, ledgers and props. This method is still widely used and the description given in Part 1 will be extended here.

The thickness of the ply decking, which may be faced, for example, with glass-reinforced plastic or similar impervious material when high quality surfaces are required, and the size of the joists will depend upon the loads the forms must carry, the spacing and span of the joists and the maximum deflection of the shuttering which may have been specified. The usual size of joist is 150×50 mm but may range from 100×50 mm to 250×75 mm. Proprietary timber I-beams of ply web and solid flanges may be used as ledgers instead of solid timber for longer spans, resulting in a reduction in the number of props required.

Methods of support   Support to the ledgers will be by individual adjustable steel props or by scaffold tubes fitted at the top with screwjacks, either of which can be finished with a flat plate or U-head as circumstances demand (see portion of floor shutter figure 10.20 D). These provide the adjustment for the initial levelling and final striking of the shutters.

The adjustable prop is shown in figure 10.21 A. It consists of an outer tube threaded at the top with a vertical slot on both sides. An inner upper tube perforated by holes is pushed up to the shutter and fine adjustment for levelling is made by a large nut which raises and lowers the tube via a pin passed through one of the holes showing through the slots. The scaffold tube prop (B) has a screwjack at the top which consists of a length of threaded rod sliding within the tube and engaging with a nut which bears on the top of the tube; this, when rotated, raises and lowers the threaded rod and thus the shutter.

Adequate bracing of all supports by horizontal and diagonal braces is essential to avoid movement. Where the floor is large in area the main bracing may be concentrated at the perimeter in the form of groups of three or four props braced together to form a 'bracing set' (figure 10.22 A) with a braced frame formed at each corner to give stiffness in both directions (B).

Proprietary steel formwork   When proprietary steel formwork is being used (figure 10.21 C) the deck panels may consist of an edge frame of light steel, with cross-members where necessary, within which is set plywood decking, giving protection to the ply edges and ensuring a close fit between the panels. The use of ply rather than steel sheet overcomes the problem of denting which can occur in steel after a number of re-uses; the panels are also lighter to handle.

The framing is perforated so that adjacent panels may be locked together and secured to supports by various types of connectors; these means of fastening permit rapid

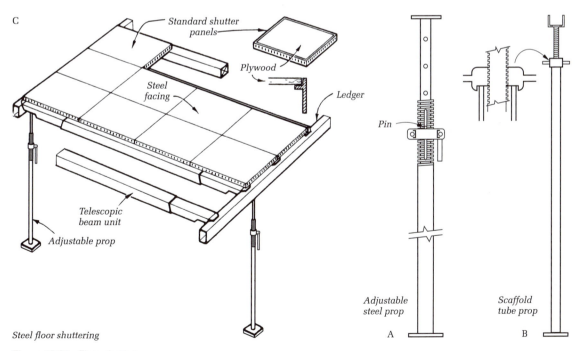

Steel floor shuttering

Figure 10.21 Floor shutters

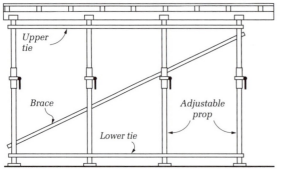

Upper tie

Brace

Adjustable prop

Lower tie

A *Bracing set*

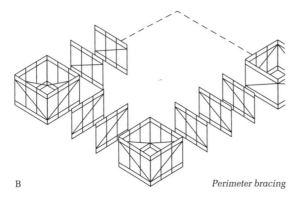

B *Perimeter bracing*

Figure 10.22 Support for floor shutters

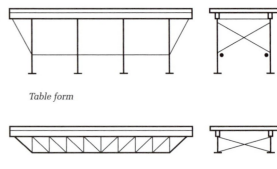

Table form

Flying form

Figure 10.23 Transportable shutters

By taking advantage of the lifting capacity of cranes erection time and labour requirements can be reduced by fabricating large floor units combining both formwork and falsework, of which there are two types:

Table form   This consists of a large shutter framed in the usual way together with its props and bracing securely attached (figure 10.23), which is raised from floor to floor by crane. It can be positioned most easily where the floors are without downstand beams, such as plate floors. These forms can be constructed with hydraulic control of the movement of the props so that striking can quickly take place over the whole of the form by the simultaneous lowering at every support.

Flying form   This also consists of a large shutter but which is carried by a pair of metal lattice girders secured to each long edge and braced laterally (figure 10.23). Very long shutters may be formed in this way. Support is by telescopic adjustable props; the jacks in the props permit levelling and striking, the telescopic action enables the form to be lowered on to rollers and moved out for the crane to raise.

Wall forms   The shutters are commonly formed of 16 mm plywood, the sheets being stiffened by vertical studs and battens at top and bottom edges (figure 10.24 A). Horizontal walings are fixed to the studs at intervals. The shutters are supported by adjustable props which have top and bottom swivel attachments bolted to the shutter and sole plate. As an alternative a steel tube with screwjacks may be used. If well strutted the two shutters are kept the thickness of the wall apart by timber cross-pieces nailed to the tops of the posts. Spacing at the bottom, as in the case of columns, is provided by a 'kicker' of concrete about 50 mm high and the exact width of the wall, which is cast on top of the foundations, or of a concrete floor. With thin walls it is an advantage first to erect one side of

erection and dismantling. The panel sizes range from 2400 × 600 mm downwards in varying lengths and widths and corner units and devices for making up dimensions are available, or infill panels of ply on bearers may be used. Lightweight steel beams some forms of which are telescopic and lightweight telescopic lattice beams which permit wide spans and reduce the number of props required in floor construction are available.

Rectangular or square grid floors and roofs (figure 5.3) present a special problem because of the large number of intersecting beams or ribs to be formed. This is usually solved by the use of square box forms or pans of metal or glass-fibre reinforced plastic or cement which are in the form of deep trays with projecting horizontal edges or lips. The pans are laid on temporary skeleton formwork with the edges touching to form the soffits of the ribs (figure 8.41). The depth of the pans varies according to the required depth of the ribs.

The metal or plastic pans can be re-used a great number of times but, as an alternative, stout, stiffened cardboard boxes or expanded polystyrene can be used as expendable forms or glass-fibre reinforced cement pans may be used as permanent shuttering.

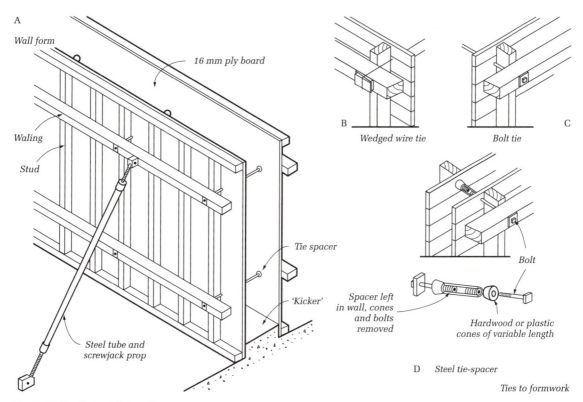

A
Wall form
16 mm ply board
Waling
Stud
Steel tube and screwjack prop
Tie spacer
'Kicker'

B
Wedged wire tie

C
Bolt tie

Bolt

Spacer left in wall, cones and bolts removed

Hardwood or plastic cones of variable length

D Steel tie-spacer

Ties to formwork

Figure 10.24 Formwork to walls

the formwork to the full height of the wall, and then to fix the reinforcement to the full height, followed by the form-work for the second side, which may be erected to the full height immediately or in lifts. If hand compacting of the concrete is to be used, the second side should be erected in successive lifts of 600 or 900 mm in height; if vibration is to be used and the thickness of the wall is sufficient, the second side may be erected to the full height before con-crete is placed. An alternative method in the case of hand compacting is to erect both sides together as the work pro-ceeds, one lift at a time.

Spacing and tying of shutters To avoid the use of large studs and an excessive amount of strutting, spacers and wire ties may be used, the spacers holding the two shutters the correct distance apart, and the ties resisting the out-ward pressure of the concrete when poured (figure 10.24 B). The wires are passed through the boarding and round the walings on each side. When the wall is cast in lifts the spacers may be of timber, which are raised as the concret-ing proceeds; but when the wall is cast in one operation, the removal of the spacers is difficult, so that the concrete spacers are used and are left in position. When the form-

work is struck the protruding ends of the wires must be cut back to at least 13 mm below the face of the wall and the holes carefully filled. As wire ties are likely to cause rust stains at the points where they are cut back, bolts are used as an alternative, being well greased or fitted with sleeves or being made slightly tapered in their length to enable them to be drawn out from the concrete when the form-work is struck (C). A number of proprietary ties are avail-able which secure the formwork without wire or spacers (D). The use of double, instead of single, walings avoids drilling, the bolts being passed through the space between the two members.

Several systems of clamps which dispense with the need for ties altogether are available for the construction of thin walls and a typical example is shown in figure 10.25.

Proprietary steel forms Figure 10.26 A shows a propriet-ary steel form for wall construction, making use of tray type panels with perforated edges which allow the use of connectors designed to facilitate rapid erection and strik-ing (B).

Proprietary systems of wall forms are also available consisting of very large ply sheets fixed to steel horizontal

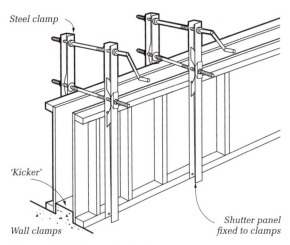

Figure 10.25 Formwork to walls

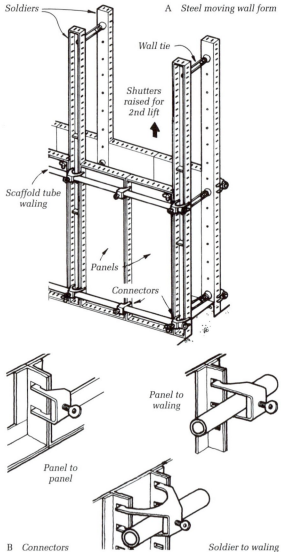

Figure 10.26 Proprietary formwork for walls

members supported by vertical steel soldiers which in turn are supported by adjustable raking props.

Slip forms or sliding shutters For the rapid construction of constant section walls it is possible to use a continuously rising form, usually known as a slip form or sliding shutter. By this means work may proceed continuously, the shutter rising from 150 to 300 mm per hour depending upon the rate of hardening of the concrete, since the cast concrete very rapidly becomes self-supporting. The form is about 900 mm or 1.20 m deep, fixed to and held apart by timber or steel frames or yokes, as shown in figure 10.27 A, B. On top of each yoke is fixed a hydraulic jack, through which passes a high tensile steel jacking rod, about 25 mm in diameter, which is cast into the walls as it rises. The jack contains a ram and a pair of upper and lower jaws which can grip the jacking rod and it works in cycles, each cycle giving a rise of about 25 mm. The jack works against the lower jaws to raise the yoke and the form with it. When the pressure is released, the upper jaws grip the rod and the lower jaws are released and raised under the action of a spring. Replacing the rod by a steel tube of larger diameter, because of its greater stiffness, permits the use of larger capacity jacks at greater yoke centres. A rarely used alternative to the hydraulic jack is the manually operated screwjack (C).

A working deck is constructed level with the top of the form, from which is usually suspended a hanging scaffold from which the concrete may be inspected and rubbed down as it leaves the shutters.

Stair forms These are constructed on the lines shown in figure 10.28. The shutter is carried on cross joists and rak-

ing ledgers. The risers, which are fixed after the reinforcement is placed, are bevelled at the bottom to permit the whole of the tread face to be trowelled. The outer ends are carried by a cut string and the wall ends by hangers secured to a board fixed to, or strutted against, the wall face. The treads are left open to permit concreting. Depending upon the width of the stair and thickness of the risers, the latter may require stiffening as shown against the outward pressure of the wet concrete when poured.

Proprietary sheet steel forms are available, either as standardised units or 'made to measure', prefabricated

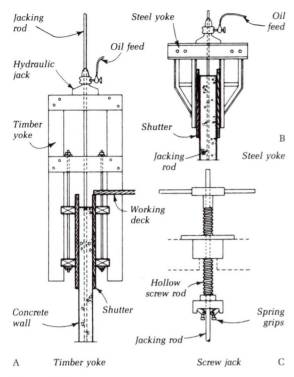

complete with soffit, strings and risers. They have the reinforcement in position and are meant to be left in place after completion. Some temporary support between the end bearings is usually necessary.

Doubly curved shell forms As explained in chapter 8, the hyperbolic paraboloid form can be developed from straight lines. The formwork can, therefore, be made up entirely from straight members. In principle, therefore, the construction is the same as for floor slabs, with decking, joists, ledgers and props. The edge beam forms are framed up, braced and supported as normal outer floor beams as shown in figure 10.30.

Shell barrel vault forms Steel forms can be used in the construction of shell vaults (see figure 10.29). Adjustable props, or screw jacks fitted to the heads of tubes, provide vertical adjustment. Mobile scaffolds running on rails may be employed, and is a useful method on very long vaults, but necessitates a clear barrel soffit. All stiffening ribs and frames to the vault must, therefore, be above the curved shell. Unless the vault is, in fact, long, striking and re-erection of the shuttering is generally quicker and more accurate. The edge beam forms are framed up, braced and supported as normal outer floor beams as shown in figure 10.30.

Figure 10.27 Sliding shutters

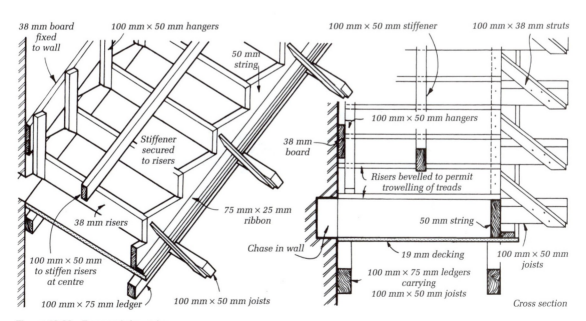

Figure 10.28 Formwork for stairs

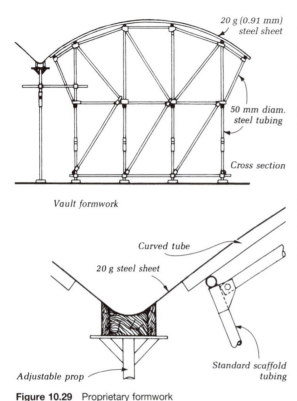

Vault formwork

Figure 10.29 Proprietary formwork

10.4.2 Treatment of formwork

The nature and treatment of the working faces of the form-work, that is, the faces in contact with the concrete, will affect the finished surface of the concrete. All working timber faces should always be treated with mould oil to prevent the concrete adhering to them and thus reduce the risk of damage when the formwork is stripped. In cases where a good key will be required on the final surface or where it is desired ultimately to expose the aggregate on the surface, a retarding liquid may be applied to the form-work. This prevents the setting of the cement at the sur-face, so that when the formwork is struck, the concrete face may be brushed down with stiff brushes to form a rough surface of exposed aggregate. Alternatively, aggre-gate transfer may be used. This consists of sticking selected aggregates to shutter liners with a suitable water-soluble adhesive. On stripping the shutters the aggregate is transferred to the concrete to which it is, by then, bonded. To produce a good smooth face, the formwork may be lined. Bare plywood will produce a smooth finish, but will leave the grain pattern on the surface. A better finish may be obtained by the use of paper-faced plywood, which will give a matt finish, or, by the use of a sealed paper, a gloss finish is possible. Plastics of various types can also be used to line the forms, either as sheeting or as rigid liners. Where a patterned surface is required, patterned tough

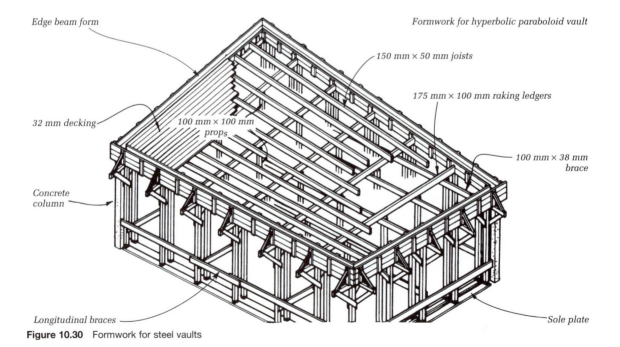

Figure 10.30 Formwork for steel vaults

rubber sheet or expanded plastic is used as a lining or glass reinforced plastic forms may be used; very deep patterning may be produced by the first two methods because the linings can be pulled away reasonably easily from the set concrete.[10]

Permanent shuttering of precast concrete may be used to provide the final finished face. Concrete pipes may be used for circular columns and concrete slabs for walls. Wood wool slabs used for thermal insulation to a wall may be used as an inside permanent shutter.

Striking times Striking times for formwork from BS 8110-1 are given in table 10.1, page 359. These are intended as a general guide only. Actual times will vary on each project according to the size of member, type of structure and day to day weather conditions. The periods given are for ordinary Portland cement concrete and in cold weather they should be increased. A guide for soffit formwork increases the periods for 7 °C by half a day for each day on which the concrete temperature was generally between 2 °C and 7 °C and by a whole day for each day on which the concrete temperature was below 2 °C.[11]

Notes

1 Recommendations on the design and construction of scaffolds are to be found in BS EN 12811-1: *Temporary Works Equipment. Scaffolds. Performance Requirements and General Design.*
2 See BS EN 39: *Loose Steel Tubes for Tube and Coupler Scaffolds. Technical Delivery Conditions.*
3 See BS 1139-1.2: *Metal Scaffolding. Tubes. Specification for Aluminium Tube.*
4 Otherwise known as planks or battens. Produced from sawn softwood, 225 mm wide × 38, 50 or 63 mm thick: 38 mm is normally used for scaffold systems, the greater thicknesses for trestle systems and access towers. See BS 2482: *Specification for Timber Scaffold Boards.*
5 The Work at Height Regulations 2005.
6 See BS EN 1004: *Mobile Access and Working Towers Made of Prefabricated Elements. Materials, Dimensions, Design Loads, Safety and Performance Requirements* and BS 1139-6: *Metal Scaffolding. Specification for Prefabricated Towers Outside the Scope of BS EN 1004, but Utilising Components from Such Systems.*
7 See also Part 3.1, *Ground Water Control, Advanced Construction Technology* by R Chudley and R Greeno, Pearson.
8 For a full coverage of excavations and other ground works see *Modern Construction and Ground Engineering Equipment and Methods*, 2nd edition, by Frank Harris, Prentice Hall, 1995.
9 *Formwork* is a general term that covers all types of mould for cast in situ concrete. The word *shuttering* is correctly applied only to the flat panels that are fixed together to make the complete formwork. Parts of the formwork such as column and beam boxes are called *forms*. Boxes for precast concrete are called *moulds*.
10 For the special requirements of 'board-marked' concrete finish and for other finishes referred to here, see *Guide to Exposed Concrete Finishes* by Michael Gage, Architectural Press, 1972.
11 See also Report 136: *Formwork Striking Times* by T Harrison, CIRIA, 1996.

Index